Cellular and Molecular Approaches in Fish Biology

Cellular and Molecular Approaches in Fish Biology

Edited by

Ignacio Fernández Monzón
Researcher, Center for Aquaculture Research, Agrarian Technological Institute of Castile and Leon, Valladolid, Spain

Jorge M.O. Fernandes
Professor, Genomics Division, Faculty of Biosciences and Aquaculture, Nord University, Bodø, Norway

ELSEVIER

ACADEMIC PRESS
An imprint of Elsevier

Academic Press is an imprint of Elsevier
125 London Wall, London EC2Y 5AS, United Kingdom
525 B Street, Suite 1650, San Diego, CA 92101, United States
50 Hampshire Street, 5th Floor, Cambridge, MA 02139, United States
The Boulevard, Langford Lane, Kidlington, Oxford OX5 1GB, United Kingdom

Library of Congress Cataloging-in-Publication Data
A catalog record for this book is available from the Library of Congress

British Library Cataloguing-in-Publication Data
A catalogue record for this book is available from the British Library

ISBN 978-0-12-822273-7

For information on all Academic Press publications
visit our website at https://www.elsevier.com/books-and-journals

Publisher: Nikki Levy
Acquisitions Editor: Anna Valutkevich
Editorial Project Manager: Lindsay Lawrence
Production Project Manager: Stalin Viswanathan
Cover Designer: Miles Hitchen

Typeset by STRAIVE, India

Contents

4. Noncoding RNAs in fish physiology and development: miRNAs as a cornerstone in gene networks

Francisco Javier Toledo Solís, Jorge M.O. Fernandes, Elena Sarropoulou, and Ignacio Fernández Monzón

5. Nutritional epigenetics

Kaja H. Skjærven, Anne-Catrin Adam, Saito Takaya, Rune Waagbø, and Marit Espe

Contributors

Numbers in parenthesis indicate the pages on which the authors' contributions begin.

Anne-Catrin Adam (161), Institute of Marine Research, IMR, Bergen, Norway

Juan F. Asturiano (1), Institute for Animal Science and Technology, Universitat Politècnica de València, Valencia, Spain

Nadine Azzam (347), The Department of Pediatrics, Division of Hematology/Oncology, Children's Hospital of Eastern Ontario Research Institute (CHEO RI), Ottawa, ON, Canada

Jason N. Berman (347), The Department of Pediatrics, Division of Hematology/Oncology, Children's Hospital of Eastern Ontario Research Institute (CHEO RI); Faculty of Medicine, University of Ottawa, Ottawa, ON, Canada

Elsa Cabrita (1), Centre of Marine Sciences (CCMAR), Faculty of Sciences and Technology, University of Algarve, Faro, Portugal

M. Leonor Cancela (381), CCMAR—Centre of Marine Sciences; FMCB—Faculty of Medicine and Biomedical Sciences, University of Algarve; ABC—Algarve Biomedical Center, Campus de Gambelas, Faro, Portugal

Alberto Cuesta (251), Immunobiology for Aquaculture Group, Department of Cell Biology & Histology, Faculty of Biology, University of Murcia, Murcia, Spain

Rose Ruiz Daniels (309), The Roslin Institute and Royal (Dick) School of Veterinary Studies, The University of Edinburgh, Edinburgh, United Kingdom

María P. Di Yorio (221), Department of Biodiversity and Experimental Biology, Faculty of Exact and Natural Sciences, University of Buenos Aires; Institute of Biodiversity and Experimental and Applied Biology (IBBEA), CONICET-University of Buenos Aires, Buenos Aires, Argentina

Marit Espe (161), Institute of Marine Research, IMR, Bergen, Norway

María Ángeles Esteban (251), Immunobiology for Aquaculture Group, Department of Cell Biology & Histology, Faculty of Biology, University of Murcia, Murcia, Spain

Jorge M.O. Fernandes (105), Genomics Division, Faculty of Biosciences and Aquaculture, Nord University, Bodø, Norway

Ignacio Fernández Monzón (105), Center for Aquaculture Research, Agrarian Technological Institute of Castile and Leon, Valladolid, Spain

Clémence Fraslin (193), The Roslin Institute, University of Edinburgh, Edinburgh, United Kingdom

Manuel Gesto (285), Section for Aquaculture, National Institute of Aquatic Resources, Technical University of Denmark, Hirtshals, Denmark

Silvia González-Rojo (69), Department of Molecular Biology, Universidad de León, Campus de Vegazana, León; Center of Biofuels and Bioproducts, Instituto Tecnológico Agrario deCastilla y León (ITACyL), Villarejo de Órbigo, Spain

Francisco A. Guardiola (251), Immunobiology for Aquaculture Group, Department of Cell Biology & Histology, Faculty of Biology, University of Murcia, Murcia, Spain

Manu Kumar Gundappa (309), The Roslin Institute and Royal (Dick) School of Veterinary Studies, The University of Edinburgh, Edinburgh, United Kingdom

María Paz Herráez (69), Department of Molecular Biology, Universidad de León, Campus de Vegazana, León, Spain

Ákos Horváth (1), Department of Aquaculture, Hungarian University of Agriculture and Life Sciences, Gödöllő, Hungary

Ross D. Houston (193), The Roslin Institute, University of Edinburgh, Edinburgh, United Kingdom

Kim Kobar (347), The Department of Pediatrics, Division of Hematology/Oncology, Children's Hospital of Eastern Ontario Research Institute (CHEO RI); Faculty of Medicine, University of Ottawa, Ottawa, ON, Canada

Vincent Laizé (381), CCMAR—Centre of Marine Sciences, Campus de Gambelas, Faro; S2 AQUA—Sustainable and Smart Aquaculture Collaborative Laboratory, avenida Parque Natural da Ria Formosa, Olhão, Portugal

Marta Lombó (69), Department of Molecular Biology, Universidad de León, Campus de Vegazana, León, Spain

Daniel J. Macqueen (309), The Roslin Institute and Royal (Dick) School of Veterinary Studies, The University of Edinburgh, Edinburgh, United Kingdom

Zoran Marinović (1), Department of Aquaculture, Hungarian University of Agriculture and Life Sciences, Gödöllő, Hungary

Nicole Melong (347), The Department of Pediatrics, Division of Hematology/Oncology, Children's Hospital of Eastern Ontario Research Institute (CHEO RI), Ottawa, ON, Canada

Diego Perojil Morata (309), The Roslin Institute and Royal (Dick) School of Veterinary Studies, The University of Edinburgh, Edinburgh, United Kingdom

Daniela I. Pérez Sirkin (221), Department of Biodiversity and Experimental Biology, Faculty of Exact and Natural Sciences, University of Buenos Aires; Institute of Biodiversity and Experimental and Applied Biology (IBBEA), CONICET-University of Buenos Aires, Buenos Aires, Argentina

Marta F. Riesco (49), Department of Molecular Biology, Cell Biology Area, University of León, León, Spain

Diego Robledo (193), The Roslin Institute, University of Edinburgh, Edinburgh, United Kingdom

Vanesa Robles (49), Department of Molecular Biology, Cell Biology Area, University of León, León, Spain

Joana T. Rosa (381), CCMAR—Centre of Marine Sciences, Campus de Gambelas, Faro; S2 AQUA—Sustainable and Smart Aquaculture Collaborative Laboratory, avenida Parque Natural da Ria Formosa, Olhão, Portugal

Elena Sarropoulou (105), Institute of Marine Biology, Biotechnology and Aquaculture, Hellenic Centre for Marine Research, Iraklion, Crete, Greece

Kaja H. Skjærven (161), Institute of Marine Research, IMR, Bergen, Norway

Saito Takaya (161), Institute of Marine Research, IMR, Bergen, Norway

Marco Tarasco (381), CCMAR—Centre of Marine Sciences, Campus de Gambelas, Faro, Portugal

Richard S. Taylor (309), The Roslin Institute and Royal (Dick) School of Veterinary Studies, The University of Edinburgh, Edinburgh, United Kingdom

Francisco Javier Toledo Solís (105), Consejo Nacional de Ciencia y Tecnología (CONACYT), Universidad de Ciencias y Artes de Chiapas, Tuxtla, Mexico

David G. Valcarce (49), Spanish Institute of Oceanography (IEO), Santander, Spain

Paula G. Vissio (221), Department of Biodiversity and Experimental Biology, Faculty of Exact and Natural Sciences, University of Buenos Aires; Institute of Biodiversity and Experimental and Applied Biology (IBBEA), CONICET-University of Buenos Aires, Buenos Aires, Argentina

Rune Waagbø (161), Institute of Marine Research, IMR, Bergen, Norway

Editor's biographies

Jorge M.O. Fernandes is a professor in genomics and molecular biology at Nord University, Norway. He has also held honorary associate professorships at the University of Messina, Italy, and the University of Tokyo, Japan. Prof. Fernandes obtained his BEng in chemical engineering with specialization in biotechnology from Instituto Superior Técnico, Portugal, where he graduated with distinction in 1997. His passion for biology led him to pursue a doctoral degree in comparative immunology at the University of St Andrews. After his PhD graduation in 2003, he took a postdoc position until 2007, studying the molecular mechanisms that regulate muscle growth in fish. Prof. Fernandes has more than 15 years of experience as an independent researcher and is a recognized expert in aquaculture genomics. His current research fields are muscle growth and innate immunity in teleosts, with focus on the epigenetic regulation (DNA methylation and miRNAs) of gene networks by environmental factors in commercially important species. Prof. Fernandes is an ERC grantee and has an established track record of high standard research, with 122 publications in international journals with peer review and an h-index of 37. He is a regular reviewer for various scientific journals and funding bodies, and has won the Nord University research prize in 2014 and 2016.

Ignacio Fernández Monzón is a researcher in fish development and molecular biology at the Agro-Technological Institute of Castilla y León (ITACyL), Spain. Dr. Fernández obtained his BS in marine sciences from the University of Vigo, Spain, in 2003. His enthusiasm and curiosity for understanding how fish grow and develop led him to get a doctoral degree. He received a fellowship to conduct his PhD research at the Institute of Agrifood Research and Technology (IRTA, Spain) in collaboration with the Institut Français de Recherche pour l'Exploitation de la Mer (IFREMER, France), receiving his PhD graduation in aquaculture at the University of Barcelona, Spain, in 2011. In 2012 he won a postdoctoral fellowship from the Portuguese Foundation for Science and Technology (FCT) at the Centre of Marine Sciences (CCMAR) Portugal. There he studied the molecular pathways governing the skeletal development using a multidisciplinary approach. Along his career, Dr. Fernández performed several short scientific missions at different internationally recognized institutions. The various research skills gained from his academic experience all over the years enabled him to win two Spanish grants of excellence in research: Beatriu de Pinós (2018) and Ramón y Cajal (2019). Currently, Dr. Fernández has coauthored 47 publications in international peer-reviewed journals. His main research interests are in (i) the development of sustainable aquaculture production; (ii) the identification, validation, and implementation of integrative biomarkers of fish physiological condition, particularly on nutritional status; and (iii) the identification of suitable predictive and diagnosis descriptors of fish quality.

Preface

The first fishes are thought to have emerged on Earth over 500 million years ago during the Cambrian Period of the Paleozoic Era and they occupy a key phylogenetic position in vertebrate evolution. Extant species range from jawless fish to elasmobranchs to the abundant and diverse group of ray-finned fishes (Actinopterygii). Fishes are the most diverse and abundant taxonomic group within vertebrates, accounting for more than half the total of all other vertebrate species combined, with more than 33,000 species of fishes found in freshwater and marine environments throughout the world. This remarkable biodiversity is accompanied by fascinating adaptations that make fish extremely attractive subjects in developmental biology, physiology, and evolution.

The continuously increasing interest in research involving fishes is also due to their ecological and economic relevance in fisheries and aquaculture, and as models in human medicine. The approaches used to study fish biology have also advanced substantially in the last two decades, particularly with the advent of genomics. This book comprises 12 exciting chapters in which experts review the state of the art of some of the most relevant fields in fish biology (e.g., reproduction, nutrition, immunity, endocrinology, welfare, genetics, epigenetics, and genome evolution) and present the methodologies currently used to help us perceive and understand the wonderful world of fishes.

Chapter 1 by Cabrita et al. examines the factors that should be considered when developing protocols for cryopreservation of different biological materials, namely sperm, germline stem cells, and even embryos. These genome resource banks can be applied in the conservation of endangered species such as the European eel (*Anguilla anguilla*), which has been intensively fished to near extinction. Cryopreservation techniques are also increasingly used for model species and in selective breeding programs for improved production of salmonids and other farmed fishes. All these new and exciting possibilities must ensure the faithful preservation of maternal and paternal genomes, as discussed by Riesco et al. in Chapter 2. After reviewing traditional methods to evaluate how cryopreservation affects DNA integrity, the authors put forward a novel approach to detect DNA lesions by quantitative PCR and illustrate it with a specific protocol to determine DNA damage in primordial germ cells of zebrafish (*Danio rerio*) after cryopreservation. Importantly, this technique can discover genome changes that are undetectable using conventional techniques to analyze DNA fragmentation. In Chapter 3, Herráez et al. show that the reliable transmission of the genome goes beyond its structural integrity. Epigenetic changes in DNA methylation and histone modifications induced by environmental factors and domestication can have a profound impact on the phenotype. Several molecular techniques can be used to profile histone epigenetic marks, global and promoter DNA methylation, in order to assess the status of sperm chromatin and its implication in the development of the progeny in several farmed fish species. Moreover, environmental toxicants can alter the epigenetic information contained in zebrafish spermatozoa, thus jeopardizing future generations. Toledo-Solís et al. (Chapter 4) explore another important mechanism of epigenetic and posttranscriptional regulation of gene expression: the noncoding RNAs. They focus on miRNAs, an important class of small RNAs that play a major role in numerous biological processes. Current methodological approaches are presented, along with several examples of their application to study the role of miRNAs in development, reproduction and sex differentiation, adaptation and response to environmental cues, metabolism, and immunocompetence and disease resistance in fish. The recent identification of circulating miRNAs in biological fluids suggests they may be used as ethical biomarkers, paving the way for groundbreaking discoveries and biotechnological applications in the near future. Chapter 5 by Skjærven et al. presents several examples of how nutrients and metabolites influence the epigenetic regulators, including DNA methylation. A practical case of a bioinformatics workflow to identify differentially methylated sites from reduced representation bisulfite sequencing data is provided. Nutritionally induced changes in phenotype can be transmitted from broodstock to offspring, but the persistence of epigenetic effects across generations is still a matter of debate. Intra- and intergenerational approaches to nutritional epigenetics are of great relevance for aquaculture, especially when linked to the concept of nutritional programming, as they can be applied to produce more robust fish. This can be used in synergy with the more established strategies to identify the causative variants in the genome sequence that are directly responsible for a difference in a trait of interest, as reviewed by Fraslin et al. (Chapter 6). Several methods for mapping causative variants are discussed in detail, from more conventional linkage analysis to genome-wide association studies to recent multimarker models. A typical step-by-step example pipeline for genome-wide association analysis is also described, along with examples of quantitative trait loci and causative variant implementation into breeding programs. The development of genome editing technologies in aquaculture species is expected to prove the association of causative variants and a specific phenotype in the coming years.

In Chapter 7, Vissio et al. discuss how double-labeling immunofluorescence can be used as an accessible and effective tool to perform neuroendocrine studies in fish. Detailed protocols (with practical tips) of double-labeling immunofluorescence and pituitary cultures are put forward and used to demonstrate how these techniques, combined with genetic tools, can reveal novel insights on the physiological action of neuropeptides and allow us to unveil new actors in the regulation of the pituitary function in teleost fish. Furthermore, this system can be used as a model to better understand the hypothalamus-pituitary relationship in tetrapods, as most of the main components and functions of the hypothalamic-pituitary axis are conserved from fish to mammals. Guardiola et al. (Chapter 8) provide an overview of mucosal immunology in fish, with a focus on gilthead seabream (*Sparus aurata*), which is a commercially important species in Europe. Mucosal surfaces not only constitute the first line of host defense against pathogens during infection, but they also play a major role during wound repair and healing. The authors summarize the modulating effects of different biotic and abiotic conditions on mucosal immunity, including dietary prebiotics and probiotics, and exposure to contaminants such as heavy metals and microplastics. The most popular methods to collect intestine, skin, and gill mucus samples are described and discussed, along with examples of the main downstream analyses of mucus samples. In Chapter 9, Gesto discusses the gaps in knowledge related to fish welfare and the difficulties in finding reliable indicators, especially when the fish are exposed to chronic disturbances of low severity. A comprehensive methodology to determine the stress status of fishes as part of the assessment of their welfare is proposed, along with technical considerations for experimental design and sampling. This approach has been applied to assess the stressor potential of different factors, including handling by chasing, alterations of water quality, presence of pollutants, and formation of social hierarchies, among others.

The great biological and genetic diversity within fishes makes this taxon an exceptional system for studies of genome evolution. Taylor et al. (Chapter 10) introduce key concepts, advances, and resources to study the evolution of fish genomes, providing a reference for the status and directions in the field. Several aspects of fish genome evolution are covered, including protein coding genes, noncoding regions, ohnologue regulation, and the epigenome. Additionally, the authors developed a detailed approach to infer the evolutionary relationships of miRNA families expanded by whole genome duplication events during teleost evolution. Chapter 11 by Melong et al. presents remarkable examples of translational research using the zebrafish model. Larval zebrafish xenotransplantation models have been developed to investigate tumor-induced angiogenesis, and liquid and solid tumor xenotransplants. Their applications as patient avatars include drug screening, novel drug discovery, radiotherapy enhancement, and stem cell transplants for regenerative therapies. Both larval and adult zebrafish xenotransplant models have proven extremely useful to bridge knowledge gaps and uncover crucial mechanisms and pathways in the fields of oncology and regenerative medicine. There is a growing interest for cell lines of fish origin, as they complement in vivo systems. The chapter by Laizé et al. (Chapter 12) reviews the current status of fish cell culture and the use of fish cells lines in various fields, such as cell biology, immunology, toxicology, biotechnology, and species conservation. An updated protocol to develop bone-derived cell lines is described along with their potential applications. These fish mineralogenic cell lines hold great potential for genetic and epigenetic studies to gain insights into cellular mechanisms and signaling pathways, as well as to accelerate drug screening for bone regeneration.

The editors believe that this exciting book will be an inspiring source for graduate students and a reference for more experienced researchers who want to learn how to implement multidisciplinary cellular and molecular approaches in their fields. We are very grateful to the different leading experts who have contributed to this book, highlighting the research issues covered here and their relevance, detailing the diverse set of methodologies, and especially the effort put on a smooth knowledge transfer for potential readers. Enjoy your reading.

Chapter 1

Technologies and strategies for ex situ conservation of aquatic organisms: The role of cryopreservation in long-term management☆

Elsa Cabrita[a], Ákos Horváth[b], Zoran Marinović[b], and Juan F. Asturiano[c]

[a]*Centre of Marine Sciences (CCMAR), Faculty of Sciences and Technology, University of Algarve, Faro, Portugal,* [b]*Department of Aquaculture, Hungarian University of Agriculture and Life Sciences, Gödöllő, Hungary,* [c]*Institute for Animal Science and Technology, Universitat Politècnica de València, Valencia, Spain*

1. Factors and steps to consider in setting up a cryopreservation protocol

1.1 Cryopreservation definition

By definition, cryopreservation is the process where biological material can be stored at very low temperatures, usually at – 196 °C, the boiling point of liquid nitrogen, for long periods. Spermatozoa freezing is a good example of the application of this procedure. What happens to cells during cooling/freezing has been extensively studied and reviewed recently elsewhere, but for further interested readers Wolkers and Oldenhof (2021) is highly recommended. The general impacts of fast or slow cooling on cells is summarized in Fig. 1. The rate at which most of the cells are cooled is proven to have a dramatic impact on its long-term viability. Cooling rate affects the formation, type, and size of both intracellular and extracellular ice crystals. It can affect the osmolarity of the extender media producing what is called a solution effect, since during temperature decrease ice tends to form, excluding the solutes to the media, which at this point is still unfrozen. Therefore while fast cooling minimizes solute concentration effects, due to uniform ice formation, it increases intracellular ice formation because

☆ In memory of Serafín Pérez Cerezales.

Cellular and Molecular Approaches in Fish Biology. https://doi.org/10.1016/B978-0-12-822273-7.00011-2

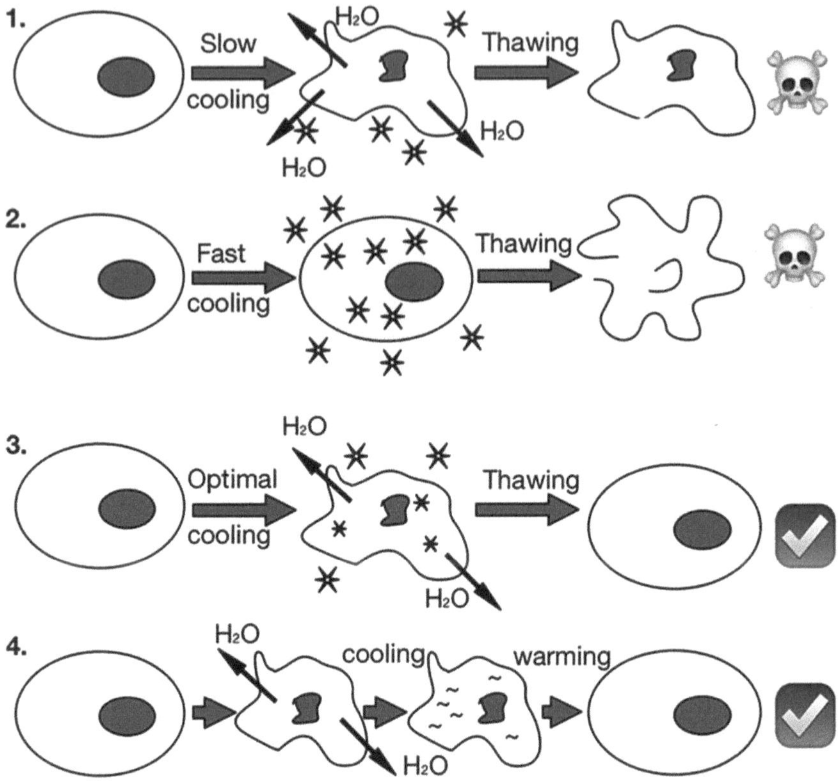

FIG. 1 Fate of cells during cryopreservation. When cells are cooled slowly (1) they are exposed to loss of water and increasing solute concentration which is lethal. When cells are cooled fast (2) they cannot lose enough water which freezes inside them causing lethal damage. During optimal cooling (3) cells lose enough water to avoid freezing in large ice crystals, yet are not exposed to solutes long enough to be lethal, thus they survive. During vitrification (4) cells lose a sufficient volume of water prior to cooling. Then they are cooled with very fast cooling rates so that the solution in them does not freeze but becomes an amorphous glassy solid. With very rapid warming, the cells survive.

water does not have enough time to move out of the cell. The opposite occurs with slow cooling, where water molecules have enough time, until the solution is completely frozen, to exchange from the inner to the outer part of the cell, reducing intracellular ice formation. Since ice crystal formation is one of the most important problems associated with cryopreservation in most biological systems using controlled cooled freezing, there is other technique avoiding this problem, known as vitrification.

1.2 Vitrification

Vitrification is the formation of an aqueous solid without crystalline ice through ultra-rapid cooling at temperatures below $-130\,^{\circ}\mathrm{C}$. In the process of vitrification, the solution attains a glassy state, meaning that in the vitrified matter chemical and physical processes are severely limited, thus biological time is effectively stopped without the formation of ice crystals (Fahy et al., 1984). This process is also known as fast nonequilibrium vitrification, as opposed to slow equilibrium freezing (Mazur, 1990). During vitrification, water molecules do not have time to arrange into a crystalline structure and turn to an amorphous state that is a snapshot of the liquid state. Successful vitrification requires dehydration of cells even prior to exposure to subzero temperatures by exposing them to high concentrations of cryoprotectants (both penetrating and nonpenetrating). Vitrified liquids are metastable, meaning that they transit to crystalline ice once their temperature is increased above the glass temperature (Franks, 1986). This means that warming rates also need to be extremely fast in order to prevent devitrification and subsequent recrystallization. Fast warming rates are currently considered even more important than rapid cooling or high cryoprotectant concentrations (Seki et al., 2014). In fact, partial vitrification is known to appear even in samples cryopreserved by conventional freezing, thereby increasing the chances of cell survival (Rall, 1987).

Vitrification of live cells and tissues is typically achieved by plunging them directly into liquid nitrogen (Fahy et al., 1984). Prior to this, cells are treated with cryoprotectants or their mixtures in order to increase the viscosity of the solution within and outside the cells by partial removal of water. Each penetrating cryoprotectant contributes to glass formation at a different concentration, and due to their toxicity, instead of single cryoprotectants, mixtures are commonly used (Cuevas-Uribe, 2011). The volume of intracellular water can also be reduced by the use of nonpenetrating cryoprotectants, leading to successful vitrification of cells. This process is sometimes misleadingly labeled as "cryoprotectant-free" vitrification (Isachenko, 2004; Merino et al., 2012), although it only means that penetrating cryoprotectants are not necessary for vitrification. Finally, the volume of vitrification solution around the cells or organs should be minimized to achieve successful glass transition.

A crucial point for the development of vitrification methods is the choice of devices used for cryopreservation of various cell or organ types. Vitrification in straws includes the use of conventional 0.25 mL straws, the specifically designed Open-Pulled Straws (OPS) with thinner walls for better heat transition, as well as the somewhat similar Cryotips. Reduction of vitrification solution volume can successfully be achieved by vitrification on electron microscope copper grids by blotting the excess vitrification solution on a filter membrane or by suspending cells in a film of vitrification solution in devices called cryoloops. Cryotips are thin and narrow polypropylene strips attached to a plastic handle that can be inserted into a protective plastic sleeve and the vitrification solution is loaded onto the strip, which is then plunged into liquid nitrogen (Chen and Yang, 2007). An interesting solution for the vitrification of tissues and organs is offered by the Needle-immersed vitrification (NIV) method, where tissue pieces of 20–80 mg or 1–2 mm in diameter are pulled on acupuncture needles and plunged into liquid nitrogen following exposure to equilibration and vitrification solutions (Wang et al., 2008a, b).

In case of fish, vitrification of fish sperm has been tested in several species (Merino et al., 2012; Cuevas-Uribe et al., 2011a,b, 2015; Kása et al., 2017, 2018). However, due to its limitations and low efficiency, this methodology is not likely to replace conventional freezing methods. Vitrification of gonadal tissue, on the other hand, offers a viable alternative to freezing. These techniques will be discussed in detail later in this chapter.

1.3 Setting up a cryopreservation protocol

To set up a cryopreservation protocol for any species, and particularly for fish sperm, it is important to bear in mind some steps that need to be considered. Here, we explain the principal steps that will be elucidated, especially with examples for each species from Sections 3.1 and 4.

The choice of the container system is the first step to consider after taking care of the proper method to collect your sperm samples (see Section 3.1), which of course is species specific (Beirão et al., 2019). Sperm samples (or other biological material) can easily be stored and identified either in straws or cryovials, commercially available, and with different volume sizes. This is important for species that produce large amounts of sperm such as most salmonids (Section 4.2) or for species that produce small quantities such as the model species (Section 4.3). The selection of one or other system is related to the amount of sperm to be cryopreserved or to the objective of the cryopreservation protocol (conservation of individual samples or bulk cryopreservation for production). The selection of the correct cryoprotectant, its concentration and dilution in the extender media are also important and may compromise the success of the procedure.

Cryoprotectants (CPAs) are used to reduce the freezing temperature of solution and to produce a dehydration process due to their high osmolarity, reducing the risks of ice formation inside the spermatozoa. However, they can also be toxic. According to Paredes (2016) and Riesco et al. (2017a), it is necessary to study the toxic effects and toxicity thresholds of the different chemical compounds in order to select the right CPA or its combinations to be used before cryopreservation, as well as the temperature of addition and equilibration (time necessary to penetrate the CPA into the cells until freezing starts), if necessary. In this regard, factors such as membrane permeability, temperature, and osmotic excursions effects need to be taken into account. Fish sperm, like that of other species, allows a fast penetration of most cryoprotectants, reducing the time of equilibration to usually the time necessary to package all samples (3–4 min). The first protocols developed in most salmonid species used a long equilibration time (30–60 min) (Cabrita et al., 1998), but nowadays it is recommended to reduce this time, since some cryoprotectants can be toxic with prolonged exposure (see Section 3.1).

The third step in a cryopreservation design will be the choice of the freezing methodology. The most common cooling agent used for the cryopreservation of fish sperm is the vapor of liquid nitrogen. Dry ice has also been used for the cryopreservation of fish sperm in various forms, e.g., as straws immersed into powdered dry ice (Yasui et al., 2008) or put on top of a block of dry ice (Lichtenstein et al., 2010). In either case, liquid nitrogen vapor and dry ice are only used for the precooling of sperm while the final temperature is attained in the storage environment, which is most often liquid nitrogen.

Several methods are available to cryopreserve samples, from the use of portable or bench top controlled cooling devices (biofreezers), where slow freezing rates (e.g., − 0.3 °C/min) and fast freezing rates (e.g., − 50 °C/min) can be attained, to other methods such as vitrification (for details, see Section 1.2). Both systems can be performed using automated equipment, which is usually expensive but accurate. Another way of freezing samples is by using a polystyrene (Styrofoam) box or a liquid nitrogen container. In this case, samples are placed at different levels from the surface of liquid nitrogen, in the case of conventional cryopreservation, or directly submerged into liquid nitrogen, producing a vitreous state (if the concentration of cryoprotectant is high enough to avoid ice crystal formation or recrystallization). A more consistent version of cryopreservation in the vapor of liquid nitrogen is the use of dry shippers for freezing fish sperm. In this case, straws are put directly into the canister positioned in the dry shipper and are allowed to cool in the vapor of liquid nitrogen absorbed into its walls. This has successfully been used in a number of fish species, especially in Brazil (Viveiros and Godinho, 2009). Other systems have been developed such as the use of − 150 °C electric ultrafreezers (Diogo et al., 2018) (for more details, see Section 4.3). This system can be advantageous over the use of liquid nitrogen, especially for samples cryopreserved in cryovials, since they do not require a constant supply of liquid nitrogen and samples are stored separately in boxes, reducing the risk of cross-contamination of the stored material (Larman et al., 2014). Using either this method or the traditional liquid nitrogen storage, it is important to avoid temperature fluctuations. Samples should continuously be stored below their glass transition temperature; temperature excursions out of this range will result in recrystallization of the ice, a process in which large ice crystals grow at the expense of smaller ones, causing a reduction of cell viability or even cell destruction (Paredes, 2016).

To evaluate the success of a cryopreservation protocol, samples are thawed and the metabolism of the cells, which at this point is in a quiescent state, needs to be restored. Sperm thawing is also a critical step. The most common method used for thawing is a water bath set at a constant temperature, where samples are submerged for some seconds to few minutes, depending on the volume of the stored material (the lower the volume the lower the time). Thawing should be done as fast as possible to avoid recrystallization of ice and to allow a homogenous recovery. In the case of fish sperm, thawing temperature is also species specific and can range from 5 °C used for spotted wolfish (*Anarhichas minor*) sperm (Santana et al., 2020) to 25 °C as in gilthead seabream (*Sparus aurata*) or Senegalese sole (*Solea senegalensis*) sperm (Cabrita et al., 2005; Riesco et al., 2017b), 35 °C in European seabass (*Dicentrarchus labrax*; Martínez-Páramo et al., 2012), or 40 °C as used by several authors for Atlantic salmon (*Salmo salar*; Yang et al., 2018).

After thawing, which is the last step of the cryopreservation protocol, samples need to be evaluated. Once again, it may depend on the purpose and on the applications of those samples, but even using basic methodologies, an assay of the freezing–thawing procedure should be performed.

1.4 Assay of freezing–thawing process

1.4.1 Cell integrity and functionality

To test the success of a sperm cryopreservation protocol, the analysis of cell membrane integrity is one of the methods most commonly used in research. The plasma membrane is composed of an amphipathic lipid bilayer that separates and protects the cell from the extracellular environment and is also responsible for the control of what enters and leaves the cell, mainly by controlling transmembrane proteins. Plasma membrane integrity and functionality are primary needs for cell survival, and several tests have been developed to assay cryopreserved spermatozoa viability—a term commonly used as a synonym of cells with a functional plasma membrane (Cabrita et al., 2010). Most tests determine plasma membrane integrity and/or selective permeability using vital dyes such as trypan blue (Lubzens et al., 1997), eosin (Lahnsteiner et al., 1996) and Hoescht 33258 (Cabrita et al., 1998). These dyes only enter the cell if the plasma membrane is altered. The dual staining SYBR-14/propidium iodide combined with a flow cytometry analysis, allows a faster and more accurate evaluation of plasma membrane integrity, and is a common method used nowadays (Cabrita et al., 2010, 2014).

Cell morphology and membrane ultrastructure are often evaluated in cryopreserved fish sperm using scanning electron microscopy (SEM) and/or transmission electron microscopy (TEM). Several studies have identified morphological and morphometric changes in head, flagella, and midpiece in species such as sharpsnout seabream (*Diplodus puntazzo*) and ocean pout (*Macrozoarces americanus*; Taddei et al., 2001; Liu et al., 2007). Other methods such as computer-assisted sperm morphology analysis (ASMA) have been applied to detect cryopreservation effects on the morphology of fish spermatozoa (Asturiano et al., 2007, 2017). This software has advantages over traditional SEM analysis because it provides an objective and faster automated evaluation of sperm morphometry, which together with statistical tools, allows the identification of spermatozoa populations resistant to cryoinjury.

Programmed cell death occurs during spermatogenesis to eliminate aged spermatozoa but other sperm management technics such as cryopreservation can also induce this phenomenon. Different methods to detect damage associated to apoptosis are available using fluorescent dyes with the conjunction of more than one probe. Annexin-V FITC (Diogo et al., 2018), YO-PRO 1, although not so specific (Beirão et al., 2010), and caspase activity using commercial available kits (Riesco et al., 2017b), are some of the probes that provide additional information on the effects of cryopreservation on spermatozoa apoptosis. These probes are used in combination with flow cytometry to quantify different stages of apoptosis (combined usually with an impermeable dye such as PI or SYTOX) or with confocal microscopy.

Despite the importance of proteins in cell metabolism and structure, this topic only recently gained more relevance in the evaluation of fish sperm cryopreservation protocols (Ciereszko et al., 2017; Dietrich et al., 2019). The analysis of protein expression using two-dimensional polyacrylamide gel electrophoresis (2-DE) together with western blot or matrix-associated laser desorption/ionization time-of-flight mass spectrometry (MALDI-TOF) for the identification of specific spots was first explored in the characterization of proteins in European seabass and gilthead seabream sperm cryopreservation by Zilli et al. (2005, 2008). These authors reported a loss of some of these components or even a redistribution between spermatozoa heads and flagella after cryopreservation (Zilli et al., 2014). Sperm cryopreservation seems to affect the integrity and expression of some proteins as well as the activity of some enzymes and their phosphorylation state, thus compromising sperm motility activation in fish sperm (Zilli et al., 2005, 2008). In recent studies, Dietrich and Ciereszko (2018) identified proteins that might be useful bioindicators of freezing resilience and thus, could be used to screen samples before cryopreservation. The authors found that in common carp (*Cyprinus carpio*) sperm, good freezability (the capacity of cells to withstand cooling temperatures) is related to high concentrations of proteins involved in the maintenance of flagella structure, membrane fluidity, control of Ca^{2+} and sperm motility, energy production, and antioxidant protection.

Plasma membrane lipid components are also important and affected by cryopreservation. Their determination could be a good indicator of sperm freezability. Several works in European seabass and gilthead seabream sperm demonstrated a reduction in cholesterol/phospholipid ratio and in some specific phospholipid content (e.g., phosphatidylcholine) as well as changes in fatty acid profile after cryopreservation (Martínez-Páramo et al., 2012; Beirão et al., 2012). This fact was somehow counteracted with the supplementation of antifreeze proteins in the extender media (Beirão et al., 2012). Protein and lipid redistribution, loss, or degradation is associated with lipid and protein peroxidation, cell leakage due to ruptures, and other oxidative events that might occur in the cells during cryopreservation.

1.4.2 Mitochondria and sperm motility

ATP levels in fish spermatozoa are only sufficient to maintain motility for some seconds to a few minutes, with mitochondria being responsible for ATP supply to the cell. Thus, the effect of cryopreservation methods on ATP reserves and mitochondria status is of utmost importance (Cabrita et al., 2010; Figueroa et al., 2019).

Current methods to analyze mitochondria status after fish sperm cryopreservation use flow cytometry combining probes such as JC-1, a lipophilic cation, to identify mitochondria displaying a normal membrane potential gradient (Cabrita et al., 2005; Figueroa et al., 2015, 2019). Oxygen consumption is another indirect way of determining mitochondrial activity (Zilli et al., 2005; Figueroa et al., 2019). ATP and ADP concentrations can be measured by luminescence, and their values are usually affected by the osmotic stress caused by cryoprotectant concentrations and solute effects during cooling (Lahnsteiner et al., 1996; De Baulny et al., 1997; Cabrita et al., 2005). Changes in the mitochondria (both structural and those related to membrane polarization) and/or decrease in cell ATP content have been registered in cryopreserved sperm of several species like gilthead seabream, striped bass (*Morone saxatilis*), rainbow trout (*Oncorhynchus mykiss*), and Atlantic salmon, causing a reduction in motility duration and sperm velocity after cryopreservation (De Baulny et al., 1997; Cabrita et al., 2005; Guthrie et al., 2014; Figueroa et al., 2019).

Motility is one of the most used parameters to evaluate semen quality after cryopreservation. Unlike in species with internal fertilization, fish sperm is immotile in the testis, being motility triggered after sperm dilution in a hyper or hypoosmotic medium. Traditionally, motility evaluation is based on a subjective classification (from 0 to 5), where the observer considers the percentage of motile sperm and the type of movement for the estimation, a method commonly used in production. Software systems (computer-assisted sperm analysis—CASA) gained relevance in research and are used for evaluating motility in cryopreserved sperm (Cabrita et al., 2010). CASA was first introduced in 1979 for mammalian sperm motility assessment (Dott and Foster, 1979), with several software versions commercially available nowadays. An open source software was developed by Wilson-Leedy and Ingermann (2007) and was recently updated to a more complex system integrating also other quality parameters (Alquézar-Baeta et al., 2019). Both commercial and open source systems have been used with fish sperm to assess the effects of different cryoprotectants (Asturiano et al., 2007; Beirão et al., 2011; Liu et al., 2007), to characterize different subpopulations of spermatozoa in fish milt samples or after cryopreservation (Gallego and Asturiano, 2018, 2019), or to characterize how broodstock feeding regimes affect gamete quality (Beirão et al., 2015; Butts et al., 2020; Martins et al., 2020).

1.4.3 DNA damage analysis

The final end of spermatozoa is the transmission of male genetic information to the offspring. Therefore the preservation of spermatozoa genomic information should be one of the priorities when a cryopreservation protocol is designed. Several studies have demonstrated that sperm cryopreservation might promote

DNA damage in fish sperm (Cabrita et al., 2005; Pérez-Cerezales et al., 2009; Riesco et al., 2017b,c). In recent years, the development of techniques for detecting different types of DNA damage has revealed the nature of cell injury and has been facilitating the improvement of freezing–thawing protocols. The most common technique to detect DNA damage in fish cryopreserved sperm, in particular DNA fragmentation, is the single cell gel electrophoresis (SCGE), also known as comet assay. This technique has been used in cryopreserved sperm from species such as Senegalese sole (Riesco et al., 2017b), European seabass (Martínez-Páramo et al., 2013), gilthead seabream and rainbow trout (Cabrita et al., 2005, Pérez-Cerezales et al., 2009), zebrafish (*Danio rerio*; Diogo et al., 2018), common carp (Shaliutina-Kolešová et al., 2020), dusky grouper (*Epinephelus marginatus*), and Portuguese oyster (*Crassostrea angulata*; Riesco et al., 2017c, 2019). This technique reveals different types of damage depending on the pH buffer. A neutral pH electrophoresis buffer detects double strand breaks while double and single strand breaks as well as alkaline labile sites are revealed with basic pH. In addition, the application of specific endonucleases that recognize oxidized bases allows their detection by increased DNA fragmentation (reviewed by Cabrita et al., 2014).

There are other assays to determine DNA fragmentation such as Terminal deoxynucleotidyl transferase-mediated dUTP-biotin end-labeling (TUNEL), but not so commonly used. Reports on Atlantic salmon, zebrafish, and European seabass sperm showed very low levels of DNA damage (9%, 1%, and 3%; respectively) when compared with samples or similar studies where comet assay was used (around 10% for European seabass and zebrafish) (Cabrita et al., 2011; Figueroa et al., 2015; Diogo et al., 2018; González-Rojo et al., 2018), therefore displaying a poor correlation with already established methods. Pérez-Cerezales et al. (2012) justified this difference in mouse sperm because the TUNEL assay probably requires more marked DNA and chromatin damage before it can detect any change, indicating lower sensitivity than the comet assay.

Cryopreservation can also produce other types of damage related with gene methylation, although it has been quite controversial the results obtained by several studies, especially because most deal with DNA global methylation. It seems that alterations in DNA methylation may be associated with chemical-specific interactions in damaged cells, rather than with the freezing by itself. Depincé et al. (2019) found a decrease in DNA global methylation in goldfish sperm when using DMSO, one of the most common cryoprotectants used, whereas no effect was seen with methanol. The same results were reported for eel sperm (Herranz-Jusdado et al., 2019a). In other species such as zebrafish, methanol did not protect so well the cells and produced hypermethylation after cryopreservation (Depincé et al., 2019). The same authors reported no effects of sperm snap-frozen (without any cryoprotectant) or on sperm exposed to cryoprotectant solutions (without freezing). Therefore it seems intriguing the main causes behind the changes in DNA methylation patterns. Moreover, more specific methods to evaluate the effects on DNA can be applied (see Chapter 2).

1.5 Cell types used for genome cryobanking

Different cell types have been used in order to find suitable protocols for fish genome cryobanking: spermatozoa, oocytes, spermatogonia, and primordial germ cells, as well as somatic cells, blastomeres, and embryos (Labbé et al., 2013). Spermatozoa are easy to collect in most of the fish species and they have a small size, a relatively simple structure and high chilling resistance. Thus they have been the objective of most of the studies with cryopreservation purposes, and used to establish commercial techniques, although it allows the preservation of only male germplasm.

On the other hand, the cryopreservation of fish oocytes is limited by their large cell volume, the presence of a chorion, the low permeability to cryoprotectants, and a high chilling sensitivity. In this case, it would allow the preservation of only female germplasm and would need the development of in vitro techniques to induce oocyte development and maturation. Although an increasing body of research is available (reviewed by Martínez-Páramo et al., 2017), fish oocyte (and embryo) cryopreservation is still far from practical applications in aquatic organisms.

Spermatogonia and primordial germ cells have been cryopreserved in several fish species (Lee et al., 2016; Marinović et al., 2019a; Robles et al., 2017; Yoshizaki et al., 2011, for specific examples see Section 3.2), and its use allows preservation of the entire genome of the individuals. However, the development of specific biotechnological tools such as transplantation is required to establish the progression of the germline (reviewed by Asturiano et al., 2017; Franěk et al., 2019a, b). Recent studies on cryobanking techniques have assayed the cryopreservation of ovarian (or testis) tissue fragments, containing early stage follicles, with the intention of using them to be fertilized after using in vitro culture methods (Higaki et al., 2017; Kawasaki et al., 2016), or as origin of oogonia (or spermatogonia) to be transplanted (reviewed by Šćekić et al., 2020).

2. Applications of cryopreservation techniques in aquatic animals

2.1 Historic aspects and current situation

2.1.1 Fish

Considering the huge diversity among the 25,000–30,000 fish species, being the largest group of vertebrates, including the high variability of the gamete (spermatozoa) morphology and biology (Mattei, 1991), it seems logical that cryopreservation methods are fish species specific (Cabrita et al., 2009b, 2014). Blaxter published in 1953 the first gamete cryopreservation protocol in aquatic species. During the following (almost) 70 years, protocols have been adapted to successfully cryopreserve sperm of a high number of fish species (Gallego and Asturiano, 2019; Tsai and Lin, 2012). However, a good part of the research effort has been centered on a limited number of fish groups. Freshwater species have received a greater attention, and specific protocols have mainly been established for salmonids, cyprinids (mostly carps and zebrafish), sturgeons, and a few siluriformes (catfishes). In general, the publication of cryopreservation protocols for sperm of marine species is historically more recent (Asturiano et al., 2017; Gallego and Asturiano, 2019). As Martínez-Páramo et al. (2017) suggested, the main reason explaining this difference between freshwater and marine species is the fact that reproduction of most marine commercial species occurs naturally in a tank, without requiring artificial fertilization and the application of cryopreservation techniques.

In summary, most of the scientific publications evidencing these advances in fish sperm cryopreservation have been published in the present century, and many of them have been included in several previous review publications (Asturiano et al., 2017; Cabrita et al., 2010; Figueroa et al., 2016a; Kopeika et al., 2007; Labbé et al., 2013; Martínez-Páramo et al., 2017; Suquet et al., 2000) and specific books (Cabrita et al., 2009b; Tiersch and Green, 2011).

2.1.2 Invertebrates

Cryopreservation techniques have also been applied to invertebrates, with the main intention of transferring its use to the aquaculture industry, but with evident application possibilities in the preservation of natural genetic resources, as well as laboratory animal models used in embryo–larvae bioassays (Fabbrocini et al., 2014). Similar to fish, invertebrate sperm and spermatogonia can be preserved, along with spermatophores, oocytes, embryos, and larvae (as it was subsequently reviewed by Lin et al., 1999; Labbé et al., 2013; Martínez-Páramo et al., 2017; Paredes et al., 2020).

Sperm has been cryopreserved in mollusks such as different oyster species (*C. angulata*, *Crassostrea gigas*, *Crassotrea virginica*, *Ostrea edulis*, *Pinctada margaritifera*, or *Pinctada fucata martensii*; reviewed by Martínez-Páramo et al., 2017), mussels (*Mytilus galloprovincialis*), abalones (*Haliotis laevigata*), as well as echinoderms (several sea urchin species; Paredes, 2016; Paredes et al., 2020), tunicates, and shrimps. A few exploratory trials to set up protocols for spermatophore cryopreservation have been carried out with cephalopods (*Illex coindetii*) and shrimps (*Penaeus monodon*), as well as gorgonian coral sperm sacs (Robles et al., 2013; Vuthiphandchai et al., 2007; Tsai et al., 2014).

Robles et al. (2009a) explained that the success of the cryopreservation of invertebrate embryos and larvae is favored by a limited embryo size, improving water and cryoprotectant exchange, a low yolk content in the oocyte, and a holoblastic cleavage, improving cryoprotectant penetration. Martínez-Páramo et al. (2017) listed 40 studies carried out in a few invertebrate species (many in Pacific oyster (*C. gigas*)) and justified by its high commercial value. Most of them aimed to define basic cryopreservation protocols (including embryonic developmental stages), the evaluation of embryo survival after thawing, and the long-term rearing performances of thawed larvae.

Larvae of other invertebrates, such as several sea urchin species (*Evechinus chloroticus*, *Hemicentrotus pulcherrimus*, *Paracentrotus lividus*, *Strongylocentrotus nudus*, etc.) were cryopreserved (extensively reviewed by Paredes, 2016). Many of them are used as biomarkers in ecotoxicology bioassays. Among penaeids, several species (as *Penaeus japonicus*, *P. stylirostris*, or *P. monodon*) have been used in the first steps of protocols for embryo and larvae cryopreservation, including the determination of the best developmental stage to be frozen, although further research is still required.

2.2 Cryopreservation applications: The use of cryobanks

The techniques for the cryopreservation of gametes from aquatic species have a high number of potential applications, ranging from aquaculture to ecology perspectives. Aquaculture purposes are the most evident way to use them, improving broodstock management at hatcheries (modifying the reproductive season for instance) or allowing the preservation of genetically selected strains coming from genetic improvement programs (Asturiano et al., 2017; Martínez-Páramo et al., 2017).

Cryopreservation techniques can assist in the management of species having problems related with reproduction, such as lack of synchronization in gamete production of males and females (as in the case of the European eel (*Anguilla anguilla*); Asturiano et al., 2004); when it is difficult to capture both sexes simultaneously (i.e., European sturgeon (*Acipenser sturio*; Williot et al., 2011)); with species producing low sperm volume (i.e., Senegalese sole; Cabrita et al., 2006); or to give time for the genetic analysis of a given individual to decide about its use in a conservation program (Horváth et al., 2012). Moreover, these techniques can be potentially applied to preserve genetic information from natural populations of (fish) species entering the domestication process and experimenting genetic modifications. The preserved original wild genotypes can be considered a phenotypic backup to have the chance of recovering deleted genes in the future.

Of course, another purpose of gamete cryopreservation protocols is their use in restocking and conservation programs, or allowing cryobanking for biodiversity maintenance (Martínez-Páramo et al., 2009b, 2017), considering the increasing number of endangered species all over the world. Moreover, cryobanking techniques are complementary for flowering reproductive biotechnologies, such as germ cell xenotransplantation. Finally, but not less important is the use of these techniques to store the increasing number of transgenic lines, knockouts, and mutant strains derived from the use of aquatic models, such as zebrafish, in studies of biotechnology, toxicology, and/or pharmacology (Da Silva et al., 2019; Tiersch et al., 2011).

Gamete cryopreservation techniques are the basis of important business in mammals (both cattle and humans), while commercial application of fish sperm cryopreservation is hindered by several factors, such as the lack of commercial-scale know-how for scaling-up to industry and practical aquaculture, offering quality assessment of products, methods, and cryobanking services. Moreover, although aquaculture is a fast growing industry, it is still in its infancy on this regard in comparison with terrestrial animal production.

However, the recent increase of genetic improvement programs has resulted in a growing offer of cryobanking services for several salmonids (where the use of Quantitative Trait Loci (QTL) is gaining importance; Yue, 2014; Tsai et al., 2015; also see Chapter 6 for an overview of concepts, methods, current state and future prospects of QTL mapping for genetic gain of aquaculture breeding programs) and other species. Some companies are already offering lines of specific products (extenders, activating solutions, etc.) for fish sperm cryopreservation, and even cryopreservation services (reviewed by Martínez-Páramo et al., 2017).

Obviously, cryobanking means added costs for companies and programs, and long-term cost–benefits must be evaluated.

The scientific community has called for the creation of national or supranational specialized centers not only to maintain collection of frozen samples, but also to assist in the design of species- and cell-specific cryopreservation protocols, improving standardization (definitions, methodologies, and reporting) and increasing the chances of transference to the industry. As Martínez-Páramo et al. (2017) suggested, the management of these banks requires technical capacity in genetics, reproductive physiology, cryobiology and data administration, as well as a close cooperation of cryobiologists with other researchers and managers.

In general, the available information on aquatic species cryobanks is quite limited, but Martínez-Páramo et al. (2017) made a detailed review of the existing cryobanks in Europe, USA, Brazil, Australia, and New Zealand, as well as the aquatic species and even the type of cells they work with. Some cryobanks have been created for aquaculture applications in a limited number of fish species (common carp; Horváth et al., 2007; tambaqui (*Colossoma macropomum*) and cachara (*Pseudoplatystoma reticulatum*), Streit Jr. et al., 2013; tench (*Tinca tinca*), wels catfish (*Silurus glanis*) or several salmonids such as rainbow- and brown trout (*Salmo trutta* m. *fario*)), combining aquaculture and biodiversity preservation (Atlantic salmon; O'Reilly and Doyle, 2007; and several sturgeon species) or directly as part of conservation programs and restocking projects (Brazilian ichthyofauna, Viveiros and Godinho, 2009; Adriatic grayling (*Thymallus thymallus*); Horváth et al., 2012; UK fish species, into the Frozen Ark project).

There are centers, as the *European Zebrafish Resource Centre* (EZRC; Leopoldshafen, Germany) or the *Zebrafish International Resource Center* (ZIRC; Oregon, USA), specialized in acting as repository of lines of model species (basically zebrafish, and medaka (*Oryzias latipes*)) with the objective of providing access to those lines for the research community.

Some examples of cryobanking for the conservation of invertebrates are the US Fish and Wildlife Service, acting as a repository of several endangered species and includes sperm samples and glochidia (larval stage) from freshwater mussels (Wayman, 2011). Meanwhile, the Taronga Conservation Society Australia maintains the Taronga CryoReserve, storing germplasm from endangered species, including the first created frozen repository of coral species from the Great Barrier Reef (Hagedorn et al., 2012a, b; Hagedorn and Spindler, 2014) to preserve its biodiversity. Finally, the Cawthron Institute (Nelson, New Zealand) develops breeding programs for mussels and oysters for the New Zealand aquaculture sector, which include sperm banking to carry out family crosses (Martínez-Páramo et al., 2017).

3. Cryopreservation of different biological materials to create Genome Resource Banks

3.1 Sperm

By far, the most commonly cryopreserved tissue/cell suspension type in aquaculture (as well as in domestic livestock) is sperm. Although it is very difficult to indicate the number of fish species whose sperm has successfully been cryopreserved, estimates between 50 and 200 species (marine and freshwater, respectively) have been reported (Tiersch et al., 2011). Regardless of the species, the key elements of cryopreservation protocols are similar with little variation in each step, as was already mentioned in Section 1.3.

Sperm of fish species is collected following hormonal stimulation of spermiation during the spawning season. Hormonal stimulation is administered in the form of an intramuscular or intraperitoneal injection of various hormone products (carp pituitary extract, gonadotropin-releasing hormones and their synthetic analogues, etc.). Spermiation is seasonal in species of the temperate climate while in case of tropical and subtropical fish it is more or less continuous. Hormonal stimulation of spermiation is not practiced in several species, including salmonids, where sperm can be collected from males during the natural or artificially induced spawning season (including changes in the light and temperature regimes). It is also unnecessary and impractical in model tropical species such as the zebrafish due to their small size and continuous sperm production.

Sperm is collected from fish by hand-stripping into a dry vessel in order to avoid activation of spermatozoa by accidental contact with water, urine, and/or other activating media. In some fish species—notably catfishes—sperm cannot be collected by stripping. In these cases, sperm is extracted from surgically removed testes or testicular sections (Legendre et al., 1996). The volume of collected sperm varies with species, from around 1 μL in the zebrafish to as much as 1000 mL in larger sturgeons. These represent a logistical challenge in cryopreservation; however, in most cases working volumes of 1–20 mL of sperm are used. Following collection, it is generally recommended to store sperm until further use at 4 °C or on melting ice. Sperm quality is determined most commonly by motility or the percentage of spermatozoa performing forward movement.

This can be done either visually or using computer-assisted sperm analysis as was described in Section 1.4.2. Additional methods to determine sperm quality include determination of viability, mitochondrial function, DNA fragmentation, or ATP levels; however, these are more time consuming than motility tests.

Prior to cryopreservation, fish sperm is always diluted in solutions called extenders. Extenders are buffered solutions of sugars, ions, or their combinations that need to be isotonic to the seminal plasma of fish to prevent their activation. They should also provide favorable conditions for cells to survive osmotic and ionic concentration changes during the process of freezing. Extenders can be common cell culture solutions such as Hanks' balanced salt solution (HBSS) that is used for the sperm of a wide variety of species, from marine sciaenids to cyprinids and medaka (Tiersch et al., 2004; Wayman et al., 1997; Yang et al., 2007, 2010). More often, however, research groups prefer to compose their own extender recipes and adapt those to the fish species that they work with. For example, an extender has been developed originally for the sperm of the grayling (Horváth et al., 2012, 2015a); however, it has been used on various other salmonids (Horváth et al., 2015b) as well as cyprinids (Lujić et al., 2017a). In many cases, extenders developed by individual research groups gain international recognition and are used by several teams.

With the exception of a few reported cases of cryoprotectant-free vitrification (Merino et al., 2012), all sperm cryopreservation protocols involve the use of penetrating cryoprotectants. Their concentration is almost invariably given as percent volume concentration (% v/v) for the sake of convenience, although, molar concentration would be more appropriate. For the freezing of fish sperm, cryoprotectant concentrations ranging from 5% to 15% (v/v) are typically used. The choice of cryoprotectants for fish sperm is quite broad, although in practice two major ones are used. By far, the most common cryoprotectant is dimethylsulfoxide (DMSO) that has been tested in a variety of species, representing completely different families and orders of fish (Cabrita et al., 2009a; Chereguini et al., 1997; Fabbrocini et al., 2000; Thirumala et al., 2006; Tsvetkova et al., 1996). Methanol has been found effective in concentrations of 5%–10% (v/v) in sturgeons (Glogowski et al., 2002), salmonids (Lahnsteiner et al., 1997), cyprinids (Horváth et al., 2003), catfishes (Tiersch et al., 1994), and percids (Bernáth et al., 2015). Other cryoprotectants tested for the cryopreservation of teleost sperm include glycerol (Blaxter, 1953), dimethyl-acetamide (DMA) (McNiven et al., 1993; Warnecke and Pluta, 2003), ethylene glycol (Jähnichen et al., 1999), and methyl glycol (Viveiros et al., 2014).

Dilution of sperm with an extender containing the cryoprotectant is considered one of the crucial steps of cryopreservation. Dilution ratios ranging from 1:1 to 1:20 have extensively been studied in various freshwater and marine fish species (Suquet et al., 2000) and have been optimized even relatively recently (Judycka et al., 2016). However, the efficiency of dilution ratios has been questioned, as concentration of spermatozoa in individual sperm samples varies. Therefore standardization of sperm concentration yielding the highest cryosurvival was suggested by several authors (Dong et al., 2007; Tan et al., 2010), which is the norm in the commercial cryopreservation of dairy bull sperm. The importance of diluting sperm to an optimal sperm concentration of $0.5–1.0 \times 10^9$ spermatozoa was demonstrated in the rainbow trout, yielding not only high postthaw motility (60%–66%), but also consistency and low variation of results (Nynca et al., 2017). Later, standardization of sperm concentration was applied to other salmonid species, as well, with similar results (Judycka et al., 2018).

Following dilution, sperm is equilibrated in the extenders before freezing. The effect of equilibration time on the outcome of cryopreservation in fish sperm has been tested in several studies and has always been found to yield optimal results as long as it was kept to less than 10–15 min (He and Woods III, 2003; Horváth and Urbányi, 2000; Stoss and Holtz, 1983). Less attention has been paid to equilibration time in later studies; however, Judycka et al. (2018) reported the use of a 15 min equilibration for the sperm of various salmonid species.

During equilibration, sperm is loaded into the container in which it is frozen. By far, the most common container type for fish sperm cryopreservation is the 0.25 and 0.5 mL straws that were originally developed by Robert Cassou in1963 (Lonergan, 2018). Other types of straws used include the 1.2 mL straws that were tested in salmonids and cyprinids (Horváth et al., 2007; Lahnsteiner et al., 1997), yielding fertilization results similar to those of 0.5 mL straws. Larger (4 or 5 mL) straws, also known as macrotubes, have also been tested in salmonids (Cabrita et al., 2001a; Lahnsteiner et al., 1997), the paddlefish (*Polyodon spathula*) (Horváth et al., 2010) and the wels catfish (Bokor et al., 2009). The general consensus is that although cryopreservation of sperm in these larger straws results in lower fertilization percentages than with the use of 0.5 mL straws, their larger volume compensates the loss of fertility. A notable exception was the case of wels catfish, where fertilization rates with sperm frozen in 4 mL straws were similar to those of fresh sperm. Current developments in fish sperm cryopreservation, however, favor the use of the conventional 0.25 or 0.5 mL straws as these are considered the industry standard for sperm cryopreservation. Other containers such as cryotubes or cryovials were also tested for the cryopreservation of fish sperm. Turbot (*Scophthalmus maximus*) and dusky grouper sperm have successfully been cryopreserved in 2 mL cryotubes, yielding similar results as 0.5 mL straws (Chereguini et al., 2003; Riesco et al., 2017c). Similarly, good results were reported with the use of 10 mL cryotubes in the sperm of the wels catfish by Bokor et al. (2019). On the other hand, fertilization rates were significantly lower than the fresh control when 1.6 or 4.5 mL cryotubes were used for the cryopreservation of tambaqui sperm (Maria et al., 2015).

Following equilibration, sperm is frozen to the temperature of cryopreserved storage using one of the methods described in Section 1.3. Freezing rates are species specific and should be set according to the chosen protocol conditions (e.g., sperm loading container). Cryopreserved sperm samples are almost universally stored in liquid nitrogen containers, either in canisters or in racks, the former being developed primarily for straws while the latter one for cryotubes (although cryotubes can also be stored in canisters attached to cryocanes). Adequate labeling as well as database development is imperative in any laboratory dealing with cryopreservation.

3.2 Germline stem cells (GSCs)

Even though sperm is the most commonly cryopreserved cell type, as previously mentioned, the greatest downside of this technology is that it preserves solely male genetic material and cannot produce live progeny alone; the presence of an ovulating female able to produce fresh eggs is needed for fertilization and the production of progeny. To preserve female genetic material, attempts were made to cryopreserve eggs or embryos. However, effective cryopreservation protocols for these cell types have not yet been developed due to their complex structure, large size and amounts of yolk material, high chilling sensitivity, and poor membrane permeability to solutes (for more details, see Section 3.3). To circumvent this problem, application of GSC (including primordial germ cells (PGCs); spermatogonial stem cells (SSCs), and oogonial stem cells (OSCs)) manipulation techniques started to gain attention. These cells are the baseline stem cells of gametogenesis, and through their cryopreservation and subsequent transplantation into suitable recipients, both male and female genetic resources can be preserved and reconstituted (reviewed by Yoshizaki and Lee, 2018; Yoshizaki and Yazawa, 2019). In short, donor GSCs are isolated from the gonads, most commonly transplanted into the body cavity of recipient larvae where they migrate toward the recipient's genital ridge, which they colonize, start to proliferate and differentiate, and ultimately give rise to donor-derived gametes (most commonly both sperm and eggs if transplanted into mixed-sex pool of larvae).

PGCs are firstly formed germline cells which begin to differentiate during early embryonic development outside of the gonad (Dosch, 2015) and subsequently, begin to migrate toward the genital ridge. After successful colonization of the gonad, they start to proliferate and differentiate into SSCs or OSCs, depending on the donor gender as specified afterward. Attempts to cryopreserve PGCs have been made during early embryonic development, during migration, as well as when PGCs were settled within the genital ridges (Franěk and Pšenička, 2020). Cryopreservation of cell suspensions containing isolated PGCs is uncommon and preference is usually given to cryopreservation of gonadal ridges or whole embryos/larvae. Genital ridges of rainbow trout (Kobayashi et al., 2007) and zebrafish (Riesco et al., 2012) were successfully cryopreserved with high PGC viability/recovery, and in the case of rainbow trout, cryopreserved PGCs transplanted into abdominal cavities of allogeneic trout larvae retained their functionality by migrating into the recipient's genital ridges, proliferating, differentiating, and ultimately giving rise to donor-derived gametes and offspring. Furthermore, cryopreservation of dechorionated 14–18-somite zebrafish embryos (Higaki et al., 2010) and dechorionated 12–14-somite loach (*Misgurnus anguillicaudatus*) embryos (Inoue et al., 2012) yielded viable PGCs postthaw, which were functional and were able to induce germline chimeras after transplantation.

Even though cryopreservation of PGCs is plausible, several factors severely hinder the application of this technology: (1) PGCs are present only in larvae of certain age, therefore finding these individuals during spawning season might be very difficult in many fish species; (2) each individual contains only a limited number of PGCs, thus large number of larvae carrying PGCs need to be collected; (3) identification and visualization of PGCs is very challenging, possible by using injections of specific fluorescent markers or by using transgenic lines but really difficult in most fish species; and (4) collection of PGCs from a large number of larvae is very laborious and time consuming. Therefore an alternative was found in SSCs and OSCs which are present in juvenile and adult individuals. In this way, a large number of early stage germ cells can be isolated from each individual in a more labor- and time-efficient manner through a simple mechanical separation or enzymatic treatment.

Cryopreservation of GSCs is similar, but less diverse, to that of sperm. Samples are usually suspended in available cell culture solutions such as L-15 or phosphate-buffered saline (PBS) and supplemented with permeating and nonpermeating cryoprotectants at various concentrations. Samples are then frozen in cryotubes at slow cooling rates (most commonly 1 °C/min) until − 80 °C, when they are plunged into liquid nitrogen. GSCs are usually frozen in one of two possible ways: within the tissue pieces or within cell suspensions which require previous enzymatic dissociation of the tissue. Marinović et al. (2017) and Pšenička et al. (2016) indicated that slightly better results are obtained when freezing tissue pieces. Indeed, freezing of tissue pieces is more time efficient as it does not require dissociation prior to cryopreservation, while dead cells present within the tissue after thawing are digested during the subsequent enzymatic dissociation. Therefore the obtained cell suspensions contain almost no dead cells, which may facilitate subsequent GSC manipulations. However, when working with species that have large gonads, fragmentation of gonads should be conducted prior to cryopreservation. Studies in common carp (Franěk et al., 2019a) and wels and African catfish (Marinović et al., unpublished data) indicate that tissue fragments should not surpass 100–150 mg.

The choice of optimal cryoprotectants when freezing GSCs is species specific. DMSO proved to be optimal in rainbow trout (Lee et al., 2013, 2016), tench (Linhartová et al., 2014; Marinović et al., 2016), goldfish (*Carassius auratus*; Marinović et al., 2017), common carp (Franěk et al., 2019a,b), zebrafish (Marinović et al., 2019a), and European eel (Šćekić et al., 2020), whereas ethylene glycol was optimal for Siberian sturgeon (*Acipenser baerii*; Pšenička et al., 2016) or methanol in Manchurian trout (*Brachymystax lenok*; Lee and

Yoshizaki, 2016). Optimal cryoprotectant concentrations are also species specific; however, a general trend is higher concentrations are optimal for cyprinids (Franěk et al., 2019a; Marinović et al., 2016), while lower concentrations are optimal for salmonids (Lee et al., 2013, 2016, Lee and Yoshizaki, 2016). As for cooling rates, lower cooling rates of approximately 1 °C/min are optimal as demonstrated in common carp (Franěk et al., 2019a), Manchurian trout (Lee and Yoshizaki, 2016), and Siberian sturgeon (Pšenička et al., 2016). In addition to permeating cryoprotectants, nonpermeating cryoprotectants, which act as osmotic stabilizers (such as sugars or protein), can be added to the cryomedium. However, sugar and protein supplementation usually have no effect when freezing tissue pieces, since the uptake and exposure of cells to these substances is questionable due to testicular structure, and tissues already contain high levels of sugars and protein within the blood vessels. Therefore, additional supplementation usually does not have a significant protective role (Franěk et al., 2019a; Marinović et al., 2019a; Šćekić et al., 2020).

In recent years, the effectiveness of testicular and ovarian tissue vitrification in preservation of GSCs was tested. Vitrification offers several advantages to freezing such as cost- and time-effectiveness, lower volumes of liquid nitrogen needed, field-friendliness, smaller size of tissue pieces used, which could be advantageous in fish with small gonads (e.g., zebrafish), and in several cases higher viability. Usually, needle-immersed vitrification (NIV; Lujić et al., 2017b; Marinović et al., 2018) or copper mesh (Seki et al., 2017) is used as tissue carriers since they enable direct exposure of tissue to liquid nitrogen with minimal volumes of cryoprotectants attached to them, therefore maximizing cooling rates and reducing the concentration of cryoprotectants needed. The only downside to direct liquid nitrogen exposure is possible cross-contamination of samples by bacteria or viruses present in the liquid nitrogen, therefore steps for sterilization should be taken, and liquid nitrogen should not be reused when utilizing these methods.

The response of GSCs to vitrification is variable. SSC viability rates of up to 70%–80% were attained in zebrafish (Marinović et al., 2019a) and brown trout (Marinović et al., unpublished data), whereas significantly lower rates of 10%–20% were obtained in common carp (Franěk et al., 2019a) and European and African catfish (Marinović et al., unpublished data). The main difference between these two groups of species was the presence of spermatozoa. Brown trout testes were immature and contained only SSCs and somatic cells, whereas mature zebrafish testes proportionally contain far less spermatozoa compared to mature common carp and catfish species, where spermatozoa constitute up to 95% of all testicular cells. As sperm vitrification is usually ineffective compared to freezing, and yields lower viability and functionality of these cells after warming/thawing (Asturiano et al., 2017; Kása et al., 2017, 2018), the high proportion of spermatozoa within common carp and catfish species may have influenced lower vitrification success. Therefore, the choice of applying freezing or vitrification of a given testicular sample in cryobanking should mostly depend on the testicular structure and the number of spermatozoa present within the tissue. Vitrification could be recommended for immature testicular tissue or tissue which contains low numbers of spermatozoa, whereas freezing would be greatly advantageous for mature testicular tissue, which is heavily dominated by spermatozoa.

The most common end point in assessing the effectiveness of GSC cryopreservation trials is determining their viability, usually assessed through trypan blue (or some other) dye exclusion test. However, viability of cells does not always reflect their functionality as demonstrated many times in sperm where sperm viability and motility are not necessarily correlated. Therefore application of functionality tests is of utmost importance when cryopreserving GSCs. The most common functionality tests are proliferation of GSCs in in vitro cell culture or their ability to colonize, proliferate, and give rise to gametes after transplantation into suitable recipients. Transplantation studies conducted with cryopreserved rainbow trout OSCs and SSCs (Lee et al., 2013, 2016), common carp OSCs and SSCs (Franěk et al., 2019a,b), zebrafish SSCs (Marinović et al., 2019a), medaka SSCs (Seki et al., 2017), and Chinese rosy bitterling (*Rhodeus ocellatus ocellatus*) SSCs (Octavera and Yoshizaki, 2020) demonstrated that cryopreserved GSCs retain their functionality after cryopreservation (either through freezing or vitrification) and are able to colonize recipient gonads and produce progeny (Lee et al., 2013; Marinović et al., 2019a; Octavera and Yoshizaki, 2020).

Lastly, as pointed out by Franěk and Pšenička (2020), GSC cryopreservation mostly depends on its downstream application. Development of GSC cryopreservation protocols for many species without further development of means for its application (either through in vitro gametogenesis or transplantation) could be regarded as "work lacking in concept." Even though development of GSC cryopreservation protocols for many species has its indisputable merit in biobanking, possibilities for downstream applications need to be evaluated, making this technology more difficult to implement in practice than sperm cryopreservation. Attention must be given to plausibility of intra- or interspecific transplantation, i.e., the presence of appropriate recipients for transplantation, as well as the existence of sterilization methodology for recipients. In some cases, cryopreservation of GSCs can only be a first step toward development of further manipulation technologies which will enable a continuous pool of GSCs, as suggested by Šćekić et al. (2020). However, development of cryopreservation protocols for new species must go hand in hand with the development of further downstream applications such as in vitro culture and transplantation, at least if we want to successfully implement this technology in aquaculture or species conservation programs.

3.3 Embryos and larvae cryopreservation

Fish embryo cryopreservation has been a challenging objective for decades. The interest in developing fish embryo cryopreservation protocols can be explained considering the advantages associated to this type of material: from a conservation point of view, ensuring the preservation of both paternal and maternal genomes; while from an aquaculture point of view, it would significantly simplify the establishment and management of genetic selection programs in fish farms (Martínez-Páramo et al., 2017).

Challenges hindering fish embryo cryopreservation are well known and could be summarized in four areas: fish embryos have low surface-to-volume ratio, large size of yolk, low membrane permeability, and high chilling sensitivity (reviewed in Martínez-Páramo et al., 2017). In this section, different approaches in fish embryo cryopreservation used during the last decades will be reported including the limited success achieved together with future perspectives. A comparison with invertebrate success will be shown.

Embryo resistance to chilling and cryopreservation has been studied in more than 20 teleost species, all reviewed in Martinez-Páramo et al. (2017) and more recently in kelp grouper (*Epinephelus moara*; Zhang et al., 2020). However, there is no doubt that zebrafish, as a model species, has been the most studied and with recent success (Khosla et al., 2017).

Embryo chilling capacity, or in other words, the effects of subzero temperatures on embryo survival, has been one of the focus of research where authors have obtained good results in different species (Fornari et al., 2014; Pessoa et al., 2015; Robles et al., 2007). However, embryo survival after cryopreservation (controlled slow freezing or vitrification) has been very limited and very few studies reported preliminary positive results (Robles et al., 2005; Zhang et al., 2020). Several efforts have been made involving the development of cryopreservation protocols, such as the choice of cryoprotectants, freezing–thawing rates, adequate embryonic developmental stages; and studying membrane permeability and cryoprotectant penetration within the embryos, aiming to provide new tools to evaluate embryo viability/survival after freezing–thawing or to provide new technologies and procedures to improve embryo cryoresistance (Khosla et al., 2017; Martínez-Páramo et al., 2017). Advances in research reported different methods to evaluate cryoprotectant penetration within the embryos using impedance spectroscopy (Zhang et al., 2006), magnetic resonance microscopy (Hagedorn et al., 1996), HPLC (Cabrita et al., 2003), and/or scanning calorimetry (Liu et al., 2001). The scarce or nonexistent embryo survival after freezing–thawing has pushed researchers to explore different methods to evaluate cryopreservation protocols combining morphological studies with molecular biology methods in order to provide more information on embryo metabolic activity at enzymatic or molecular level (Desai et al., 2011; Robles et al., 2004). The use of techniques to overcome permeability barriers and introduce cryoprotectants inside embryo compartments was investigated using microinjection (Janik et al., 2000), ultrasound (Wang et al., 2008a, b), or other delivery systems to introduce antifreeze proteins into the embryos (Martínez-Páramo et al., 2009a; Robles et al., 2007), or even altering fish embryo membranes with aquaporin 3 to increase permeability (Hagedorn et al., 2002).

Contrarily to fish species, invertebrate embryos or larvae cryopreservation yielded better results, favored by some biological characteristics such as a limited embryo size, improving water and cryoprotectant exchange, a low yolk oocyte content and a holoblastic cleavage, improving cryoprotectant penetration (Robles et al., 2009a). Since the pioneering work published by Renard (1991), more than 40 studies have been published in a few invertebrate species (Martínez-Páramo et al., 2017; Paredes, 2016).

The increasing number of studies devoted to invertebrate embryo cryopreservation is explained by the recent improvement of farming techniques, including breeder selection and conservation purpose (Daly et al., 2018, Paredes, 2016). Embryo or larvae cryopreservation studies have mainly focused on Pacific oyster because of its high commercial value. Whatever the species, the survival remains limited ($<1\%$) but promising with reports on F1 larvae obtained from cryopreserved breeders (Suquet et al., 2014). When there is no possibility to follow survival for long periods of time, one of the methods used to assess the quality of postthaw larvae is the percentage of motile larvae. The decrease of larval movement velocity observed using a CASA device is suggested to be a quick and more reliable estimation of the quality of thawed Pacific oyster larvae (Suquet et al., 2012). However, studying the long-term rearing performances of thawed larvae is a prerequisite to the development of invertebrate cryobanking. According to (Paniagua-Chavez et al. (2000)), the survival of postthaw oysters was not different from the control after 4 months of development. Close to 3 years after larvae thawing, the growing-out and reproductive capacities of Pacific oysters were similar to those observed for unfrozen ones (Suquet et al., 2014). Both examples suggest an absence of genome alterations of thawed larvae, allowing subsequent development of these oysters and their use in cryobanks.

With the new technologies of vitrification, combined with laser warming and gold nanorods, already used in some invertebrate species (e.g., coral larvae; Daly et al., 2018) and model fish species (e.g., zebrafish embryos; Khosla et al., 2017), it is foreseen that this investigation will provide new insights on the cryopreservation of embryos and larvae from different aquatic organisms.

4. Examples of sperm cryopreservation in specific species

4.1 Endangered species

4.1.1 The European eel case study

The European eel is one of the 19 species included in the genus *Anguilla*. This is a demanded species with high market values, intensively fished at all its life stages, including glass, yellow, and silver eels. Moreover, wild glass eels are the base of the eel aquaculture, which supplies most of the consumed eels. The intense decline of its populations during the last decades, due to several factors, caused its inclusion as "Critically Endangered" on the Red List of Threatened Species, by the International Union for Conservation of Nature (IUCN). In 2007 the EU categorized the European eel as critically endangered and established measures for the recovery of the stock, while the Convention on International Trade in Endangered Species of Wild Fauna and Flora (CITES) included it in Appendix II, prohibiting the international trade of this species. Thus the conservation status of eel species justifies the needs for taking actions such as development of its reproduction in captivity and the control of fisheries.

Eel species require the use of hormonal treatments to induce sperm production (Gallego et al., 2012; Mylonas et al., 2017; Pérez et al., 2000) or the manipulation of environmental factors (Asturiano, 2020; Rozenfeld et al., 2019). The development of protocols for the cryopreservation of Japanese (*Anguilla japonica*) and European eels sperm started in the early 2000s (reviewed by Herranz-Jusdado et al., 2019c). Cryopreservation of Japanese eel sperm was first achieved by Tanaka et al. (2002a). Briefly, they designed an extender (containing NaCl, $NaHCO_3$, and soya lecithin) and added DMSO at 10% (v/v) as cryoprotectant. To prevent the activation of sperm motility due to the hypertonicity caused by DMSO, they included $NaHCO_3$ in the freezing medium (Tanaka et al., 2002b). Moreover, they defined the cryovials volume (2 mL) and the freezing and thawing conditions. They reported postthaw sperm motility values around 40%, but using thawed sperm in fertilization trials they found that the hatchability of the fertilized eggs was lower in comparison with eggs fertilized with fresh sperm.

Years later, parallel works of Müller et al. (2017) and Koh et al. (2017) assayed different extenders, cryoprotectants, cryovial volumes, and freezing–thawing conditions, defining alternative methods for the Japanese eel. In the first case, the embryos hatched using thawed sperm samples showed lower survival and higher malformation rates than those obtained with fresh sperm, indicating that the protocol was still suboptimal. Better results were obtained with Koh et al. (2017) method, which was used by Nomura et al. (2018) to establish a large-scale cryopreservation protocol for Japanese eel sperm in fertilization programs. They adapted the cooling rate to use 5 mL straws, and the cryoprotectants and extenders were as described by Koh et al. (2017). The fertilization trials did not show any difference in egg hatching or survival rates between cryopreserved and fresh sperm. Moreover, the morphology of larvae produced from cryopreserved sperm was like that of larvae from fresh sperm, and the larvae were further grown into normal Japanese glass eels.

Following the cryopreservation method described by Tanaka et al. (2002a), Asturiano et al. (2003, 2004) published the first method to cryopreserve European eel sperm. After testing different extenders, best results in terms of postthawing spermatozoa motility were obtained with the P1 medium (in mM: 125 NaCl, 20 $NaHCO_3$, 30 KCl, 2.5 $MgCl_2$, 1 $CaCl_2$, with pH adjusted to 8.5), isoionic to the seminal plasma of European eel. All the tested freezing media contained 10% DMSO as cryoprotectant, and different sperm dilution ratios were also tested. The freezing was done in 0.25 mL straws placed for 10 min, 5 cm above liquid nitrogen surface before plunging them into, and the thawing was carried out by submerging the straws in water at 20 °C during 45 s.

Shortly after, Müller et al. (2004) developed a quite different cryopreservation protocol for the sperm of the European species using a modified Kurokura solution as extender and 10% methanol as cryoprotectant. They got similar results to those obtained with Asturiano et al.'s protocol. Succeeding this study, Szabó et al. (2005) conducted a series of experiments to test different extenders and cryoprotectants (DMSO and methanol), getting their best results using Tanaka extender with DMSO (10%) or methanol (10%).

On the other hand, Asturiano's lab carried out studies to test the effect of DMSO, methanol, and other cryoprotectants, with different dilution ratios and freezing medium supplementation with fetal bovine serum (FBS) or L-α-phosphatidylcholine, on European eel spermatozoa motility, viability, and morphology (Asturiano et al., 2007; Garzón et al., 2008; Marco-Jiménez et al., 2006). Although valuable results were obtained using DMSO, the use of this cryoprotectant increased the medium osmolality, resulting in inducing spermatozoa motility activation and premature ATP consumption. To avoid this problem, Peñaranda et al. (2009) tested different combinations of pH and concentrations of $NaHCO_3$ (previously used in the Japanese eel sperm cryopreservation protocols; Tanaka et al., 2002b). Thus an improved medium based on P1, but containing 100 mM $NaHCO_3$ and pH 6.5, partially prevented the activation effect of DMSO. Moreover, they improved the method using a 1:2 (sperm:freezing medium) dilution, and the 0.25 mL straws were cooled 1.6 cm above liquid nitrogen surface for 5 min before being immersed into liquid nitrogen. Using this protocol, they obtained postthaw spermatozoa motility values close to 40% that was considered as enough for fertilization trials. In fact, using this last protocol, Asturiano et al. (2016) successfully used cryopreserved sperm in fertilization trials, producing viable offspring, although with a lower percentage of fertilized eggs than that observed with fresh sperm, and only a few larvae found at 55 h after fertilization. Similarly, Müller et al. (2018) successfully used cryopreserved European eel sperm to fertilize Japanese eel eggs, obtaining hybrid larvae. Their cryopreservation method included the use of a modified Tanaka's solution as extender and methanol 10% as cryoprotectant.

Although both protocols succeeded in fertilization trials, their numerous differences made necessary a comparison, looking for its standardization. Thus, the Hungarian and the Spanish groups conducted together a joint study, comparing their protocols using the same sperm samples (Herranz-Jusdado et al., 2019a). In summary, they found that the protocol using methanol, initially developed by the Hungarian group (Müller et al., 2017; Szabó et al., 2005), was better in terms of higher spermatozoa viability and motility, and lower methylation effects on the spermatozoa DNA than the protocol with DMSO, developed by Asturiano's lab. Very recently, Herranz-Jusdado et al. (2019b) modified the Hungarian protocol to scale up the volume of sperm cryopreserved using 2 and

5 mL cryotubes, by adapting the cooling rate. Moreover, adding egg yolk to the extender solution, they obtained motility values over 50%, being the highest motility ever reported in cryopreserved European eel sperm.

Kása et al. (2017) developed a vitrification protocol for European eel sperm, consisting in a sperm:diluent ratio of 1:1, with 40% cryoprotectant (20% methanol and 20% propylene glycol), and 10% FBS using Cryotops of 2 μL as cooling device. The percentage of motile spermatozoa after thawing was low in comparison to conventional sperm cryopreservation.

4.2 Aquacultured species

4.2.1 The salmonid species case study

Salmonid fish include some of the most common cultured species in the world, both in marine and freshwater environments, such as the Atlantic salmon or the rainbow trout. Cryopreservation techniques have been developed for these species since the 1960s (Hodgins and Ridgway, 1964), and they probably have the most abundant scientific literature on sperm freezing among all groups of teleosts. Cryopreservation of salmonid sperm is plausibly the closest to industrial application, and products such as extenders are readily available for these species in commercial trade. Private companies are also known to specialize in the cryopreservation of salmonid sperm and offering it as a commercial service. Salmonids also include rare species that are either endangered or are of local interest such as the huchen (*Hucho hucho*) or the Mediterranean brown trout (*Salmo cettii*) for which sperm cryopreservation has also been reported (Di Iorio et al., 2019; Nynca et al., 2015). For these reasons, we are going to concentrate on the developments of the last decade (2010–20, Table 1) as this period brought significant changes in research that facilitate the adoption of cryopreservation methods to the practice.

Methods developed in the time period mentioned before are closely linked to a given research group. Thus, the choice of extenders can include well-known recipes such as that of Erdahl and Graham (1987) used by Pérez-Cerezales et al. (2010) and Fernández-Díez and Herráez (2018), or the Cortland medium reported by Truscott et al. (1968) and used by Figueroa et al. (2016a, b, 2019) and Díaz et al. (2019). Other teams prefer to use their own extenders which can be as simple as a 0.15 M glucose solution (Ciereszko et al., 2014; Judycka et al., 2018; Nynca et al., 2015, 2017) or multicomponent such as those by Lahnsteiner et al. (2011) or Kutluyer et al. (2014). Almost all extenders employ sugars (typically glucose, sucrose, or trehalose) in varying concentrations, both as nonpenetrating cryoprotectants and for osmoregulation. The addition of various antioxidants did not have a beneficial effect on the fertilizing capacity of salmonid sperm (Kutluyer et al., 2014; Lahnsteiner et al., 2011), although it has reduced oxidative stress in some species (Sandoval-Vargas et al., 2021). The choice of cryoprotectants has been reduced to methanol and DMSO in concentrations ranging from 7% to 10% (v/v). This decade saw the emergence of

TABLE 1 Summary table of protocols used for the cryopreservation of salmonid species in the period between 2010 and 2020. Only the protocols yielding the best results are shown.

Reference	Species	Extender	Cryoprotectant
Ciereszko et al. (2013)	Whitefish (*Coregonus lavaretus*)	0.225 M glucose	7.5% methanol
Ciereszko et al. (2014)	Rainbow trout (*Oncorhynchus mykiss*)	0.15 M glucose	7.5% methanol
Di Iorio et al. (2019)	Mediterranean brown trout (*Salmo cettii*)	300 mM glucose, 10% egg yolk (extender A), 75 mM NaCl, 70 mM KCl, 2 mM $CaCl_2$, 1 mM $MgSO_4$, 20 mM TRIS, 10% egg yolk (extender B)	10% DMSO
Dziewulska and Domagała (2013)	Sea trout (*Salmo trutta* m. *trutta*)	0.3 M glucose, 10% hen egg yolk	10% methanol
Figueroa et al. (2016a, b)	Atlantic salmon (*Salmo salar*)	Cortland medium	1.3 M DMSO
Horváth et al. (2012)	Adriatic grayling (*Thymallus thymallus*)	200 mM glucose, 40 mM KCl, 30 mM Tris, pH 8.0	10% methanol
Judycka et al. (2016)	Rainbow trout (*O. mykiss*)	175–200 mM trehalose	10% methanol
Judycka et al. (2018)	Brown trout (*S. trutta* m. *fario*), sea trout (*S. trutta* m. *trutta*), brook trout (*Salvelinus fontinalis*), Atlantic salmon (*S. salar*)	0.15 M glucose	7.5% methanol
Kutluyer et al. (2014)	Rainbow trout (*O. mykiss*)	6.52 mg/mL NaCl, 0.8 mg/mL KCl, 2 mg/mL $NaHCO_3$, 2 mg/mL glucose, 4 mg/mL BSA, 7.5% egg yolk	10% DMSO
Lahnsteiner et al. (2011)	Rainbow trout (*O. mykiss*), brook trout (*S. fontinalis*)	103 mM NaCl, 40 mM KCl, 1 mM $CaCl_2$, 0.8 mM $MgSO_4$, 20 mM Hepes (pH 7.8), 1.5% BSA; 7% egg yolk, 0.5% sucrose	10% methanol
Lee-Estevez et al. (2019)	Atlantic salmon (*S. salar*)	Strofish®, 0.3 M glucose, 2% BSA	1.3 M DMSO
Nynca et al. (2015)	huchen (*Hucho hucho*), grayling (*T. thymallus*)	0.15 M glucose	7.5% methanol
Nynca et al. (2017)	Rainbow trout (*O. mykiss*)	0.15 M glucose	7.5% methanol
Pérez-Cerezales et al. (2010)	Rainbow trout (*O. mykiss*)	Erdhal and Graham extender	7% DMSO

Dilution ratio or sperm concentration	Container	Cooling conditions	Thawing conditions	Representative results
1:3, 1:1, 3:1	0.25 mL straws	LN_2 vapor, 3 cm, 3 min	40 °C, 4 s	21.6%–27.5% postthaw motility
1:5	0.25 mL straws	LN_2 vapor, 3 cm, 3 min	40 °C, 5 s	84.4% hatch
1:3	0.25 mL straws	LN_2 vapor, 5 cm, 10 min	30 °C, 10 s	42.6% postthaw motility, 36.5% eyed eggs, 32.5% hatch
1:3	0.25 mL straws	LN_2 vapor, 4 cm, 3 min	40 °C, 5–6 s	10.2% motility, up to 94% eyed eggs
1:3	0.5 mL straws	LN_2 vapor, 2 cm, 10 min	35 °C, 9 s	50.1% postthaw motility, 81.6% fertilization
1:1	0.5 mL straws	LN_2 vapor, 3 cm, 3 min	40 °C, 13	Up to 82% hatch
1:11.5	0.25 mL straws	LN_2 vapor, 3 cm, 5 min	40 °C, 5 s	up to 79% eyed eggs, up to 78.6% hatch
Brown trout and sea trout: 3.0×10^9/mL Brook trout: 2.0×10^9/mL Atlantic salmon: 4.0×10^9/mL	0.5 mL straws	LN_2 vapor, 3 cm, 5 min	40 °C, 10 s	Postthaw motility, brown trout: 60%–70%, sea trout: 70%, brook trout: 56%, Atlantic salmon: 63%
1:10	0.25 mL straws	LN_2 vapor, 3 cm, 10 min	40 °C, 5 s	Up to 90.0% fertilization, up to 80.1% hatch
1:3	0.5 mL straws	LN_2 vapor, 1.5 cm for rainbow trout, 2.5 cm for brook trout, 10 min	25 °C, 30 s	Up to 92% eyed eggs in the rainbow trout, up to 76.3% eyed eggs in brook trout
1:3	0.5 mL straws	LN_2 vapor, 2 cm, 10 min	37 °C, 7 s	15.1% postthaw motility, 46.6% membrane integrity
1:5	0.25 mL straws	LN_2 vapor, 3 cm, 5 min	40 °C, 5 s	Eyed eggs, huchen: 88%, grayling: 50%; hatch, huchen: 76%–80%
0.5–1.0 $\times 10^9$/mL	0.5 mL straws	LN_2 vapor, 3 cm, 5 min	40 °C, 10 s	60%–66% postthaw motility
1:3	0.5 mL straws	LN_2 vapor, 2 cm, 10 min	25 °C, 30 s	37%–41% postthaw cell viability, 60% eyed eggs

methanol as a highly effective cryoprotectant for salmonid sperm following the groundbreaking work of Lahnsteiner et al. (1997).

Significant progress has been made toward the standardization of cryopreservation methods for salmonids. The team of Ciereszko has optimized the final (postdilution) concentration of the extender and cryoprotectant to 0.15 M glucose and 7.5% methanol by adjusting them to a standard concentration of spermatozoa in one straw instead of the traditional dilution ratios of sperm to extender (Judycka et al., 2018; Nynca et al., 2017). Using a uniform sperm concentration has been the norm in the cryopreservation of sperm from domestic livestock; however, this has not been adapted to fish sperm. Standardization of sperm concentration has been facilitated by the use of automated cell counters instead of the slower and more cumbersome process of counting cells in a hemocytometer. Optimal cell concentrations for the cryopreservation of sperm from various salmonids are species specific and result in significantly reduced variation among individual samples expressed in postthaw motility.

Progress has also been made for the optimization of the conditions of fertilization with salmonid sperm. In rainbow trout, sperm:egg ratios as low as 6×10^5:1 resulted in fertilization and hatch percentages that were not significantly different from the control (84.4% with frozen sperm and 87% with fresh sperm, Ciereszko et al., 2014). This was in stark contrast with previous reports of effective ratios as high as 3×10^6:1 (Babiak et al., 2001; Stoss and Holtz, 1981). Later, sperm:egg ratios of 3×10^5:1 were found optimal for the brook trout (*Salvelinus fontinalis*, Nynca et al., 2014), 1.5×10^5:1 for the Adriatic grayling, and 5×10^5:1 for the huchen (Nynca et al., 2015). In Adriatic grayling, this ratio has been reduced to as low as 5×10^4:1 in the work of Horváth et al. (2015a). The time period elapsed between thawing and fertilization with cryopreserved sperm (postthaw storage time) has also been a crucial point in salmonid sperm cryopreservation as earlier studies suggested that as little as 30 s to 2 min already significantly reduced the fertilizing capacity of sperm (Lahnsteiner et al., 1996; Stoss and Holtz, 1981). We have found that application of methanol as cryoprotectant (as opposed to DMSO) allowed storage of thawed sperm for 1 h in rainbow trout and 10 min in marble trout (*Salmo marmoratus*) without loss of fertilizing capacity. Postthaw storage of Adriatic grayling and brown trout sperm for up to 1 h yielded similar fertilization percentages as the fresh control (Horváth et al., 2015b). These findings were also confirmed in brook trout (Nynca et al., 2014).

Current trends in the investigation of cryopreserved sperm quality go beyond the traditional assessment of fertilizing capacity. Nusbaumer et al. (2019) found that although the fertilizing capacity of brown trout sperm was not affected by cryopreservation, it led to significantly impaired larval growth, which was not dependent on the individual males. A transcriptomic study on rainbow trout larvae hatched from fertilization with fresh or cryopreserved sperm has determined that cryogenic damage to sperm DNA also led to differentially expressed genes affecting metabolic processes as well as cellular differentiation, signaling, and trafficking among others (Fernández-Díez and Herráez, 2018). These findings urge investigations into the DNA-damaging properties of any cryopreservation protocol before being adopted into aquaculture practice.

4.3 Model species

During the last years, zebrafish and other ornamental fish have become more important as model species and the creation of sperm cryobanks has become an important demand in some of these species. Some constraints are still present, although in the last decade several studies have emerged, especially in zebrafish (Table 2). In contrast to commercial fish species, research model fishes are characterized by small body sizes, producing limited sperm volume (1–2 μL; Yang and Tiersch, 2009), which restricts the possibility to develop several studies to characterize properly the sperm quality before and after cryopreservation. This fact, together with the significant heterogeneity of sperm samples, creates problems for male selection and sample quality identification for cryopreservation. Like for other fishes, sperm cryopreservation of model species, such as zebrafish, medaka, or Southern platyfish (*Xiphophorus maculatus*), follows several procedures to be stored properly (Da Silva et al., 2019; Diogo et al., 2019a, b; Martínez-Páramo et al., 2017; Robles et al., 2009b; Yang and Tiersch, 2009). The choice of sample collection method, sample processing, freezing and storage methods, thawing, and quality assessment are critical, especially when very small volumes of sperm are at risk and no errors can be committed. In this section, we will focus on zebrafish sperm since it has been the species with more advancements on its cryopreservation in the last years.

Collection of sperm is usually performed in MS222 anesthetized fish using a glass capillary filled with Hanks balanced salt solution (HBSS), or directly with a very small micropipette tip (10 μL), followed by a dilution in HBSS to avoid sperm dehydration, since the volumes collected are very small (Diogo et al., 2019a, b). This last method works very well since the volume of sperm collected can be measured before dilution. Other methods such as testis ablation have been used in the past (reviewed in Robles et al., 2009b), but it required the sacrifice of the animals and the quality of samples was poorer, since apart from sperm there are other germ cell populations during testis processing. Sample processing includes the suspension of sperm in the extender, sperm concentration adjustment, and quality assessment before starting any process of cryopreservation. According to Diogo et al. (2019a, b), only samples with higher concentration than 3×10^7 cells/mL, and total motility of $>50\%$, should be considered for cryostorage.

Sample packaging can influence the cooling rate during freezing, storage efficiency after freezing, sample identification, and biosecurity. In zebrafish and medaka, glass capillary tubes or cryovials were used in sperm cryopreservation (Aoki et al., 1997; Diogo et al., 2018, Table 2). To standardize protocols with potential for high-throughput automation, 0.25 mL French straws were

TABLE 2 Cronological order of zebrafish sperm cryopreservation protocols developed since Harvey's work (1982).

Reference	Method of sperm extraction	Extender + CPA	Sperm container	Cooling rate	Thawing method	Postthaw analysis
Harvey et al. (1982)	Stripping	Ginsburg + MeOH + skim milk	Capillary	16 °C/mim (dry ice)	Room temperature	Motility estimation, hatching rate
Yang et al. (2007)	Testis dissection	HBSS + DMA/ DMSO/glycerol/ MeOH	0.25 mL French straws	10-20 °C/min (biofreezer)	40 °C, 5 s	Motility estimation, cell viability
(Morris et al., 2009)	Testis dissection	BSMIS + DMA	Cryovials	Dry ice	37 °C, 10-15 s	Motility estimation, fertilization rate
Draper and Moens (2009)	Stripping	Ginsburg/HBSS + MeOH + skim milk	Cryovials	Dry ice	33 °C, 8–10 s	Fertilization rate
Carmichael et al. (2009)	Stripping/testis dissection	HBSS + MeOH	Cryovials	Dry ice	–	–
Hagedorn et al. (2012a, b)	Testis dissection	HBSS + DMA/ DMSO/MeOH	0.25 mL French straws	10 °C/min (biofreezer)	30 °C, 20–40 s	Motility estimation and cell viability
Bai et al. (2013)	Testis dissection	HBSS + DMSO	0.25 mL French straws	25 °C/min until −30 °C; 5 °C/min until −80 °C (biofreezer)	40 °C, 7 s	Motility by CASA and cell viability
Wang et al. (2015)	Testis dissection	HBSS + DMSO	0.25 mL French straws	25 °C/min until −30 °C; 5 °C/min until −80 °C (biofreezer)	40 °C, 7 s	Motility by CASA, cell viability, ROS, MDA, ATP, MMP, hatching rate, larval survival

Yang et al. (2016)	Testis dissection	HBSS/Glucose + MeOH	0.25 mL French straws	10 °C/min (biofreezer)	40 °C, 5 s	Motility estimation, cell viability, fertilization rate, embryo survival
Diogo et al. (2018)	Stripping	HBSS + DMF	Cryovials	20 °C/min (biofreezer); 66 °C/min (ultrafreezer)	40 °C, 8 s	Motility by CASA, cell viability, apoptosis, fertilization rate, embryo survival, hatching rate
Matthews et al. (2018)	Stripping	E400 + MeOH	Cryovials	5-35 °C/min (biofreezer/dry ice)	38 °C, 10–15 s	Motility by CASA, fertilization rate
Da Silva et al. (2019)	Stripping	Grayling extender + MeOH	0.25 mL French straws	Dry ice	40 °C, 5 s	Fertilization rate

HBSS, Hank's balanced salt solution; MeOH, methanol; DMSO, dimethylsulfoxide, DMA, dimethylacetamide; DMF, dimethylformamide; CASA, computer-assisted sperm analysis; ROS, reactive oxygen species; MDA, malondialdehyde (lipid peroxidation); MMP, mitochondrial membrane potential; E400: 130 mM KCL, 50 mM NaCl, 2 mM $CaCl_2$, 1 mM $MgSO_4$, 10 mM D-(+)-Glucose, and 30 mM HEPES-KOH (pH 7.9, 400 mmol/kg).

chosen for packaging the sperm of medaka and *Xiphophorus* fishes (Yang and Tiersch, 2009; Yang et al., 2010). Cryopreservation of zebrafish sperm using cryovials yielded very good results when combined with fast freezing rates. This controlled cooling can be achieved with programmed biofreezers using cooling rates for zebrafish sperm that are higher than − 50 °C/min. Other simpler methods have been used in the past by placing samples on dry ice or suspension in liquid nitrogen vapor (Table 2) that can be used in field situations. Programmable freezers are typically used to produce accurate and repeatable cooling rates, especially for sperm that are sensitive to small variations in cooling rate during freezing. For example, for medaka sperm, a change of 5 °C/min in cooling rate yielded a significant change in postthaw sperm motility (Yang et al., 2010). Therefore strict control of cooling rate can assure quality and uniformity of cryopreserved sperm. Recently, Diogo et al. (2018) developed a method to freeze and directly store samples in a − 150 °C biofreezer with good results. However, the traditional method to store zebrafish sperm, as other fish sperm species, is in the liquid nitrogen. In the case of zebrafish, an important aspect when using cryovials is the possibility of frozen samples to float, especially when small volumes are packaged in lightweight containers with a large air space. To avoid this, there are goblets with caps or cryotube supports that individually hold and fix each tube.

Postthaw quality assessment methods for model fishes include motility estimations or the use of CASA (described in Section 1.3), evaluation of cell constituents by flow cytometry (Daly and Tiersch, 2011), and DNA damage analysis using comet assay, all the analytical techniques that consume very few microliters of sperm. Procedures are not very different from the ones previously described in Section 1.4. Ultimately, fertilization with cryopreserved sperm and hatching rate are the methods that can attest for a successful cryopreservation procedure. After fertilization, eggs can be incubated at 28.5 °C in petri dishes with daily water removal. Diogo et al. (2019b) showed that assessment of embryonic survival can be done at 24 h postfertilization yielding the same results as hatching, reducing the time needed to verify fertilization success with cryopreserved sperm. Fertilization rates are variable with cryopreserved zebrafish sperm (60%–80%, at morula stage, Diogo et al., 2018; Yang et al., 2016) and is reduced to 4%–6% at 24 h postfertilization or hatching (Diogo et al., 2018). These data show again the high variability between samples, experiments, males and highlights the importance of several aspects that need to be taken into consideration such as age, body weight, body length, broodstock husbandry conditions (e.g., feeding and spawning setup), male stripping frequency, motility activation conditions, etc. (Diogo et al., 2019a, b; Martins et al., 2020).

5. Conclusions and future perspectives

A large number of studies have been performed during the last decades to design cryopreservation protocols for ex situ conservation of aquatic organisms.

Research on germplasm cryobanking in aquatic species has embraced diverse cell types, including sperm, oocytes, somatic cells, spermatogonia, and PGCs, besides cryopreservation of embryos and larvae. Fish sperm cryopreservation has been the focus of research mainly due to the handicaps presented by other cell types such as oocytes or embryos, which still need technology development to guarantee enough cryoprotectant distribution into all compartments for a successful cryopreservation. Studies on cryopreservation of GSCs can constitute an alternative for the cryopreservation of both paternal and maternal genomes. However, more work needs to be carried out in the development and optimization of some reproductive biotechnological tools, such as transplantation, to restore an individual or a species, since up to the moment there are few applications. Cryopreservation methods have been developed for various purposes: to assist reproduction in aquaculture, to be applied in selective breeding programs, or species conservation actions. The aquaculture industry has been showing technological advancements in this area accompanied by an increasing interest on sperm cryopreservation. However, the application of cryopreservation methods as part of a standard methodology used by fish farming industry is still limited to sperm of few species of interest (mostly Atlantic salmon) in comparison to the cryopreservation industry serving domestic livestock farming. To overcome this limitation, cryobiologists are doing a huge effort to reduce ambiguities in results, reporting and in maintaining more collaborations with the industry to promote knowledge transfer. Some of these efforts have contributed to the application of this technology in a wider number of species in recent years, but still for particular aspects (e.g., selective breeding). Some of the areas in which future research will contribute with significant advances are (i) the identification of the key parameters defining sample susceptibility to cryopreservation, in order to establish criteria for sample selection. In this aspect, sperm (seminal plasma and spermatozoa) proteome analysis or the relevance of specific mRNAs transported into fish spermatozoa may represent key elements in research development to achieve this goal. Another aspect is (ii) the use of different approaches for increasing sperm pre- and postfreezing/thawing quality. This could be achieved through breeder nutrition, supplementation of sperm extenders with particular compounds, and with a more exhaustive quality analysis. Finally, (iii) given the high potential of sperm cryopreservation in gene banking and genetic selection, the analysis and evaluation of the preservation and transmission of a faithful genome must be guaranteed. For that it is important to study specific genes that can be altered by cryopreservation and look for their methylation changes and how they can alter progeny. Epigenetic risks associated with permeant cryoprotectants must be investigated.

Acknowledgments

We would like to thank the financial support of the Spanish Ministry of Science Innovation and Universities through project EELGONIA (RTI2018-096413-B-I00). In Portugal the work was financed by Interreg POCTEP through project VENUS (0139_VENUS_5_E), ASSEMBLE+ JRA2 CRYOMAR project (H2020-INFRAIA-2016-2017, N 730984), and European Regional Development Fund, Interreg Atlantic Area (EBB-EAPA_501/2016 project). This study was also supported by Portuguese national funds from Foundation for Science and Technology (FCT) through project UID/Multi/04326/2021 and from the operational programs CRESC Algarve 2020 and COMPETE 2020. In Hungary, the work was supported by the Ministry of Innovation and Technology within the framework of the Thematic Excellence Programme 2020, Institutional Excellence Subprogramme (TKP2020-IKA-12), the NKFIH projects K129127 and FK124585, Fisheries Fund for Renewable Fisheries provided by the EU and Hungary, as well as the Stipendium Hungaricum Scholarship Programme (grant 106360).

References

Alquézar-Baeta, C., Gimeno-Martos, S., Miguel-Jiménez, S., Santolaria, P., Yániz, J., Palacín, I., Casao, A., Cebrián-Pérez, J.A., Muiño-Blanco, T., Pérez-Pe, R., 2019. OpenCASA: a new open-source and scalable tool for sperm quality analysis. PLoS Comput. Biol. 15 (1), e1006691.

Aoki, K., Okamoto, M., Tatsumi, K., Ishikawa, Y., 1997. Cryopreservation of medaka spermatozoa. Zoolog. Sci. 14, 641–644.

Asturiano, J.F., 2020. Improvements on the reproductive control of the European eel. In: Yoshida, M., Asturiano, J.F. (Eds.), Reproduction in Aquatic Animals. From Basic Biology to Aquaculture Technology. Springer Nature, pp. 293–320. ISBN: 978-981-15-2290-1 (eBook). Chapter 14.

Asturiano, J.F., Pérez, L., Marco-Jiménez, F., Olivares, L., Vicente, J.S., Jover, M., 2003. Media and methods for the cryopreservation of European eel (*Anguilla anguilla*) sperm. Fish Physiol. Biochem. 28, 501–502.

Asturiano, J.F., Pérez, L., Garzón, D.L., Marco-Jiménez, F., Peñaranda, D.S., Vicente, J.S., Jover, M., 2004. Physio-chemical characteristics of seminal plasma and development of media and methods for the cryopreservation of European eel sperm. Fish Physiol. Biochem. 30, 283–293.

Asturiano, J.F., Marco-Jiménez, F., Peñaranda, D.S., Garzón, D.L., Pérez, L., Vicente, J.S., Jover, M., 2007. Effect of sperm cryopreservation on the European eel sperm viability and spermatozoa morphology. Reprod. Domest. Anim. 42, 162–166.

Asturiano, J.F., Sørensen, S.R., Pérez, L., Lauesen, P., Tomkiewicz, J., 2016. First production of larvae using cryopreserved sperm. Effects of preservation temperature and cryopreservation on European eel sperm fertilization capacity. Reprod. Domest. Anim. 51, 485–491.

Asturiano, J.F., Cabrita, E., Horváth, Á., 2017. Progress, challenges and perspectives on fish gamete cryopreservation: a mini-review. Gen. Comp. Endocrinol. 245, 69–76.

Babiak, I., Glogowski, J., Goryczko, K., Dobosz, S., Kuzminski, H., Strzezek, J., Demianowicz, W., 2001. Effect of extender composition and equilibration time on fertilization ability and enzymatic activity of rainbow trout cryopreserved spermatozoa. Theriogenology 56, 177–192.

Bai, C., Wang, X., Lu, G., Wei, L., Liu, K., Gao, H., Huang, C., Dong, Q., 2013. Cooling rateoptimization for zebrafish sperm cryopreservation using a cryomicroscope coupled with SYBR14/PI dual staining. Cryobiology 67, 117–123.

Beirão, J., Pérez-Cerezales, S., Martínez-Páramo, S., Herráez, M.P., 2010. Detection of early damage of sperm cell membrane in gilthead seabream (*Sparus aurata*) with the nuclear stain YO-PRO 1. J. Appl. Ichthyol. 26, 794–796.

Beirão, J., Cabrita, E., Pérez-Cerezales, S., Martínez-Páramo, S., Herráez, M.P., 2011. Effect of cryopreservation on fish sperm subpopulations. Cryobiology 62, 22–31.

Beirão, J., Zilli, L., Vilella, S., Cabrita, E., Schiavone, R., Herráez, M.P., 2012. Improving sperm cryopreservation with antifreeze proteins: effect on gilthead seabream (*Sparus aurata*) plasma membrane lipids. Biol. Reprod. 86 (59), 1–9.

Beirão, J., Soares, F., Pousão-Ferreira, P., Diogo, P., Dias, J., Dinis, M.T., Herráez, M.P., Cabrita, E., 2015. The effect of enriched diets on *Solea senegalensis* sperm quality. Aquaculture 435, 187–194.

Beirão, J., Boulais, M., O'Brien, J.K., Peixoto, S., Robeck, T.R., Cabrita, E., 2019. Sperm handling in aquatic animals for artificial reproduction. Theriogenology 133, 161–178.

Bernáth, G., Bokor, Z., Kása, E., Várkonyi, L., Hegyi, Á., Kollár, T., Urbányi, B., Żarski, D., Radóczi, L.F.J., Horváth, Á., 2015. Comparison of two different methods in the cryopreservation of Eurasian perch (*Perca fluviatilis*) sperm. Cryobiology 70, 76–78.

Blaxter, J.H.S., 1953. Sperm storage and cross-fertilization of spring and autumn spawning herring. Nature 26, 1189–1190.

Bokor, Z., Urbányi, B., Horváth, L., Horváth, Á., 2009. Commercial-scale cryopreservation of wels catfish (*Silurus glanis*) semen. Aquacult. Res. 41, 1549–1551.

Bokor, Z., Bernáth, G., Várkonyi, L., Molnár, J., Láng, Z.L., Tarnai-Király, Z., Solymosi, E., Urbányi, B., 2019. The applicability of large-scale sperm cryopreservation in wels catfish (*Silurus glanis*) optimized for hatchery practice. Aquaculture 506, 337–340.

Butts, I.A.E., Hilmarsdóttir, G.S., Zadmajid, V., Gallego, V., Holst, L.K., Støttrup, J., Jacobsen, C., Krüger-Johnsen, M., Politis, S., Asturiano, J.F., Holst, L.K., Tomkiewicz, J., 2020. Dietary amino acids impact sperm performance traits for a catadromous fish, *Anguilla anguilla* reared in captivity. Aquaculture 518, 734602.

Cabrita, E., Alvarez, R., Anel, L., Rana, K., Herráez, M.P., 1998. Sublethal damage during cryopreservation of rainbow trout sperm. Cryobiology 37 (3), 245–253.

Cabrita, E., Robles, V., Alvarez, R., Herráez, M.P., 2001a. Cryopreservation of rainbow trout sperm in large volume straws: application to large scale fertilization. Aquaculture 201, 301–314.

Cabrita, E., Robles, V., Chereguini, O., de Paz, P., Anel, L., Herraez, M.P., 2003. Dimethyl sulfoxide influx in turbot embryos exposed to a vitrification protocol. Theriogenology 60, 463–473.

Cabrita, E., Robles, V., Rebordinos, L., Sarasquete, C., Herráez, M.P., 2005. Evaluation of DNA damage in rainbow trout (*Oncorhynchus mykiss*) and gilthead sea bream (*Sparus aurata*) cryopreserved sperm. Cryobiology 50, 144–153.

Cabrita, E., Soares, F., Dinis, M.T., 2006. Characterization of Senegalese sole, *Solea senegalensis*, male broodstock in terms of sperm production and quality. Aquaculture 261, 967–975.

Cabrita, E., Engrola, S., Conceição, L.E.C., Pousão-Ferreira, P., Dinis, M.T., 2009a. Successful cryopreservation of sperm from sex-reversed dusky grouper, *Epinephelus marginatus*. Aquaculture 287, 152–157.

Cabrita, E., Robles, V., Herráez, M.P., 2009b. Methods in Reproductive Aquaculture: Marine and Freshwater Species. CRC Press, Boca Raton, FL, USA. 549 pp.

Cabrita, E., Sarasquete, C., Martínez-Páramo, S., Robles, V., Beirão, J., Pérez-Cerezales, S., Herráez, M.P., 2010. Cryopreservation of fish sperm: applications and perspectives. J. Appl. Ichthyol. 26, 623–635.

Cabrita, E., Ma, S., Diogo, P., Martínez-Páramo, S., Sarasquete, C., Dinis, M.T., 2011. The influence of certain aminoacids and vitamins on post-thaw fish sperm motility, viability and DNA fragmentation. Anim. Reprod. Sci. 125, 189–195.

Cabrita, E., Martínez-Páramo, S., Gavaia, P.J., Riesco, M.F., Valcarce, D.G., Sarasquete, C., Herráez, M.P., Robles, V., 2014. Factors enhancing fish sperm quality and emerging tools for sperm analysis. Aquaculture 432, 389–401.

Carmichael, C., Westerfield, M., Varga, Z.M., 2009. Cryopreservation and in vitro fertilization at the Zebrafish International Resource Center. In: Lieschke, G.J., Oates, A.C., Kawakami, K. (Eds.), Zebrafish: Methods and Protocols. Humana Press, Totowa, NJ, pp. 45–65.

Chen, S.U., Yang, Y.S., 2007. Vitrification of oocytes: various procedures. In: Tucker, M.J., Liebermann, J. (Eds.), Vitrification in Assisted Reproduction. Informa Healthcare, London, UK, pp. 129–144.

Chereguini, O., Cal, R.M., Dreanno, C., De Baulny, B.O., Suquet, M., Maisse, G., 1997. Short-term storage and cryopreservation of turbot (*Scophthalmus maximus*) sperm. Aquat. Living Resour. 10, 251–255.

Chereguini, O., Garcia de la Banda, I., Herrera, M., Martinez, C., De la Hera, M., 2003. Cryopreservation of turbot *Scophthalmus maximus* (L.) sperm: Fertilization and hatching rates. Aquacult. Res. 34, 739–747.

Ciereszko, A., Dietrich, G.J.J., Nynca, J., Liszewska, E., Karol, H., Dobosz, S., 2013. The use of concentrated extenders to improve the efficacy of cryopreservation in whitefish spermatozoa. Aquaculture 408–409, 30–33.

Ciereszko, A., Dietrich, G.J., Nynca, J., Dobosz, S., Zalewski, T., 2014. Cryopreservation of rainbow trout semen using a glucose-methanol extender. Aquaculture 420–421, 275–281.

Ciereszko, A., Dietrich, M.A., Nynca, J., 2017. Fish semen proteomics- New opportunities in fish reproductive research. Aquaculture 472, 81e92.

Cuevas-Uribe, R., 2011. A General Approach for Vitrification of Fish Sperm. LSU Doctoral Dissertations, p. 3597. http://digitalcommons.lsu.edu/gradschool_dissertations/3597.

Cuevas-Uribe, R., Leibo, S.P., Daly, J., Tiersch, T.R., 2011a. Production of channel catfish with sperm cryopreserved by rapid non-equilibrium cooling. Cryobiology 63, 186–197.

Cuevas-Uribe, R., Yang, H., Daly, J., Savage, M.G., Walter, R.B., Tiersch, T.R., 2011b. Production of F_1 offspring with vitrified sperm from a live-bearing fish, the green swordtail *Xiphophorus hellerii*. Zebrafish 8, 167–179.

Cuevas-Uribe, R., Chesney, E.J., Daly, J., Tiersch, T.R., 2015. Vitrification of sperm from marine fish: effect on motility and membrane integrity. Aquacult. Res. 46, 1770–1784.

Da Silva, C.C., Kollár, T., Csenki-Bakos, Z., Fernandes, B.H.V., Horváth, Á., Carvalho, L.R., 2019. An easy method for cryopreservation of zebrafish (*Danio rerio*) sperm. Zebrafish 16 (3), 321–323.

Daly, J., Tiersch, T.R., 2011. Flow cytometry for the assessment of sperm quality in aquatic species. In: Tiersch, T.R., Green, C. (Eds.), Cryopreservation in Aquatic Species, second ed. World Aquaculture Society, Baton Rouge, Louisiana, pp. 201–207.

Daly, J., Zuchowicz, N., Lendo, C.I.N., Khosla, K., Lager, C., Henley, E.M., Bischof, J., Kleinhans, F.W., Lin, C., Peters, E.C., Hagedorn, M., 2018. Successful cryopreservation of coral larvae using vitrification and laser warming. Sci. Rep. 8, 15714.

De Baulny, B.O., LeVern, Y., Kerboeuf, D., Maisse, G., 1997. Flow cytometric evaluation of mitochondrial activity and membrane integrity in fresh and cryopreserved rainbow trout (*Oncorhynchus mykiss*) spermatozoa. Cryobiology 34, 141–149.

Depincé, A., Gabory, A., Dziewulska, K., Le Bail, P., Jammes, H., Labbé, C., 2019. DNA methylation stability in fish spermatozoa upon external constraint: Impact of fish hormonal stimulation and sperm cryopreservation. Mol. Reprod. Dev. 87, 124–134.

Desai, K., Spikings, E., Zhang, T., 2011. Effect of chilling on *sox2*, *sox3* and *sox19a* gene expression in zebrafish (*Danio rerio*) embryos. Cryobiology 63, 96–103.

Di Iorio, M., Esposito, S., Rusco, G., Roncarati, A., Miranda, M., Gibertoni, P.P., Cerolini, S., Iaffaldano, N., 2019. Semen cryopreservation for the Mediterranean brown trout of the Biferno River (Molise- Italy): comparative study on the effects of basic extenders and cryoprotectants. Sci. Rep. 9 (9), 9703. https://doi.org/10.1038/s41598-019-45006-4.

Díaz, R., Lee-Estevez, M., Quiñones, J., Dumorné, K., Short, S., Ulloa-Rodríguez, P., Valdebenito, I., Sepúlveda, N., Farías, J.G., 2019. Changes in Atlantic salmon (*Salmo salar*) sperm morphology and membrane lipid composition related to cold storage and cryopreservation. Anim. Reprod. Sci. 204, 50–59.

Dietrich, M.A., Ciereszko, A., 2018. Proteomic characterization of fresh spermatozoa and supernatant after cryopreservation in relation to freezability of carp (*Cyprinus carpio* L) semen. PLoS One 13 (3), e0192972.

Dietrich, M.A., Nynca, J., Ciereszko, A., 2019. Proteomic and metabolomic insights into the functions of the male reproductive system in fishes. Theriogenology 132, 182e200.

Diogo, P., Martins, G., Quinzico, I., Nogueira, R., Gavaia, P.J., Cabrita, E., 2018. Electric ultrafreezer (−150 °C) as an alternative for zebrafish sperm cryopreservation and storage. Fish Physiol. Biochem., 1–13.

Diogo, P., Martins, G., Eufrásio, A., Silva, T., Cabrita, E., Gavaia, P.J., 2019a. Selection criteria of zebrafish male donors for sperm cryopreservation. Zebrafish 16, 189–196.

Diogo, P., Martins, G., Nogueira, R., Marreiros, A., Gavaia, P.J., Cabrita, E., 2019b. Cryoprotectants synergy improve zebrafish sperm cryopreservation and offspring skeletogenesis. Cryobiology 91, 115–127.

Dong, Q., Huang, C., Tiersch, T.R., 2007. Control of sperm concentration is necessary for standardization of sperm cryopreservation in aquatic species: evidence from sperm agglutination in oysters. Cryobiology 54, 87–98.

Dosch, R., 2015. Next generation mothers: maternal control of germline development in zebrafish. Crit. Rev. Biochem. Mol. Biol. 50, 54–68.

Dott, M.H., Foster, G.C.A., 1979. Estimation of sperm motility in semen, on a membrane slide, by measuring the area change frequency with an image analyzing computer. J. Reprod. Fertil. 55, 161–166.

Draper, B.W., Moens, C.B., 2009. A high-throughput method for zebrafish sperm cryopreservation and *in vitro* fertilization. J. Vis. Exp., 1–5. https://doi.org/10.3791/1395.

Dziewulska, K., Domagała, J., 2013. Spermatozoa concentration influences cryopreservation success in sea trout (*Salmo trutta* m. *trutta* L.). Theriogenology 80, 659–664.

Erdahl, A.W., Graham, E.F., 1987. Fertility of teleost semen as affected by dilution and storage in a seminal plasma-mimicking medium. Aquaculture 60, 311–321.

Fabbrocini, A., Lavadera, S.L., Rispoli, S., Sansone, G., 2000. Cryopreservation of seabream (*Sparus aurata*) spermatozoa. Cryobiology 40, 46–53.

Fabbrocini, A., D'Adamo, R., Pelosi, S., Oliveira, L.F.J., Silvestri, F., Sansone, G., 2014. Gamete cryobanks for laboratory research: Developing a rapid and easy to perform protocol for the cryopreservation of the sea urchin *Paracentrotus lividus* (Lmk, 1816) spermatozoa. Cryobiology 69, 149–156.

Fahy, G.M., MacFarlane, D.R., Angell, C.A., Meryman, H.T., 1984. Vitrification as an approach to cryopreservation. Cryobiology 21, 407–426.

Fernández-Díez, C., Herráez, M.P., 2018. Changes in transcriptomic profile of trout larvae obtained with frozen sperm. Aquaculture 492, 306–320.

Figueroa, E., Merino, O., Risopatrón, J., Isachenko, V., Sánchez, R., Effer, B., Isachenko, E., Farias, J.G., Valdebenito, I., 2015. Effect of seminal plasma on Atlantic salmon (*Salmo salar*) sperm vitrification. Theriogenology 83, 238–245.

Figueroa, E., Valdebenito, I., Farias, J.G., 2016a. Technologies used in the study of sperm function in cryopreserved fish spermatozoa. Aquacult. Res. 47 (6), 1691–1705.

Figueroa, E., Valdebenito, I., Merino, O., Ubilla, A., Risopatrón, J., Farias, J.G., 2016b. Cryopreservation of Atlantic salmon *Salmo salar* sperm: effects on sperm physiology. J. Fish Biol. 89, 1537–1550.

Figueroa, E., Lee-Estevez, M., Valdebenito, I., Watanabe, I., Oliveira, R.P.S., Romero, J., Castillo, R.L., Farias, J.G., 2019. Effects of cryopreservation on mitochondrial function and sperm quality in fish. Aquaculture 511. https://doi.org/10.1016/j.aquaculture.2019.06.004, 634190.

Fornari, D.C., Ribeiro, R.P., Streit, D.J., Godoy, L.C., Neves, P.R., de Oliveira, D., Sirol, R.N., 2014. Effect of cryoprotectants on the survival of cascudo preto (*Rhinelepis aspera*) embryos stored at −8 °C. Zygote 22, 58–63.

Franěk, R., Pšenička, M., 2020. Cryopreservation of germ stem cells in fish. In: Betsy, J., Kumar, S. (Eds.), Cryopreservation of Fish Gametes. Springer, Singapore.

Franěk, R., Marinović, Z., Lujić, J., Urbányi, B., Fučíková, M., Kašpar, V., Pšenička, M., Horváth, Á., 2019a. Cryopreservation and transplantation of common carp spermatogonia. PLoS One 14 (4), e0205481.

Franěk, R., Tichopád, T., Steinbach, C., Xie, X., Lujić, J., Marinović, Z., Horváth, Á., Kašpar, V., Pšenička, M., 2019b. Preservation of female genetic resources of common carp through oogonial stem cell manipulation. Cryobiology 87, 78–85.

Franks, F., 1986. Metastable water at subzero temperatures. J. Microsc. 141, 243–249.

Gallego, V., Asturiano, J.F., 2018. Sperm motility in fish: technical applications and perspectives through CASA-Mot systems. Reprod. Fertil. Dev. 30 (6), 820–832.

Gallego, V., Asturiano, J.F., 2019. Fish sperm motility assessment as a tool for aquaculture research, a historical approach. Rev. Aquac. 11, 697–724.

Gallego, V., Mazzeo, I., Vílchez, M.C., Peñaranda, D.S., Carneiro, P.C.F., Pérez, L., Asturiano, J.F., 2012. Study of the effects of thermal regime and alternative hormonal treatments on the reproductive performance of European eel males (*Anguilla anguilla*) during induced sexual maturation. Aquaculture 354–355, 7–16.

Garzón, D.L., Peñaranda, D.S., Pérez, L., Marco-Jiménez, F., Espert, X., Müller, T., Jover, M., Asturiano, J.F., 2008. Effects of pH, sodium bicarbonate, cryoprotectants and foetal bovine serum on the cryopreservation of European eel sperm. Reprod. Domest. Anim. 43, 99–105.

Glogowski, J., Kolman, R., Szczepkowski, M., Horváth, Á., Urbányi, B., Sieczyński, P., Rzemienecki, A., Domagała, J., Demianowicz, W., Kowalski, R., Ciereszko, A., 2002. Fertilization rate of Siberian sturgeon (*Acipenser baeri*, Brandt) milt cryopreserved with methanol. Aquaculture 211, 367–373.

González-Rojo, S., Fernandez-Díez, C., Lombo, M., Herráez, M.P., 2018. Distribution of DNA damage in the sperm nucleus: a study of zebrafish as a model of histone-packaged chromatin. Theriogenology 122, 109–115.

Guthrie, H.D., Welch, G.R., Woods III, L.C., 2014. Effects of frozen and liquid hypothermic storage and extender type on calcium homeostasis in relation to viability and ATP content in striped bass (*Morone saxatilis*) sperm. Theriogenology 81, 1085–1091.

Hagedorn, M., Spindler, R., Comizzoli, P., 2014. The reality, use and potential for cryopreservation of coral reefs. In: Holt, W.V., Brown, J.L. (Eds.), Reproductive Sciences in Animal Conservation. Springer, New York, pp. 317–329.

Hagedorn, M., Hsu, E.W., Pilatus, U., Wildt, D.E., Rall, W.R., Blackband, S.J., 1996. Magnetic resonance microscopy and spectroscopy reveal kinetics of cryoprotectant permeation in a multicompartmental biological system. Proc. Natl. Acad. Sci. U. S. A. 93, 7454–7459.

Hagedorn, M., Lance, S.L., Fonseca, D.M., Kleinhans, F.W., Artimov, D., Fleischer, R., Hoque, A.T., Hamilton, M.B., Pukazhenthi, B.S., 2002. Altering fish embryos with aquaporin-3: an essential step toward successful cryopreservation. Biol. Reprod. 67, 961–966.

Hagedorn, M., McCarthy, M., Carter, V.L., Meyers, S., 2012a. Oxidative stress in zebrafish (*Danio rerio*) sperm. PLoS One 7, 2–12.

Hagedorn, M., van Oppen, M.J.H., Carter, V., Henley, M., Abrego, D., Puill-Stephan, E., Negri, A., Heyward, A., MacFarlane, D., Spindler, R., 2012b. First frozen repository for the Great Barrier Reef coral created. Cryobiology 65, 157–158.

Harvey, B., Norman, K.R., Ashwood-Smith, M.J., 1982. Cryopreservation of zebrafish spermatozoa using methanol. Can. J. Zool. 60, 1867–1870.

He, S., Woods III, L.C., 2003. The effects of osmolality, cryoprotectant and equilibration time on striped bass *Morone saxatilis* sperm motility. J. World Aquac. Soc. 34, 255–265.

Herranz-Jusdado, J.G., Gallego, V., Morini, M., Rozenfeld, C., Pérez, L., Kása, E., Kollár, T., Depincé, A., Labbé, C., Horváth, Á., Asturiano, J.F., 2019a. Comparison of European eel sperm cryopreservation protocols with standardization as a target. Aquaculture 498, 539–544.

Herranz-Jusdado, J.G., Gallego, V., Rozenfeld, C., Morini, M., Pérez, L., Asturiano, J.F., 2019b. European eel sperm storage: optimization of short-term protocols and cryopreservation of large volumes. Aquaculture 506, 42–50.

Herranz-Jusdado, J.G., Gallego, V., Morini, M., Rozenfeld, C., Pérez, L., Müller, T., Horváth, Á., Ohta, H., Asturiano, J.F., 2019c. Eel sperm cryopreservation: an overview. Theriogenology 133, 210–215.

Higaki, S., Eto, Y., Kawakami, Y., Yamaha, E., Kagawa, N., Kuwayama, M., Nagano, M., Katagiri, S., Takahashi, Y., 2010. Production of fertile zebrafish (*Danio rerio*) possessing germ cells (gametes) originated from primordial germ cells recovered from vitrified embryos. Reproduction 139, 733–740.

Higaki, S., Shimada, M., Kawamoto, K., Todo, T., Kawasaki, T., Tooyama, I., Fujioka, Y., Sakai, N., Takada, T., 2017. *In vitro* differentiation of fertile sperm from cryopreserved spermatogonia of the endangered endemic cyprinid honmoroko (*Gnathopogon caerulescens*). Sci. Rep. 7, 42852.

Hodgins, H.O., Ridgway, G.J., 1964. Recovery of viable salmon spermatozoa after fast-freezing. Progress. Fish Cult. 26, 95.

Horváth, Á., Miskolczi, E., Urbányi, B., 2003. Cryopreservation of common carp sperm. Aquat. Living Resour. 16, 457–460.

Horváth, Á., Urbányi, B., 2000. The effect of cryoprotectants on the motility and fertilizing capacity of cryopreserved African catfish *Clarias gariepinus* (Burchell 1822) sperm. Aquac. Res. 31, 317–324.

Horváth, Á., Miskolczi, E., Mihálffy, S., Osz, K., Szabó, K., Urbányi, B., 2007. Cryopreservation of common carp (*Cyprinus carpio*) sperm in 1.2 and 5 ml straws and occurrence of haploids among larvae produced with cryopreserved sperm. Cryobiology 54, 251–257.

Horváth, A., Urbányi, B., Wang, C., Onders, R.J., Mims, S.D., 2010. Cryopreservation of paddlefish sperm in 5 ml straws. J. Appl. Ichthyol. 26, 715–719.

Horváth, Á., Jesenšek, D., Csorbai, B., Bokor, Z., Raboczki, É., Kaczkó, D., Bernáth, G., Hoitsy, G., Urbányi, B., Bajec, S.S., Snoj, A., 2012. Application of sperm cryopreservation to hatchery practice and species conservation: a case of the Adriatic grayling (*Thymallus thymallus*). Aquaculture 358-359, 213–215.

Horváth, Á., Bokor, Z., Bernáth, G., Csenki, Z., Gorjan, A., Herráez, M.P., Urbányi, B., Jesenšek, D., 2015a. Very low sperm–egg ratios result in successful fertilization using cryopreserved sperm in the Adriatic grayling (*Thymallus thymallus*). Aquaculture 435, 75–77.

Horváth, Á., Labbé, C., Jesenšek, D., Hoitsy, G., Bernáth, G., Kaczkó, D., Bokor, Z., Urbányi, B., 2015b. Post-thaw storage of sperm from various salmonid species. J. Appl. Ichthyol. 31, 119–124.

Inoue, D., Fujimoto, T., Kawakami, Y., Yasui, G.S., Yamaha, E., Arai, K., 2012. Vitrification of primordial germ cells using whole embryos for gene-banking in loach, *Misgurnus anguillicaudatus*. J. Appl. Ichthyol. 28, 919–924.

Isachenko, V., 2004. Cryoprotectant-free cryopreservation of human spermatozoa by vitrification and freezing in vapor: effect on motility, DNA integrity, and fertilization ability. Biol. Reprod. 71, 1167–1173.

Jähnichen, H., Warnecke, D., Trölsch, E., Kohlmann, K., Bergler, H., Pluta, H.J., 1999. Motility and fertilizing capability of cryopreserved *Acipenser ruthenus* L. sperm. J. Appl. Ichthyol. 15, 204–206.

Janik, M., Kleinhans, F.W., Hagedorn, M., 2000. Overcoming a permeability barrier by microinjecting cryoprotectants into zebrafish embryos (*Brachydanio rerio*). Cryobiology 41, 25–34.

Judycka, S., Cejko, B.I., Dryl, K., Dobosz, S., Grudniewska, J., Kowalski, R.K., 2016. The effect of supplementation of a trehalose-based extender with KCl on rainbow trout (*Oncorhynchus mykiss*) sperm freezability and post-thaw motility. Aquaculture 465, 303–310.

Judycka, S., Nynca, J., Liszewska, E., Dobosz, S., Grudniewska, J., Ciereszko, A., 2018. Optimal sperm concentration in straws and final glucose concentration in extender are crucial for improving the cryopreservation protocol of salmonid spermatozoa. Aquaculture 486, 90–97.

Kása, E., Bernáth, G., Kollár, T., Żarski, D., Lujić, J., Marinović, Z., Bokor, Z., Hegyi, Á., Urbányi, B., Vílchez, M.C., Morini, M., Peñaranda, D.S., Pérez, L., Asturiano, J.F., Horváth, Á., 2017. Development of sperm vitrification protocols for freshwater fish (Eurasian perch, *Perca fluviatilis*) and marine fish (European eel, *Anguilla anguilla*). Gen. Comp. Endocrinol. 245, 102–107.

Kása, E., Lujić, J., Marinović, Z., Kollár, T., Bernáth, G., Bokor, Z., Urbányi, B., Lefler, K.K., Jesenšek, D., Horváth, Á., 2018. Development of sperm vitrification protocols for two endangered salmonid species: the Adriatic grayling, *Thymallus thymallus*, and the marble trout, *Salmo marmoratus*. Fish Physiol. Biochem. 44, 1499–1507.

Kawasaki, T., Siegfried, K.R., Sakai, N., 2016. Differentiation of zebrafish spermatogonial stem cells to functional sperm in culture. Development 143, 566–574.

Khosla, K., Wang, Y., Hagedorn, M., Qin, Z., Bischof, J., 2017. Gold nanorod induced warming of embryos from the cryogenic state enhances viability. ACS Nano 11, 7869–7878.

Kobayashi, T., Takeuchi, Y., Takeuchi, T., Yoshizaki, G., 2007. Generation of viable fish from cryopreserved primordial germ cells. Mol. Reprod. Dev. 74, 207–213.

Koh, I.C.C., Hamada, D., Tsuji, Y.A., Okuda, D., Nomura, K., Tanaka, H., Ohta, H., 2017. Sperm cryopreservation of Japanese eel, *Anguilla japonica*. Aquaculture 473, 487–492.

Kopeika, E., Kopeika, J., Zhang, T., 2007. Cryopreservation of fish sperm. Methods Mol. Biol. 368, 203–217.

Kutluyer, F., Kayim, M., Öğretmen, F., Büyükleblebici, S., Tuncer, P.B., 2014. Cryopreservation of rainbow trout *Oncorhynchus mykiss* spermatozoa: effects of extender supplemented with different antioxidants on sperm motility, velocity and fertility. Cryobiology 69, 462–466.

Labbé, C., Robles, V., Herráez, M.P., 2013. Cryopreservation of gametes for aquaculture and alternative cell sources for genome preservation. In: Allan, G., Burnell, G. (Eds.), Advances in Aquaculture Hatchery Technology. Woodhead Publishing, pp. 76–116. ISSN: 0213-3911.

Lahnsteiner, F., Patzner, R., Weismann, T., 1996. Semen cryopreservation of salmonid fishes: influence of handling parameters on the postthaw fertilization rate. Aquacult. Res. 27, 659–671.

Lahnsteiner, F., Weismann, T., Patzner, R.A., 1997. Methanol as cryoprotectant and the suitability of 1.2 ml and 5 ml straws for cryopreservation of semen from salmonid fishes. Aquacult. Res. 28, 471–479.

Lahnsteiner, F., Mansour, N., Kunz, F.A., 2011. The effect of antioxidants on the quality of cryopreserved semen in two salmonid fish, the brook trout (*Salvelinus fontinalis*) and the rainbow trout (*Oncorhynchus mykiss*). Theriogenology 76, 882–890.

Larman, M.G., Hashimoto, S., Morimoto, Y., Gardner, D.K., 2014. Cryopreservation in ART and concerns with contamination during cryobanking. Reprod. Med. Biol. 13, 107–117.

Lee, S., Yoshizaki, G., 2016. Successful cryopreservation of spermatogonia in critically endangered Manchurian trout (*Brachymystax lenok*). Cryobiology 72, 165–168.

Lee, S., Iwasaki, Y., Shikina, S., Yoshizaki, G., 2013. Generation of functional eggs and sperm from cryopreserved whole testes. Proc. Natl. Acad. Sci. U. S. A. 110, 1640–1645.

Lee, S., Katayama, N., Yoshizaki, G., 2016. Generation of juvenile rainbow trout derived from cryopreserved whole ovaries by intraperitoneal transplantation of ovarian germ cells. Biochem. Biophys. Res. Commun. 478, 1478–1483.

Lee-Estevez, M., Herrera, L., Díaz, R., Beltrán, J., Figueroa, E., Dumorné, K., Ulloa-Rodríguez, P., Short, S., Risopatrón, J., Valdebenito, I., Farías, J., 2019. Effects of cryopreservation on cAMP-dependent protein kinase and AMP-activated protein kinase in Atlantic salmon (*Salmo salar*) spermatozoa: relation with post-thaw motility. Anim. Reprod. Sci. 209, 106133.

Legendre, M., Linhart, O., Billard, R., 1996. Spawning and management of gametes, fertilized eggs and embryos in Siluroidei. Aquat. Living Resour. 9, 59–80.

Lichtenstein, G., Elisio, M., Miranda, L.A., 2010. Development of sperm cryopreservation techniques in pejerrey *Odontesthes bonariensis*. Aquaculture 306, 357–361.

Lin, T.T., Chao, N.H., Tung, H.T., 1999. Factors affecting survival of cryopreserved oyster (*Crassostrea gigas*) embryos. Cryobiology 39, 192–196.

Linhartová, Z., Rodina, M., Huralp, H., Gazo, I., Saito, T., Pšenička, M., 2014. Isolation and cryopreservation of early stages of germ cells of tench (*Tinca tinca*). Czech J. Anim. Sci. 59, 381–390.

Liu, X.H., Zhang, T., Rawson, D.M., 2001. Differential scanning calorimetry studies of intraembryonic freezing and cryoprotectant penetration in zebrafish (*Danio rerio*) embryos. J. Exp. Zool. 290, 299–310.

Liu, Q.H., Li, J., Zhang, S.C., Xiao, Z.Z., Ding, F.H., Yu, D.D., Xu, X.Z., 2007. Flow cytometry and ultrastructure of cryopreserved red seabream (*Pagrus major*) sperm. Theriogenology 67, 1168–1174.

Lonergan, P., 2018. Review: historical and futuristic development in bovine semen technology. Animal 12, S4–S18.

Lubzens, E., Daube, N., Pekarsky, I., Magnus, Y., Cohen, A., Yusefovich, F., Feigin, P., 1997. Carp (*Cyprinus carpio* L.) spermatozoa cryobanks - strategies in research and application. Aquaculture 155, 13–30.

Lujić, J., Bernáth, G., Marinović, Z., Radojković, N., Simić, V., Ćirković, M., Urbányi, B., Horváth, Á., 2017a. Fertilizing capacity and motility of tench *Tinca tinca* (L., 1758) sperm following cryopreservation. Aquacult. Res. 48, 102–110.

Lujić, J., Marinović, Z., Sušnik Bajec, S., Djurdjevič, I., Kása, E., Urbányi, B., Horváth, Á., 2017b. First successful vitrification of salmonid ovarian tissue. Cryobiology 76, 154–157.

Marco-Jiménez, F., Garzón, D.L., Peñaranda, D.S., Pérez, L., Viudes-de-Castro, M.P., Vicente, J.S., Jover, M., Asturiano, J.F., 2006. Cryopreservation of European eel (*Anguilla anguilla*) spermatozoa: effect of dilution ratio, foetal bovine serum supplementation, and cryoprotectants. Cryobiology 53, 51–57.

Maria, A.N., Carvalho, A.C.M., Araújo, R.V., Santos, J.P., Carneiro, P.C.F., Azevedo, H.C., 2015. Use of cryotubes for the cryopreservation of tambaqui fish semen (*Colossoma macropomum*). Cryobiology 70, 109–114.

Marinović, Z., Lujić, J., Kása, E., Bernáth, G., Urbányi, B., Horváth, Á., 2017. Cryosurvival of isolated testicular cells and testicular tissue of tench *Tinca tinca* and goldfish *Carassius auratus* following slow-rate freezing. Gen. Comp. Endocrinol. 245, 77–83.

Marinović, Z., Lujić, J., Kása, E., Csenki, Z., Urbányi, B., Horváth, Á., 2018. Cryopreservation of zebrafish spermatogonia by whole testes needle immersed ultra-rapid cooling. J. Vis. Exp., e56118.

Marinović, Z., Li, Q., Lujić, J., Iwasaki, Y., Csenki, Z., Urbányi, B., Yoshizaki, G., Horváth, Á., 2019a. Preservation of zebrafish genetic resources through testis cryopreservation and spermatogonia transplantation. Sci. Rep. 9, 13861.

Martínez-Páramo, S., Barbosa, V., Pérez-Cerezales, S., Robles, V., Herráez, M.P., 2009a. Cryoprotective effects of antifreeze proteins delivered into zebrafish embryos. Cryobiology 58, 128–133.

Martínez-Páramo, S., Pérez-Cerezales, S., Gómez-Romano, F., Blanco, G., Sánchez, J.A., Herráez, M.P., 2009b. Cryobanking as tool for conservation of biodiversity: effect of brown trout sperm cryopreservation on the male genetic potential. Theriogenology 71, 594–604.

Martínez-Páramo, S., Diogo, P., Dinis, M.T., Herráez, M.P., Sarasquete, C., Cabrita, E., 2012. Sea bass sperm freezability is influenced by motility variables and membrane lipid composition but not by membrane integrity and lipid peroxidation. Anim. Reprod. Sci. 131, 211–218.

Martínez-Páramo, S., Diogo, P., Dinis, M.T., Soares, F., Sarasquete, C., Cabrita, E., 2013. Effect of two sulfur-containing amino acids, taurine and hypotaurine in European sea bass (*Dicentrarchus labrax*) sperm cryopreservation. Cryobiology 66 (3), 333–338.

Martínez-Páramo, S., Horváth, Á., Labbé, C., Zhang, T., Robles, V., Herráez, P., Suquet, M., Adams, S., Viveiros, A., Tiersch, T.R., Cabrita, E., 2017. Cryobanking of aquatic species. Aquaculture 472, 156–177.

Martins, G., Diogo, P., Santos, T., Cabrita, E., Pinto, W., Dias, J., Gavaia, P.J., 2020. Microdiet formulation with phospholipid modulate zebrafish skeletal development and reproduction. Zebrafish 17, 27–37.

Mattei, X., 1991. Spermatozoon ultrastructure and its systematic implications in fishes. Can. J. Zool. 69, 3038–3055.

Matthews, J.L., Murphy, J.M., Carmichael, C., Yang, H., Tiersch, T., Westerfield, M., Varga, Z.M., 2018. Changes to extender, cryoprotective medium, and *in vitro* fertilization improve zebrafish sperm cryopreservation. Zebrafish 15, 279–290.

Mazur, P., 1990. Equilibrium, quasi-equilibrium, and nonequilibrium freezing of mammalian embryos. Cell Biophys. 17, 53–92.

McNiven, M.A., Gallant, R.K., Richardson, G.F., 1993. Dimethyl-acetamide as a cryoprotectant for rainbow trout spermatozoa. Theriogenology 40, 943–948.

Merino, O., Sánchez, R., Risopatrón, J., Isachenko, E., Katkov, I.I., Figueroa, E., Valdebenito, I., Mallmann, P., Isachenko, V., 2012. Cryoprotectant-free vitrification of fish (*Oncorhynchus mykiss*) spermatozoa: first report. Andrologia 44, 390–395.

Morris, J.P., Hagen, A., Kanki, J.P., 2009. Zebrafish sperm cryopreservation with N,N-dimethylacetamide. In: Cabrita, E., Robles, V., Herráez, M.P. (Eds.), Methods in Reproductive Aquaculture: Marine and Freshwater Species, Biology Series. CRC Press (Taylor and Francis group, Boca Raton, FL, pp. 329–337.

Müller, T., Urbányi, B., Váradi, B., Binder, T., Horn, P., Bercsényi, M., Horváth, Á., 2004. Cryopreservation of sperm of farmed European eel *Anguilla anguilla*. J. World Aquac. Soc. 35, 225–231.

Müller, T., Matsubara, H., Kubara, Y., Horváth, Á., Asturiano, J.F., Urbányi, B., 2017. Japanese eel (*Anguilla japonica* Temminck & Schlegel, 1846) propagation using cryopreserved sperm samples. J. Appl. Ichthyol. 33, 550–552.

Müller, T., Matsubara, H., Kubara, Y., Horváth, Á., Kolics, B., Taller, J., Stéger, V., Kovács, B., Horváth, L., Asturiano, J.F., Peñaranda, D.S., Urbányi, B., 2018. Testing cryopreserved European eel sperm for hybridization (*A. japonica* × *A. anguilla*). Theriogenology 113, 153–158.

Mylonas, C.C., Duncan, N.J., Asturiano, J.F., 2017. Hormonal manipulations for the enhancement of sperm production in cultured fish and evaluation of sperm quality. Aquaculture 472, 21–44.

Nomura, K., Chong, I., Koh, C.C., Lio, R., Okuda, D., Kazeto, Y., Tanaka, H., Ohta, H., 2018. Sperm cryopreservation protocols for the large-scale fertilization of japanese eel using a combination of large volume straws and low sperm dilution ratio. Aquaculture 496, 203–210.

Nusbaumer, D., Marques da Cunha, L., Wedekind, C., 2019. Sperm cryopreservation reduces offspring growth. Proc. R. Soc. Lond. B Biol. Sci. 286, 20191644. https://doi.org/10.1098/rspb.2019.1644.

Nynca, J., Dietrich, G.J., Dobosz, S., Zalewski, T., Ciereszko, A., 2014. Effect of postthaw storage time and sperm-to-egg ratio on fertility of cryopreserved brook trout sperm. Theriogenology 83, 253–256.

Nynca, J., Dietrich, G.J., Grudniewska, J., Dobosz, S., Liszewska, E., Krzyś, M., Różyński, R., Ciereszko, A., 2015. Efficient method for cryopreservation of European huchen (*Hucho hucho* L.) and grayling (*Thymallus thymallus* L.) semen. Aquaculture 435, 146–151.

Nynca, J., Judycka, S., Liszewska, E., Dobosz, S., Ciereszko, A., 2017. Standardization of spermatozoa concentration for cryopreservation of rainbow trout semen using a glucose-methanol extender. Aquaculture 477, 23–27.

Octavera, A., Yoshizaki, G., 2020. Production of Chinese rosy bitterling offspring derived from frozen and vitrified whole testis by spermatogonial transplantation. Fish Physiol. Biochem. 46, 1431–1442.

O'Reilly, P., Doyle, R., 2007. Live gene banking of endangered populations of Atlantic salmon. In: The Atlantic Salmon. Blackwell Publishing Ltd, pp. 425–469.

Paniagua-Chaves, C.G., Buchanan, J.T., Supan, J.E., Tiersch, T.R., 2000. Cryopreservation of sperm and larvae of the Eastern oyster. In: Tiersch, T.R, Mazik, P.M. (Eds.), Cryopreservation in Aquatic Species. World Aquaculture Society, Baton Rouge, LA, pp. 230–239.

Paredes, E., 2016. Biobanking of a marine invertebrate model organism: the sea urchin. J. Mar. Sci. Eng. 4, 7.

Paredes, E., Heres, P., Anjos, C., Cabrita, E., 2020. Cryopreservation of marine invertebrates: from sperm to complex larval stages. In: Wolkers, W.F., Oldenhof, H. (Eds.), Cryopreservation and Freeze-Drying Protocols. Springer, pp. 413–425. ISBN 978-1-4939-2193-5.

Peñaranda, D.S., Pérez, L., Gallego, V., Jover, M., Asturiano, J.F., 2009. Improvement of European eel sperm cryopreservation method by preventing spermatozoa movement activation caused by cryoprotectants. Cryobiology 59, 119–126.

Pérez, L., Asturiano, J.F., Tomás, A., Zegrari, S., Barrera, R., Espinós, F.J., Navarro, J.C., Jover, M., 2000. Induction of maturation and spermiation in the male European eel: assessment of sperm quality throughout treatment. J. Fish Biol. 57, 1488–1504.

Pérez-Cerezales, S., Martínez-Páramo, S., Cabrita, E., Martínez-Pastor, F., de Paz, P., Herráez, M.P., 2009. Evaluation of oxidative DNA damage promoted by storage in sperm from sex-reversed rainbow trout. Theriogenology 71, 605–613.

Pérez-Cerezales, S., Martínez-Páramo, S., Beirão, J., Herráez, M.P., 2010. Evaluation of DNA damage as a quality marker for rainbow trout sperm cryopreservation and use of LDL as cryoprotectant. Theriogenology 74, 282–289.

Pérez-Cerezales, S., Miranda, A., Gutiérrez-Adán, A., 2012. Comparison of four methods to evaluate sperm DNA integrity between mouse caput and cauda epididymidis. Asian J. Androl. 14 (2), 335–337.

Pessoa, N.O., Galvão, J.A.S., de Souza Filho, F.G.M., de Sousa, M.L.N.M., Sampaio, C.M.S., 2015. Cooling of pirapitinga (*Piaractus brachypomus*) embryos stored at − 10 oC. Zygote 23 (3), 453–459.

Pšenička, M., Saito, T., Rodina, M., Dzyba, B., 2016. Cryopreservation of early stage Siberian sturgeon *Acipenser baerii* germ cells, comparison of whole tissue and dissociated cells. Cryobiology 72, 119–122.

Rall, W.F., 1987. Factors affecting the survival of mouse embryos cryopreserved by vitrification. Cryobiology 24, 387–402.

Renard, P., 1991. Cooling and freezing tolarances in embryos of the Pacific oyster, *Crassostrea gigas*: Methanol and sucrose effects. Aquaculture 92, 43–57.

Riesco, M.F., Martínez-Pastor, F., Chereguini, O., Robles, V., 2012. Evaluation of zebrafish (*Danio rerio*) PGCs viability and DNA damage using different cryopreservation protocols. Theriogenology 77, 122–130.

Riesco, M.F., Félix, F., Matias, D., Joaquim, S., Suquet, M., Cabrita, E., 2017a. First study in cryopreserved sperm of *Crassostrea angulata*. Gen. Comp. Endocrinol. 245, 108–115.

Riesco, M.F., Oliveira, C., Soares, F., Gavaia, P.J., Dinis, M.T., Cabrita, E., 2017b. S*olea senegalensis* sperm cryopreservation: new insights on sperm quality. PLoS One 12 (10), e0186542.

Riesco, M.F., Raposo, C., Engrola, S., Martínez-Páramo, S., Mira, S., Cunha, M.E., Cabrita, E., 2017c. Improvement of the cryopreservation protocols for the dusky grouper, *Epinephelus marginatus*. Aquaculture 470, 207–213.

Riesco, M.F., Félix, F., Matias, D., Joaquim, S., Suquet, M., Cabrita, E., 2019. Comparative study on cellular and molecular responses in oyster sperm revealed different susceptibilities to cryopreservation. Aquaculture 498, 223–229.

Robles, V., Cabrita, E., de Paz, P., Cunado, S., Anel, L., Herraez, M.P., 2004. Effect of a vitrification protocol on the lactate dehydrogenase and glucose-6-phosphate dehydrogenase activities and the hatching rates of zebrafish (*Danio rerio*) and turbot (*Scophthalmus maximus*) embryos. Theriogenology 61, 1367–1379.

Robles, V., Cabrita, E., Fletcher, G.L., Shears, M.A., King, M.J., Herraez, M.P., 2005. Vitrification assays with embryos from a cold tolerant sub-arctic fish species. Theriogenology 64, 1633–1646.

Robles, V., Barbosa, V., Herraez, M.P., Martinez-Paramo, S., Cancela, M.L., 2007. The antifreeze protein type I (AFP I) increases seabream (*Sparus aurata*) embryos tolerance to low temperatures. Theriogenology 68, 284–289.

Robles, V., Cabrita, E., Acker, J.P., Herraez, M.P., 2009a. Embryo cryopreservation: what we know until now? In: Cabrita, E., Robles, V., Herráez, M.P. (Eds.), Methods in Reproductive Aquaculture. C.R.C. Press, Boca Raton, pp. 261–294.

Robles, V., Cabrita, E., Herráez, M.P., 2009b. Germplasm cryobanking in zebrafish and other aquarium model species. Zebrafish 6, 281–293.

Robles, V., Martínez-Pastor, F., Petronic, G., Riesco, M.F., Bozzano, A., Villanueva, R., 2013. Cryobiology of cephalopod (Illex Coindetti) spermatophores. Cryobiology 66 (3), 288–294.

Robles, V., Riesco, M.F., Psenicka, M., Saito, T., Valcarce, D.G., Cabrita, E., Herráez, M.P., 2017. Biology of teleost primordial germ cells (PGCs) and spermatogonia: biotechnological applications. Aquaculture 472, 4–20.

Rozenfeld, C., García-Carpintero, V., Pérez, L., Gallego, V., Herranz-Jusdado, J.G., Tveiten, H., Johnsen, H.K., Fontaine, R., Weltzien, F.A., Cañizares, J., Asturiano, J.F., Peñaranda, D.S., 2019. Cold seawater induces early maturational stages in the BPG axis of European eel males. BMC Genomics 20 (1), 597. https://doi.org/10.1186/s12864-019-5969-6.

Sandoval-Vargas, L., Jiménez, M.S., González, J.R., Villalobos, E.F., Cabrita, E., Valdebenito-Isler, I., 2021. Oxidative stress and use of antioxidants in fish semen cryopreservation. Rev. Aquac. 13, 365–387.

Santana, J., Cabrita, E., Babiak, I., Eggen, B., Beirão, J., 2020. Step by step optimization of a sperm cryopreservation protocol for spotted wolfish *(Anarhichas minor* Olafsen, 1772). Theriogenology 149, 16–24.

Šćekić, I., Marinović, Z., Lujić, J., Müller, T., Kitanović, N., Urbányi, B., Horváth, Á., 2020. A novel strategy for conservation of European eel (*Anguilla anguilla*) genetic resources: cryopreservation of ovarian stem cells. Cryobiology 95, 151–156.

Seki, S., Jin, B., Mazur, P., 2014. Extreme rapid warming yields high functional survivals of vitrified 8-cell mouse embryos even when suspended in a half-strength vitrification solution and cooled at moderate rates to -196 °C. Cryobiology 68, 71–78.

Seki, S., Kusano, K., Lee, S., Iwasaki, Y., Yagisawa, M., Ishida, M., Hiratsuka, T., Sasado, T., Naruse, K., Yoshizaki, G., 2017. Production of the medaka derived from vitrified whole testes by germ cell transplantation. Sci. Rep. 7, 43185.

Shaliutina-Kolešová, A., Ashtiani, S., Xian, M., Nian, R., 2020. Seminal plasma fractions can protect common carp (*Cyprinus carpio*) sperm during cryopreservation. Fish Physiol. Biochem. 46, 1461–1468.

Stoss, J., Holtz, W., 1981. Cryopreservation of rainbow trout (*Salmo gairdneri*) sperm: I. Effect of thawing solution, sperm density and interval between thawing and insemination. Aquaculture 22, 97–104.

Stoss, J., Holtz, W., 1983. Cryopreservation of rainbow trout (*Salmo gairdneri*) sperm: IV. The effect of DMSO concentration and equilibration time on sperm survival, sucrose and KCl as extender components and the osmolality of the thawing solution. Aquaculture 32, 321–330.

Streit Jr., D.P., Fornari, D.C., Povh, J.A., de Godoy, L.C., Lopes de Oliveira, C.A., Kawakami de Resende, E., Ribeira Pereiro, R., 2013. Germplasm banking and its role on development of the genetic improvement program for fish in Brazil. In: Martínez-Páramo, S., Oliveira, C., Dinis, M.T. (Eds.), 4th International Workshop on the Biology of Fish Gametes. Faro, Portugal, Albufeira, Portugal, pp. 140–141.

Suquet, M., Dreanno, C., Fauvel, C., Cosson, J., Billard, R., 2000. Cryopreservation of sperm in marine fish. Aquacult. Res. 31, 231–243.

Suquet, M., Le Mercier, A., Rimond, F., Mingant, C., Haffray, P., Labbe, C., 2012. Setting tools for the early assessment of the quality of thawed Pacific oyster (*Crassostrea gigas*) D-larvae. Theriogenology 78, 462–467.

Suquet, M., Labbé, C., Puyo, S., Mingant, C., Quittet, B., Boulais, M., Queau, I., Ratiskol, D., Diss, B., Haffray, P., 2014. Survival, growth and reproduction of cryopreserved larvae from a marine invertebrate, the Pacific oyster (*Crassostrea gigas*). PLoS One 9, e93486.

Szabó, G., Müller, T., Bercsényi, M., Urbányi, B., Kucska, B., Horváth, A., 2005. Cryopreservation of European eel (*Anguilla anguilla*) sperm using different extenders and cryoprotectants. Acta Biol. Hung. 56, 173–175.

Taddei, A.R., Barbato, F., Abelli, L., Canese, S., Moretti, F., Rana, K.J., Fausto, A.M., Mazzini, M., 2001. Is cryopreservation a homogeneous process? Ultrastructure and motility of untreated, prefreezing, and postthawed spermatozoa of *Diplodus puntazzo* (Cetti). Cryobiology 42, 244–255.

Tan, E., Yang, H., Tiersch, T.R., 2010. Determination of sperm concentration for small-bodied biomedical model fishes by use of microspectrophotometry. Zebrafish 7, 233–240.

Tanaka, S., Zhang, H., Horie, N., Yamada, Y., Okamura, A., Utoh, T., Mikawa, N., Oka, H.P., Kurokura, H., 2002a. Long-term cryopreservation of sperm of Japanese eel. J. Fish Biol. 60, 139–146.

Tanaka, S., Zhang, H., Yamada, Y., Okamura, A., Horie, N., Utoh, T., Mikawa, N., Oka, H.P., Kurokura, H., 2002b. Inhibitory effect of sodium bicarbonate on the motility of sperm of Japanese eel. J. Fish Biol. 60, 1134–1141.

Thirumala, S., Campbell, W.T., Vicknair, M.R., Tiersch, T.R., Devireddy, R.V., 2006. Freezing response and optimal cooling rates for cryopreserving sperm cells of striped bass, *Morone saxatilis*. Theriogenology 66, 964–973.

Tiersch, T.R., Green, C.C. (Eds.), 2011. Cryopreservation in Aquatic Species, second Edition. World Aquaculture Society, Baton Rouge, Louisiana, USA.

Tiersch, T.R., Goudie, C.A., Carmichael, G.J., 1994. Cryopreservation of channel catfish sperm: storage in cryoprotectants, fertilization trials, and growth of channel catfish produced with cryopreserved sperm. Trans. Am. Fish. Soc. 123, 580–586.

Tiersch, T.R., Figiel, C., Wayman, W., Williamson, J.H., Gorman, O., Carmichael, G., 2004. Cryopreservation of sperm from the endangered Colorado pikeminnow. N. Am. J. Aquac. 66, 8–14.

Tiersch, T.R., Yang, H., Hu, E., 2011. Outlook for development of high-throughput cryopreservation for small-bodied biomedical model fishes. Comp. Biochem. Physiol. C 154, 76–81.

Truscott, B., Idler, D.R., Hoyle, R.J., Freeman, H.C., 1968. Sub-zero preservation of Atlantic salmon sperm. J. Fish. Res. Board Can. 25, 363–372.

Tsai, S., Lin, C., 2012. Advantages and applications of cryopreservation in fisheries science. Braz. Arch. Biol. Technol. 55, 425–434.

Tsai, S., Kuit, V., Lin, Z.G., Lin, C., 2014. Application of a functional marker for the effect of cryoprotectant agents on gorgonian coral (*Junceella juncea* and *J. fragilis*) sperm sacs. Cryo Letters 35 (1), 1–7.

Tsai, H.Y., Hamilton, A., Guy, D.R., Tinch, A.E., Bishop, S.C., Houston, R.D., 2015. The genetic architecture of growth and fillet traits in farmed Atlantic salmon (*Salmo salar*). BMC Genet. 16, 51.

Tsvetkova, L.I., Cosson, J., Linhart, O., Billard, R., 1996. Motility and fertilizing capacity of fresh and frozen-thawed spermatozoa in sturgeons *Acipenser baeri* and *A. ruthenus*. J. Appl. Ichthyol. 12, 107–112.

Viveiros, A.T., Godinho, H.P., 2009. Sperm quality and cryopreservation of Brazilian freshwater fish species: a review. Fish Physiol. Biochem. 35, 137–150.

Viveiros, A.T.M., Nascimento, A.F., Leal, M.C., Gonçalves, A.C.S., Orfão, L.H., Cosson, J., 2014. Methyl glycol, methanol and DMSO effects on post-thaw motility, velocities, membrane integrity and mitochondrial function of *Brycon orbignyanus* and *Prochilodus lineatus* (Characiformes) sperm. Fish Physiol. Biochem. 41, 193–201.

Vuthiphandchai, V., Nimrat, S., Kotcharat, S., Bart, A.N., 2007. Development of a cryopreservation protocol for long-term storage of black tiger shrimp (*Penaeus monodon*) spermatophores. Theriogenology 68 (8), 1192–1199.

Wang, R., Guan, M., Rawson, D., Zhang, T., 2008a. Ultrasound enhanced methanol penetration of zebrafish (*Danio rerio*) embryos measured by permittivity changes using impedance spectroscopy. Eur. Biophys. J. 37, 1039–1044.

Wang, Y., Xiao, Z., Li, L., Fan, W., Li, S.-W., 2008b. Novel needle immersed vitrification: a practical and convenient method with potential advantages in mouse and human ovarian tissue cryopreservation. Hum. Reprod. 23, 2256–2265.

Wang, G., Kang, N., Gong, H., Luo, Y., Bai, C., Chen, Y., Ji, X., Huang, C., Dong, Q., 2015. Upregulation of uncoupling protein Ucp2 through acute cold exposure increases post-thaw sperm quality in zebrafish. Cryobiology 71, 464–471.

Warnecke, D., Pluta, H.J., 2003. Motility and fertilizing capacity of frozen/thawed common carp (*Cyprinus carpio* L.) sperm using dimethyl-acetamide as the main cryoprotectant. Aquaculture 215, 167–185.

Wayman, W., 2011. Germplasm repository programs in the United States Fish and Wildlife Service. In: Tiersch, T.R., Green, C.C. (Eds.), Cryopreservation in Aquatic Species, second ed. World Aquaculture Society, Baton Rouge, Louisiana, pp. 780–782.

Wayman, W.R., Thomas, R.G., Tiersch, T.R., 1997. Refrigerated storage and cryopreservation of black drum (*Pogonias chromis*) spermatozoa. Theriogenology 47, 1519–1529.

Williot, P., Rochard, E., Desse-Berset, N., Kirschbaum, F., Gessner, J. (Eds.), 2011. Biology and conservation of the European sturgeon *Acipenser sturio* L. 1758: the reunion of the European and Atlantic sturgeons. Springer-Verlag, Berlin Heidelberg.

Wilson-Leedy, J.G., Ingermann, R.L., 2007. Development of a novel CASA system based on open source software for characterization of zebrafish sperm motility parameters. Theriogenology 67, 661–672.

Wolkers, W.F., Oldenhof, H., 2021. Cryopreservation and Freeze-Drying Protocols. Springer, p. 742. ISBN 978-1-4939-2193-5.

Yang, H., Tiersch, T.R., 2009. Current status of sperm cryopreservation in biomedical research fish models: Zebrafish, medaka, and *Xiphophorus*. Comp. Biochem. Phys. C 149, 224–232.

Yang, H., Carmichael, C., Varga, Z.M., Tiersch, T.R., 2007. Development of a simplified and standardized protocol with potential for high-throughput for sperm cryopreservation in zebrafish *Danio rerio*. Theriogenology 68, 128–136.

Yang, H., Norris, M., Winn, R., Tiersch, T.R., 2010. Evaluation of cryoprotectant and cooling rate for sperm cryopreservation in the euryhaline fish medaka *Oryzias latipes*. Cryobiology 61, 211–219.

Yang, H., Daly, J., Carmichael, C., Matthews, J., Varga, Z.M., Tiersch, T., 2016. A procedure-spanning analysis of plasma membrane integrity for assessment of cell viability in sperm cryopreservation of zebrafish *Danio rerio*. Zebrafish 13 (2), 144–151.

Yang, H., Hu, E., Buchanan, J.T., Tiersch, T.R., 2018. A strategy for sperm cryopreservation of Atlantic salmon, *Salmo salar*, for remote commercial-scale high-throughput processing. J. World Aquac. Soc. 49 (1), 96–112.

Yasui, G.S., Arias-Rodriguez, L., Fujimoto, T., Arai, K., 2008. Simple and inexpensive method for cryopreservation of fish sperm combining straw and powdered dry ice. Cryo Lett. 29, 383–390.

Yoshizaki, G., Lee, S., 2018. Production of live fish derived from frozen germ cells via germ cell transplantation. Stem Cell Res. 29, 103–110.

Yoshizaki, G., Yazawa, R., 2019. Application of surrogate broodstock technology in Aquaculture. Fish. Sci. 85, 429–437.

Yoshizaki, G., Fujinuma, K., Iwasaki, Y., Okutsu, T., Shikina, S., Yazawa, R., Takeuchi, Y., 2011. Spermatogonial transplantation in fish: a novel method for the preservation of genetic resources. Comp. Biochem. Phys. D 6, 55–61.

Yue, G.H., 2014. Recent advances of genome mapping and marker-assisted selection in aquaculture. Fish Fish. 15, 376–396.

Zhang, T., Wang, R.Y., Bao, Q.-Y., Rawson, D.M., 2006. Development of a new rapid measurement technique for fish embryo membrane permeability studies using impedance spectroscopy. Theriogenology 66, 982–988.

Zhang, J., Tian, Y., Li, J., Wu, Y., Li, Z., Cheng, M., Wang, L., Ma, W., Zhai, J., 2020. Optimization of vitrification factors for embryo cryopreservation of kelp grouper (*Epinephelus moara*). Theriogenology 15, 390–399.

Zilli, L., Schiavone, R., Zonno, V., Rossano, R., Storelli, C., Vilella, S., 2005. Effect of cryopreservation on sea bass sperm proteins. Biol. Reprod. 72, 1262–1267.

Zilli, L., Schiavone, R., Storelli, C., Vilella, S., 2008. Effect of cryopreservation on phosphorylation state of proteins involved in sperm motility initiation in sea bream. Cryobiology 57, 150–155.

Zilli, L., Beirão, J., Schiavone, R., Herráez, M.P., Gnoni, A., Vilella, S., 2014. Comparative proteome analysis of cryopreserved flagella and head plasma membrane proteins from sea bream spermatozoa: effect of antifreeze proteins. PLoS One 9 (6), e99992.

Further reading

Cabrita, E., Anel, L., Herráez, M.P., Herraéz, M.P., 2001b. Effect of external cryoprotectants as membrane stabilizers on cryopreserved rainbow trout sperm. Theriogenology 56, 623–635.

Marinović, Z., Scekic, I., Lujić, J., Herranz-Jusdado, J.G., Urbányi, B., Müller, T., Asturiano, J.F., Horváth, Á., 2019b. Cryopreservation of European eel (*Anguilla anguilla*) spermatogonia. Cryobiology 91, 146–147 (Abstract S29).

Chapter 2

Molecular approaches on DNA damage evaluation after primordial germ cell cryopreservation in zebrafish

Marta F. Riesco[a], David G. Valcarce[b], and Vanesa Robles[a]
[a]*Department of Molecular Biology, Cell Biology Area, University of León, León, Spain,* [b]*Spanish Institute of Oceanography (IEO), Santander, Spain*

1. Introduction

Primordial germ cells (PGCs) are the precursors of gametes. These diploid cells are found in a small number in the embryo, but they have aroused great interest in fish for several reasons. On the one hand, they could be used as a target for sterilization purposes (Wong and Zohar, 2015; Saito et al., 2018). On the other hand, their preservation and transplantation could be used as a strategy to improve reproduction of valuable species through surrogate production (Goto and Saito, 2019) or as a tool for gene banking of endangered or genetically modified fish species (Yoshizaki and Lee, 2018; Mayer, 2019; Saito et al., 2019). Surrogate technology consists of transplanting donor-derived germ cells in a recipient surrogate fish of a different strain or species. This procedure was initially achieved by Takeuchi and colleagues in 2004 (Takeuchi et al., 2004). In recent years, this technique has reached a high impact in the aquaculture field since it could offer a solution for many problems (Robles et al., 2017; Yoshizaki and Yazawa, 2019) such as: (i) the reduction of breeding time by using a recipient species with a short generation time to produce gametes of a species with long generation times (e.g., Siberian sturgeon, *Acipenser baerii* (Pšenička et al., 2015)), (ii) management of species with reproductive failure (e.g., Senegalese sole, *Solea senegalensis* (Pacchiarini et al., 2014)), or for those species with large size (e.g., yellowtail, *Seriola quinqueradiata* (Morita et al., 2012) and bluefin tuna, *Thunnus orientalis* (Yazawa et al., 2010)), or (iii) genetic improvement of fish species by selection of superior traits from a single selected fish and transferring their PGCs into many recipient fish, among others.

Cellular and Molecular Approaches in Fish Biology. https://doi.org/10.1016/B978-0-12-822273-7.00009-4

Due to these reasons, scientists have made big efforts in optimizing protocols for PGC characterization (Ramasamy et al., 2006; Nagasawa et al., 2013; Li et al., 2016; Zhang et al., 2017; Duangkaew et al., 2019), isolation (Takeuchi et al., 2002; Kobayashi et al., 2004), cryopreservation (Kawakami et al., 2012; Riesco et al., 2012), and in vitro culture (Fan et al., 2008) of this special type of cells. Interestingly, it has been demonstrated that the transplantation of a single PGC could be enough for successful chimera formation and offspring generation (Saito et al., 2008). Furthermore, another study proposed the use of in vitro generation of PGCs from cultured blastomeres (Riesco et al., 2014) as a potential tool for overcoming limitations caused by the small number of PGCs in the embryo.

Whether PGCs would be used for surrogate production or gene banking purposes, the availability of successful methods for PGC storage is required, and cryopreservation seems to be the most adequate method. In this sense, different cryopreservation protocols have been described for germ cell storage with transplantation purposes in different fish species such as Nile tilapia, *Oreochromis niloticus* (Lacerda et al., 2013); medaka, *Oryzias latipes* (Seki et al., 2013); rainbow trout, *Oncorhynchus mykiss* (Yoshizaki et al., 2011); and zebrafish, *Danio rerio* (Riesco et al., 2012). However, there are several damages that cryopreservation may cause in cells, including the production of reactive oxygen species (ROS) (Len et al., 2019) that could compromise DNA integrity (Tran et al., 2018). It is interesting to notice that good cell viability values after cryopreservation do not necessarily mean the absence of important damages at different levels. Considering that these cells must give rise to gametes, the preservation of their DNA integrity is crucial, making advisable DNA evaluation after PGC cryopreservation.

2. Conventional DNA evaluation methods

Oxidative stress triggered by cryopreservation is one of the main mechanisms responsible for DNA damage, and different approaches have been employed in the evaluation of this damage: (i) 8-hydroxy-2′-deoxyguanosine (8OHdG) detection, (ii) studies based on DNA packaging failures (related to histones and protamines), and (iii) analyses of DNA fragmentation.

In the DNA, the nitrogen bases, particularly guanine, are the main targets of ROS attack, generating 8OHdG, which is considered as the hallmark of oxidative DNA damage (Li et al., 2004). The bond between guanine and the ribose units is weakened, destabilizing the DNA structure and triggering strand breakages (Aitken and De Iuliis, 2007). In this sense, different methods have been described to detect the effect of ROS on DNA fragmentation (Cabrita et al., 2014; Dutta et al., 2019; Javed et al., 2019). Specifically, 8OHdG has been quantified using different strategies such as HPLC coupled to electrochemical detection (Kao et al., 2008; Cambi et al., 2013). Nevertheless, these methods have been accepted with some skepticism because the DNA isolation and digestion steps performed during this type of analyses can provoke spontaneous 8-OHdG formation themselves (Badouard et al., 2008). For this reason, emerging methodologies based on antibody detection or binding proteins to reveal 8OHdG have been recently employed (Kao et al., 2008; De Iuliis et al., 2009; Zribi et al., 2011), providing more promising results. Specifically, in fish germ cells, 8OHdG has been employed as a DNA damage measure by antibody detection in different fish species such as rainbow trout (González-Rojo et al., 2014), Atlantic salmon (*Salmo salar*; Lloyd and Phillips, 1999), and Japanese eel (*Anguilla japonica*; Celino et al., 2012). However, these techniques are laborious and they imply different optimization steps, requiring considerable efforts and specialized equipment. More importantly, these approaches reflect only a partial status of damaged genomic sequences.

In addition to 8-OHdG formation, ROS are capable of producing abasic sites, cross-linking, and double lesions that finally result in nitrogen bases being modified and specially, in strand breakages and DNA fragmentation. In this sense, the most widely used method to quantify the percentage of fragmented chromatin in fish sperm is the comet assay (Zilli et al., 2003; Cabrita et al., 2005; Martínez-Páramo et al., 2009; Shaliutina et al., 2012). This technique was initially described by Ostling and Johanson (1984) and consists of the study of the different electrophoretic migration patterns of DNA fragments. Briefly, cells are embedded in an agarose microgel on a slide and subsequently lysed to expose and unwind the DNA. If the DNA contains breaks, DNA fragments are released and migrate toward the anode, forming a comet-like tail while the nonfragmented DNA remains in the comet's head. After staining with a DNA intercalator, the fluorescent intensity of the tail is quantified by a specific software (e.g., Tritek Comet Score Freeware v.1.5, USA), providing a DNA fragmentation rate. In addition, another advantage of this technique is its versatility since it allows the differential identification of single or double strand breaks with modification of the pH lysis solution. Due to these specific characteristics, this technique has been widely used in fish sperm samples from different species such as zebrafish (Cabrita et al., 2011; Lombó et al., 2019), Nile tilapia, goldfish (*Carassius auratus*; Kutluyer et al., 2015), gilthead seabream (*Sparus aurata*; Cabrita et al., 2005), Senegalese sole (Beirão et al., 2008), or rainbow trout (Pérez-Cerezales et al., 2010), among others. Specifically, in zebrafish PGCs, this technique has been applied by our group to confirm the absence of DNA damage after cryopreservation in order to establish an optimized storage protocol for these cells (Riesco et al., 2012). In addition, other types of DNA lesions have been detected using this technique combined with enzymatic digestion with specific endonucleases in rainbow trout sperm (Pérez-Cerezales et al., 2009). This additional step allowed cutting the strand in modified bases for the quantification of oxidized cytosines and guanosines, respectively, in nonfragmented positions of the DNA strand. To summarize, the comet assay is the most widely employed technique to measure DNA fragmentation in fish sperm as a consequence of its main characteristics: inexpensive, high sensitivity, and low cell requirements to perform the analyses. Nevertheless, this assay possesses some constraints that limit its application, such as the lack of standardized and applicable protocols among different species and laboratories, which makes it difficult to understand and compare the results obtained from different studies (Simon et al., 2017). In the case of sperm DNA, due to its high packing degree of an incomplete chromatin decondensation, this technique could not always allow the breaks to be revealed, providing sometimes inaccurate results. Particular technical issues should be considered such as the overlapping of comet tails and the presence of small tail fragments that can difficult their visualization or may be lost. The assay is also laborious and has a high level of interlaboratory variation. For this reason, several concerns concerning its accuracy, specificity, sensitivity, as well as its reproducibility still need to be addressed (Simon et al., 2017).

Concerning techniques based on the differential migration of chromatin fragments, the sperm chromatin dispersion test (SCD) is widely employed in mammalian species and was first described in human sperm (Fernández et al., 2003). The SCD test is based on the principle that sperm with fragmented DNA fails to produce the characteristic halo of dispersed DNA loops that is observed in sperm with nonfragmented DNA. Like in the comet assay, due to the specific differences in the chromatin packaging pattern, species-specific protocols for chromatin decondensation should be optimized, limiting its application. Nonetheless, some studies have been published in fish sperm, specifically in tench (*Tinca tinca*), where a high correlation with the comet assay was reported (López-Fernández et al., 2009). In order to improve the accuracy, fragmentation can also be assessed more specifically using the TUNEL assay. This assay has been commonly employed for the detection of apoptosis, but it will also effectively reveal DNA fragmentation, being caused by apoptosis or by any other process. The TUNEL assay detects DNA breakage in cells by incorporating fluorescent labeled nucleotide to the single- and double-stranded DNA ends. DNA fragmentation triggers the presence of free strand ends and the fluorescence quantity in the nucleus is proportional to fragmented DNA. Fluorescence emission can be assessed by fluorescence microscopy or flow cytometry, increasing the accuracy and allowing the evaluation of a high number of cells. In fish sperm, TUNEL was employed to find differences among freezing protocols in gilthead seabream and European seabass (*Dicentrarchus labrax*) sperm (Cabrita et al., 2011). This technique has also been applied to analyze sperm DNA damage in unfertile males in Nile tilapia (Fang et al., 2018).

Another DNA damage assay compatible with flow cytometry commonly employed in mammals is the sperm chromatin structure assay (SCSA) developed in 1980 (Evenson et al., 1980; Evenson, 2016). With this method the susceptibility to suffer DNA breaks can be evaluated indirectly with DNA denaturation status (induced by acids or heat treatment) by fluorescent dye acridine orange staining. This metachromatic dye turns to green when it is intercalated into double-stranded DNA, while it turns to red fluorescence when stacked on single-stranded DNA. The percentage of red sperm represents the DNA fragmentation index of the sample (Evenson, 2016). A variety of standardized kits have been commercialized demonstrating its wide application in mammalian species due to SCSA test having only one protocol that is now accepted internationally. Nevertheless, there are no reports on their use for fish sperm quality evaluation. This fact could be explained by the different DNA/proteins packaging among species that require specific conditions to successfully perform the analyses (Simon et al., 2017).

To summarize, the use of these techniques might imply a large variability within obtained results between species due to the different DNA packing degree among species and cell types (e.g., spermatozoa have a highly compacted DNA); the use of one standardized protocol is not always possible and these variations concerning packing degree can significantly influence the final result, decreasing their accuracy. More importantly, these approaches deliver only partially quantifiable values and could be considered as tedious, requiring extensive optimization efforts.

3. DNA lesion detection by qPCR: A novel approach

Although all these techniques reviewed in Section 2 are effective in the detection of chromatin fragmentation, they do not provide specific information and only give a general vision of chromatin damage and its integrity. In addition, some specific DNA lesions can occur that do not trigger DNA fragmentation, which these previously described techniques do not detect (as oxidized bases, abasic sites, or thymidine dimers, generated during cryopreservation). The qPCR methods of analysis (described later) are based on the ability of certain DNA lesions to delay and block the polymerase progression along DNA template (Ayala-Torres et al., 2000; San Gabriel et al., 2006; Santos et al., 2006; Rothfuss et al., 2010). More importantly, the global DNA analyses performed by the traditional techniques do not discriminate among different genome regions such as nuclear and mitochondrial genomes. Concerning mitochondrial genome analyses, substantial evidence suggests that mitochondrial DNA (mtDNA) presents a higher rate of damage than the nuclear DNA (nDNA) of the same biological sample, most notably ROS-mediated damage leading to mitochondrial genomic defects (Salazar and Van Houten, 1997; Yakes and Van Houten, 1997; Santos et al., 2003, 2006). This fact could be explained by the absence of chromatin organization and the generation of secondary ROS caused by alterations in the electron transport chain (Yakes and Van Houten, 1997). Considering that cryopreservation is widely known by its effect on ROS production, this mtDNA genome could be more prone to suffer oxidative damage and should be specifically studied. However, these mtDNA isolation approaches based on global DNA analyses present some constraints, such as the DNA isolation method provoked extensive DNA oxidation that could interfere with the analysis (Furda et al., 2014). The multiple steps of mtDNA isolation protocol (nDNA-free) could trigger high levels of damage (Michael Anson et al., 2000; Furda et al., 2014). Thus one of the main strengths of the long PCR assay is that it permits measuring mtDNA damage directly from total cellular DNA, removing the mtDNA isolation step from nDNA. This fact allows the comparison between mtDNA and nDNA damage in the same samples, making it particularly appropriate for this use. The qPCR technique can be applied to study any specific gene or DNA region that can be amplified to compare the rate of damage among regions that are hypothesized to be more quickly repaired than others. This method is based on the decrease in polymerase amplification produced by the DNA lesions, delaying the Ct (threshold cycle) (Fig. 1, panel 1, Experimental design). These data can be translated to a certain number of lesions using the formula published by Rothfuss (Rothfuss et al., 2010). This qPCR-based method was originally described to quantify DNA lesions in the mtDNA after the exposure to hydrogen peroxide in human cells. This novel approach was used for the first time by our group to quantify DNA lesions in PGCs to compare the susceptibility to suffer DNA damage after cryopreservation of particular chromatin regions (nuclear and mitochondrial) in different fish species, such as zebrafish (Riesco and Robles, 2012) and gilthead seabream (Cartón-García et al., 2013) (Fig. 1, panel 2, qPCR results). Our results confirmed the higher susceptibility of germ cell mtDNA to undergoing DNA damage after cryopreservation according with previously published data (Riesco and Robles, 2012; Cartón-García et al., 2013) (Fig. 1 panel 2, qPCR results).

Moreover, after these previous studies we could confirm that an absence of detectable DNA fragmentation after cryopreservation is not always correlated with an absence of DNA damage (Riesco et al., 2012) (Fig. 1). Global evaluation of DNA integrity could be masking some specific DNA lesions (Riesco and Robles, 2013). Applying this novel method of DNA evaluation by qPCR, we could confirm if optimized germ cell cryopreservation protocols, in terms of viability and DNA fragmentation (analyzed by Comet Assay), can provoke other type of DNA lesions even when fragmentation is absent (Riesco et al., 2012). Using this method, the effect of cryopreservation on specific PGC-related genes in zebrafish germ cells has been evaluated (Riesco and Robles, 2013), finding some specific genes with high sensibility to suffer damage during cryopreservation. This type of approach was transferred to other fish species, such as gilthead seabream (Cartón-García et al., 2013), rainbow trout (González-Rojo et al., 2014), and even to human sperm for the evaluation of a cryopreservation protocol, obtaining similar results (Valcarce et al., 2013a, b).

Our results allowed us to confirm that the limitations of the global DNA integrity analysis can be overcome by employing qPCR methods. This novel assay provides information on a specific target (a gene or a particular chromatin region) but can be also applied to multiple genes or sequences, offering, in this case, a more global and accurate assessment of DNA damage. Moreover, the qPCR-based method provides some additional advantages such as its sensitivity, the low quantity of DNA required (1–2 ng) to perform the analysis, and

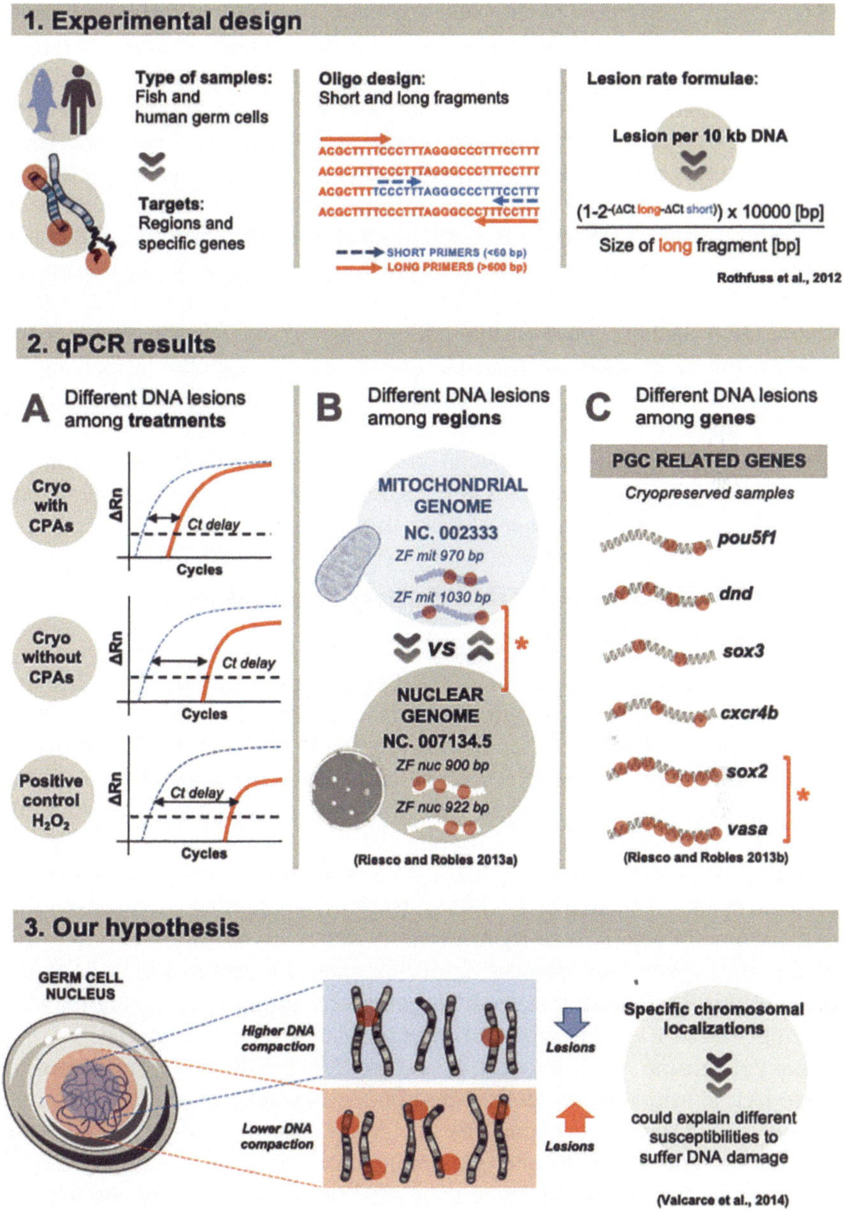

FIG. 1 (1) Schematic graphical abstract of the qPCR technique application in DNA damage evaluation: type of sample, targets, primer design, and formulae being Δ *Ct* long: *Ct* treated − *Ct* control of long amplicons and Δ *Ct* short: *Ct* treated − *Ct* control of short amplicons for each gene or region (2) main results obtained comparing treatments: cryopreserved with cryoprotectants (CPAs), cryopreserved without CPAs and peroxide treated (positive control) (step A), genome regions (nuclear and mitochondrial fragments) (step B), and specific genes crucial for PGC survival, development, and migration (step C) after germ cell cryopreservation, and (3) new hypothesis derived from our results.

the possibility to compare nuclear and mitochondrial DNA in the same sample (Furda et al., 2014). Nevertheless, in the last years, some limitations related to qPCR-based methods in DNA evaluation have been also identified: (i) it has been demonstrated the existence of a supercoiled circular form of mtDNA that prevents DNA polymerase accession, being necessary to include some additional denaturation step (Furda et al., 2014); and (ii) since this technique detects all lesions related to DNA polymerase blocking, whether the causative agents of DNA injuries trigger lesions associated to DNA fragmentation (strand breaks) or not (e.g., abasic sites and thymidine dimers) (Termini, 2000), the percentage of nonquantifiable lesions by qPCR methods is minimal although the specific induced DNA damage could be masked.

To summarize, it is recommended to use methods for global DNA integrity evaluation in combination with other analysis with higher sensitivity, such as the qPCR approach, to obtain a detailed study of the potential DNA damage. These exhaustive methods take special relevance in PGCs, since a single thawed PGC could give rise to all gametes after its transplantation to a host embryo, and we should guarantee the absence of DNA damage on it.

4. Our contribution to DNA damage evaluation in cryopreserved germ cells

One of the main advantages of qPCR-based approaches in DNA damage evaluation consists of the possibility to analyze the vulnerability of specific DNA regions or genes to suffer lesions. The main achievements of our studies were to confirm the existence of different susceptibilities to suffer DNA damage after cryopreservation in specific regions and genes in germ cells (PGCs and spermatozoa) from different fish species (zebrafish and gilthead seabream) (Riesco and Robles, 2012, 2013; Cartón-García et al., 2013) and in human sperm (Valcarce et al., 2013a, b). Previous findings pointed out that some chromatin regions are more prone to suffer cryoinjuries due to a low efficiency of chromatin remodeling during spermiogenesis (De Iuliis et al., 2009). In this vulnerability state, ROS generated during the freezing/thawing can trigger DNA fragmentation and other types of lesions (8-oxo-7,8-dihydro-2′-deoxyguanosine (8-oxodG), 8-Oxo-7,8-dihydro-2′-deoxyadenosine (8-oxodA), abasic sites, or thymidine dimers; Sikorsky et al., 2007). These previous findings were explained by our group after describing the potential different susceptibilities to undergo damage among genes belonging to different nuclear territories (Valcarce et al., 2013a, b). In this study we demonstrated that those genes located in regions with low degree of compaction were also more susceptible to undergoing damage after cryopreservation (Riesco and Robles, 2013). Sperm chromatin is divided into three structural domains: (i) the main proportion, associated with protamines for a high degree of protection; (ii) other minor part associated with histones, with a low degree of compaction (Gineitis et al., 2000; Zalenskaya et al., 2000) and therefore more exposed to suffer damage; and (iii) the sperm nuclear matrix, in the nucleus periphery and with a low degree of packing/protection. In our studies, we evidenced that the genes more affected by cryopreservation and peroxide treatment (employed as positive control of oxidative damage in our studies) were located in these lower compacted zones, close to the nuclear matrix (Valcarce et al., 2013a, b). Attending to our findings, considering the different degree of lesions that can occur and could be masked when performing a global DNA damage analysis among regions and specific genes, an accurate DNA lesions evaluation is critical (Fig. 1 panel 3, Our hypothesis).

Finally, in this chapter we present a detailed method for the evaluation of DNA damage after zebrafish PGC cryopreservation using the qPCR method.

5. Specific protocol for DNA damage evaluation in zebrafish PGCs after cryopreservation

5.1 PGC isolation

- Manually excise genital ridges (GRs) containing PGCs from earlier than 24 hours post-fertilization (hpf) in mechanically dechorionized embryos, using fine watchmaker's forceps under a dissecting microscope (Nikon, Tokyo, Japan), as described by Kobayashi and colleagues (Kobayashi et al., 2004). Collect the excised GRs and place them in modified Leibovitz medium (L15) supplemented with 5% fetal bovine serum (FBS) in a culture dish at room temperature (RT).
- Dissociate GR from embryos with collagenase 0.1% (in Hank's premix solution) by pipetting. To facilitate PGC identification and isolation, zebrafish vas:EGFP transgenic line (Krøvel and Olsen, 2002) is recommended in order to ensure PGC visualization by their bright fluorescence (see Note 1) and relatively large size. Conduct the experimental procedure in triplicates, using at least 10 GRs in each replicate (approximately 200 PGCs in each replicate).

Note 1

PGC visualization in live embryos has been limited to laboratory fish models (medaka and zebrafish) and species with aquaculture importance (e.g., rainbow trout) where different transgenic lines have been described (Yoshizaki et al., 2000). For other fish species, PGCs can be transiently labeled by embryonic injection of mRNA constructions containing green fluorescence protein gene (GFP) and the 3′-untranslated region (3′-UTR) of a maternal germ gene such as vasa. These protocols have also been described in marine species (Lin et al., 2012; Zhou et al., 2019).

5.2 Experimental groups

5.2.1 PGC cryopreservation

Expose PGCs to cryoprotectants dimethyl sulfoxide (DMSO; 5 M), ethylene glycol (EG; 1 M), and polyvinylpyrrolidone (PVP; 4%), in a three stepwise manner using multiwell plates following the next steps:

- First, incubate PGC suspension (at least 100 cells) in a pretreatment solution with DMSO 2 M and EG 0.5 M in Hank's premix solution for 10 min.
- Then, expose samples to DMSO 5 M and EG 1 M in Hank's premix for 2 min. In the last step, add external cryoprotectant agent (PVP) and incubate for 2 min.
- Place PGCs without cryoprotectants in Hank's premix solution instead of the respective cryoprotectants following the same steps. To avoid PGC lost during cryopreservation protocols, these cells can be microencapsulated after cryoprotectant exposition (see Note 2).

Note 2

Cell microencapsulation: enzymatically digest cell suspensions (collagenase 0.1% in Hank's premix) by pipetting. After that, dilute PGC suspension in L15 with 1.5% (w/v) sodium alginate (Sigma-Aldrich Co., St. Louis, MO, USA). Dissolve the solution in phosphate-buffered saline (PBS) to reach a final concentration of 0.75%. Suck cell suspension through a needle attached to a 1-mL syringe into a 60-mm plastic dish containing 1.5% (w/v) calcium chloride dissolved in PBS. The cell suspension suffers solidification immediately upon contact with the calcium chloride solution forming a high-viscosity microgel. Gently shake microgels and allow to react with calcium ions for 30 s. After freezing/thawing, place microcapsules in sodium citrate solution and vortex to release cells from the microcapsules.

- Load 0.5 mL straws with cells embedded in cryoprotectant solution after the equilibrium time (see Note 3) using a micropipette.

Note 3

Equilibrium time is a critical step considering that during this time cryoprotectants can penetrate cells (see Chapter 1). After a variable period of time, the cryoprotectants and water equilibrate to reach similar intracellular and extracellular concentrations. This step should be optimized for each cryopreservation protocol, depending on the cryoprotectant solution and exposition temperature.

- After loading, expose straws horizontally to liquid nitrogen vapors (2 cm over the surface) in the cooling rack. Close the box to reduce nitrogen evaporation and keep the straws in this atmosphere for 20 min. Other alternative devices can be employed to replicate the freezing rates such programmable ultra-freezers (please see in this regard Chapter 1).

- Plunge the straws into liquid nitrogen, transfer and keep the straws into the liquid nitrogen tank storage. Other alternative freezing methods could be employed such an ultrafast freezing: vitrification, providing satisfactory results (see Note 4).

Note 4

After cryoprotectant exposure, with a pipette aspirate 10 µL of PGCs embedded in the cryoprotectant solution (microdrop) and deposit it over a metal surface partially submerged in liquid nitrogen. Microdrops will freeze immediately and then transfer them, with precooled forceps, in a cryovial to be plunged into liquid nitrogen for their storage. This protocol has been widely described in Riesco et al. (2012). Alternative cryopreservation protocols could be employed (for more information and alternatives, please see Chapter 1).

- For thawing, carefully remove the straws from the liquid nitrogen and thaw them in a water bath for 6 s at 63 °C. Ensure they are completely thawed, cut the cotton plug with scissors to allow the suspension to be released and wash it with Hank's premix solution.

5.2.2 Without cryoprotectant treatment

Perform the same PGC freezing and thawing protocol replacing the cryoprotectant solution by PBS to obtain this experimental group.

5.2.3 Hydrogen peroxide treatment

Hydrogen peroxide treatment is employed as a positive control for DNA damage. Expose PGCs to hydrogen peroxide (H_2O_2) 1% in L15 for 30 min at room temperature (RT) in a multiwell plate, and wash twice with L15 to remove H_2O_2 (see Note 5).

Note 5

Alternative protocols and reagents for DNA damage induction could be employed as positive controls. Moreover, different concentrations and exposure times could be tested. These alternative protocols should be first evaluated by traditional techniques of DNA fragmentation (such as Comet Assay) in order to confirm their effectiveness as DNA damage inductors.

5.2.4 Fresh samples

Immediately process fresh PGCs obtained from the same cellular suspension.

5.3 DNA isolation

Extract DNA from zebrafish PGC of different experimental groups (fresh, cryopreserved, frozen without cryoprotectants, and H_2O_2-treated samples as a positive control of DNA damage). For DNA extraction, we recommend the DNA Blood and Tissue Kit (Qiagen, Spain) and following the instructions described by the manufacturer, although other DNA extraction procedures and commercial kits might be used (see Note 6). Determine DNA quantity and purity using a Nanodrop spectrophotometer (Thermofisher, Spain). The isolated DNA should show a high purity (A260/280 > 1.7). Store isolated DNA at 4 °C. For quantitation, dilute to a range of around 20 ng/μL. Diluted samples can be kept at 4 °C for up to several weeks before use.

Note 6

DNA template integrity is crucial for the satisfactory amplification of long PCR targets. Procedures that involve phenol extraction should be avoided due to potential introduction of artifactual DNA oxidation and for this reason, the DNA extraction kit from Qiagen is recommended, providing templates of relatively high molecular weight, with high stability, yielding comparable amplification during long-term storage and highly reproducible yield among samples.

5.4 Primer design

- Select sequences of the studied genes from a database. In the case of other type of DNA regions within nuclear and mitochondrial genomes, perform the same steps than in specific gene analyses. Special care should be taken in fish due to the existence of paralog genes derived from whole genome duplication (WGD) events (Van De Peer et al., 2009; Voldoire et al., 2017; see also Chapter 10). In this case, the paralog-specific amplification when using the designed primers should be confirmed by sequencing.
- Design two pairs of primers for each studied gene/DNA region. The first set of primers should generate a longer amplicon of around 800–1000 bp, and the second should create a shorter one (< 100 bp). This short fragment must be designed to be within the larger amplicon to serve as a negative control of DNA lesions. In these short fragments, the probability of lesion occurrence is minimal and can be served as an internal normalization for technical validation (Fig. 1, panel 1, Experimental design). Primers should have the following characteristics: 19–25 bases long, G + C content not higher than 70%, and melting temperature (*Tm*) between 58 °C and 62 °C. The designed primers should be evaluated for secondary structures using an appropriate software (e.g., Primer Express, Thermofisher, Spain) since the formation of primer-dimers and hairpins can interfere with qPCR reaction. For primer validation, it is necessary to establish the optimal melting temperature (*Tm*). Check optimal primer melting temperatures (*Tm*) by conventional PCR in a gradient thermocycler in order to confirm the product size and optimize the annealing temperature. Carry out a standard PCR cycle adapted for the expected product length following the manufacture instructions. Finally, the amplification of one unique band should be verified by gel electrophoresis

to select the optimal *Tm* for qPCR assays (see Note 7). Confirmation of specific amplification by Sanger sequencing is highly recommended.

Note 7

As a part of the primer validation step, it is completely necessary to evaluate the PCR products by agarose gel electrophoresis to verify the amplicon size and to assure that no other nonspecific products are generated. The percentage of agarose should be adjusted to separate long (1%–1.5% *w/v*) and small fragments (above 3% *w/v*) to check their specific lengths.

5.5 qPCR assay

- Once the primers are selected and validated, finding the optimal reaction conditions is the next step. Concerning long fragments, a specific amplification kit for long amplicons is need, for example: Fast Start DNA Master plus SYBR Green I (Roche, Germany) (see Note 8). For long fragments, the reaction mixture contains 4 μl of 5 × Fast Start DNA Master plus SYBR Green I (Roche, Germany), 1 μl of each 5 μM forward and reverse primer, 0.4 μl of 50 × ROX passive reference dye, template DNA (3 ng), and molecular grade water up to 20 μL. For short fragments, employ a routine qPCR amplification kit. In this case, the reaction mixture might consist of 10 μl of 2 × SYBR Green PCR, 1 μl each of 5 μM forward and reverse primer, 3 ng of template DNA and molecular grade water up to 20 μL. Perform a final extension step at 72 °C for 10 s (in small amplicons) or 50 s (in large amplicons).

Note 8

A specific amplification kit should be employed, in our case, we use Fast Start DNA Master plus SYBR Green I (Roche, Germany) with the following specifications: each lot is tested for performance in qPCR using three templates: (i) a GC-rich template, (ii) an AT-rich template, (iii) and long templates.

- Carry out qPCR amplification and fluorescence detection in a commercial qPCR thermocycler according to guidelines provided.
- Calculate crossing point values using the specific software of qPCR thermocycler. Include technical triplicates per sample and a nontemplate control for each set of primers.
- Verify the linear relationship between the threshold cycle and serial DNA dilutions. For each dilution, a standard qPCR protocol should be performed in triplicate for all the primer pairs included in the experiment and Ct values registered. Design a standard curve by plotting the log of the starting quantity of the template against their respective Ct value. Calculate the equation of the linear regression line and the Pearson's correlation coefficient (*r*) to evaluate whether the qPCR assay is optimized (see Note 9).

Note 9

The efficiency of a PCR measures the rate at which the polymerase converts the different components of the PCR reaction (dNTPs, oligonucleotides, and template cDNA) to amplicon. An efficiency of 100% of the reaction is achieved when the increase of amplicon per cycle is twofold. The acceptable efficiency range (90%–110%) and that higher or lower values could indicate different problems with the qPCR such as presence of polymerase inhibitors, nonoptimal annealing temperature, polymerase inactivation, wrong primer design, and/or presence of secondary structures or primer dimers.

5.6 Lesion rate calculation

- Calculate the difference in *Ct* value (Δ *Ct*) determined by qPCR analysis between each treatment (cryopreserved, freezing without cryoprotectants, and H_2O_2 treated) and the control samples (fresh PGCs). Repeat this calculation for each long and short fragment for each gene/specific region. In this way, a set of "Δlong" and "Δshort" values will be created to be introduced in the formula described by Rothfuss (Rothfuss et al., 2010).
- Replace the reported values in the formula:
  Lesion rate [Lesion per 10 kb DNA] = $(1-2^{-(\Delta Ctlong - \Delta Ctshort)}) \times 10{,}000$ [bp]/ size of long fragment [bp].

6. Summary and future perspectives

In this chapter, we have presented a DNA integrity evaluation protocol to be applied after cryopreservation in order to detect some specific lesions that could remain undetectable using the conventional techniques to analyze DNA fragmentation (Comet Assay, TUNEL, SCSA, etc.). This specific approach takes a special relevance in PGCs (that contain maternal and paternal genomes) and has been employed as a gene bank resource for surrogate techniques in fish. Considering that one single PGC will generate the gametes into the recipient embryo after its transplantation and genital ridge migration, the absence of DNA damage should be guaranteed. The application of surrogate technologies is on the rise in the field of aquaculture and new progress related to PGC preservation and their evaluation might support it. However, different approaches based on DNA modifications, such as genome-wide DNA methylation, should be considered. These methylation events have been previously observed by our group after germ cell cryopreservation in specific PGC biomarkers triggering a decrease of some important transcript levels (Riesco and Robles, 2013). However, these epigenetic events and their consequences will be described in detail in Chapter 3. To summarize, these new techniques based on the detection of molecular alterations caused by cryopreservation should be studied in detail and at different levels in order to ensure the total safety of the protocol.

Acknowledgments

The authors acknowledge the REPROSTRESS project (PID2019-108509RB-100) and Juan de la Cierva Formación Programme (FJC2018-037566-I).

References

Aitken, R.J., De Iuliis, G.N., 2007. Origins and consequences of DNA damage in male germ cells. Reprod. Biomed. Online, 727–733. https://doi.org/10.1016/S1472-6483(10)60676-1. Reproductive Healthcare Ltd.

Ayala-Torres, S., et al., 2000. Analysis of gene-specific DNA damage and repair using quantitative polymerase chain reaction. Methods 22 (2), 135–147. https://doi.org/10.1006/meth.2000.1054. Academic Press Inc.

Badouard, C., et al., 2008. Determination of new types of DNA lesions in human sperm. Zygote 16 (1), 9–13. https://doi.org/10.1017/S0967199407004340.

Beirão, J., et al., 2008. Cellular damage in spermatozoa from wild-captured *Solea senegalensis* as detected by two different assays: comet analysis and Annexin V-Fluorescein staining. J. Appl. Ichthyol. 24 (4), 508–513. https://doi.org/10.1111/j.1439-0426.2008.01144.x. John Wiley & Sons, Ltd.

Cabrita, E., et al., 2005. Evaluation of gilthead sea bream, *Sparus aurata*, sperm quality after cryopreservation in 5ml macrotubes. Cryobiology 50 (3), 273–284. https://doi.org/10.1016/j.cryobiol.2005.02.005.

Cabrita, E., et al., 2011. The influence of certain aminoacids and vitamins on post-thaw fish sperm motility, viability and DNA fragmentation. Anim. Reprod. Sci. 125 (1–4), 189–195. https://doi.org/10.1016/j.anireprosci.2011.03.003.

Cabrita, E., et al., 2014. Factors enhancing fish sperm quality and emerging tools for sperm analysis. Aquaculture. https://doi.org/10.1016/j.aquaculture.2014.04.034.

Cambi, M., et al., 2013. Development of a specific method to evaluate 8-hydroxy, 2-deoxyguanosine in sperm nuclei: relationship with semen quality in a cohort of 94 subjects. Reproduction 145 (3), 227–235. https://doi.org/10.1530/REP-12-0404.

Cartón-García, F., et al., 2013. Quantification of lesions in nuclear and mitochondrial genes of *Sparus aurata* cryopreserved sperm. Aquaculture 402–403, 106–112. https://doi.org/10.1016/j.aquaculture.2013.03.034.

Celino, F.T., et al., 2012. Proliferating spermatogonia are susceptible to reactive oxygen species attack in Japanese eel (*Anguilla japonica*). Biol. Reprod. 87 (3). https://doi.org/10.1095/biolreprod.112.099887.

De Iuliis, G.N., et al., 2009. DNA damage in human spermatozoa is highly correlated with the efficiency of chromatin remodeling and the formation of 8-hydroxy-2′-deoxyguanosine, a marker of oxidative stress. Biol. Reprod. 81 (3), 517–524. https://doi.org/10.1095/biolreprod.109.076836.

Duangkaew, R., et al., 2019. Characterization and expression of a vasa homolog in the gonads and primordial germ cells of the striped catfish (*Pangasianodon hypophthalmus*). Theriogenology 131, 61–71. https://doi.org/10.1016/j.theriogenology.2019.01.022. Elsevier Inc.

Dutta, S., Majzoub, A., Agarwal, A., 2019. Oxidative stress and sperm function: a systematic review on evaluation and management. Arab. J. Urol., 87–97. https://doi.org/10.1080/2090598X.2019.1599624. Taylor and Francis Ltd.

Evenson, D.P., 2016. The Sperm Chromatin Structure Assay (SCSA®) and other sperm DNA fragmentation tests for evaluation of sperm nuclear DNA integrity as related to fertility. Anim. Reprod. Sci. 169, 56–75. https://doi.org/10.1016/j.anireprosci.2016.01.017. Elsevier B.V.

Evenson, D.P., Darzynkiewicz, Z., Melamed, M.R., 1980. Comparison of human and mouse sperm chromatin structure by flow cytometry. Chromosoma 78 (2), 225–238. https://doi.org/10.1007/BF00328394. Springer-Verlag.

Fan, L., et al., 2008. Zebrafish primordial germ cell cultures derived from vasa::RFP transgenic embryos. Stem Cells Dev. 17 (3), 585–597. https://doi.org/10.1089/scd.2007.0178.

Fang, X., et al., 2018. Nuclear progestin receptor (Pgr) knockouts resulted in subfertility in male tilapia (*Oreochromis niloticus*). J. Steroid Biochem. Mol. Biol. 182, 62–71. https://doi.org/10.1016/j.jsbmb.2018.04.011. Elsevier Ltd.

Fernández, J.L., et al., 2003. The sperm chromatin dispersion test: a simple method for the determination of sperm DNA fragmentation. J. Androl. 24 (1), 59–66. https://doi.org/10.1002/j.1939-4640.2003.tb02641.x. John Wiley & Sons, Ltd.

Furda, A., et al., 2014. Quantitative PCR-based measurement of nuclear and mitochondrial DNA damage and repair in mammalian cells. Methods Mol. Biol. 1105, 419–437. https://doi.org/10.1007/978-1-62703-739-6_31. Humana Press Inc.

Gineitis, A.A., et al., 2000. Human sperm telomere-binding complex involves histone H2B and secures telomere membrane attachment. J. Cell Biol. 151 (7), 1591–1597. https://doi.org/10.1083/jcb.151.7.1591. The Rockefeller University Press.

González-Rojo, S., et al., 2014. Differential gene susceptibility to sperm DNA damage: analysis of developmental key genes in trout. PLoS One 9 (12). https://doi.org/10.1371/journal.pone.0114161. Public Library of Science.

Goto, R., Saito, T., 2019. A state-of-the-art review of surrogate propagation in fish. Theriogenology 133, 216–227. https://doi.org/10.1016/j.theriogenology.2019.03.032. Elsevier Inc.

Javed, A., Talkad, M.S., Ramaiah, M.K., 2019. Evaluation of sperm DNA fragmentation using multiple methods: a comparison of their predictive power for male infertility. Clin. Exp. Reprod. Med. 46 (1), 14–21. https://doi.org/10.5653/cerm.2019.46.1.14. Korean Society for Reproductive Medicine.

Kao, S.H., et al., 2008. Increase of oxidative stress in human sperm with lower motility. Fertil. Steril. 89 (5), 1183–1190. https://doi.org/10.1016/j.fertnstert.2007.05.029.

Kawakami, Y., et al., 2012. Cryopreservation of green fluorescent protein (GFP)-labeled primordial germ cells with GFP fused to the 3′ untranslated region of the nanos gene by vitrification of Japanese eel (*Aguilla japonica*) somite stage embryos. J. Anim. Sci. 90 (12), 4256–4265. https://doi.org/10.2527/jas.2011-4884.

Kobayashi, T., et al., 2004. Isolation of highly pure and viable primordial germ cells from rainbow trout by GFP-dependent flow cytometry. Mol. Reprod. Dev. 67 (1), 91–100. https://doi.org/10.1002/mrd.20003. John Wiley & Sons, Ltd.

Krøvel, A.V., Olsen, L.C., 2002. Expression of a vas::EGFP transgene in primordial germ cells of the zebrafish. Mech. Dev. 116 (1–2), 141–150. https://doi.org/10.1016/S0925-4773(02)00154-5. Elsevier.

Kutluyer, F., Öğretmen, F., İnanan, B.E., 2015. Effects of semen extender supplemented with l-methionine and packaging methods (straws and pellets) on post-thaw goldfish (*Carassius auratus*) sperm quality and DNA damage. CryoLetters 36.

Lacerda, S.M.S.N., et al., 2013. Germ cell transplantation as a potential biotechnological approach to fish reproduction. Fish Physiol. Biochem. 39 (1), 3–11. https://doi.org/10.1007/s10695-012-9606-4.

Len, J.S., Koh, W.S.D., Tan, S.X., 2019. The roles of reactive oxygen species and antioxidants in cryopreservation. Biosci. Rep. 39 (8), 20191601. https://doi.org/10.1042/BSR20191601. Portland Press Ltd.

Li, S.-Z., et al., 2016. Molecular characterization and expression pattern of a germ cell marker gene dnd in gibel carp (*Carassius gibelio*). Gene 591 (1), 183–190. https://doi.org/10.1016/j.gene.2016.07.027. Elsevier B.V.

Li, K., Shang, X., Chen, Y., 2004. High-performance liquid chromatographic detection of lipid peroxidation in human seminal plasma and its application to male infertility. Clin. Chim. Acta 346 (2), 199–203. https://doi.org/10.1016/j.cccn.2004.03.013.

Lin, F., et al., 2012. Transient and stable GFP expression in germ cells by the vasa regulatory sequences from the red Seabream (*Pagrus major*)'. Int. J. Biol. Sci. 8 (6), 882–890. https://doi.org/10.7150/ijbs.4421.

Lloyd, D.R., Phillips, D.H., 1999. Oxidative DNA damage mediated by copper(II), iron(II) and nickel(II) Fenton reactions: evidence for site-specific mechanisms in the formation of double-strand breaks, 8-hydroxydeoxyguanosine and putative intrastrand cross-links. Mutat. Res. 424 (1–2), 23–36. https://doi.org/10.1016/S0027-5107(99)00005-6.

Lombó, M., et al., 2019. Genetic and epigenetic alterations induced by bisphenol A exposure during different periods of spermatogenesis: from spermatozoa to the progeny. Sci. Rep. 9 (1), 1–13. https://doi.org/10.1038/s41598-019-54368-8. Nature Research.

López-Fernández, C., et al., 2009. Rapid rates of sperm DNA damage after activation in tench (*Tinca tinca*: Teleostei, Cyprinidae) measured using a sperm chromatin dispersion test. Reproduction 138 (2), 257–266. https://doi.org/10.1530/REP-09-0105.

Martínez-Páramo, S., et al., 2009. Cryoprotective effects of antifreeze proteins delivered into zebrafish embryos. Cryobiology 58 (2), 128–133. https://doi.org/10.1016/j.cryobiol.2008.11.013.

Mayer, I., 2019. The role of reproductive sciences in the preservation and breeding of commercial and threatened teleost fishes. Adv. Exp. Med. Biol., 187–224. https://doi.org/10.1007/978-3-030-23633-5_7. Springer New York LLC.

Michael Anson, R., Hudson, E., Bohr, V.A., 2000. Mitochondrial endogenous oxidative damage has been overestimated. FASEB J. 14 (2), 355–360. https://doi.org/10.1096/fasebj.14.2.355. Wiley.

Morita, T., et al., 2012. Production of donor-derived offspring by allogeneic transplantation of spermatogonia in the yellowtail (*Seriola quinqueradiata*). Biol. Reprod. 86 (6). https://doi.org/10.1095/biolreprod.111.097873.

Nagasawa, K., et al., 2013. Identification and migration of primordial germ cells in Atlantic salmon, *Salmo salar*: Characterization of Vasa, Dead End, and Lymphocyte antigen 75 genes. Mol. Reprod. Dev. 80 (2), 118–131. https://doi.org/10.1002/mrd.22142. Wiley-Blackwell.

Ostling, O., Johanson, K.J., 1984. Microelectrophoretic study of radiation-induced DNA damages in individual mammalian cells. Biochem. Biophys. Res. Commun. 123 (1), 291–298. https://doi.org/10.1016/0006-291X(84)90411-X.

Pacchiarini, T., Sarasquete, C., Cabrita, E., 2014. Development of interspecies testicular germ-cell transplantation in flatfish. Reprod. Fertil. Dev. 26 (5), 690–702. https://doi.org/10.1071/RD13103. CSIRO.

Pérez-Cerezales, S., et al., 2009. Evaluation of oxidative DNA damage promoted by storage in sperm from sex-reversed rainbow trout. Theriogenology 71 (4), 605–613. https://doi.org/10.1016/j.theriogenology.2008.09.057.

Pérez-Cerezales, S., et al., 2010. Fertilization capacity with rainbow trout DNA-damaged sperm and embryo developmental success. Reproduction 139 (6), 989–997. https://doi.org/10.1530/REP-10-0037. BioScientifica.

Pšenička, M., et al., 2015. Isolation and transplantation of sturgeon early-stage germ cells. Theriogenology 83 (6), 1085–1092. https://doi.org/10.1016/j.theriogenology.2014.12.010. Elsevier Inc.

Ramasamy, S., et al., 2006. Zebrafish Staufen1 and Staufen2 are required for the survival and migration of primordial germ cells. Dev. Biol. 292 (2), 393–406. https://doi.org/10.1016/j.ydbio.2006.01.014. Academic Press Inc.
Riesco, M.F., et al., 2012. Evaluation of zebrafish (*Danio rerio*) PGCs viability and DNA damage using different cryopreservation protocols. Theriogenology 77 (1), 122–130.e2. https://doi.org/10.1016/J.THERIOGENOLOGY.2011.07.024. Elsevier.
Riesco, M.F., et al., 2014. *In vitro* generation of zebrafish PGC-like cells. Biol. Reprod. 91 (5), 114. https://doi.org/10.1095/biolreprod.114.121491.
Riesco, M.F., Robles, V., 2012. Quantification of DNA damage by q-PCR in cryopreserved zebrafish Primordial Germ Cells. J. Appl. Ichthyol. 28 (6), 925–929. https://doi.org/10.1111/jai.12089. John Wiley & Sons, Ltd (10.1111).
Riesco, M.F., Robles, V., 2013. Cryopreservation causes genetic and epigenetic changes in zebrafish genital ridges. PLoS One 8 (6), e67614. https://doi.org/10.1371/journal.pone.0067614. Edited by L. Orban. Public Library of Science.
Robles, V., et al., 2017. Biology of teleost primordial germ cells (PGCs) and spermatogonia: Biotechnological applications. Aquaculture 472, 4–20. https://doi.org/10.1016/J.AQUACULTURE.2016.03.004. Elsevier.
Rothfuss, O., Gasser, T., Patenge, N., 2010. Analysis of differential DNA damage in the mitochondrial genome employing a semi-long run real-time PCR approach. Nucleic Acids Res. 38 (4), e24. https://doi.org/10.1093/nar/gkp1082.
Saito, T., et al., 2008. Xenogenesis in teleost fish through generation of germ-line chimeras by single primordial germ cell transplantation1. Biol. Reprod. 78 (1), 159–166. https://doi.org/10.1095/biolreprod.107.060038. Oxford Academic.
Saito, T., et al., 2018. Elimination of primordial germ cells in sturgeon embryos by ultraviolet irradiation. Biol. Reprod. 99 (3), 556–564. https://doi.org/10.1093/biolre/ioy076. Oxford University Press.
Saito, T., et al., 2019. Production of germ-line chimeras in zebrafish. Methods Mol. Biol., 327–341. https://doi.org/10.1007/978-1-4939-9009-2_20. Humana Press Inc.
Salazar, J.J., Van Houten, B., 1997. Preferential mitochondrial DNA injury caused by glucose oxidase as a steady generator of hydrogen peroxide in human fibroblasts. Mutat. Res., 139–149. https://doi.org/10.1016/S0921-8777(97)00047-5. Elsevier B.V.
San Gabriel, M., Zhang, X., Zini, A., 2006. Estimation of human sperm gene-specific deoxyribonucleic acid damage by real-time polymerase chain reaction analysis. Fertil. Steril. 85 (3), 797–799. https://doi.org/10.1016/j.fertnstert.2005.11.027.
Santos, J.H., et al., 2003. Cell sorting experiments link persistent mitochondrial DNA damage with loss of mitochondrial membrane potential and apoptotic cell death. J. Biol. Chem. 278 (3), 1728–1734. https://doi.org/10.1074/jbc.M208752200.
Santos, J.H., et al., 2006. Quantitative PCR-based measurement of nuclear and mitochondrial DNA damage and repair in mammalian cells. Methods Mol. Biol. 314, 183–199. https://doi.org/10.1385/1-59259-973-7:183.
Seki, S., et al., 2013. 070 Production of donor-derived offspring by allogenic transplantation of cryopreserved spermatogonia in medaka. Cryobiology 67 (3), 417. https://doi.org/10.1016/j.cryobiol.2013.09.076. Elsevier BV.
Shaliutina, A., et al., 2012. Spermatozoa motility and variation in the seminal plasma proteome of Eurasian perch (*Perca fluviatilis*) during the reproductive season. Mol. Reprod. Dev. 79 (12), 879–887. https://doi.org/10.1002/mrd.22126.
Sikorsky, J.A., et al., 2007. DNA damage reduces Taq DNA polymerase fidelity and PCR amplification efficiency. Biochem. Biophys. Res. Commun. 355 (2), 431–437. https://doi.org/10.1016/j.bbrc.2007.01.169. Academic Press.

Simon, L., Emery, B.R., Carrell, D.T., 2017. Review: diagnosis and impact of sperm DNA alterations in assisted reproduction. Best Pract. Res. Clin. Obstet. Gynaecol., 38–56. https://doi.org/10.1016/j.bpobgyn.2017.07.003. Bailliere Tindall Ltd.

Takeuchi, Y., Yoshizaki, G., Takeuchi, T., 2004. Surrogate broodstock produces salmonids. Nature 430 (7000), 629–630. https://doi.org/10.1038/430629a.

Takeuchi, Y., et al., 2002. Mass isolation of primordial germ cells from transgenic rainbow trout carrying the green fluorescent protein gene driven by the vasa gene promoter. Biol. Reprod. 67 (4), 1087–1092. https://doi.org/10.1095/biolreprod67.4.1087. Society for the Study of Reproduction.

Termini, J., 2000. Hydroperoxide-induced DNA damage and mutations. Mutat. Res. Fundam. Mol. Mech. Mutagen. 450 (1–2), 107–124. https://doi.org/10.1016/S0027-5107(00)00019-1. Elsevier.

Tran, M., et al., 2018. Cryopreservation promotes sperm DNA damage through oxidative stress [38N]. Obstet. Gynecol. 131 (1), 162S. https://doi.org/10.1097/01.aog.0000533134.61739.52. Ovid Technologies (Wolters Kluwer Health).

Valcarce, D.G., et al., 2013a. Analysis of DNA damage after human sperm cryopreservation in genes crucial for fertilization and early embryo development. Andrology 1 (5), 723–730. https://doi.org/10.1111/j.2047-2927.2013.00116.x.

Valcarce, D.G., et al., 2013b. Effect of cryopreservation on human sperm messenger RNAs crucial for fertilization and early embryo development. Cryobiology 67 (1), 84–90. https://doi.org/10.1016/j.cryobiol.2013.05.007.

Van De Peer, Y., Maere, S., Meyer, A., 2009. The evolutionary significance of ancient genome duplications. Nat. Rev. Genet., 725–732. https://doi.org/10.1038/nrg2600.

Voldoire, E., et al., 2017. Expansion by whole genome duplication and evolution of the sox gene family in teleost fish. PLoS One 12 (7). https://doi.org/10.1371/journal.pone.0180936. Public Library of Science.

Wong, T.T., Zohar, Y., 2015. Production of reproductively sterile fish: a mini-review of germ cell elimination technologies. Gen. Comp. Endocrinol. 221, 3–8. https://doi.org/10.1016/j.ygcen.2014.12.012. Academic Press Inc.

Yakes, F.M., Van Houten, B., 1997. Mitochondrial DNA damage is more extensive and persists longer than nuclear DNA damage in human cells following oxidative stress. Proc. Natl. Acad. Sci. U. S. A. 94 (2), 514–519. https://doi.org/10.1073/pnas.94.2.514.

Yazawa, R., et al., 2010. Chub mackerel gonads support colonization, survival, and proliferation of intraperitoneally transplanted xenogenic germ cells. Biol. Reprod. 82 (5), 896–904. https://doi.org/10.1095/biolreprod.109.081281.

Yoshizaki, G., Lee, S., 2018. Production of live fish derived from frozen germ cells via germ cell transplantation. Stem Cell Res. 29, 103–110. https://doi.org/10.1016/j.scr.2018.03.015. Elsevier B.V.

Yoshizaki, G., Yazawa, R., 2019. Application of surrogate broodstock technology in aquaculture. Fish. Sci. 85 (3), 429–437. https://doi.org/10.1007/s12562-019-01299-y. Springer.

Yoshizaki, G., et al., 2000. Germ cell-specific expression of green fluorescent protein in transgenic rainbow trout under control of the rainbow trout vasa-like gene promoter. Int. J. Dev. Biol. 44 (3), 323–326. https://doi.org/10.1387/ijdb.10853829. UPV/EHU Press.

Yoshizaki, G., et al., 2011. Spermatogonial transplantation in fish: a novel method for the preservation of genetic resources. Comp. Biochem. Physiol. Part D Genomics Proteomics, 55–61. https://doi.org/10.1016/j.cbd.2010.05.003. Elsevier Inc.

Zalenskaya, I.A., Bradbury, E.M., Zalensky, A.O., 2000. Chromatin structure of telomere domain in human sperm. Biochem. Biophys. Res. Commun. 279 (1), 213–218. https://doi.org/10.1006/bbrc.2000.3917. Academic Press Inc.

Zhang, J., et al., 2017. MiR-202-5p is a novel germ plasm-specific microRNA in zebrafish. Sci. Rep. 7 (1), 7055. https://doi.org/10.1038/s41598-017-07675-x.

Zhou, L., et al., 2019. Visualization of turbot (*Scophthalmus maximus*) primordial germ cells *in vivo* using fluorescent protein mediated by the 3′ untranslated region of nanos3 or vasa gene. Marine Biotechnol. 21 (5), 671–682. https://doi.org/10.1007/s10126-019-09911-z. Springer New York LLC.

Zilli, L., et al., 2003. Evaluation of DNA damage in *Dicentrarchus labrax* sperm following cryopreservation. Cryobiology 47 (3), 227–235. https://doi.org/10.1016/j.cryobiol.2003.10.002. Academic Press Inc.

Zribi, N., et al., 2011. Sperm DNA fragmentation and oxidation are independent of malondialdheyde. Reprod. Biol. Endocrinol. 9 (1), 47. https://doi.org/10.1186/1477-7827-9-47. BioMed Central.

Chapter 3

The role of epigenetics in fish biology and reproduction: An insight into the methods applied to aquaculture

María Paz Herráez[a], Marta Lombó[a,*], and Silvia González-Rojo[a,b,*]
[a]*Department of Molecular Biology, Universidad de León, Campus de Vegazana, León, Spain,*
[b]*Center of Biofuels and Bioproducts, Instituto Tecnológico Agrario deCastilla y León (ITACyL), Villarejo de Órbigo, Spain*

1. Introduction: Concept and importance of epigenetics in terms of evolution

Since the central dogma of biology was stated by Francis Crick in 1957, who established how the information sequentially flows in a biological system from the nucleic acids to the proteins (Crick, 1970), a tremendous research effort has been made in order to understand how the transfer of this information is precisely controlled. Any eukaryotic cell contains within the nucleus a great deal of information encoding for a high number of proteins, whose expression requires a strict spatiotemporal regulation in order to allow a proper functionality. The process of embryo development clearly illustrates how the selective and coordinated expression of specific genes is the key event driving cell differentiation: a unique cell originates different cell lineages containing the same genetic program. The study of the genetic code, protein synthesis, and experimental embryology were burgeoning research fields in the mid-20th century, so it was in this context where the modern term epigenetic was coined. The embryologist Conrad Waddington named as epigenetics "the branch of biology which studies the causal interactions between genes and their products which bring the phenotype into being". Waddington described the process of differentiation as a one-way flow in which an undifferentiated cell rolls down in a branched landscape; the choice of every branch in the path downstream is mediated by

*Equal contribution.

Cellular and Molecular Approaches in Fish Biology. https://doi.org/10.1016/B978-0-12-822273-7.00008-2

environmental and intracellular factors and will determine the fate of the cell (Waddington, 1957). With this description, the influence of intra and extracellular events on gene expression and silencing was incorporated, linking them to the emergence of the phenotype.

From that moment onward, the term epigenetics has evolved in order to incorporate the molecular mechanisms governing the process and several definitions have emerged. At the end of the 1980s, Holliday proposed as definition "the nuclear inheritance which is not based on changes in DNA sequence" (Holliday, 1987). Later on, epigenetics was described as "the structural adaptation of chromosomal regions so as to register, signal or perpetuate altered activity states" (Bird, 2007). Felsenfeld (2014) redefined epigenetic as "the study of mitotically and/or meiotically inherited changes in gene expression that are not produced by alterations of DNA sequence". More recently, Best et al. (2018) incorporated the link between this processes and the specific phenotype, defining epigenetics as "the study of factors that heritably regulate the spatio-temporal genome expression that underlies the emergence of physiological phenotypes" and, shortly after, Cavalli and Heard (2019) proposed that it is "the study of molecules and mechanisms that can perpetuate alternative gene activity states in the context of the same DNA sequence". The last definition avoided to encompass the inheritance of these changes, to include the persistence of gene activity in the absence of division in postmitotic cells such as adult neurons. Each definition is focused on one particular aspect, but all of them consider epigenetics as the study of the molecular processes that interfere with the gene expression process—independently of the DNA sequence—in a stable, inheritable, and reversible way.

Three mechanisms are recognized as the molecular mediators of epigenetics: DNA methylation, histone covalent modifications, and the presence of coding and noncoding RNAs (Feil and Fraga, 2012). DNA methylation and histone modifications affect chromatin structure and accessibility, interfering with the transcription machinery, and they will be the focus of this chapter, whereas RNA mechanisms are mainly related to the posttranscriptional regulation and will be the focus of another chapter (Chapter 4) in this book. The intra- and/or extracellular events are able to influence gene expression by switching on/off different signaling pathways to which the three epigenetic mechanisms can respond. Reprograming the original epigenetic signature of a given cell/gene to face a changing environment could represent a homeostatic and adaptive response, providing the basis for the physiological plasticity. Moreover, the epigenetic code can be inherited during mitosis, spreading their phenotypical consequences in the daughter cells. In addition, some of the epigenetic marks are also conserved during meiosis, thus allowing the transmission of epigenetic changes to future generations if the germline cells are affected (transgenerational inheritance) (Soubry et al., 2014). Even if some mechanisms are activated during gametogenesis and embryo development to prevent a bulky inheritance of epigenetic marks, some cases of transgenerational inheritance have been confirmed (both mechanisms and cases will be described later). The transmission of specific epigenetic signatures to the progeny represents a mechanism of inheritance of the environmental pressure consequences, in which the unexposed individuals could suffer phenotypic changes promoted by their ancestors contact to specific stimulus or by their lifestyle. This mechanism might well allow a fast and reversible adaptive response to a changing environment, which can affect a limited number of generations or, on the contrary, it can turn into stable phenotypic changes, as suggested during the last decade by the apparent relationship between some epigenetic signatures and mutations on the DNA sequence (Cavalli and Heard, 2019). The long-term stabilization of such epigenetic changes would require a revision of the evolution concepts from a pure Darwinist point of view, to incorporate the epigenetic mechanism as an additional via of heredity, as it was deeply analyzed by Skinner (2015).

This chapter will explore the DNA methylation and histone covalent modifications in fish, summarizing how these chemical modifications take place, how they are affected by physiological or environmental factors, particularly those related to farming practices, and finally reviewing the methods that can be applied to their study.

2. Epigenetic mechanisms

2.1 DNA covalent modifications

The DNA helix can be modified by the covalent addition of methyl and hydroxymethyl groups to a nucleotide base. In vertebrates, this modification occurs at cytosines, the presence of some quantities of methyl-adenosine being still controversial (Zhang et al., 2020). Cytosine is commonly methylated in the 5th carbon of the base (5mC) (Liu et al., 2019; Seidl, 2017), mostly when it is bound to guanines by phosphate residues at the known CpG sites, where the "C" and "G" stand for cytosine and guanine, respectively, and "p" indicates a phosphodiester bond between them (Ramsahoye et al., 2000). Genomic areas statistically enriched in CpG are named as CpG islands (CGIs), their number and density being very variable among fish species, as demonstrated by Han and Zhao (2008). CGIs located upstream of the transcription start sites (TSS) are particularly affected by methylation: CpG unmethylation in these locations represents a permissive state and it is generally associated with active gene transcription (Smith and Meissner, 2013), whereas methylation is commonly associated with genome silencing. High levels of DNA methylation are involved in female X chromosome inactivation, in silencing of alleles during genomic parental imprinting (still nondemonstrated in fish), in the repression of retroviral elements, and in suppressing transcription during tissue-specific gene expression (Jones, 2012). The relationship between the methylation rate and the transcriptional activity is more controversial in the coding regions or gene bodies. While in slowly dividing and in nondividing cells, both methylation in the first exon and in the rest of the gene bodies are also related to gene repression, in dividing cells the higher the percentage of methylation, the higher the level of the gene expression (Moore et al., 2013).

The DNA strand undergoes methylation and demethylation processes. Methylation occurs during DNA replication, to copy in the new strand the methylation pattern of the template and maintain the methylation profile at mitosis, or to incorporate new methylation marks in the helix during development and differentiation (de novo methylation). These processes are catalyzed by DNA methyltransferases (DNMTs), which transfer a methyl group from S-adenyl methionine (SAM) to 5C (Feil and Fraga, 2012). Two DNMT families have been identified in mammals: DNMT1, in charge of DNA methylation maintenance, and DNMT3, responsible for de novo methylation (Lyko, 2018). As reviewed by Best et al. (2018), the methylation landscape is more diverse in fish due to the high number of paralogues resulting from the gene duplication events occurring during fish evolution. A number of maintenance and de novo methyltransferases, orthologues to the mammalian DNMT1 and DNMT3 family members, have been described in zebrafish (*Danio rerio*) (Campos et al., 2012), rainbow trout (*Oncorhynchus mykiss*), and Atlantic salmon (*Salmo salar*) (Best et al., 2018), whose functional relevance still remains unclear. Indeed, the lack of a paralogue of the mammalian DNMT3-like (DNMT3L), involved in the methylation of imprinted genes, has led to speculate that the monoallelic maternal or paternal expression could occur exclusively in eutherian mammals (Chédin, 2011).

DNA demethylation can be accomplished by passive or active mechanisms. The inactivity of DNMT1 during genome replication implies that the new strand is not methylated, passively diluting the DNA methylation level in each cell division (Moore et al., 2013). The most common active demethylation requires a sequential process, which starts when the methyl group is hydroxylated to 5-hydroxymethylcytosine (5hmC) by ten-eleven translocation enzymes (TETs). Then, the hydroxymethyl cytosine, after some chemical modifications, is removed by the DNA repair machinery, restoring an unmethylated cytosine at place (reviewed by Kohli and Zhang (2013)). Similar TET-dependent demethylation processes have been identified in fish (Kamstra et al., 2015).

2.2 Histone modifications

Histones are the protein component of the nucleosomes, the basic unit of the chromatin in somatic cells. The nucleosomes are composed of an octamer of two units of each one of the core histones H2A, H2B, H3, and H4, wrapped by two loops of DNA involving 145–147 base pairs. The linker histone H1 stabilizes the structure among nucleosomes. Additional noncanonical histone variants can be incorporated into this structure, contributing to the dynamic chromatin activity (Berger, 2007) but, for clarity, we will focus on the canonical core histones. The eight histones in each nucleosome have protruding N-terminal tails, enriched in lysine and arginine residues, which actively interact with proteins related to chromatin condensation, structure, and activity. The tails are susceptible to suffer a number of posttranslational modifications, which result in altered binding affinities, allowing to regulate recruitment of nonhistone effector proteins and modifying the transcriptional activity (Lawrence et al., 2016). Such modifications have been named as "the histone code" (Strahl and Allis, 2000), whose complexity is determined by the large amount of reversible processes (acetylation, methylation, phosphorylation, ubiquitination, sumoylation, ADP ribosylation, propionylation, and butyrylation) they are subjected to (Kebede et al., 2015). The enzymes catalyzing those reactions, both the writers (adding the modifications) and the erasers (recovering the "naked" state), are often components of protein complexes involved in chromatin organization and regulation of expression, such as polycomb or trithorax complexes (PRC and TRX) (Noordermeer and Duboule, 2013). Each specific histone modification (occurring at diverse amino acid residues) represents a specific "epigenetic mark". Methylation and acetylation are the most extensively studied and best understood modifications. Histone methylation is catalyzed by histone methyltransferases (HMTs) and it is erased by histone demethylases (HDMs), taking place at lysine and arginine residues, mainly in histones H3 and H4. More than one methyl group can be added to each residue, so lysines can become mono, di, or trimethylated (Hyun et al., 2017). The number and location of the methyl groups at the lysine residues determines a different role, either as activator or repressor of gene transcription (Zhang et al., 2015). As an example, the addition of a single methyl group at the lysine 9 or the lysine 27 of the histone H3 (marks H3K9me and H3K27me) are permissive modifications, whereas trimethylation of the same lysines (H3K9me3 and H3K27me3) is associated with gene repression (Dong and Weng, 2013). The trimethylation at the fourth lysine residue of histone H3 (H3K4me3) is characteristic in the active and low condensed euchromatin. To further add complexity, histones at the nucleosomes of the same gene promoter can be differentially modified, coexisting permissive and repressive marks in the so-called bivalent domains, which are particularly represented during embryo development (Brykczynska et al., 2010). Tails of all the four core histones are susceptible to acetylation, neutralizing the positive charges of lysine residues and promoting a more relaxed and accessible chromatin state. Acetylation marks in histones H3 and H4 are frequently associated with the transcriptional activity (Gates et al., 2017) and are considered permissive signatures. Histone acetylation is mediated by histone acetyltransferases (HATs) and the erasure of acetyl groups is catalyzed by histone deacetylases (HDACs) (Eberharter and Becker, 2002). As reviewed by Best et al. (2018), posttranslational modifications of histones seem to play a similar role in fish than in other vertebrates, and the genes encoding for the enzymes involved in histone dynamics are conserved throughout teleost evolution.

2.3 Noncoding RNAs

The third epigenetic mechanism, independent from the chromatin structure and organization, is related to the role of noncoding RNAs (ncRNAs), a set of RNAs that do not encode for proteins. ncRNAs represent an essential mechanism of gene expression and chromatin structure regulation (Wei et al., 2017). These RNAs can be grouped according to their size in long (lncRNAs) or small noncoding RNAs (sncRNAs), which comprise microRNAs (miRNAs), small interference-RNAs (siRNAs), and piwi-interacting RNAs (piRNAs) (Stefani and Slack, 2008), among others.

The best ncRNAs described as epigenetic modulators of normal and disease development are the miRNAs. They are approximately 22-nt sncRNAs that bind to 3′-UTR (untranslated region) of mRNAs (Cannell et al., 2008), through complementary base pair binding, interfering with its stability or its translation. The ability of miRNAs to increase mRNA degradation and to inhibit translation of specific transcripts, demonstrated in teleost by Bazzini et al. (2005), is involved in several developmental and reproductive processes, such as tissue determination and maintenance (Best et al., 2018), germline development, spermatogenesis, and oogenesis (Robles et al., 2019). Furthermore, miRNAs play a crucial role in the transmission of environmental-induced epimutations to the next generations, as reviewed by Champroux et al. (2018). Other sncRNAs are implied in the regulation of reproduction, development, and disease. As an example, piRNAs are primarily involved in silencing transposable elements in zebrafish gametes, contributing to the process of gametogenesis (Houwing et al., 2007).

lncRNAs display a much wider range of actions, activating or inhibiting gene expression by the interaction with different key players such as mRNA, DNA, microRNA, histone modifiers, RNA-binding proteins, and chromatin (Luk et al., 2014). They closely interact with other epigenetic mechanisms, guiding many of the proteins that catalyze DNA methylation and posttranscriptional histone modifications. Functional roles for lncRNAs have been established, among others, in the control of spermatogenesis (Joshi and Rajender, 2020; Luk et al., 2014), lipid metabolism (Muret et al., 2019), or pathological processes (Boltaña et al., 2016).

3. The role of epigenetics in farmed fish species

Even if vertebrates share the main epigenetic mechanisms responsible for the modifications in the chromatin, some aspects must be deeper explored in fish. The study of DNA methylation has revealed some differences when comparing to the mammalian pattern that has been summarized by Gavery and Roberts (2017) in (i) the very likely absence of gene imprinting, (ii) the higher global DNA methylation respect to the one described in mammals, and (iii) the maintenance of the DNA methylation pattern during early embryo development (Jiang et al., 2013; Potok et al., 2013), which is particularly intriguing and will be detailed later. An additional particular aspect of fish is related to the whole genome duplications suffered during evolution (Jatllon et al., 2004), which require the regulation of multiallele expression using specific dosage compensation mechanisms that could, to some extent, rely on epigenetics.

As it has been explained, epigenetic events are mediators between the genome and the phenotype, and they can be influenced by environmental signals. Fish and amphibians are considered excellent biomarkers of environmental quality because of their enhanced sensitivity to environmental changes (temperature, oxygen levels, presence of contaminants, etc.). Every day, more and more evidences indicate that these factors interact with the epigenetic activity (Hammond et al., 2016; Lombó et al., 2019a; Wang et al., 2016), interfering with the physiological performance of living organisms. A well-known process illustrating how the environment meets physiology and development is the control of sex differentiation and sexual reversion in fish. The use of epigenetic approaches has demonstrated that these processes are mediated by changes in the methylation profile on specific promoters as a result of environmental (reviewed by Piferrer et al. (2019) and Ortega-Recalde et al. (2020)) or even social clues (Todd et al., 2019). Therefore the epigenetic events are of utmost importance in aquaculture. In fact, many of the farming practices such as the process of domestication itself, the nutrition with artificial diets, or even the reproductive strategies used to increase the productivity can promote epigenetic changes resulting in modified performances. The identification of the epigenetic signatures related to specific factors can provide both the farmers and the researchers with molecular markers of fitness.

3.1 Domestication

Rearing fish in captivity, both in commercial and restocking aquaculture, may imply changes in phenotypic traits. Moreover, selection by phenotype is a common practice in aquaculture in order to obtain the more efficient and/or appreciated production. The basis of the process has been related to a passive or active genetic selection. Nevertheless, recent studies with salmonids (Gavery et al., 2018), Nile tilapia (*Oreochromis niloticus*) (Konstantinidis et al., 2020) and European sea bass (*Dicentrarchus labrax*) (Anastasiadi and Piferrer, 2019), demonstrated that, in the absence of genetic variations, DNA methylome and hydroxymethylome changed in a single generation of domestication. The genetic and epigenetic study of steelhead trout (*O. mykiss*) populations from natural and hatchery origin revealed an identical genetic profile, but there are around 100 of differentiated methylated regions in red blood cells and sperm (Gavery et al., 2018). In tilapia, the observed alteration in the extent of DNA hydroxymethylation was accompanied by intense transcriptome changes affecting the expression of more than 2000 genes involved in immune-, growth-, and neuronal-related pathways (Konstantinidis et al., 2020). The study of epimutations in domesticated European sea bass demonstrated that early domesticated populations undergo changes in DNA methylation during gastrulation which have phenotypic consequences related to the "domestication syndrome," such as jaw malformations. Moreover, considering that methylated cytosines are much more prone to mutations, mainly cytosine-to-thymine polymorphisms, these epigenetic changes can become persistent epimutations in subsequent generations by permanent integration into the genomes (Anastasiadi and Piferrer, 2019). The stress (Uren Webster et al., 2018) and the environmental impoverishment (Berbel-Filho et al., 2020) linked to intensive farming are also sources of epigenetic reprogramming that could have effects even on fish behavior (Berbel-Filho et al., 2020).

3.2 Nutrition

Since the effects of maternal nutrition on the progeny performance were linked to epigenetic modifications, both in mammals (Susiarjo and Bartolomei, 2014; Vickers, 2014) and fish (Adam et al., 2018), the epigenetic changes triggered by the diet are among the most reported in the literature. Nutrients and diet components definitively affect the epigenetic activity, the DNA methylation being the more studied process. The increasing use of substitution diets in fish feeding, including more plant-based than fish-based ingredients, has fueled several studies in farmed and model species. Global hypomethylation of DNA and hypoacetylation of H3K9 were noticed in the liver of rainbow trout fed with a high-carbohydrate diet (Marandel et al., 2016). The inclusion of plant proteins in the diet of zebrafish during 42 days induced important changes in the global methylation rate of the mid intestine, affecting genes involved in immunity, neurogenesis, or antioxidant defense system (Dhanasiri et al., 2020). Such a clear interaction among nutrients, epigenetics, and phenotype implies that parental nutrition affects the progeny performance and that larval nutrition modulates further fitness at the adulthood. A deeper insight into the mechanisms involved can be used to design a convenient nutritional program, as suggested by the study performed in gilthead seabream (*Sparus aurata*) by Perera et al. (2020). They demonstrated that feeding the broodstock with increased amounts of linoleic acid (ALA) led to the hypermethylation of the stearoyl-CoA desaturase promoter in the descendants, preserving the fatty acid unsaturation in the liver in spite of feeding a diet with a low content of polyunsaturated fatty acids. The fact that beneficial epimutations can be targeted to genes of interest opens the door to a new field of epigenetic edition that is still at its infancy.

3.3 Reproduction

Reproductive practices in fish farming are additional sources of epigenetic variation. Induction of sex maturation, sterilization, sex reversal, artificial fertilization, short- and long-term gametes preservation, surrogate production, ploidy manipulation, and nuclear transfer from somatic to germ cells are some of the processes involving intensive environmental and intracellular signaling, which can promote dramatic changes in the epigenetic signatures of both somatic and germinal cells. As remarked by Labbé and colleagues (2017), who reviewed the epigenetic effects of some of these factors on the germline cells, this is still an almost unexplored field in fish. The impact of artificial reproductive technologies has been carefully analyzed in human and domestic mammals, but a bigger research effort is still required in teleosts. The promoted changes can affect the reproductive performance of progenitors, having phenotypic effects on the F1 progeny (intergenerational transmission) or on a higher number of generations if they are transmitted via the germline cells (transgenerational inheritance) (Soubry et al., 2014). Some authors even proposed the incorporation of epigenotypes in the process of broodstock selection for those quantitative traits clearly related to particular DNA methylation signatures (see revision by Gavery and Roberts (2017)). Nevertheless, the success of an epigenetic selection would rely on the epigenetic dynamics during life cycle, which has been only explored in some model species, as will be summarized later on for zebrafish.

3.4 Rearing temperature

The phenotypic effects of rearing temperature, during the whole life cycle or at specific developmental stages, can affect development, growth rate, or sex determination among other processes. Recent studies have identified epigenetic events related to some of these phenomena. The activity of epigenetic enzymes can be deeply affected by temperature, promoting differential epigenetic signatures. According to Varriale (2014) there is a correlation in vertebrates between the global DNA methylation and the body temperature, cold-living vertebrates showing a hypermethylated state when compared with warm-living species. Nevertheless, the cold acclimation could follow a more complicated pattern as it was shown using zebrafish fibroblast in vitro, which suffered an increased whole genome methylation after 5 days of exposure to cold, but a decreased methylation after a long-term cold exposure (30 days) (Han et al., 2016). Embryo development is a particularly sensitive stage in which a thermal imprinting can be established. As an example, muscle development at adulthood can be affected by embryo rearing temperature in Atlantic salmon: the incubation of embryos at 8 °C promoted a significantly higher muscle development and growth rate during adulthood than the incubation at 4 °C. The enhanced muscle development is related to a reduced methylation of the myogenin promoter at higher temperatures and to changes in the expression levels of the DNA methyltransferase genes *dnmt1*, *dnmt3a*, and *dnmt3b* (Burgerhout et al., 2017). Similar effects of rearing temperature were observed in Senegalese sole (*Solea senegalensis*) at larval stages, exhibiting lower promoter methylation and higher expression of myogenin when reared at 21 °C rather than at 15 °C (Campos et al., 2013).

Sex differentiation and, consequently, sex ratio, are documented events affected by rearing temperature at specific developmental stages. Navarro-Martín et al. (2011) established that methylation of the aromatase promoter in European sea bass varies with temperature. Aromatase converts androgens into estrogens, promoting feminization, but high temperatures during embryo development entail hypermethylation at its promoter, downregulating aromatase expression at the gonads and leading to masculinization. Since this epigenetic mechanism was linked to the well-known relationship between water temperature and sex ratio in lower vertebrates, similar mechanisms have been described in other species dependent on environmental determination of sex. Moreover, experiments performed with zebrafish showed that the bias to male development promoted by high temperatures persisted in the F1 of some males exposed to heat, which showed altered testicular epigenome (Valdivieso et al., 2020).

4. Epigenetic mechanisms in model species: A case study in zebrafish

Throughout the life of organisms, the epigenome undergoes two major epigenetic reprogramming waves: during the determination and migration of primordial germ cells (PGCs) and, after fertilization, during the first steps of embryo development (Hales et al., 2011).

In 1981, Streisinger et al. (1981) first proposed the use of zebrafish as model organism. From this moment onward, a large number of molecular tools, mutants, and transgenic zebrafish lines have been created. Moreover, zebrafish genome was fully sequenced at the Sanger Institute and it was shown to contain approximately 26,000 genes and over 1.4 billion base pairs on 25 pairs of chromosomes. The comparison to the human reference genome revealed that approximately 70% of human genes have at least one obvious zebrafish orthologue (Howe et al., 2013). Last but not least, the transparency and the small size of zebrafish embryos—which allows them to meet oxygenation and nutrient supply by passive diffusion during the first week of development and, therefore, to survive even under pathological states—make them a good species for the study of embryo development, among other processes. The severe health conditions induced in embryos by certain experimental procedures do not prevent researchers from monitoring, while they would be probably lethal in other vertebrate species (Singleman and Holtzman, 2012). Therefore a wide variety of mutants have been created, including those relative to epigenetic processes (Mudbhary and Sadler, 2011).

4.1 The epigenetic events involved in gametogenesis

In mammals, all epigenetic marks in PGCs are erased before gametogenesis in order to get rid of the epimutations acquired by the germline cells, thus avoiding their transmission to following generations and allowing the gametes to acquire a new epigenetic pattern during differentiation (Jammes et al., 2011; McCarrey, 2014). Throughout this epigenetic reprogramming, PGCs undergo a genome-wide methylation erasure that leads to the "epigenetic ground state". On the contrary, a novel study about zebrafish PGC DNA methylome, from 4 to 24h postfertilization (hpf), has demonstrated the retention of paternal epigenetic configuration in these cells and it has also identified that the expression of several zebrafish germline markers is independent of DNA methylation-mediated regulation (Skvortsova et al., 2019b). Although this work confirms that genome-wide DNA methylation erasure does not occur in zebrafish germline from blastula stages to genital ridge colonization, the possibility that the methylome reprogramming might take place before 4hpf or later during spermatogenesis and oogenesis should be still taken into account. Anyway, the lack of epigenetic erasure during fish development could have implications in the intergenerational transmission of epigenetic signatures and of some acquired phenotypic traits.

Once the settlement and proliferation of PGCs takes place, germ cells suffer an intense epigenetic remodeling to acquire the sex-specific patterns that further define the female and male germline. Regarding oogenesis, the meiotic arrest that occurs in primary oocytes is linked to an increase in histone acetylation, whereas the accomplishment of meiosis is associated with the reduction of histone acetylation (Duffié and Bourc'his, 2013). In the case of zebrafish oocytes, it has been established that their nuclear DNA has approximately 75% of methylated CpGs while its mitochondrial DNA is essentially unmethylated (Potok et al., 2013). On the other hand, the meiotic phase of spermatogenesis implies a gradual gene silencing that involves an increase in DNA methylation, a decrease of activating histone marks (H3K4me3), and an increase in silencing marks (H3K27me3 and H3K9me3) (Carrell and Hammoud, 2010). In fact, zebrafish spermatozoa are characterized by their highly methylated DNA (91%–95% of methylated CpGs) (Potok et al., 2013). However, some hypomethylated regions of sperm DNA are associated with activating histone marks (H3K4me3, H3K4me2, and H4K16ac) to allow the expression of crucial genes prior to mid-blastula transition (MBT) (Wu et al., 2011), the moment when the embryonic transcription is activated. Therefore these authors proved that developmental genes are packed in both bivalent and multivalent marks, in which DNA hypomethylation coexists with activating and repressing histone marks. This peculiar epigenetic profile avoids the expression of certain genes during male germline formation and allows the activation of others in the embryo (Carrell, 2011).

4.2 Epigenetic changes during embryo development

Epigenetic regulation of chromatin structure is essential for gene activation and repression during embryo development. As stated before, the second round of epigenetic reprogramming occurs in early embryos, soon after fertilization, to erase all the marks established during gametogenesis (Smallwood and Kelsey, 2012).

In mammals, a passive loss of global DNA methylation, mainly affecting the male pronucleus, occurs soon after fertilization up to the blastocyst stage.

At the onset of gastrulation, the global genome remethylation starts to allow the loss of cellular pluripotency and, therefore, to enable the determination of cellular lineages. The highly methylated pattern of spermatozoa is diluted in the zygote, especially in retroelements, since hypomethylated status of the oocyte is reflected on the zygote (Smith et al., 2012). In contrast, zebrafish DNA suffers a moderate demethylation after fertilization and DNA remethylation starts sooner than in mammals. Although the DNA of the oocyte is also hypomethylated in zebrafish, this maternal pattern is maintained in the embryos only until 16-cell stage. In fact, Jiang et al. (2013) demonstrated that the global methylation level at this stage of development overlaps the mean value of oocyte and sperm DNA methylation (80% and 91%, respectively). Then, the embryonic epigenome is progressively methylated and, by the MBT stage, the methylome of zebrafish embryos is almost identical to that of the sperm (Potok et al., 2013). Due to the gradual reset of maternal DNA methylation pattern, the methylation profile in most gene promoters of MBT embryos is also very similar to that of sperm (Lindeman et al., 2010). For instance, genes involved in embryo development (*hox* clusters) or in germline function (*vasa*, *piwi*, and *dazl*) are hypermethylated in oocytes and hypomethylated in both sperm and MBT embryo (Potok et al., 2013). From MBT to 24 hpf, thousands of differentially methylated regions (DMRs) were identified by Lee et al. (2015), most of them being located in intergenic regions (outside gene promoters, CpG islands, and island shores) where they surprisingly act as functional developmental enhancers.

Histone modifications also vary throughout zebrafish embryo development and they depend on the sperm pattern as well. In fact, genes enriched in activating marks in spermatozoa, usually related to cell cycle and metabolism, are transcribed in the embryos before MBT, whereas those genes lacking enhancing marks are expressed at later stages (Wu et al., 2011). Since zygotic genome activation takes place at MBT, the epigenetic regulation is a key process during this period. In fact, Lindeman and colleagues (Lindeman et al., 2010, 2011) categorized the genes at MBT according to the histone modifications present in their promoters. The genes of the first group were those enriched in the active marks H3K9ac, H4ac, and H3K4me3, including *pou5f1*, *sox3*, and the housekeeping *bactin2*. The second group contained genes with activating (H3K9ac, H4ac, and H3K4me3) and repressing (H3K27me3) marks, such as *sox2* and *klf4*. Genes of the third group harbor histone methylation (H3K4me3 and H3K27me3) but no acetylation, such as *nnr* and *otx1b*. In the fourth group, the histone profile of gene promoters contained both activating (H3K4me3) and repressive (H3K27me3 together with H3K9me3) marks, as it was observed in *vasa*. In light of these results, all genes in zebrafish MBT embryos are enriched in H3K4me3 regardless of their expression (Lindeman et al., 2010). Lately, Murphy et al. (2018) have described the existence of "placeholder" nucleosomes, containing histones H2A.Z and H3K4me1, which occupy all hypomethylated DNA in both zebrafish sperm and early embryos. Upon genome activation, placeholders become either active (in housekeeping genes) or repressive (in developmental genes) marks.

5. Techniques applied for the evaluation of DNA methylation

5.1 Evaluation of global DNA methylation

Global status of DNA methylation can be assessed by direct measurements using high-performance separation technologies or by indirect quantification using other methods. The former methodology employs high-performance liquid chromatography (HPLC) (usually in combination with tandem mass spectrometry, LC-MS/MS) or capillary electrophoresis (HPCE) (Fraga et al., 2002). In terms of cost-effectiveness, direct measurement approaches require expensive equipment and large amounts of DNA sample. After enzymatic or acidic hydrolysis of DNA, free nucleotides are released and can be quantified by these techniques. LC-MS/MS allows quantitative analysis of the global DNA methylation and identifies not only 5mC but also other derivatives such as 5-hydroxymethylcytosine (5hmC), 5-formylcytosine (5fC), or 5-carboxylcytosine (5caC) (Chowdhury et al., 2017).

Regarding indirect technologies, there are a wide variety of methods, including luminometric methylation assay (LUMA) or enzyme-linked immunosorbent assay (ELISA) (Skvortsova et al., 2019a). The LUMA technique was first described by Karimi et al. (2006) and consists of two parallel restriction digest reactions, followed by pyrosequencing of the digested DNA, as we will describe in detail later on.

The evaluation of global DNA methylation by ELISA is a commercially available option in the form of a variety of kits. Even though it is prone to a high variability, ELISA offers a quick and easy evaluation of the DNA landscape (Kurdyukov and Bullock, 2016). The basis of the technique consists of capturing the DNA on a surface and DNA methylation is detected by subsequent incubation steps with specific antibodies. Firstly, a primary antibody against 5mC is used. Next, a labeled secondary antibody is incubated and, finally, the level of DNA methylation is revealed by colorimetric or fluorometric detectors (for a detailed description of an immunohistochemical (IHC) technique protocol, please see Chapter 7). Repetitive DNA elements are often used for estimating global DNA methylation. Specifically, ELISA is often employed to determine the methylation levels of long interspersed nucleotide element-1 (LINE-1) retrotransposons. The technique is rather easy, the DNA being mainly fragmented and hybridized to a biotinylated LINE-1 probe. The product is immobilized to a streptavidin coated plate and after washing and blocking, DNA methylation is revealed by an anti-5mC antibody. A secondary antibody (usually anti-IgG HRP conjugated) is used and detection is carried out using chemiluminescence reagents (Kurdyukov and Bullock, 2016). Alternatively, LINE-1 methylation status can be quantified by bisulfite conversion of DNA, followed by PCR amplification. This methodology is also discussed later.

It is worth mentioning the new genome-wide technologies for assessing global DNA methylation, such as whole-genome bisulfite sequencing (WGBS), nucleosome occupancy and methylation sequencing (NOMe-Seq), methylated

DNA immunoprecipitation (MeDIP), or methyl-Cpg binding domain-based capture and sequencing (MBDCap-Seq). All of them employ next-generation sequencing (NGS) and third generation sequencing (TGS; providing a larger fragment single molecule reads) platforms, offering a single-nucleotide resolution. However, these technologies represent an expensive option and need highly qualified people for resolving technical issues and handling data analysis (Skvortsova et al., 2019a). Reduced representation bisulfite sequencing (RRBS) deserves special attention due to its cost-effectiveness to generate genome-wide DNA methylation maps. Basically, DNA is fragmented using restriction endonucleases (usually *MspI*) and DNA fragments are selected and ligated to a methylated adapter, which facilitates the subsequent PCR amplification and sequencing (Chatterjee et al., 2013; Meissner et al., 2005; Taylor et al., 2010). Today, the focus is set on single-cell level and technologies have evolved to avoid the differences in methylation levels to be masked by the population heterogeneity. In this way, new approaches have appeared such as single-cell bisulfite sequencing (scBS-Seq), single-cell nucleosome occupancy and methylation sequencing (scNOME-Seq), or single-cell reduced-representation bisulfite sequencing (scRRBS) among others (Guo et al., 2015; Skvortsova et al., 2019a).

These techniques applied for the evaluation of DNA methylation have been used in different aquatic species, to assess the alteration of methylation levels by dietary regime, rearing temperatures, thermal environmental changes, or by different environmental toxicants or radiation. The most popular options have been ELISA-based methods, which have shown the DNA methylation status in gonads of adult rare minnow (*Gobiocypris rarus*), zebrafish, or medaka (*Oryzias latipes*) (Bertucci et al., 2020; Liu et al., 2014; Teng et al., 2020) after exposure to different toxicants. However, other alternatives have been thoroughly used in zebrafish, such as the LUMA technology (Laing et al., 2016), high-performance liquid chromatography (Bouwmeester et al., 2016; Gombeau et al., 2016; Lombó et al., 2015; Xing et al., 2014), or the WGBS, MeDIP, and RRBS approaches (Adam et al., 2019; Chatterjee et al., 2014; Dhanasiri et al., 2020; Falisse et al., 2018; Olsvik et al., 2019; Qian et al., 2019). Moreover, RRBS or cytosine methylation assayed by pyrosequencing has been applied to other aquaculture species to evaluate DNA methylation status after altering natural thermal conditions in European grayling (*Thymallus thymallus*) or in Atlantic salmon (Sävilammi et al., 2020; Burgerhout et al., 2017).

5.1.1 Luminometric methylation assay (LUMA): A brief overview

For the LUMA technique, the use of isoschizomer endonucleases, restriction enzymes whose target sequence is identical, leaving a 5′-overhang, is necessary. The most usual enzymes are *HpaII* and *MspI*. Both enzymes recognize CCGG sequence. However, the former does not cut the sequence if the internal cytosine of the recognition sequence is methylated (C^{m}CGG), whereas the latter cuts the recognition sequence independently of its methylation status (Karimi et al.,

2006). The main steps of the technique are summarized as follows (Delaney et al., 2015; Ho and Tang, 2007; Karimi et al., 2006, 2011; Kurdyukov and Bullock, 2016; Mohsen et al., 2006):

First—DNA sample: it is crucial to obtain a high-quality DNA. DNA integrity could be visually checked by agarose gel electrophoresis, without any smear below the main high-molecular weight band, and/or numerically by the DNA integrity number (DIN) through specific gel image analysis. Purity of DNA is also ensured by spectrophotometric analysis, to discard any protein and/or RNA contamination.

Second—Enzyme restriction digestion of the DNA sample: genomic DNA is prepared in duplicate to be subjected to two reactions with restriction enzymes: one for *HpaII* and one for *MspI*. *EcoRI* is included in both reactions as internal control for the normalization between runs and for DNA input. *EcoRI* recognizes GAATTC sequence and it is insensitive to CpG methylation, leading a 5′-AATT overhang after the reaction. In contrast, *HpaII* and *MspI* yield a 5′-CG overhang. These ends are filled in a polymerase extension in the subsequent pyrosequencing assay.

Third—Pyrosequencing assay (Fig. 1): this methodology incorporates nucleotides in orderly manner throughout four steps (step one—dATPαS, step

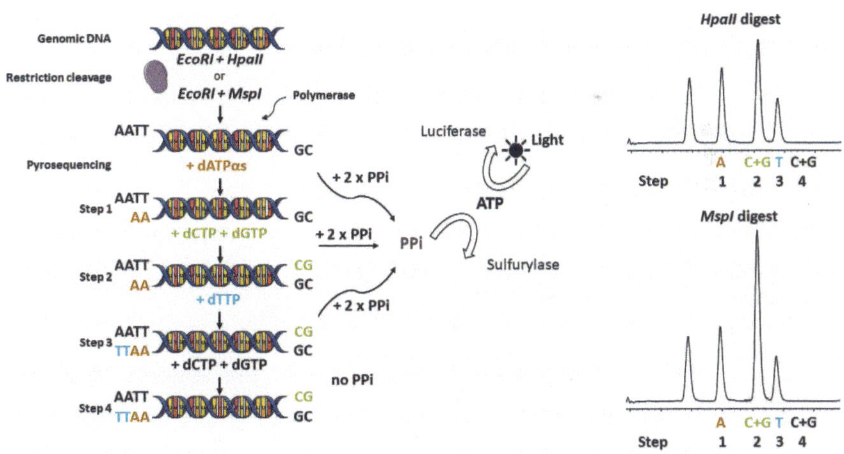

FIG. 1 Schematic representation for analysis of global DNA methylation by luminometric methylation assay (LUMA). Genomic DNA is subjected to enzyme restriction digestion combining restriction enzymes: *EcoRI/HpaII* or *EcoRI/MspI*. Both reactions lead a 5′-AATT overhang (*EcoRI*) or 5′-CG overhangs (*HpaII*—sensitive to methylation; or *MspI*—insensitive to methylation). These ends are filled by pyrosequencing in a four-step reaction. Pyrosequencing chromatograms are obtained for the two restriction enzymes combinations: A and T peaks correspond to steps 1 and 3, respectively, reflecting the *EcoRI* digestion and should be equal. C + G peak (in green color in the figure) represents the detected signal from step 2, after *HpaII* or *MspI* cleavage. The second C + G peak (appeared in *black* color in the chromatogram) corresponds to step 4, functions as an internal control, and should be close to zero. *HpaII/MspI* ratio at C + G peaks of step 2 determines the methylation status. In unmethylated samples, this value is close to 1, whereas in methylated samples the ratio is close to zero.

two—dGTP+dCTP, step three—dTTP, and step four—dCTP+dGTP). After each nucleotide incorporation by DNA polymerase, an inorganic pyrophosphate is released and converted to ATP by ATP sulfurylase. The reaction contains luciferin that is converted by luciferase in presence of ATP to oxyluciferin, which produces visible light being detected by a CCD camera. The amount of detected light is proportional to the dNTP incorporation in the overhangs. Peaks detected after dATPαS and dTTP incorporation correspond to the amount of *EcoRI* restriction, whereas peaks detected after the incorporation of dCTP and dGTP in the second step represent the amount of *HpaII* or *MspI* digestion.

Fourth—Data analysis: once the pyrosequencing assay is finished, data analysis is performed to obtain the *HpaII/MspI* ratio as a value of the degree of DNA methylation:

$$\frac{HpaII(C+G)}{EcoRI(A)} \frac{MspI(C+G)}{EcoRI(A)}$$

$$\frac{HpaII}{MspeI} \to \begin{matrix} \text{max.} = 1.0(\textit{Unmethylated samples}) \\ \text{min.} = 0(\textit{Fully methylated samples}) \end{matrix}$$

The lower the ratio, the higher the amount of methylation.

5.2 Analysis of gene promoter methylation

This method is usually performed after the evaluation of global DNA methylation in order to have a deeper insight into the levels of methylation in target DNA regions.

5.2.1 Bisulfite conversion and nested PCR

The evaluation of DNA methylation at specific regions requires the treatment of the extracted genomic DNA with sodium bisulfite to selectively convert nonmethylated cytosines into uracils (Fig. 2).

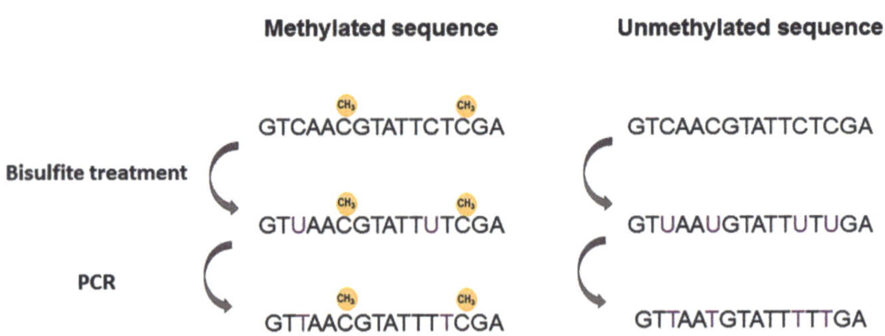

FIG. 2 Scheme of bisulfite conversion and bisulfite-converted sequence. Bisulfite conversion of template DNA in which unmethylated cytosines are transformed into uracils and, after the PCR, they become thymines.

Once DNA is bisulfite converted, a nested PCR is performed to improve the sensitivity and the specificity of the reaction. It consists of two rounds of successive amplifications in which the first set of primers anneal to bigger sequences that include the regions where the second set is going to attach. For this purpose, total PCR is usually performed from 1 μL bisulfite-converted DNA and products diluted 1:20 (v:v) are used as templates for the second amplification. To check the efficiency of the reactions and the lack of carryover contaminations, electrophoresis on agarose (1.8% (*w*/v)) gel is usually performed before further steps are done.

5.2.2 Sequencing

There are two main methods to analyze the methylation of CpGs of each sample. The first method, which involves the cloning of PCR products, is considered the "gold but old" of specific DNA methylation analysis. Despite the fact that it provides the researcher with reliable and accurate information, it is much more tedious and slower than the newly developed methods. After cloning the PCR products, DNA is extracted from the agarose gel. These consecutive steps can be easily carried out using commercial kits (Fang et al., 2013). Once DNA is extracted, it has to be sequenced. Both the alignment with the original sequences and the assessment of specific DNA methylation can be performed using specific software, such as FinchTV (Geospiza) and CpG viewer, respectively.

The second method is based on the pyrosequencing technology, previously explained, which provides more rapid and high-throughput means of detecting the methylation levels at individual loci. In this case, the amplicons obtained from the nested PCR are denaturated and biotinylated. This biotinylated strand is captured on streptavidin-coated sepharose beads. Then, pyrosequencing primers are annealed to the retained DNA and the assay starts with the dispensation of nucleotides in a pyrosequencer. Eventually, the peak heights on the resulting pyrogram represent the ratio of cytosine to thymine at each evaluated CpG site, showing the proportion of methylated DNA (Delaney et al., 2015).

5.2.3 CpG island identification and primer design

For this protocol, the first step is to check whether the region of interest harbors a CpG island. This information can be obtained from the Genome Browser (University of California Santa Cruz; https://genome.ucsc.edu/).

Once the sequence is selected, it has to be in silico bisulfite converted to design the primer sets used in the nested PCR. For the conversion step, the sequence can be submitted in the online version of MethPrimer (https://www.urogene.org/methprimer/), whereas the Primer Design and Search Tool is very useful for obtaining the forward and reverse primer sequences (http://bisearch.enzim.hu/?m=temp). Generally, the primers for the amplification of bisulfite-converted sequences have between 26 and 32 nucleotides. The reverse primer used in the second round of amplification must be biotinylated to allow the caption of the DNA sequence in the pyrosequencing station. Regarding the amplicon size, it must be around 500 bp (base pair) for the first reaction and around 200 bp for the second round of amplification.

As for pyrosequencing, the primers can be designed through GeneScript (https://www.genscript.com). In this case, the template DNA is the sequence obtained in the second PCR, and only forward primers (of approximately 10 bp) are required. To avoid failures during the annealing process, the product sizes must be smaller than 40 bp but including the maximum number of CpGs.

It is noteworthy that, when designing primers from a bisulfite-converted sequence, if there is a potentially methylated cytosine in a CpG island it has to be represented as Y in the primer sequence, instead of C.

6. Techniques applied for the evaluation of histone modifications

6.1 Global evaluation of epigenetic modifications in histones

As stated before, posttranslational modifications of the N-terminal tail of histones include acetylation, methylation, ubiquitination, and phosphorylation, among others. The study of histone methylation and acetylation using different approaches has been widely expanded throughout aquatic species during the last years. Proteomic methods are suitable for unveiling posttranslational modifications of histones, mass spectrometry being an accurate tool for characterization and quantification of global levels of specific histone modifications (Chimento et al., 2013; Kimura, 2013; Partolina et al., 2017; Quinn and Simeonov, 2011; Villar-Garea et al., 2008). Despite its specificity, mass spectrometry requires expensive equipment and it is inappropriate for routine screening applications. Therefore, chromatin immunoprecipitation (ChIP) has become a promise in the ongoing growth of histone evaluation. ChIP allows the study of the interaction of proteins with specific chromosomal loci. The basis of the technique consists of the precipitation of DNA using antibodies against specific histones modifications, after the isolation and shearing of chromatin (Kimura, 2013; Zachary Ostrow et al., 2015). The obtained DNA could be subjected to different analysis according to the target loci to be studied. For the global assessment of histone modifications, microarray analysis or deep sequencing could be useful, rendering the sequences of the immunoprecipitated DNA and an epigenome map (Kimura, 2013; Nakato and Sakata, 2020). Moreover, a more specific evaluation could be performed by qPCR, if the target loci are known, as we will describe later in detail.

Other methodologies frequently employed to assess global status of histone modifications also include the use of specific antibodies, western blot, ELISA, and immunostaining being widely used. Currently, ELISA is one of the most common chosen approaches to generate a preliminary overview since it is easy and quick to perform, and it is a cost-effective technique. This assay implies the immobilization of the histones on a coated plate and, then, the detection and quantification of the specific histone modification is performed using antibodies and colorimetric or fluorometric developing reagents (Quinn and Simeonov, 2011). There are a wide variety of commercial kits for both histone acetylation and histone methylation assessment. Other variants immobilize nucleosomes for monitoring the level of histone modifications, such as NU-ELISA (nucleosomes-ELISA) (Dai et al., 2013). As it occurs with the DNA methylation evaluation, single cell epigenomics applied to histone modifications is currently pioneering the field. Nevertheless, approaches that allow the study of cellular heterogeneity are being developed, such as single-cell ChIP-Seq (scChIP-seq) (Nakato and Sakata, 2020) or mass cytometry (Cheung et al., 2018).

6.1.1 Analysis of global histone H3 acetylation by ELISA

To have an overview of global levels of histone H3 acetylation it could be very useful to perform an ELISA-based assay. In this section, we will explain a specific method applied to zebrafish testes, but easily expandable to other species and tissues once dissociated cells are obtained. It is necessary to note that several commercial kits offer reagents for tissue dissociation and histone purification. However, our experience shows that they are inappropriate for fish species in most cases. For that reason, we briefly explain a tissue dissociation protocol suitable for zebrafish testes.

First—Tissue dissociation: a pool of testes from 4 males are dissociated in Leibovitz's medium (L-15) supplemented with 0.2% (*w*/v) collagenase, 10 mM HEPES, 0.5% (*w*/v) bovine serum albumin, and 1 U/mL DNase I. Tissues are incubated at 28 °C for 2 h, with occasional pipette mixing or using pestles. After dissociation, the cell suspension is filtered using 0.75 μm filters and washed twice with phosphate buffered saline (PBS). Cells are centrifuged at 1000*g* for 5 min at room temperature (RT) between the two washing steps.

Second—Histone extraction: after washing the cell pellet, 1 mL of hypotonic lysis buffer is added, containing 10 mM Tris-HCl (pH 8.0), 1 mM KCl, 1.5 mM $MgCl_2$, and 1 mM DTT. Cells are incubated for 30 min on ice and, after that, they are centrifuged at 10,000*g* for 10 min at 4 °C. The obtained nuclear pellets are incubated in 0.4 N HCl overnight at 4 °C, to ensure histone solubilization. Next, histones are precipitated overnight at 4 °C, using trichloroacetic acid (TCA, 33% (*v*/v)). Histones are then washed twice with acetone and centrifuged at 16,000*g* for 10 min at 4 °C between washing steps. Finally, total histones are solubilized in 30 μL of molecular biology water and they can be used for subsequent analysis or stored at − 80 °C.

Third—Quantification of histone concentration: this step could be easily performed due to the wide availability of commercial kits, from those involving bicinchoninic acid (BCA) to those based on Coomassie dye (Bradford assay).

Next steps—General protocol for histone H3 acetylation measurement: once histones are extracted and their concentration is known, the rest of the protocol is common among commercial kits, with slight differences. It is important to adjust protein concentration in the range of 200 ng/μL to 400 ng/μL in order to start the assay with 5 μL (initial quantity of histones: 1–2 μg), which are spread out over the well surface. A blank (histone resuspension buffer) and a positive control need to be included. Generally, the positive control consists of a calibration curve using serial dilutions of an acetylated histone H3 control (commercially available or included in the kit), the concentration ranging from 2 to 30 ng/μL. This curve will serve for the quantification of histone acetylation level, as it will relate absorbance with the quantity (in ng) of the acetylated histones. After spreading the blank, the acetylated histone control, and the studied samples on the wells, the steps to be followed are as follows:

First—Blocking step: to minimize unspecific binding of the primary antibody to those sites which do not harbor acetylated histones.
Second—Washing step: well surface is washed three times with a washing buffer (e.g., PBS) to remove the rests of the previous reagent.
Third—Primary antibody incubation: the histones attached to the plate are incubated with a primary antibody (or capture antibody), generally on an orbital shaker, for at least 1 h at RT.
Fourth—Washing step: once again, well surface is washed three times.
Fifth—Secondary antibody incubation: after washing the well surface to remove the primary antibody not bound to the target, a secondary antibody (or detection antibody) is added. This secondary antibody can be conjugated to different labels, but usually it consists of an enzyme label, visualized by chromogenic reactions.
Sixth—Washing step: to separate the unbound secondary antibody, the plate surface is washed again.
Seventh—Developing step: according to the conjugated labeled secondary antibody, the developing solution could vary. If an enzyme is conjugated to the secondary antibody (for example, streptavidin-HRP), its substrate will be added in this step. This substrate is usually a chromogenic molecule, which suffers the enzyme reaction and produces a visible color during the reaction. By monitoring the color production of the standard (positive control) and the one of samples, a calibration curve can be obtained and therefore, the levels of histone acetylation are quantified. A microplate reader is necessary to measure the absorbance of the color at the specific wavelength.

6.1.2 Analysis of the enzymes in charge of histone modifications

Posttranslational modifications of histones are catalyzed by different enzymes as it has been previously described. Not only the information about global or specific modification of histones is useful, but also the state of the enzymes in charge of these histone modifications. Following the analysis of histone acetylation, it is feasible to measure the activity of the enzymes responsible for that modification. In this section, we will explain how to quantify the activity of histone acetyltransferases (HAT), whose imbalance could display some effects, related to improper nucleosome assembly, unsuitable chromatin folding, altered DNA damage repair, etc. which could alter certain cellular processes such as apoptosis, oxidative stress, or inflammation (Lee and Grant, 2018; Marmorstein and Trievel, 2009).

There are different commercially available kits, the vast majority offering an indirect analysis of HAT activity since they quantify the acetyl group donor, detecting the generation of free CoA (or CoA-SH). Other methods detect the reaction product, quantifying the amount of acetylated histone substrate. Here, we will give an overview of the preparation of nuclear extracts from zebrafish testes, which could be used directly on a commercial kit:

First—Nuclear extracts: testes or other tissue have to be dissociated to obtain a cell suspension. Testes are dissociated as described in the first step of section 5.1.1. Cell pellets are resuspended in 50 µL of lysis buffer 1 (10 mM HEPES pH 7.5, 10 mM KCl, 0.1 mM EDTA, 1 mM DTT, 0.5% (*v*/v) NP-40, and 1 × protease inhibitor (a general composition of this cocktail could be: 1 mM AEBSF, 800 nM aprotinin, 50 µM bestatin, 15 µM E64, 20 µM leupeptin, 10 µM pepstatin A, and 5 mM EDTA, or commercially available)), for 20 min on ice. After that, cells are pelleted by centrifugation at 12,000*g* for 10 min at 4 °C. The supernatant corresponds to cytoplasmic extracts that can be stored at − 80 °C for further analysis. The nuclei pellet is resuspended in 25 µL of lysis buffer 2 (20 mM HEPES pH 7.5, 400 mM NaCl, 1 mM EDTA, 1 mM DTT, and 1 × protease inhibitor cocktail). The high osmolality of this solution supports the nuclei content release to the aqueous phase. The incubation time last 30 min on ice. After that, the solution is sonicated three times for 10 s each and centrifuged at 12,000*g* for 15 min at 4 °C, the nuclear extract being collected from the supernatant. Nuclear extracts could be used immediately or stored at − 80 °C for further analysis.

Second—Quantification of protein concentration: as described in section 5.1.1.

Third—Histone acetyltransferase activity assay: this evaluation is based on the reaction catalyzed by HAT (Fig. 3). Firstly, well surface is coated with HAT substrate at a given concentration. After that, the nuclear extracts are added with acetyl-CoA, necessary for HAT activity. To quantify the activity of HAT, a calibration curve with acetylated histones at a given concentration needs to be performed. After the reaction time, three washing steps are carried out to remove any remaining HAT or cofactors used in the catalytic reaction. From this point onward, the rest of the protocol is a typical ELISA, assessing the quantity of acetylated histones, a direct rating of HAT activity. Briefly, a primary antibody (or capture antibody) against acetylated

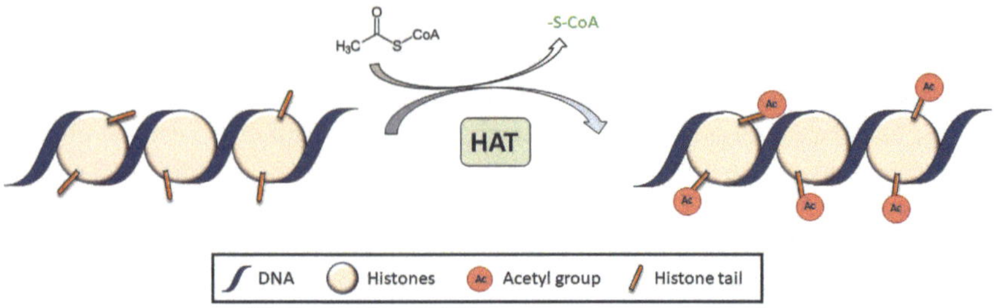

FIG. 3 Reaction catalyzed by histone acetyltransferase (HAT) in the DNA strand. HAT catalyzed the acetylation of histone tails at specific amino acids (mainly lysine) in the nucleosome, by the addition of an acetyl group from the cofactor acetyl coenzyme A.

histone is used and later a secondary antibody will detect the first one. This secondary antibody is generally conjugated to a label that catalyzes a chromogenic reaction. This reaction can be monitored in a microplate reader at a specific wavelength, measuring the absorbance of the different wells and consequently, assessing the quantity of acetylated histones, a direct rating of HAT activity.

6.2 Specific measurements of histone epigenetic marks

The different histone marks can be evaluated by immunostaining techniques using specific antibodies.

6.2.1 *Whole-mount immunostaining*

In order to assess the histone marks throughout the early development of fish, whole-mount immunostaining is recommended, since once cell differentiation occurs, the patterns of histone methylation/acetylation could vary within the different cell types of the embryos or the larvae.

Here we describe a protocol in which zebrafish embryos and larvae are used as model species (Lombó et al., 2019a, b). It is important to note that for fish species with bigger sizes of embryos and larvae, the timing for the incubation steps and the concentrations of antibodies should be previously optimized.

First—Fixation and depigmentation: for each sample, a pool of at least 5 embryos/larvae are collected in tubes with PBS. Then, fixation is carried out using 4% (*w*/v) paraformaldehyde (PFA) overnight at 4 °C under constant agitation. In order to avoid the autofluorescence of larvae that are already pigmented, an extra step with 6% (*v*/v) H_2O_2 has to be performed at RT until pigmentation vanished. Before next steps are done, samples are washed three times with PBS.

Second—Permeabilization: to ensure that the antibody goes through the cell and nuclear lipid bilayers, embryos/larvae are incubated with pure methanol at −20 °C. Next, they are rinsed three times in 1% tris-buffered saline

(19.8 mM Tris hydrochloride, 0.15 M NaCl (pH 7.6)) plus 1% (*v*/v) Triton X-100 (TBST) at RT for 5 min each time.

Third—Blocking and primary antibody incubation: so as to prevent the non-specific binding of antibodies, an incubation with the blocking solution is usually carried out before adding the primary antibody. This blocking solution usually contains bovine serum albumin (BSA), cold water fish gelatin, goat serum, nonfat dry milk, or casein. In this case, zebrafish embryos/larvae are rinsed in a solution containing 3% (*w*/v) BSA in 1% (*v*/v) TBST for 1–2 h at RT. Embryos/larvae are incubated with primary antibodies at the corresponding dilution in blocking solution for 2 days at 4 °C under constant agitation. For the negative controls, embryos/larvae must be incubated under the same conditions with blocking solution.

Fourth—Detection of primary antibodies and contrast of nuclei: to detect the binding of primary antibodies, embryos/larvae are rinsed in PBS plus the corresponding fluorescence-conjugated secondary antibodies (against the species in which primary antibodies have been made) overnight at 4 °C. Eventually, nuclei are stained with 180 μM diamidino-2-phenylindole (DAPI) for 8 min.

Fifth—Analysis of histone acetylation levels: whole embryos/larvae must be kept in PBS with an antifade mounting until further observation in a confocal microscope, using an ibidi chamber. A representative number of images from the embryos/larvae are taken to ensure that, at least, 200 nuclei can be analyzed from each sample. Relative quantity of histone marks is easily quantified using Fiji ImageJ software.

6.2.2 *Analysis of histone acetylation in gene promoters by ChIP-qPCR*

The levels of histone acetylation at specific loci (usually on gene promoters) can be assessed by combining the ChIP with the quantitative PCR (qPCR). The protocol used for this technique, using zebrafish as a model species, has been developed by Santos-Pereira et al. (2019):

First—Crosslinking of embryos: the number of embryos depends on the size of the species and the stage of development. To dechorionate one thousand zebrafish embryos at 24 hpf, a solution of 0.3 mg/ml pronase diluted in egg water (0.038 mM $CaCO_3$, 0.446 mM $NaHCO_3$, 1.025 mM sea salt, and 0.005% (vol/vol) methylene blue; pH 7.4) can be used. Then, embryos should be washed with egg water and fixed in 1% (*w*/v) formaldehyde in 200 mM phosphate buffer for 10 min at RT to cross-link the proteins to the DNA. Glycine at a final concentration of 125 mM for 5 min is added to quench the formaldehyde. Next, embryos/larvae are rinsed three times in ice-cold PBS. After this step, the samples may be kept at − 80 °C if needed.

Second—Chromatin fragmentation and immunoprecipitation: the frozen embryos are rinsed in 5 mL cell lysis buffer (10 mM Tris-HCl pH 7.5, 10 mM

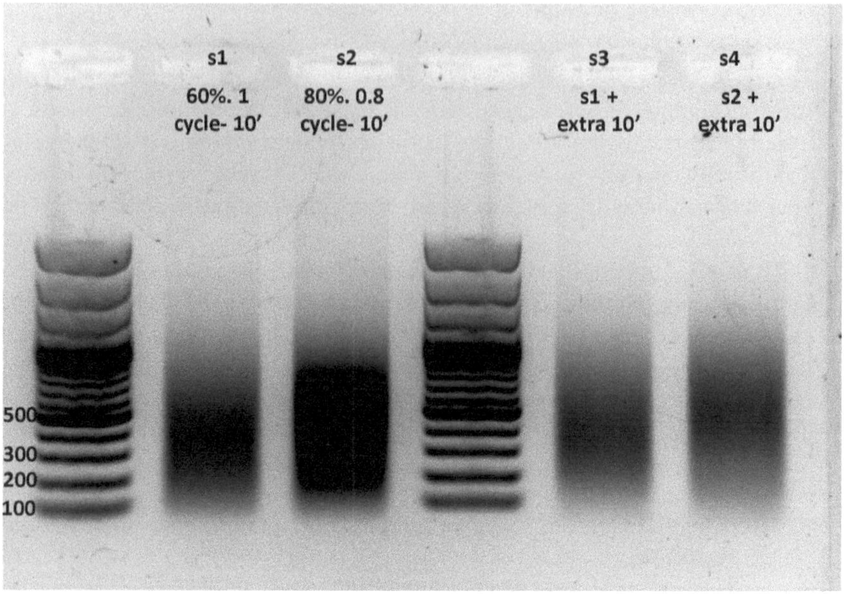

FIG. 4 Representative 2% agarose gel electrophoresis of the ChIP DNA from zebrafish embryos at different conditions of sonication (s1, s2, s3, and s4). The smears represent the fragmented DNA of different sizes. A ladder marking a range from 100 to 500bp is used to determine the fragment sizes.

NaCl, 0.3% NP-40, and 1 × Roche Complete protease inhibitors cocktail). A Potter homogenizer is used in ice for 30min. Then, embryos are transferred to a 15mL Falcon tube and centrifuged at 1200*g* for 5min at 4 °C. The pellets are resuspended in 1320μL of nuclear lysis buffer (50mM Tris-HCl pH 7.5, 10mM EDTA, 1% SDS, and 1 × Roche Complete protease inhibitors cocktail) for 5min on ice and, to block this lysis step, 2680μL of ChIP dilution buffer (16.7mM Tris-HCl pH 7.5, 1.2mM EDTA, 167mM NaCl, 0.01% SDS, 1.1% Triton-X100) are added. Then, chromatin is sonicated. The cycle amplification and the time of sonication might depend on the type of equipment used. Samples are centrifuged at 18,000*g* for 5min at 4 °C and 20μL of supernatant is used to extract the DNA and check if the chromatin is correctly sonicated in a 2% agarose gel (Fig. 4).

At this step, the chromatin is divided in aliquots of 200μL that can be frozen at − 80 °C until further use. To proceed with the immunoprecipitation, 20μL of chromatin must be reserved as input chromatin, whereas the rest of aliquots are incubated with the corresponding amount of antibody overnight at 4 °C and under agitation.

Third—Incubation with the beads and de-crosslinking: Twenty microliters of protein G Dynabeads (Invitrogen) is washed twice in 1mL of ChIP dilution buffer. Next, protein G Dynabeads are resuspended in 50μL of this buffer and added to each chromatin sample. Incubation is performed for 1 h at 4 °C under rotation. To eliminate the nonprecipitated chromatin, a sequence of washing steps is carried out using a magnetic rack: wash buffer 1 (20 mM Tris-HCl pH 7.5, 2 mM EDTA, 150 mM NaCl, 1% SDS, 1% Triton-X100), then in wash buffer 2 (20 mM Tris-HCl pH 7.5, 2 mM EDTA, 500 mM NaCl, 0.1% SDS, 1% Triton-X100), wash buffer 3 (10 mM Tris-HCl pH 7.5, 1 mM EDTA, 250 mM LiCl, 1% NP-40, 1% Na-deoxycholate), and 10 mM Tris-HCl pH 8. The samples are rinsed twice in 1 mL each solution and incubated for 5 min on ice before proceeding with the next wash. Eventually, chromatin is eluted twice in 100 μL of elution buffer (50 mM $NaHCO_3$ pH 8.8, 1% SDS) and 180 μL of this buffer is added to the 20 μL of input DNA. Incubation is carried out during 15 min for each elution step.

The chromatin de-crosslinking is performed by incubating the eluted chromatin and the input DNA with 10 μL of 4 M NaCl and 1 μL of 10 mg/mL proteinase K at 65 °C for 6 h. Finally, DNA is purified using a PCR purification kit.

Fourth—qPCR and data analysis: first of all, primers for the promoter regions of the genes containing the histone marks must be designed using the database Genome Browser (University of California Santa Cruz; https://genome.ucsc.edu/). Once temperatures and amplification efficiency are optimized, qPCRs can be performed using immunoprecipitated chromatin as template, whereas input DNA is used as internal control. For each reaction, 5 μL of 1× SYBR Green Master mix (Applied Biosystems), 1 μL of 5 μM primers, and 1 μL of immunoprecipitated chromatin and 3 μL of bi-distilled water are used. Regarding the input DNA, serial 1:10 dilutions are done for each sample and each primer set to obtain a calibration curve. Inputs are adjusted using the logarithm of the dilution, and this corrected input value is used to calculate the ΔCt ($Ct_{\text{corrected input DNA}} - Ct_{\text{immunoprecipitated chromatin}}$). Additionally, a primer set amplifying an intergenomic region, to which the antibody for histone marks should not attach, can be used as negative control. In this case, we calculate the $\Delta\Delta Ct$ ($(Ct_{\text{corrected input DNA}} - Ct_{\text{immunoprecipitated chromatin}})_{\text{target region}} - (Ct_{\text{corrected input DNA}} - Ct_{\text{immunoprecipitated chromatin}})_{\text{negative control}}$). Percent of input can be obtained using both $2^{-\Delta}Ct$ and $2^{-\Delta\Delta}Ct$ allowing the evaluation of significant differences among treatments.

7. Application of the epigenetic methods to our research

This research group, led by Dr. Paz Herráez, has been looking into the paternal contribution to embryo development. For many years we have been assessing the status of sperm chromatin and its implication in the development of the progeny in several fish farming species (brown trout (*Salmo trutta*), rainbow trout, European sea bass, gilthead seabream, and Senegalese sole) (Beirão et al., 2011a, b; Cabrita et al., 2005; Fernández-Díez et al., 2015, 2016; González-Rojo et al., 2014a, b; Pérez-Cerezales et al., 2010, 2011). More recently, we started studying how environmental toxicants, such as bisphenol A (BPA), are able to alter the information contained in the spermatozoa (DNAs, RNAs, and epigenetic marks), thus jeopardizing the health of future generations. To carry out this research line we introduced the use of zebrafish as model species, since it presents several advantages for the toxicological studies.

Regarding the paternal factor, we have exposed zebrafish males to BPA during early stages of spermatogenesis, analyzing the sperm DNA methylation by ultra-performance liquid chromatography-mass spectrometer (UPLC-MS; Lombó et al., 2015), and also during the whole spermatogenesis, assessing the levels of 5mC in testicular cells by flow cytometry (González-Rojo et al., 2019). The results showed that exposure during the early steps of male germline formation does not affect the DNA methylation of spermatozoa, whereas exposure to 2000 μg/L BPA also during meiotic phase and spermiogenesis leads to an increase in the levels of 5mC of haploid and tetraploid cells. Given the importance of histone posttranslational modifications in the spermatozoa, we also looked into the levels of histone acetylation (H3K9ac, H3K14ac, H3K27ac, and H4K12ac) as well as of histone methylation (H3K27me3) by immunocytochemistry. In this case, results provided a conclusive evidence of BPA epigenotoxicity: sperm from males exposed to 100 and 2000 μg/L showed higher levels of histone acetylation and a decrease in histone methylation (González-Rojo et al., 2019; Lombó et al., 2019a). Moreover, we used this methodology to go deeper into the mechanisms leading to histone acetylation changes promoted by BPA in testes. For that reason, we set a tissue culture system to create a toxic in vitro environment, allowing us to show a reversion of histone acetylation changes promoted by BPA (González-Rojo et al., 2019). In testicular cells, the use of the method described before for measuring the activity of epigenetic enzymes proved that the rise in histone acetylation levels was triggered by a greater activity of HAT enzymes (González-Rojo et al., 2019). In addition, we have performed several epigenetic assays in the progeny of BPA-exposed males to have a deeper insight into the paternal inheritance of BPA epigenotoxic effects. Concerning DNA methylation, male exposure to BPA did not affect the global levels of 5mC at MBT (also analyzed by UPL-MS) nor the methylation of some gene promoters at 24 hpf (assessed by bisulfite conversion and pyrosequencing) (Lombó and Herráez, 2021). In contrast, the analysis of histone epigenetic marks by whole-mount immunostaining revealed that the increase in H3K9ac and H3K27ac in sperm of exposed males was inherited by their F1 embryos (Lombó et al., 2019a). Moreover, ChIP-qPCR was performed to check if the increase in H3K9ac specifically affects the promoters of genes involved in early development, both in larvae from exposed males and in larvae directly exposed to BPA during embryo development (Lombó et al., 2019b).

In conclusion, the use of a wide variety of accurate molecular techniques has allowed us to describe how the exposure to environmental toxicants affects the male germline and, what is more, to establish the mechanisms by which paternal exposure impacts the development of future generations.

References

Adam, A.C., Helvik Skjærven, K., Whatmore, P., Moren, M., Lie, K.K., 2018. Parental high dietary arachidonic acid levels modulated the hepatic transcriptome of adult zebrafish (*Danio rerio*) progeny. PLoS One 13. https://doi.org/10.1371/journal.pone.0201278.

Adam, A.C., Lie, K.K., Whatmore, P., Jakt, L.M., Moren, M., Skjærven, K.H., 2019. Profiling DNA methylation patterns of zebrafish liver associated with parental high dietary arachidonic acid. PLoS One 14. https://doi.org/10.1371/journal.pone.0220934.

Anastasiadi, D., Piferrer, F., 2019. Epimutations in Developmental Genes Underlie the Onset of Domestication in Farmed European Sea Bass. Mol. Biol. Evol. 36, 2252–2264. https://doi.org/10.1093/molbev/msz153.

Bazzini, A.A., Lee, M.T., Giraldez, A.J., 2005. Ribosome profiling shows that mi R-430 reduces translation before causing mRNA decay in zebrafish. Science 336, 233–237. https://doi.org/10.1126/science.1215704.

Beirão, J., Cabrita, E., Pérez-Cerezales, S., Martínez-Páramo, S., Herráez, M.P., 2011a. Effect of cryopreservation on fish sperm subpopulations. Cryobiology 62, 22–31. https://doi.org/10.1016/j.cryobiol.2010.11.005.

Beirão, J., Soares, F., Herráez, M.P., Dinis, M.T., Cabrita, E., 2011b. Changes in *Solea senegalensis* sperm quality throughout the year. Anim. Reprod. Sci. 126, 122–129. https://doi.org/10.1016/j.anireprosci.2011.04.009.

Berbel-Filho, W.M., Berry, N., Rodríguez-Barreto, D., Rodrigues Teixeira, S., Garcia de Leaniz, C., Consuegra, S., 2020. Environmental enrichment induces intergenerational behavioural and epigenetic effects on fish. Mol. Ecol. 29, 2288–2299. https://doi.org/10.1111/mec.15481.

Berger, S.L., 2007. The complex language of chromatin regulation during transcription. Nature 447 (7143), 407–412. https://doi.org/10.1038/nature05915.

Bertucci, E.M., Mason, M.W., Camus, A.C., Rhodes, O.E., Parrott, B.B., 2020. Chronic low dose irradiation alters hepatic transcriptional profiles, but not global DNA methylation in medaka (*Oryzias latipes*). Sci. Total Environ. 729, 138680. https://doi.org/10.1016/j.scitotenv.2020.138680.

Best, C., Ikert, H., Kostyniuk, D.J., Craig, P.M., Navarro-Martin, L., Marandel, L., Mennigen, J.A., 2018. Epigenetics in teleost fish: from molecular mechanisms to physiological phenotypes. Comp. Biochem. Physiol. Part - B Biochem. Mol. Biol. 224, 210–244. https://doi.org/10.1016/j.cbpb.2018.01.006.

Bird, A., 2007. Perceptions of epigenetics. Nature 447, 396–398. https://doi.org/10.1038/nature05913. 4477143.

Boltaña, S., Valenzuela-Miranda, D., Aguilar, A., Mackenzie, S., Gallardo-Escárate, C., 2016. Long noncoding RNAs (lnc RNAs) dynamics evidence immunomodulation during ISAV-infected Atlantic salmon (*Salmo salar*). Sci. Rep. 6. https://doi.org/10.1038/srep22698.

Bouwmeester, M.C., Ruiter, S., Lommelaars, T., Sippel, J., Hodemaekers, H.M., van den Brandhof, E.-J., Pennings, J.L.A., Kamstra, J.H., Jelinek, J., Issa, J.-P.J., Legler, J., van der Ven, L.T.M., 2016. Zebrafish embryos as a screen for DNA methylation modifications after compound exposure. Toxicol. Appl. Pharmacol. 291, 84–96. https://doi.org/10.1016/j.taap.2015.12.012.

Brykczynska, U., Hisano, M., Erkek, S., Ramos, L., Oakeley, E.J., Roloff, T.C., Beisel, C., Schübeler, D., Stadler, M.B., Peters, A.H.F.M., 2010. Repressive and active histone methylation mark distinct promoters in human and mouse spermatozoa. Nat. Struct. Mol. Biol. 17, 679–687. https://doi.org/10.1038/nsmb.1821.

Burgerhout, E., Mommens, M., Johnsen, H., Aunsmo, A., Santi, N., Andersen, Ø., 2017. Genetic background and embryonic temperature affect DNA methylation and expression of myogenin and muscle development in Atlantic salmon (*Salmo salar*). PLoS One 12. https://doi.org/10.1371/journal.pone.0179918, e0179918.

Cabrita, E., Robles, V., Rebordinos, L., Sarasquete, C., Herráez, M.P., 2005. Evaluation of DNA damage in rainbow trout (*Oncorhynchus mykiss*) and gilthead sea bream (*Sparus aurata*) cryopreserved sperm. Cryobiology 50, 144–153. https://doi.org/10.1016/j.cryobiol.2004.12.003.

Campos, C., Valente, L.M.P., Conceição, L.E.C., Engrola, S., Fernandes, J.M.O., 2013. Temperature affects methylation of the myogenin putative promoter, its expression and muscle cellularity in *Senegalese sole* larvae. Epigenetics 8. https://doi.org/10.4161/epi.24178.

Campos, C., Valente, L.M.P., Fernandes, J.M.O., 2012. Molecular evolution of zebrafish dnmt3 genes and thermal plasticity of their expression during embryonic development. Gene 500, 93–100. https://doi.org/10.1016/j.gene.2012.03.041.

Cannell, I.G., Kong, Y.W., Bushell, M., 2008. How do microRNAs regulate gene expression? Biochem. Soc. Trans. 36, 1224–1231.

Carrell, D.T., 2011. Epigenetic marks in zebrafish sperm: insights into chromatin compaction, maintenance of pluripotency, and the role of the paternal genome after fertilization. Asian J. Androl. 13, 620–621. https://doi.org/10.1038/aja.2011.37.

Carrell, D.T., Hammoud, S.S., 2010. The human sperm epigenome and its potential role in embryonic development. Mol. Hum. Reprod. 16, 37–47. https://doi.org/10.1093/molehr/gap090.

Cavalli, G., Heard, E., 2019. Advances in epigenetics link genetics to the environment and disease. Nature 571 (7766), 489–499. https://doi.org/10.1038/s41586-019-1411-0.

Champroux, A., Cocquet, J., Henry-Berger, J., Drevet, J.R., Kocer, A., 2018. A decade of exploring the mammalian sperm epigenome: paternal epigenetic and transgenerational inheritance. Front. Cell Dev. Biol. 6, 50. https://doi.org/10.3389/fcell.2018.00050.

Chatterjee, A., Ozaki, Y., Stockwell, P.A., Horsfield, J.A., Morison, I.M., Nakagawa, S., 2013. Mapping the zebrafish brain methylome using reduced representation bisulfite sequencing. Epigenetics 8, 979–989. https://doi.org/10.4161/epi.25797.

Chatterjee, A., Stockwell, P.A., Horsfield, J.A., Morison, I.M., Nakagawa, S., 2014. Base-resolution DNA methylation landscape of zebrafish brain and liver. Genom. Data 2, 342–344. https://doi.org/10.1016/j.gdata.2014.10.008.

Chédin, F., 2011. The DNMT3 family of mammalian de novo DNA methyltransferases. In: Progress in Molecular Biology and Translational Science. Elsevier B.V, pp. 255–285, https://doi.org/10.1016/B978-0-12-387685-0.00007-X.

Cheung, P., Vallania, F., Dvorak, M., Chang, S.E., Schaffert, S., Donato, M., Rao, A.M., Mao, R., Utz, P.J., Khatri, P., Kuo, A.J., 2018. Single-cell epigenetics—chromatin modification atlas unveiled by mass cytometry. Clin. Immunol. 196, 40–48. https://doi.org/10.1016/j.clim.2018.06.009.

Chimento, A., Casaburi, I., Bartucci, M., Patrizii, M., Dattilo, R., Avena, P., Ando, S., Pezzi, V., Sirianni, R., 2013. Selective GPER activation decreases proliferation and activates apoptosis in tumor Leydig cells. Cell Death Dis. 4. https://doi.org/10.1038/cddis.2013.275, e747.

Chowdhury, B., Cho, I.H., Irudayaraj, J., 2017. Technical advances in global DNA methylation analysis in human cancers. J. Biol. Eng. https://doi.org/10.1186/s13036-017-0052-9.

Crick, F., 1970. Central dogma of molecular biology. Nature 227, 561–563. https://doi.org/10.1038/227561a0.

Dai, B., Giardina, C., Rasmussen, T.P., 2013. Quantitation of nucleosome acetylation and other histone posttranslational modifications using microscale NU-ELISA. Methods Mol. Biol. 981, 167–176. https://doi.org/10.1007/978-1-62703-305-3_13.

Delaney, C., Garg, S.K., Yung, R., 2015. Analysis of DNA methylation by pyrosequencing. In: Methods in Molecular Biology. Humana Press Inc, pp. 249–264, https://doi.org/10.1007/978-1-4939-2963-4_19.

Dhanasiri, A.K.S., Chen, X., Dahle, D., Siriyappagouder, P., Fæste, C.K., Fernandes, J.M.O., 2020. Dietary inclusion of plant ingredients induces epigenetic changes in the intestine of zebrafish. Epigenetics 15 (10), 1035–1051. https://doi.org/10.1080/15592294.2020.1747777.

Dong, X., Weng, Z., 2013. The correlation between histone modifications and gene expression. Epigenomics 5, 113–116. https://doi.org/10.2217/epi.13.13.

Duffié, R., Bourc'his, D., 2013. Parental epigenetic asymmetry in mammals. In: Current Topics in Developmental Biology. Academic Press Inc, pp. 293–328, https://doi.org/10.1016/B978-0-12-416027-9.00009-7.

Eberharter, A., Becker, P.B., 2002. Histone acetylation: A switch between repressive and permissive chromatin. Second in review on chromatin dynamics. EMBO Rep. 3, 224–229. https://doi.org/10.1093/embo-reports/kvf053.

Falisse, E., Ducos, B., Stockwell, P.A., Morison, I.M., Chatterjee, A., Silvestre, F., 2018. DNA methylation and gene expression alterations in zebrafish early-life stages exposed to the antibacterial agent triclosan. Environ. Pollut. 243, 1867–1877. https://doi.org/10.1016/j.envpol.2018.10.004.

Fang, X., Corrales, J., Thornton, C., Scheffler, B.E., Willett, K.L., 2013. Global and gene specific DNA methylation changes during zebrafish development. Comp. Biochem. Physiol. B Biochem. Mol. Biol. 166, 99–108. https://doi.org/10.1016/j.cbpb.2013.07.007.

Feil, R., Fraga, M.F., 2012. Epigenetics and the environment: emerging patterns and implications. Nat. Rev. Genet. 13, 97–109. https://doi.org/10.1038/nrg3142.

Felsenfeld, G., 2014. A brief history of epigenetics. Cold Spring Harb. Perspect. Biol. 6. https://doi.org/10.1101/cshperspect.a018200.

Fernández-Díez, C., González-Rojo, S., Lombó, M., Herráez, M.P., 2016. Impact of sperm DNA damage and oocyte-repairing capacity on trout development. Reproduction 152, 57–67. https://doi.org/10.1530/rep-16-0077.

Fernández-Díez, C., González-Rojo, S., Montfort, J., Le Cam, A., Bobe, J., Robles, V., Pérez-Cerezales, S., Herráez, M.P., 2015. Inhibition of zygotic DNA repair: transcriptome analysis of the off spring in trout (*Oncorhynchus mykiss*). Reproduction 149, 101–111. https://doi.org/10.1530/REP-14-0382.

Fraga, M.F., Uriol, E., Diego, L.B., Berdasco, M., Esteller, M., Cañal, M.J., Rodríguez, R., 2002. High-performance capillary electrophoretic method for the quantification of 5-methyl 2′-deoxycytidine in genomic DNA: application to plant, animal and human cancer tissues. Electrophoresis 23, 1677–1681. https://doi.org/10.1002/1522-2683(200206)23:11<1677::AID-ELPS1677>3.0.CO;2-Z.

Gates, L.A., Shi, J., Rohira, A.D., Feng, Q., Zhu, B., Bedford, M.T., Sagum, C.A., Jung, S.Y., Qin, J., Tsai, M.J., Tsai, S.Y., Li, W., Foulds, C.E., O'Malley, B.W., 2017. Acetylation on histone H3 lysine 9 mediates a switch from transcription initiation to elongation. J. Biol. Chem. 292, 14456–14472. https://doi.org/10.1074/jbc.M117.802074.

Gavery, M.R., Nichols, K.M., Goetz, G.W., Middleton, M.A., Swanson, P., 2018. Characterization of genetic and epigenetic variation in sperm and red blood cells from adult hatchery and natural-origin steelhead, *Oncorhynchus mykiss*. G3 (Bethesda) 8, 3723–3736. https://doi.org/10.1534/g3.118.200458.

Gavery, M.R., Roberts, S.B., 2017. Epigenetic considerations in aquaculture. PeerJ 2017. https://doi.org/10.7717/peerj.4147.

Gombeau, K., Pereira, S., Ravanat, J.L., Camilleri, V., Cavalie, I., Bourdineaud, J.P., Adam-Guillermin, C., 2016. Depleted uranium induces sex- and tissue-specific methylation patterns in adult zebrafish. J. Environ. Radioact. 154, 25–33. https://doi.org/10.1016/j.jenvrad.2016.01.004.

González-Rojo, S., Fernández-Díez, C., Guerra, S.M., Robles, V., Herraez, M.P., 2014a. Differential gene susceptibility to sperm DNA damage: analysis of developmental key genes in trout. PLoS One 9, 1–21. https://doi.org/10.1371/journal.pone.0114161.

González-Rojo, S., Fernández-Díez, C., Guerra, S.M., Robles, V., Herraez, M.P., 2014b. Differential gene susceptibility to sperm DNA damage: analysis of developmental key genes in trout. PLoS One 9 (12). https://doi.org/10.1371/journal.pone.0114161, e114161.

González-Rojo, S., Lombó, M., Fernández-Díez, C., Herráez, M.P., 2019. Male exposure to bisphenol a impairs spermatogenesis and triggers histone hyperacetylation in zebrafish testes. Environ. Pollut. 248, 368–379. https://doi.org/10.1016/j.envpol.2019.01.127.

Guo, H., Zhu, P., Guo, F., Li, X., Wu, X., Fan, X., Wen, L., Tang, F., 2015. Profiling DNA methylome landscapes of mammalian cells with single-cell reduced-representation bisulfite sequencing. Nat. Protoc. 10, 645–659. https://doi.org/10.1038/nprot.2015.039.

Hales, B.F., Grenier, L., Lalancette, C., Robaire, B., 2011. Epigenetic programming: from gametes to blastocyst. Birth Defects Res. A Clin. Mol. Teratol. https://doi.org/10.1002/bdra.20781.

Hammond, S.A., Nelson, C.J., Helbing, C.C., 2016. Environmental influences on the epigenomes of herpetofauna and fish. Biochem. Cell Biol. 94, 95–100. https://doi.org/10.1139/bcb-2015-0111.

Han, B., Li, W., Chen, Z., Xu, Q., Luo, J., Shi, Y., Li, X., Yan, X., Zhang, J., 2016. Variation of DNA methylome of zebrafish cells under cold pressure. PLoS One 11. https://doi.org/10.1371/journal.pone.0160358, e0160358.

Han, L., Zhao, Z., 2008. Comparative analysis of CpG Islands in four fish genomes. Comp. Funct. Genomics 2008, 1–6. https://doi.org/10.1155/2008/565631. 565631.

Ho, S.M., Tang, W.Y., 2007. Techniques used in studies of epigenome dysregulation due to aberrant DNA methylation: an emphasis on fetal-based adult diseases. Reprod. Toxicol. 23 (3), 267–282. https://doi.org/10.1016/j.reprotox.2007.01.004.

Holliday, R., 1987. The inheritance of epigenetic defects. Science 238, 163–170. https://doi.org/10.1126/science.3310230.

Houwing, S., Kamminga, L.M., Berezikov, E., Cronembold, D., Girard, A., van den Elst, H., Filippov, D.V., Blaser, H., Raz, E., Moens, C.B., Plasterk, R.H.A., Hannon, G.J., Draper, B.W., Ketting, R.F., 2007. A role for Piwi and pi RNAs in germ cell maintenance and transposon silencing in zebrafish. Cell 129, 69–82. https://doi.org/10.1016/j.cell.2007.03.026.

Howe, K., Clark, M.D., Torroja, C.F., Torrance, J., Berthelot, C., Muffato, M., Collins, J.E., Humphray, S., McLaren, K., Matthews, L., McLaren, S., Sealy, I., Caccamo, M., Churcher, C., Scott, C., Barrett, J.C., Koch, R., Rauch, G.-J., White, S., Chow, W., Kilian, B., Quintais, L.T., Guerra-Assunção, J.A., Zhou, Y., Gu, Y., Yen, J., Vogel, J.-H., Eyre, T., Redmond, S., Banerjee, R., Chi, J., Fu, B., Langley, E., Maguire, S.F., Laird, G.K., Lloyd, D., Kenyon, E., Donaldson, S., Sehra, H., Almeida-King, J., Loveland, J., Trevanion, S., Jones, M., Quail, M., Willey, D., Hunt, A., Burton, J., Sims, S., McLay, K., Plumb, B., Davis, J., Clee, C., Oliver, K., Clark, R., Riddle, C., Elliot, D., Eliott, D., Threadgold, G., Harden, G., Ware, D., Begum, S., Mortimore, B., Mortimer, B., Kerry, G., Heath, P., Phillimore, B., Tracey, A., Corby, N., Dunn, M., Johnson, C., Wood, J., Clark, S., Pelan, S., Griffiths, G., Smith, M., Glithero, R., Howden, P., Barker, N., Lloyd, C., Stevens, C., Harley, J., Holt, K., Panagiotidis, G., Lovell, J., Beasley, H., Henderson, C., Gordon, D., Auger, K., Wright, D., Collins, J., Raisen, C., Dyer, L., Leung, K., Robertson, L., Ambridge, K., Leongamornlert, D., McGuire, S., Gilderthorp, R., Griffiths, C., Manthravadi, D., Nichol, S., Barker, G., Whitehead, S., Kay, M., Brown, J., Murnane, C., Gray, E., Humphries, M., Sycamore, N., Barker, D., Saunders, D., Wallis, J., Babbage, A., Hammond, S., Mashreghi-Mohammadi, M., Barr, L., Martin, S., Wray, P., Ellington, A., Matthews, N., Ellwood, M., Woodmansey, R., Clark, G., Cooper, J.D., Cooper, J., Tromans, A., Grafham, D., Skuce, C., Pandian, R., Andrews, R., Harrison, E., Kimberley, A., Garnett, J., Fosker, N., Hall, R., Garner, P., Kelly, D., Bird, C., Palmer, S., Gehring, I., Berger, A., Dooley, C.M., Ersan-Ürün, Z., Eser, C., Geiger, H., Geisler, M., Karotki, L., Kirn, A., Konantz, J., Konantz, M., Oberländer, M., Rudolph-Geiger, S., Teucke, M., Lanz, C., Raddatz, G., Osoegawa, K., Zhu, B., Rapp, A., Widaa, S., Langford, C., Yang, F., Schuster, S.C., Carter, N.P., Harrow, J., Ning, Z., Herrero, J., Searle, S.M.J., Enright, A., Geisler, R., Plasterk, R.H.A., Lee, C., Westerfield,

M., de Jong, P.J., Zon, L.I., Postlethwait, J.H., Nüsslein-Volhard, C., Hubbard, T.J.P., Roest Crollius, H., Rogers, J., Stemple, D.L., 2013. The zebrafish reference genome sequence and its relationship to the human genome. Nature 496, 498–503. https://doi.org/10.1038/nature12111.

Hyun, K., Jeon, J., Park, K., Kim, J., 2017. Writing, erasing and reading histone lysine methylations. Exp. Mol. Med. 49, e324. https://doi.org/10.1038/emm.2017.11.

Jammes, H., Junien, C., Chavatte-Palmer, P., 2011. Epigenetic control of development and expression of quantitative traits. In: Reproduction, Fertility and Development, pp. 64–74, https://doi.org/10.1071/RD10259.

Jatllon, O., Aury, J.M., Brunet, F., Petit, J.L., Stange-Thomann, N., Maucell, E., Bouneau, L., Fischer, C., Ozouf-Costaz, C., Bernot, A., Nicaud, S., Jaffe, D., Fisher, S., Lutfalla, G., Dossat, C., Segurens, B., Dasilva, C., Salanoubat, M., Levy, M., Houdet, N., Castellano, S., Anthouard, V., Jubin, C., Castelli, V., Katinka, M., Vacherie, B., Blémont, C., Skalli, Z., Cattolico, L., Poulain, J., De Berardinis, V., Cruaud, C., Dupart, S., Brottler, P., Coutanceau, J.P., Gouzy, J., Parra, G., Lardier, G., Chapple, C., McKernan, K.J., McEwan, P., Bosak, S., Kellis, M., Volff, J.N., Gulgó, R., Zody, M.C., Mesirov, J., Lindblad-Toh, K., Birren, B., Nusbaum, C., Kahn, D., Robinson-Rechavi, M., Laudet, V., Schachter, V., Quétler, F., Saurin, W., Scarpeill, C., Wincker, P., Lander, E.S., Weissenbach, J., Roest Crollius, H., 2004. Genome duplication in the teleost fish *Tetraodon nigroviridis* reveals the early vertebrate proto-karyotype. Nature 431, 946–957. https://doi.org/10.1038/nature03025.

Jiang, L., Zhang, J., Wang, J.J., Wang, L., Zhang, L., Li, G., Yang, X., Ma, X., Sun, X., Cai, J., Zhang, J., Huang, X., Yu, M., Wang, X., Liu, F., Wu, C.I., He, C., Zhang, B., Ci, W., Liu, J., 2013. Sperm, but not oocyte, DNA methylome is inherited by zebrafish early embryos. Cell 153, 773–784. https://doi.org/10.1016/j.cell.2013.04.041.

Jones, P.A., 2012. Functions of DNA methylation: islands, start sites, gene bodies and beyond. Nat. Rev. Genet. 13 (7), 484–492. https://doi.org/10.1038/nrg3230.

Joshi, M., Rajender, S., 2020. Long non-coding RNAs (lncRNAs) in spermatogenesis and male infertility. Reprod. Biol. Endocrinol. 18 (1), 103. https://doi.org/10.1186/s12958-020-00660-6.

Kamstra, J.H., Aleström, P., Kooter, J.M., Legler, J., 2015. Zebrafish as a model to study the role of DNA methylation in environmental toxicology. Environ. Sci. Pollut. Res. 22, 16262–16276. https://doi.org/10.1007/s11356-014-3466-7.

Karimi, M., Johansson, S., Stach, D., Corcoran, M., Grandér, D., Schalling, M., Bakalkin, G., Lyko, F., Larsson, C., Ekström, T.J., 2006. LUMA (LUminometric methylation assay)-a high throughput method to the analysis of genomic DNA methylation. Exp. Cell Res. 312, 1989–1995. https://doi.org/10.1016/j.yexcr.2006.03.006.

Karimi, M., Luttropp, K., Ekström, T.J., 2011. Global DNA methylation analysis using the luminometric methylation assay. Methods Mol. Biol. 791, 135–144. https://doi.org/10.1007/978-1-61779-316-5_11.

Kebede, A.F., Schneider, R., Daujat, S., 2015. Novel types and sites of histone modifications emerge as players in the transcriptional regulation contest. FEBS J. 282, 1658–1674. https://doi.org/10.1111/febs.13047.

Kimura, H., 2013. Histone modifications for human epigenome analysis. J. Hum. Genet. 58 (7), 439–445. https://doi.org/10.1038/jhg.2013.66.

Kohli, R.M., Zhang, Y., 2013. TET enzymes, TDG and the dynamics of DNA demethylation. Nature 502 (7472), 472–479. https://doi.org/10.1038/nature12750.

Konstantinidis, I., Sætrom, P., Mjelle, R., Nedoluzhko, A.V., Robledo, D., Fernandes, J.M.O., 2020. Major gene expression changes and epigenetic remodelling in Nile tilapia muscle after just one generation of domestication. Epigenetics, 1–16. https://doi.org/10.1080/15592294.2020.1748914.

Kurdyukov, S., Bullock, M., 2016. DNA methylation analysis: choosing the right method. Biology (Basel) 5 (1), 3. https://doi.org/10.3390/biology5010003.

Labbé, C., Robles, V., Herraez, M.P., 2017. Epigenetics in fish gametes and early embryo. Aquaculture 472, 93–106. https://doi.org/10.1016/J.AQUACULTURE.2016.07.026.

Laing, L.V., Viana, J., Dempster, E.L., Trznadel, M., Trunkfield, L.A., Uren Webster, T.M., van Aerle, R., Paull, G.C., Wilson, R.J., Mill, J., Santos, E.M., 2016. Bisphenol A causes reproductive toxicity, decreases dnmt 1 transcription, and reduces global DNA methylation in breeding zebrafish (*Danio rerio*). Epigenetics 11, 526–538. https://doi.org/10.1080/15592294.2016.1182272.

Lawrence, M., Daujat, S., Schneider, R., 2016. Lateral thinking: how histone modifications regulate gene expression. Trends Genet. 32 (1), 42–56. https://doi.org/10.1016/j.tig.2015.10.007.

Lee, C.Y., Grant, P.A., 2018. Role of histone acetylation and acetyltransferases in gene regulation. In: Toxicoepigenetics: Core Principles and Applications. Elsevier, pp. 3–30, https://doi.org/10.1016/B978-0-12-812433-8.00001-0.

Lee, H.J., Lowdon, R.F., Maricque, B., Zhang, B., Stevens, M., Li, D., Johnson, S.L., Wang, T., 2015. Developmental enhancers revealed by extensive DNA methylome maps of zebrafish early embryos. Nat. Commun. 6, 6315. https://doi.org/10.1038/ncomms7315.

Lindeman, L.C., Andersen, I.S., Reiner, A.H., Li, N., Aanes, H., Østrup, O., Winata, C., Mathavan, S., Müller, F., Aleström, P., Collas, P., 2011. Prepatterning of developmental gene expression by modified histones before zygotic genome activation. Dev. Cell 21, 993–1004. https://doi.org/10.1016/j.devcel.2011.10.008.

Lindeman, L.C., Winata, C.L., Håvard, A., Sinnakaruppan, M., Aleström, P., Collas, P., 2010. Chromatin states of developmentally-regulated genes revealed by DNA and histone methylation patterns in zebrafish embryos. Int. J. Dev. Biol. 54, 803–813. https://doi.org/10.1387/ijdb.103081ll.

Liu, Z.-Y., Xing, J.-F., Chen, W., Luan, M.-W., Xie, R., Huang, J., Xie, S.-Q., Xiao, C.-L., 2019. MDR: an integrative DNA N6-methyladenine and N4-methylcytosine modification database for Rosaceae. Hortic. Res. 6, 78. https://doi.org/10.1038/s41438-019-0160-4.

Liu, Y., Yuan, C., Chen, S., Zheng, Y., Zhang, Y., Gao, J., Wang, Z., 2014. Global and cyp 19a1a gene specific DNA methylation in gonads of adult rare minnow *Gobiocypris rarus* under bisphenol A exposure. Aquat. Toxicol. 156, 10–16. https://doi.org/10.1016/j.aquatox.2014.07.017.

Lombó, M., Fernández-díez, C., González-rojo, S., Herráez, M.P., 2019a. Genetic and epigenetic alterations induced by bisphenol A exposure during different periods of spermatogenesis : from spermatozoa to the progeny. Sci. Rep. 1–13. https://doi.org/10.1038/s41598-019-54368-8.

Lombó, M., Fernández-Díez, C., González-Rojo, S., Navarro, C., Robles, V., Herráez, M.P., 2015. Transgenerational inheritance of heart disorders caused by paternal bisphenol A exposure. Environ. Pollut. 206, 667–678. https://doi.org/10.1016/j.envpol.2015.08.016.

Lombó, M., González-Rojo, S., Fernández-Díez, C., Herráez, M.P., 2019b. Cardiogenesis impairment promoted by bisphenol A exposure is successfully counteracted by epigallocatechin gallate. Environ. Pollut. 246, 1008–1019. https://doi.org/10.1016/j.envpol.2019.01.004.

Luk, A.C.S., Chan, W.Y., Rennert, O.M., Lee, T.L., 2014. Long noncoding RNAs in spermatogenesis: insights from recent high-throughput transcriptome studies. Reproduction 147 (5), R131–R141. https://doi.org/10.1530/REP-13-0594.

Lyko, F., 2018. The DNA methyltransferase family: a versatile toolkit for epigenetic regulation. Nat. Rev. Genet. 19 (2), 81–92. https://doi.org/10.1038/nrg.2017.80.

Marandel, L., Lepais, O., Arbenoits, E., Véron, V., Dias, K., Zion, M., Panserat, S., 2016. Remodelling of the hepatic epigenetic landscape of glucose-intolerant rainbow trout (*Oncorhynchus mykiss*) by nutritional status and dietary carbohydrates. Sci. Rep. 6, 32187. https://doi.org/10.1038/srep32187.

Marmorstein, R., Trievel, R.C., 2009. Histone modifying enzymes: structures, mechanisms, and specificities. Biochim. Biophys. Acta, Gene Regul. Mech. 1789 (1), 58–68. https://doi.org/10.1016/j.bbagrm.2008.07.009.

McCarrey, J.R., 2014. Distinctions between transgenerational and non-transgenerational epimutations. Mol. Cell. Endocrinol. 398, 13–23. https://doi.org/10.1016/j.mce.2014.07.016.

Meissner, A., Gnirke, A., Bell, G.W., Ramasahoye, B., Lander, E.S., Jaenisch, R., 2005. Reduced representation bisulfite sequencing for comparative high-resolution DNA methylation analysis. Nucleic Acids Res. 33, 5868–5877. https://doi.org/10.1093/nar/gki901.

Mohsen, K., Johansson, S., Ekström, T.J., Karimi, M., 2006. Using LUMA: a luminometric-based assay for global DNA-methylation. Epigenetics 1, 45–48. https://doi.org/10.4161/epi.1.1.2587.

Moore, L.D., Le, T., Fan, G., 2013. DNA methylation and its basic function. Neuropsychopharmacology 38, 23–38. https://doi.org/10.1038/npp.2012.112.

Mudbhary, R., Sadler, K.C., 2011. Epigenetics, development, and cancer: zebrafish make their ARK. Birth Defects Res. C Embryo Today 93 (2). https://doi.org/10.1002/bdrc.20207, 21671358.

Muret, K., Désert, C., Lagoutte, L., Boutin, M., Gondret, F., Zerjal, T., Lagarrigue, S., 2019. Long noncoding RNAs in lipid metabolism: literature review and conservation analysis across species. BMC Genomics 20 (1), 882. https://doi.org/10.1186/s12864-019-6093-3.

Murphy, P.J., Wu, S.F., James, C.R., Wike, C.L., Cairns, B.R., 2018. Placeholder nucleosomes underlie germline-to-embryo DNA methylation reprogramming. Cell 172, 993–1006.e13. https://doi.org/10.1016/j.cell.2018.01.022.

Nakato, R., Sakata, T., 2020. Methods for ChIP-seq analysis: a practical workflow and advanced applications. Methods 187, 44–53. https://doi.org/10.1016/j.ymeth.2020.03.005.

Navarro-Martín, L., Viñas, J., Ribas, L., Díaz, N., Gutiérrez, A., Di Croce, L., Piferrer, F., 2011. DNA methylation of the gonadal aromatase (cyp 19a) promoter is involved in temperature-dependent sex ratio shifts in the European sea bass. PLoS Genet. 7. https://doi.org/10.1371/journal.pgen.1002447, e1002447.

Noordermeer, D., Duboule, D., 2013. Chromatin architectures and Hox gene collinearity. In: Current Topics in Developmental Biology. Academic Press Inc, pp. 113–148, https://doi.org/10.1016/B978-0-12-416027-9.00004-8.

Olsvik, P.A., Whatmore, P., Penglase, S.J., Skjærven, K.H., Anglès d'Auriac, M., Ellingsen, S., 2019. Associations between behavioral effects of bisphenol A and DNA methylation in zebrafish embryos. Front. Genet. 10, 184. https://doi.org/10.3389/fgene.2019.00184.

Ortega-Recalde, O., Goikoetxea, A., Hore, T.A., Todd, E.V., Gemmell, N.J., 2020. The genetics and epigenetics of sex change in fish. Annu. Rev. Anim. Biosci. 8, 47–69. https://doi.org/10.1146/annurev-animal-021419-083634.

Partolina, M., Thoms, H.C., MacLeod, K.G., Rodriguez-Blanco, G., Clarke, M.N., Venkatasubramani, A.V., Beesoo, R., Larionov, V., Neergheen-Bhujun, V.S., Serrels, B., Kimura, H., Carragher, N.O., Kagansky, A., 2017. Global histone modification fingerprinting in human cells using epigenetic reverse phase protein array. Cell Death Dis. 3, 16077. https://doi.org/10.1038/cddiscovery.2016.77.

Perera, E., Turkmen, S., Simó-Mirabet, P., Zamorano, M.J., Xu, H., Naya-Català, F., Izquierdo, M., Pérez-Sánchez, J., 2020. Stearoyl-CoA desaturase (scd1a) is epigenetically regulated by broodstock nutrition in gilthead sea bream (*Sparus aurata*). Epigenetics 15, 536–553. https://doi.org/10.1080/15592294.2019.1699982.

Pérez-Cerezales, S., Gutiérrez-Adán, A., Martínez-Páramo, S., Beirão, J., Herráez, M.P., 2011. Altered gene transcription and telomere length in trout embryo and larvae obtained with DNA cryodamaged sperm. Theriogenology 76, 1234–1245. https://doi.org/10.1016/j.theriogenology.2011.05.028.

Pérez-Cerezales, S., Martínez-Páramo, S., Beirão, J., Herráez, M.P., 2010. Evaluation of DNA damage as a quality marker for rainbow trout sperm cryopreservation and use of LDL as cryoprotectant. Theriogenology 74, 282–289. https://doi.org/10.1016/j.theriogenology.2010.02.012.
Piferrer, F., Anastasiadi, D., Valdivieso, A., Sánchez-Baizán, N., Moraleda-Prados, J., Ribas, L., 2019. The model of the conserved epigenetic regulation of sex. Front. Genet. 10, 857. https://doi.org/10.3389/fgene.2019.00857.
Potok, M.E., Nix, D.A., Parnell, T.J., Cairns, B.R., 2013. Reprogramming the maternal zebrafish genome after fertilization to match the paternal methylation pattern. Cell 153, 759–772. https://doi.org/10.1016/j.cell.2013.04.030.
Qian, Y., Ji, C., Yue, S., Zhao, M., 2019. Exposure of low-dose fipronil enantioselectively induced anxiety-like behavior associated with DNA methylation changes in embryonic and larval zebrafish. Environ. Pollut. 249, 362–371. https://doi.org/10.1016/j.envpol.2019.03.038.
Quinn, A.M., Simeonov, A., 2011. Methods for activity analysis of the proteins that regulate histone methylation. Curr. Chem. Genom. 5 (Suppl 1), 95–105. https://doi.org/10.2174/1875397301005010095.
Ramsahoye, B.H., Biniszkiewicz, D., Lyko, F., Clark, V., Bird, A.P., Jaenisch, R., 2000. Non-CpG methylation is prevalent in embryonic stem cells and may be mediated by DNA methyltransferase 3a. Proc. Natl. Acad. Sci. U. S. A. 97, 5237–5242. https://doi.org/10.1073/pnas.97.10.5237.
Robles, V., Valcarce, D.G., Riesco, M.F., 2019. Non-coding RNA regulation in reproduction: their potential use as biomarkers. Non-coding RNA Res. 4 (2), 54–62. https://doi.org/10.1016/j.ncrna.2019.04.001.
Santos-Pereira, J.M., Gallardo-Fuentes, L., Neto, A., Acemel, R.D., Tena, J.J., 2019. Pioneer and repressive functions of p63 during zebrafish embryonic ectoderm specification. Nat. Commun. 10, 1–13. https://doi.org/10.1038/s41467-019-11121-z.
Sävilammi, T., Papakostas, S., Leder, E.H., Vøllestad, L.A., Debes, P.V., Primmer, C.R., 2020. Cytosine methylation patterns suggest a role of methylation in plastic and adaptive responses to temperature in European grayling (*Thymallus thymallus*) populations. Epigenetics, 1–18. https://doi.org/10.1080/15592294.2020.1795597.
Seidl, M.F., 2017. Adenine N6-methylation in diverse fungi. Nat. Genet. 49 (6), 823–824. https://doi.org/10.1038/ng.3873.
Singleman, C., Holtzman, N.G., 2012. Analysis of postembryonic heart development and maturation in the zebrafish, *Danio rerio*. Dev. Dyn. 241, 1993–2004. https://doi.org/10.1002/dvdy.23882.
Skinner, M.K., 2015. Environmental epigenetics and a unified theory of the molecular aspects of evolution: a neo-Lamarckian concept that facilitates neo-Darwinian evolution. Genome Biol. Evol. 7, 1296–1302. https://doi.org/10.1093/gbe/evv073.
Skvortsova, K., Stirzaker, C., Taberlay, P., 2019a. The DNA methylation landscape in cancer. Essays Biochem. 63 (6), 797–811. https://doi.org/10.1042/EBC20190037.
Skvortsova, K., Tarbashevich, K., Stehling, M., Lister, R., Irimia, M., Raz, E., Bogdanovic, O., 2019b. Retention of paternal DNA methylome in the developing zebrafish germline. Nat. Commun. 10, 3054. https://doi.org/10.1038/s41467-019-10895-6.
Smallwood, S.A., Kelsey, G., 2012. *De novo* DNA methylation: a germ cell perspective. Trends Genet. 28 (1), 33–42. https://doi.org/10.1016/j.tig.2011.09.004.
Smith, Z.D., Chan, M.M., Mikkelsen, T.S., Gu, H., Gnirke, A., Regev, A., Meissner, A., 2012. A unique regulatory phase of DNA methylation in the early mammalian embryo. Nature 484, 339–344. https://doi.org/10.1038/nature10960.
Smith, Z.D., Meissner, A., 2013. DNA methylation: roles in mammalian development. Nat. Rev. Genet. 14, 204–220. https://doi.org/10.1038/nrg3354.

Soubry, A., Hoyo, C., Jirtle, R.L., Murphy, S.K., 2014. A paternal environmental legacy: evidence for epigenetic inheritance through the male germ line. Bioessays 36, 359–371. https://doi.org/10.1002/bies.201300113.

Stefani, G., Slack, F.J., 2008. Small non-coding RNAs in animal development. Nat. Rev. Mol. Cell Biol. 9 (3), 219–230. https://doi.org/10.1038/nrm2347.

Strahl, B.D., Allis, C.D., 2000. The language of covalent histone modifications. Nature 403, 41–45. https://doi.org/10.1038/47412.

Streisinger, G., Walker, C., Dower, N., Knauber, D., Singer, F., 1981. Production of clones of homozygous diploid zebra fish (*Brachydanio rerio*). Nature 291, 293–296. https://doi.org/10.1038/291293a0.

Susiarjo, M., Bartolomei, M.S., 2014. You are what you eat, but what about your DNA? Science 345 (6198), 733–734. https://doi.org/10.1126/science.1258654.

Taylor, K.H., Shi, H., Caldwell, C.W., 2010. Next generation sequencing: advances in characterizing the methylome. Genes (Basel) 1, 143–165. https://doi.org/10.3390/genes1020143.

Teng, M., Chen, X., Wang, C., Song, M., Zhang, J., Bi, S., Wang, C., 2020. Life cycle exposure to propiconazole reduces fecundity by disrupting the steroidogenic pathway and altering DNA methylation in zebrafish (*Danio rerio*). Environ. Int. 135, 105384. https://doi.org/10.1016/j.envint.2019.105384.

Todd, E.V., Ortega-Recalde, O., Liu, H., Lamm, M.S., Rutherford, K.M., Cross, H., Black, M.A., Kardailsky, O., Marshall Graves, J.A., Hore, T.A., Godwin, J.R., Gemmell, N.J., 2019. Stress, novel sex genes, and epigenetic reprogramming orchestrate socially controlled sex change. Sci. Adv. 5 (7). https://doi.org/10.1126/sciadv.aaw7006, eaaw7006.

Uren Webster, T.M., Rodriguez-Barreto, D., Martin, S.A.M., Van Oosterhout, C., Orozco-terWengel, P., Cable, J., Hamilton, A., Garcia De Leaniz, C., Consuegra, S., 2018. Contrasting effects of acute and chronic stress on the transcriptome, epigenome, and immune response of Atlantic salmon. Epigenetics 13, 1191–1207. https://doi.org/10.1080/15592294.2018.1554520.

Valdivieso, A., Ribas, L., Monleón-Getino, A., Orbán, L., Piferrer, F., 2020. Exposure of zebrafish to elevated temperature induces sex ratio shifts and alterations in the testicular epigenome of unexposed offspring. Environ. Res. 186, 109601. https://doi.org/10.1016/j.envres.2020.109601.

Varriale, A., 2014. DNA methylation, epigenetics, and evolution in vertebrates: facts and challenges. Int. J. Evol. Biol. 2014, 1–7. https://doi.org/10.1155/2014/475981.

Vickers, M.H., 2014. Early life nutrition, epigenetics and programming of later life disease. Nutrients 6, 2165–2178. https://doi.org/10.3390/nu6062165.

Villar-Garea, A., Israel, L., Imhof, A., 2008. Analysis of histone modifications by mass spectrometry. Curr. Protoc. Protein Sci. 92 (1). https://doi.org/10.1002/0471140864.ps1410s51, e54.

Waddington, C.H., 1957. The Strategy of the Genes; A Discussion of some Aspects of Theoretical Biology. Allen & Unwin.

Wang, S.Y., Lau, K., Lai, K.P., Zhang, J.W., Tse, A.C.K., Li, J.W., Tong, Y., Chan, T.F., Wong, C.K.C., Chiu, J.M.Y., Au, D.W.T., Wong, A.S.T., Kong, R.Y.C., Wu, R.S.S., 2016. Hypoxia causes transgenerational impairments in reproduction of fish. Nat. Commun. 7, 12114. https://doi.org/10.1038/ncomms12114.

Wei, J.W., Huang, K., Yang, C., Kang, C.S., 2017. Non-coding RNAs as regulators in epigenetics. Oncol. Rep. 37, 3–9. https://doi.org/10.3892/or.2016.5236.

Wu, S.-F., Zhang, H., Cairns, B.R., 2011. Genes for embryo development are packaged in blocks of multivalent chromatin in zebrafish sperm. Genome Res. 21, 578–589. https://doi.org/10.1101/gr.113167.110.

Xing, H., Wang, C., Wu, H., Chen, D., Li, S., Xu, S., 2014. Effects of atrazine and chlorpyrifos on DNA methylation in the brain and gonad of the common carp. Comp. Biochem. Physiol., Part C: Toxicol. Pharmacol. 168, 11–19. https://doi.org/10.1016/j.cbpc.2014.11.002.

Zachary Ostrow, A., Viggiani, C.J., Aparicio, J.G., Aparicio, O.M., 2015. ChIP-Seq to analyze the binding of replication proteins to chromatin. Methods Mol. Biol. 1300, 155–168. https://doi.org/10.1007/978-1-4939-2596-4_11.

Zhang, T., Cooper, S., Brockdorff, N., 2015. The interplay of histone modifications—writers that read. EMBO Rep. 16, 1467–1481. https://doi.org/10.15252/embr.201540945.

Zhang, M., Yang, S., Nelakanti, R., Zhao, W., Liu, G., Li, Z., Liu, X., Wu, T., Xiao, A., Li, H., 2020. Mammalian ALKBH1 serves as an N 6-mA demethylase of unpairing DNA. Cell Res. 30, 197–210. https://doi.org/10.1038/s41422-019-0237-5.

Further reading

Lombó, M., Herráez, M.P., 2021. Paternal Inheritance of Bisphenol A Cardiotoxic Effects: The Implications of Sperm Epigenome. Int. J. Mol. Sci. 20, 2125. https://doi.org/10.3390/ijms22042125.

Chapter 4

Noncoding RNAs in fish physiology and development: miRNAs as a cornerstone in gene networks

Francisco Javier Toledo Solís[a], Jorge M.O. Fernandes[b], Elena Sarropoulou[c], and Ignacio Fernández Monzón[d]

[a]*Consejo Nacional de Ciencia y Tecnología (CONACYT), Universidad de Ciencias y Artes de Chiapas, Tuxtla, Mexico,* [b]*Genomics Division, Faculty of Biosciences and Aquaculture, Nord University, Bodø, Norway,* [c]*Institute of Marine Biology, Biotechnology and Aquaculture, Hellenic Centre for Marine Research, Iraklion, Crete, Greece,* [d]*Center for Aquaculture Research, Agrarian Technological Institute of Castile and Leon, Valladolid, Spain*

1. Noncoding RNAs

Since the discovery of the first small noncoding RNAs (ncRNAs) by both Lee et al. (1993) and Wightman et al. (1993), the scientific community made a huge research effort to unveil the biological implications of these molecules. For years, the central dogma proposed by Crick, where DNA replication, DNA transcription to mRNA, and mRNA translation to proteins, explained any biological process and/or species evolution (Crick, 1970). Currently, it is estimated that only 2% of the mammalian DNA encodes for proteins, while the other remaining DNA are noncoding sequences (The ENCODE Consortium, 2004). This 98%, previously known as "junk" DNA, has been progressively characterized during the last decades. Some noncoding and nontranscribed DNA sequences are known to harbor basic information to regulate the transcription of protein-coding genes (e.g., promoters, enhancers, insulators, etc.; reviewed by Riethoven (2010)). Other sequences, although not coding for proteins, are indeed constitutively transcribed and essential to assist on protein-coding genes translation (e.g., ribosomal RNAs (rRNAs) and transfer RNAs (tRNAs); Ramakrishnan, 2002). Many other ncRNAs control key biological processes such as mRNA posttranscriptional regulation, degradation and stability, silence retrotransposon elements, and/or govern the transcription of mRNAs through epigenetic mechanisms (see later). The identification, quantification, and functional characterization of these ncRNAs, revealing their central role on the complex and tightly

Cellular and Molecular Approaches in Fish Biology. https://doi.org/10.1016/B978-0-12-822273-7.00012-4

regulated processes sustaining life, were only possible thanks to an everlasting improvement of biotechnological approaches at the cellular and molecular levels. Particularly, with the advent of the high-throughput sequencing technologies, the notion that the protein-coding region was sufficient to explain the complexity of higher multicellular organisms was progressively turned down.

Among the ncRNAs, two main classes can be considered based on their function: structural and regulatory RNAs (Zhang et al., 2019a). Although both are required for the normal proliferation, differentiation, function, and viability of the cells, structural (or housekeeping) ncRNAs are normally constitutively expressed and include tRNAs, rRNAs, small nuclear RNAs (snRNAs), small nucleolar RNAs (snoRNAs), telomerase RNA (TERC), tRNA-derived fragments (tRF), and tRNA halves (tiRNA). In contrast, regulatory ncRNAs are differentially transcribed depending on the cell cycle, developmental stage, and/or in response to diverse endogenous/exogenous stimuli. They can be further categorized in distinct groups depending on their sequence length: mainly in microRNA (miRNA), small interfering RNA (siRNA), piwi-interacting RNA (piRNA), and long noncoding RNAs (lncRNA), including enhancer RNA (eRNA) and/or circular RNA (circRNA), among others. LncRNAs have been mechanistically characterized and may be also distinguished by their biogenesis loci in intergenic lncRNAs (lincRNAs), intronic lncRNAs, enhancer RNAs, and/or antisense RNAs (Fang and Fullwood, 2016).

The main features, biogenesis, and mechanisms of action of the best-known classes of regulatory ncRNAs are briefly described as follows. Small interfering RNAs (siRNAs) are linear double-stranded RNA (21–23 nt) perfectly base paired, containing a mRNA target sequence (sense strand) and its complement (antisense active strand; McManus and Sharp, 2002). SiRNAs are Dicer endonuclease processed and loaded onto the RNA-induced silencing complex (RISC) to silence mRNA targets. They have been used as biotechnological approaches to treat human diseases, as a gene knockdown method for functional analysis in developmental biology, or even as a strategy to control viral diseases in farmed fish (e.g., Shinya et al., 2013; Wittrup and Lieberman, 2015). MiRNAs are the most extensively studied small ncRNAs. The originally transcribed pre-miRNAs are forming distinctive hairpin structures and are further sequentially processed by Drosha and Dicer into the mature (functional) miRNAs. They are around 18–24 nt single-stranded molecules that finally interact with Argonaute (Ago) proteins to form the RISC and pair with the target mRNAs (reviewed in Ha and Kim, 2014). The biogenesis and mechanisms of action of miRNAs will be described in detail in the next section.

P-element-induced wimpy testis (PIWI)-interacting RNAs (piRNAs) are 24–31 nt length single-strand sequences, whose primary role is to silence transposable elements (TEs) in the germline of animals (recently reviewed in Ramat and Simonelig, 2021). The insertion or transposition of TEs within the genome represents a threat to its integrity and may induce gene disruption. Curiously, the genes from the piRNA pathway have evolved very rapidly in teleosts, likely to adapt to the higher diversity of transposons in these fish species (Yi et al.,

2014). In addition to silence TEs, recent studies provided evidence of piRNAs having a gene regulation function. Through base pairing with mRNA targets, piRNAs can cleavage, stabilize, and/or activate their translation, central to the evolutionary conserved maternal mRNA decay during the maternal-to-zygotic transition (Ramat and Simonelig, 2021).

LncRNAs (> 200 nt) are quite heterogeneous regarding their localization (from the nucleus to cytoplasm) or the presence of a polyA tail (reviewed in Fatica and Bozzoni, 2014), and showed a relative low degree of conservation between mammals and fish (Basu et al., 2013). LncRNAs are central in gene expression control during both developmental and differentiation processes. At the nuclear level, through repressive heterochromatin formation and/or by acting as decoys for transcription factors, and in the cytoplasm by modulating mRNA translation through direct complementary pairing with mRNA targets and/or by sequestering specific miRNAs. Particular classes of lncRNA are the eRNAs and the circRNAs. eRNAs gene regulation is through chromatin environment alteration and the interaction with transcriptional regulators, but they also control RNA processing and translation (reviewed in Sartorelli and Lauberth, 2020). They are recognized by having a histone methylation signature. More particularly, eRNAs are originated from DNA regions extremely rich in the co-occurrence of histone H3 lysine 4 monomethylation (H3K4me1) and histone H3 lysine 27 acetylation (H3K27ac). In contrast, circRNAs have their 3′ and 5′ ends not free but covalently joined, which make them different from other lncRNAs and extremely resistant to deadenylation, decapping, and degradation (reviewed in Kristensen et al., 2019). They exert important biological functions acting as microRNA or protein inhibitors ("sponges"), including in fish species (Rbbani et al., 2021).

2. Biogenesis and mechanisms of action of miRNAs

The most studied class of regulatory small ncRNAs is the miRNAs. Their biogenesis, functions, and mechanisms of action have beeen thoroughly reviewed (see Bartel, 2018; Treiber et al., 2019; Dexheimer and Cochella, 2020), where the most relevant findings in humans and particular model organisms (e.g., mouse (*Mus musculus*), *Caenorhabditis elegans*, housefly (*Musca domestica*), and/or zebrafish (*Danio rerio*)) have been compiled.

MiRNAs, together with DNA methylation and histone covalent modifications, are important molecular mediators of epigenetic regulation. MiRNAs principally regulate mRNA translation through the binding of their seed region (2–8 nt of miRNA from the 5′ end) to the 3′UTR (most commonly), 5′UTR, or coding region of the target mRNA. Translation repression can be induced by mRNA cleavage, deadenylation and decay, or through the blockage of either initiation or elongation steps. Through these mechanisms, including the (direct or indirect) modulation of DNA methylation (Collins et al., 2011; Wang et al., 2017a), miRNAs are known to be a central part of the gene regulation process behind each biological function of economic and ecological interest, including development and body patterning, muscle growth, fat deposition, immune response and/or stress resistance, among others. Indeed, the metazoan miRNAs can be thought of as the sculptors of their transcriptome (Bartel, 2018).

In general, the biogenesis of miRNAs is almost fully conserved in vertebrates and includes canonical and noncanonical pathways (reviewed in Bartel, 2018; Treiber et al., 2019; Dexheimer and Cochella, 2020). The general (canonical) biogenesis of miRNAs in animals is briefly described here and depicted in Fig. 1. Depending on the origin, miRNAs can be monocistronic, polycistronic, or intronic, the latter known as miRtrons and being processed differently by the mRNA splicing machinery rather than by the microprocessor complex (Berezikov et al., 2007). MiRNAs are transcribed by RNA polymerase II (Pol II) as capped and polyadenylated primary miRNAs (pri-miRNAs) (Treiber et al., 2019; Fig. 1, upper right corner). After transcription, produced pri-miRNA can be subjected or not to posttranscriptional modification (inhibiting or enhancing its processing) before being further processed into single hairpins termed precursor miRNAs (pre-miRNAs) by a complex of nuclear proteins known as microprocessor. In addition to several less well-understood auxiliary factors, microprocessor is mainly composed of the RNase III enzyme Drosha and the double-stranded RNA (dsRNA)-binding protein (dsRBP) DiGeorge critical region 8 (Dgcr8). Afterward, pre-miRNAs are exported to the cytoplasm through direct interaction with the export receptor exportin 5 (Exp5; Fig. 1, middle right side). There, pre-miRNAs are further processed by Dicer, another RNase III-type enzyme, that cuts both strands near the loop to generate the miRNA duplex, containing the 5p and 3p miRNA pair (before known as guide and passenger strands (Bartel, 2018)). Recently, a reevaluation of the roles performed by Drosha, Dicer, and Exp5 in miRNA biogenesis showed as many miRNAs are detectable in Dicer-knockout cells, suggesting that canonical miRNAs can undergo biogenesis through alternative pathways. The effect of Exp5-knockout in miRNA biogenesis was also found to be modest, evidencing the existence of alternative export routes. An example of noncanonical miRNAs is that requiring Dicer but not Drosha for their biogenesis. For example, in vertebrates, miR-451, essential for erythrocyte maturation, needs Drosha but not Dicer. After Drosha cleavage, the sequence of the pre-miR-451 hairpin is too short to be cleaved by Dicer, but it is processed by Argonaute 2 (Ago2; reviewed in Bartel (2018)).

In the canonical pathway, the double-stranded miRNA is loaded into an RNA-induced silencing complex (RISC), formed by the miRNA duplex and a member of the Ago protein family, which selects one strand (5p or 3p) to become the mature miRNA and discards the other (Fig. 1, bottom right corner). Mature miRNAs loaded into the RISC are very stable and can then be released/exported to the bloodstream and transported either associated to lipoproteins or included as cargos in exosomes to be uptaken by other cells, as another system of cell-cell communication (Turchinovich et al., 2016). Instead, in the cytoplasm, mature miRNAs loaded onto Ago proteins can interact with mRNAs, inducing its deadenylation, cleavage, or translational repression (Fig. 1, center). The proportion of mRNAs suffering decay or translational repression is developmental cell context

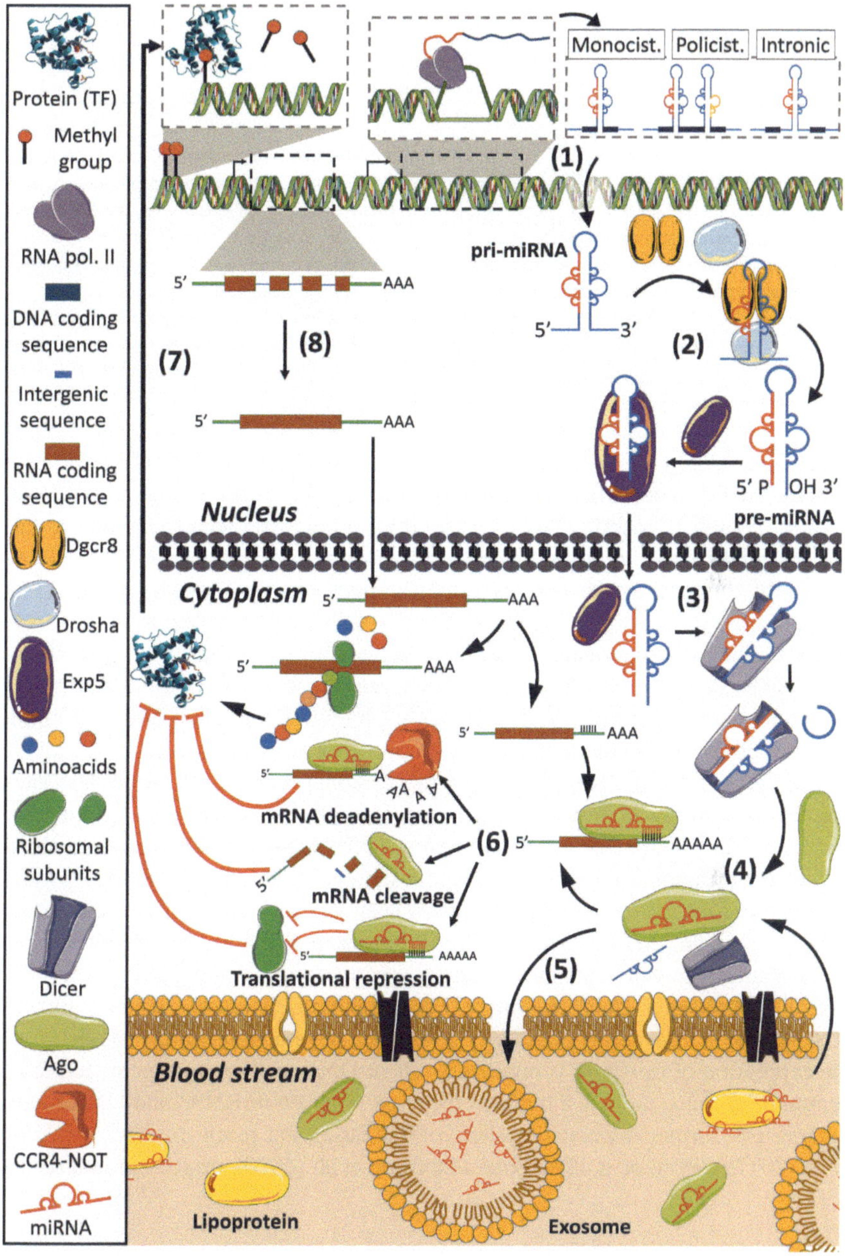

FIG. 1 Biogenesis of canonical miRNAs and mechanisms of action. The following main steps are depicted: (1) The biogenesis of monocistronic, polycistronic, or intronic miRNAs started with their transcription by RNA polymerase II (Pol II) and produced as capped and polyadenylated primary miRNAs (pri-miRNAs) (upper right corner); (2) pri-miRNAs are processed into single hairpins termed precursor miRNAs (pre-miRNAs) by a complex of nuclear proteins known as microprocessor, mainly composed by Drosha and the double-stranded RNA (dsRNA)-binding protein (dsRBP) DiGeorge critical region 8 (Dgcr8) (upper right corner); (2–3) after, pre-miRNAs are exported to the cytoplasm through the export receptor exportin 5 (Exp5) (middle-right); (3) pre-miRNAs are then processed by Dicer, cutting both strands near the loop to generate the miRNA duplex (middle-right); (4) the double-stranded miRNA is loaded into an RNA-induced silencing complex (RISC),

(Continued)

FIG. 1, CONT'D formed by the miRNA duplex and a member of the Argonaute (Ago) protein family, which selects one strand to become the mature miRNA (middle-right); (5) mature miRNAs loaded into the RISC can be released/exported to the blood stream and transported either associated to lipoproteins or included in exosomes as cargos to be up taken by other cells, as another system of cell-cell communication (bottom right corner); (6) in the cytoplasm, mature miRNAs loaded onto AGO proteins can interact with mRNAs to be translated, inducing its deadenylation by Carbon Catabolite Repression-Negative On TATA-less (CCR4-NOT), cleavage or translational repression (center); (7) silencing of mRNAs by miRNAs can induce an altered methylation profile of the DNA (left); and thus (8) reduced mRNA transcription rate (upper left corner). *(Modified from Treiber, T., Treiber, N., Meister, G., 2019. Regulation of microRNA biogenesis and its crosstalk with other cellular pathways. Nat. Rev. Mol. Cell Biol. 20, 5–20. https://doi.org/10.1038/s41580-018-0059-1.)*

dependent. Decay of mRNAs dominates in all postembryonic cells examined (Bartel, 2018). In zebrafish, early embryos, poly(A)-tail shortening of an mRNA does not change its stability but decreases translational efficiency. In contrast, during gastrulation and in postembryonic cells, shortening the poly(A)-tail reduces its stability but does not change its translational efficiency (Subtelny et al., 2014).

Although translational silencing of mRNAs is normally through a not fully complementary base pairing of the miRNA seed region (bases 2–8 of miRNA), nucleotides 2–5 are the most critical for target association (Betel et al., 2010). Nevertheless, further pairing with the mRNA by miRNA nucleotides 9–12 might induce a conformational change of RISC, allowing a better position of the active site of the Ago protein for target slicing. Additional regulatory and binding sites are also known, such as a 3′-compensatory site, but seems to only account for less than 1% of the preferentially conserved miRNA sites in mammalian mRNAs (reviewed in Bartel, 2018). In any case, translational silencing of mRNAs by miRNAs can induce further and deeper effects such as altering the DNA methylation (Fig. 1, left side) and thus, mRNA transcription rate (Fig. 1, upper left corner). For instance, the miR-290 cluster targets retinoblastoma-like 2 protein (Rbl2), known to represses transcription of DNA methyltransferases (Dnmts; Benetti et al., 2008; Sinkkonen et al., 2008). Moreover, several studies investigated how the expression of miRNAs is regulated, where DNA methylation seems to play a central role. Thus, there is a mutual cross-talk between miRNAs and DNA methylation. For further interested readers in this issue, the recent reviews by Wang et al. (2017a), Fuso et al. (2020), and Divisato et al. (2020) are recommended.

3. Methodological approaches applied in miRNA research

Different methodologies have allowed researchers to expand our understanding of how miRNAs are involved in different biological processes. While the level and dynamic expression of each miRNA, as well as the tissue/cell type where it is expressed, clearly provide the first evidence of a potential role in a particular biological process and/or defined cell context (homeostasis versus disease), a bioinformatic prediction of their mRNA targets would offer further insights on the underlying mechanism of such function. This is consistent with the potential genomic inference that we can perform considering that this miRNA has been previously characterized in a particular closely related species. Nevertheless, since both miRNA and mRNA sequences can diverge to some extent, even when only different teleost species were compared (Desvignes et al., 2021), functional validation of the targets should be conducted. Here, in addition to providing a brief overview of the different technologies that have been (or can be) applied in miRNA quantification, a general protocol of miRNA library preparation to guide researchers in this scientific topic is presented (Figs. 2 and 3).

3.1 Methodological approaches to quantify miRNA expression

Both narrow (quantitative polymerase chain reaction (qPCR)) and wider (microarray or small RNA sequencing (RNA-Seq)) approaches can be used to determine the level of expression of a single/few or hundreds of miRNAs in the cell/tissue/organism of interest, respectively (Fig. 2). The selection of one or another approach usually depends on the objective of the research: qPCR to characterize the expression of particular miRNAs and microarray or small RNA-Seq to either discover new miRNA signatures and/or have a full picture of how the miRNome is modulated. Nevertheless, and although the costs of high-throughput analysis (even for nonmodel species) are becoming more affordable, the available budget might also direct our analytical approach to less expensive options. In this sense, a combined approach might be recommended, applying a small RNA-Seq procedure on pooled biological samples to infer the most interesting (and probably differentially expressed) miRNAs that subsequently will be quantified by qPCR procedures in separated biological samples. An example of such a convenient approach is the characterization of altered

	qPCR	ddPCR	Microarray	RNA-Seq
Model organism	✓	✓	✓	✓
Non-model organism	✓	✓	✗	✓
Low expressed miRNAs	✓	✓	✗	✓
High-throughput	✗	✗	✓	✓
Pri-miRNAs	✓	✓	✓	✓
Pre-miRNAs	✓	✓	✓	✓
Mature miRNAs	✓	✓	✓	✓
isomiRs	✗	✗	✗	✓
Capacity to distinguish	✗	✗	✗	✓

FIG. 2 Comparative panel of the different methods to quantify miRNA expression. Comparative analysis was based on the ability of each method to use small noncoding RNAs from model species (those where a full or substatial genomic information is available) or nonmodel species (only few genomic sequences are available), if the intended goal is to evaluate the gene expression of low abundant miRNAs, if a high-throughput result is obtained, if it is able to specifically quantify the abundance of specific pri-miRNAs, pre-miRNAs, mature miRNAs, and isomiRs; and/or if it is able to distinguish between them. *ddPCR*, droplet digital PCR; *qPCR*, quantitative PCR.

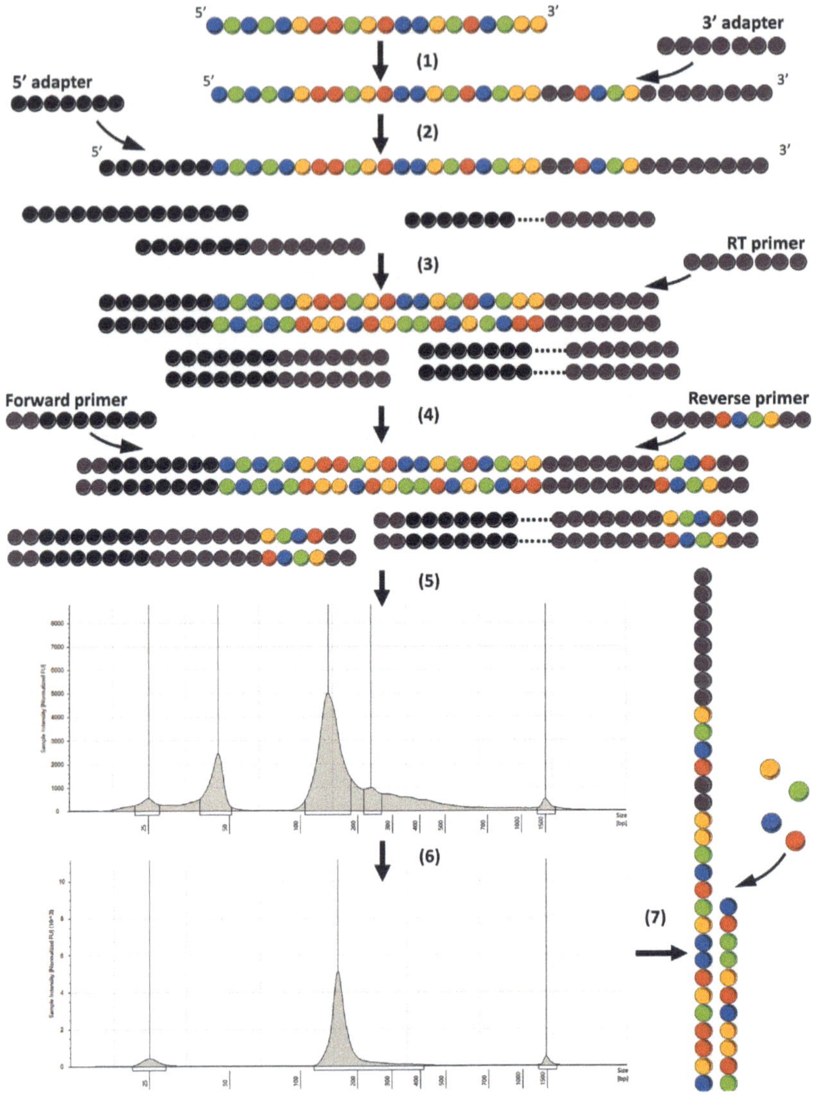

FIG. 3 General scheme of a small RNA-Seq library preparation. Extracted RNA is first submitted to a 3′ adapter ligation (1). Afterward, a 5′ adapter ligation is performed (2). Attention: at this stage, reaction should be optimized in order to reduce the probability of getting adapter dimers and subsequently ligation bias. A reverse transcription (RT) of RNA molecules with 3′ adapter is performed with a RT primer complementary to the 3′ adapter (3). An amplification of the number of molecules is done with forward and reverse PCR primers complementary to 5′ and 3′ adapters, with the 3′ adapter reverse primer adding an index for multiplexed barcode sequencing of different samples (4). Attention: at this stage, reaction should be optimized regarding the number of PCR cycles to reduce the probability of getting PCR bias. The small RNA-Seq library produced is quantified and the size profile characterized (5) with a 2100 Bioanalyzer. The result of adapter dimers (~50 bp products, left peak) and adapter ligation to longer noncoding RNAs (>200 bp products; right peak) led to its amplification, in addition to the intended products (small RNAs; central peak, ~200 bp products). To evaluate the amplified products and isolate the specific amplicons to be sequenced, a size selection procedure using TBE 10% gels is performed (6). A quality control of size selection procedure as well as an accurate quantification is done with a 2100 Bioanalyzer, right before pooling libraries and sequencing (7). *Gray* and *black circles*, not specified nucleotides; *Red, yellow, green,* and *blue circles* represent specific A, T, C, and G nucleotides.

expression of miRNAs in the fish muscle tissue when embryogenesis occurred at different temperatures (Campos et al., 2014).

Independently of the approach to be conducted, total RNA extraction through sample lysis with phenol:chloroform or guanidine isothiocyanate, and final precipitation with isopropanol is a common procedure (Kure et al., 2013; Li et al., 2017). However, a kind of isolation/enrichment procedure of the miRNAs extracted from the sample of interest, perhaps with a commercial kit such as miRVana miRNA Isolation kit (Brzuzan et al., 2012; Riesco et al., 2019), miRNAiso plus kit (Li et al., 2019), Small RNA kit (Gao et al., 2019a), or miRNeasy Serum/Plasma Kit (Fernández et al., 2019; Oliveira et al., 2020), is always recommended. Nevertheless, researchers should be aware that different results might be obtained depending on the kit used (Brown et al., 2018). Afterward, an accurate size profiling followed by miRNA quantification should be performed. Since the obtained range of concentrations of isolated miRNAs can be low, particularly from liquid biopsies, the use of NanoDrop (Thermo Scientific) is not recommended (Chu and Nabity, 2019). Instead, using a TapeStation (Agilent), 2100 Bioanalyzer (Agilent), or Qubit (Invitrogen) equipment is much more convenient, with the Bioanalyzer and TapeStation providing extra information regarding the size distribution of the isolated small RNAs. Then, the extracted small ncRNA population should be submitted to a reverse transcription (RT). For qPCR for example, different commercially available options are mainly based on an RNA polyadenylation step or the use of a poly (T) adapter before the cDNA synthesis (e.g., PrimeScript miRNA cDNA Synthesis Kit (Li et al., 2019) and TianGen miRcute Plus miRNA First-Strand cDNA Synthesis Kit (Sun et al., 2019a)), a miRNA first-strand cDNA synthesis by stem-loop primers (Li et al., 2019; Guan et al., 2020), or on the technology of Locked Nucleic Acid (LNA) (mirCury LNA RT kit; Campos et al., 2014) for example, with the latter allowing an accurate quantification of low expressed miRNAs.

Nowadays, researchers can choose among a diverse set of qPCR methods for miRNA expression analysis, including the use of universal or miRNA-specific qPCR primers and/or probes. This issue has been recently reviewed by Forero et al. (2019) where advantages and disadvantages have been analyzed and discussed. Of note, most of these methods are mainly designed to quantify mature miRNAs instead of studying pri- and pre-miRNAs. In contrast, small RNA-Seq can quantify them all, as well as potential sequence variants termed isomiRs. There is evidence that isomiRs are functionally distinct and differentially regulated in teleosts. For example, out of 70 different miR-1-3p isomiRs identified in Senegalese sole (*Solea senegalensis*), only one was consistently upregulated in fish treated with the Kisspeptin protein (Oliveira et al., 2020).

The use of probes (e.g., Taqman) instead of (most economic) SYBR Green dye, as well as the use of stem-loop RT primers, might lead to higher qPCR amplification specificity. Unfortunately, the qPCR multiplexing method, a very convenient cost-effective procedure, is compromised in miRNAs when using universal primers. Nevertheless, as mRNAs, both absolute and relative quantitative expressions can be performed, the latter requiring a proper normalization for a precise quantification (reviewed in Schwarzenbach et al., 2015). For this purpose, several normalization procedures have been applied in fish species, normally using the small ncRNA U6 snRNA (Gao et al., 2019a; Sun et al., 2019a), 5S rRNA (Li et al., 2019), miR-92-3p (Riesco et al., 2019), or piR-41105 (Fernández et al., 2019) as internal controls. Lately, the use of an exogenous control such as the UniSp6 or the cell-miR-39 mimic (Zhu et al., 2018) has been propagated, although it is more related to the normalization of the miRNA extraction procedure among experimental samples. A particularly interesting approach, mainly for low expressed miRNAs, is the absolute quantification of single or few miRNAs by droplet digital PCR (ddPCR). This method is particularly applied for quantifying circulating miRNAs (Miotto et al., 2014). Nevertheless, whatever the system and/or method used, authors should always provide the minimum information required for the publication of qPCR experiments (Bustin et al., 2009; The dMIQE Group, 2020).

Other much wider approaches used for quantifying miRNAs are microarrays and small RNA-Seq (reviewed in Pritchard et al., 2012). Several microarray systems have been developed for miRNAs, including the classic dual-channel approach (labeling with Cy3 and Cy5) or the more innovative oligonucleotide probes printed onto a slide with the posterior incorporation of labeled dATPs and based on the binding of two fluorescent LNA-DNA oligonucleotide probes (Pritchard et al., 2012; Hsieh et al., 2018). The use of microarrays is still a cost-effective approach, although mainly limited to species with extensive genomic information.

The most recent (and nowadays, commonly used) technology to get a wider profile of all the miRNAs involved in a particular cell/tissue or biological process is the small RNA-Seq. The constant evolution and development of OMIC analytical techniques, particularly from next-generation sequencing (NGS; Mutz et al., 2013), has opened unlimited possibilities to expand our understanding of the gene networks underlying each biological process. RNA-Seq has a higher throughput, is more specific and precise, and can be applied to a wider dynamic range than high-throughput proteomic analysis (Qian et al., 2014). Indeed, when compared with microarrays, small RNA-Seq has emerged as the preferred method for miRNA profiling, offering high sensitivity, single-nucleotide resolution, and allowing the analysis of a great number of samples in parallel (Coenen-Stass et al., 2018). However, to get reliable results, this technology has some requirements. Good RNA quality, purity, and integrity are fundamental. Higher integrity of RNA has been associated with greater numbers of uniquely mapped reads and reads mapped to genes (Gallego et al., 2014). If using total RNA, the RNA integrity can be evaluated using the 28S to 18S rRNA ratio (or RNA integrity number (RIN) values), with ratios greater than 8 highly desirable to warrant accurate sequencing and identification of miRNAs. If the sample to be sequenced is depleted of rRNAs (e.g., from biofluids or coming from an enriching small RNA protocol), RIN values cannot be calculated. In this case, the inspection of the size profile of small RNAs might allow a fair evaluation of their integrity.

A general pipeline of the different steps to be performed from isolated miRNAs to miRNA sequencing and counting (including differential expression analysis) is detailed as follows. Once the population of small ncRNAs has been extracted and quantified, successful library preparation is the first step (specific instructions might vary depending on the commercial library preparation kit used; Fig. 3). In brief, two sequential adapter-ligation (3′- and 5′-adapters) processes should be conducted to allow first-strand synthesis and PCR amplification with RT and PCR primers, respectively. Although adapter-ligation procedures have been improved, they still might be a source of bias in the final results. Posterior to first-strand synthesis, a PCR amplification procedure is performed to amplify, purify and identify to which sample each sequenced miRNA is belonging to. In this sense, a specific barcoding sequence (from 4 to 8 nt) is included within the primers to identify the sample where the sequenced miRNA comes from PCR amplification procedure can also introduce some bias, so the number of PCR cycles (from 10 to 24) should be adjusted to the minimum required. Since adapters dimers, as well as adapter ligations to the other RNAs present in the sample, might occur, a quality check (using a Bioanalyzer) is needed to confirm the size distribution and the amount of amplified material. As a result of a successful library prep, those sequences around 200 bp size are the intended miRNA population to be sequenced, and should be isolated through a size selection process. After the size selection procedure, the size, purity, and concentration of sequences from each sample should be checked again, just before normalizing and pooling the libraries to be sequenced.

After sequencing, the bioinformatic analysis can be performed using particular pipelines (see Conesa et al. (2016) for a general example). It should start with quality control (QC) check of the raw data (reviewed in Sheng et al. (2017) for the most common sequencing platform used, the Illumina). In brief, the most common parameters to assess in the raw data are (1) the total number of reads sequenced, recommended to be higher than 5 million per sample to get good coverage, or at least numbers should be homogeneous between samples (Sun et al., 2014); (2) the base quality score given in Phred scale, to be expected between 30 and 40 (values below 20 are usually considered low quality), giving the probability of the base being incorrectly assigned; (3) nucleotide distribution by cycle, the 4 nt (A, T, C, and G) should have a stable distribution across all cycles; and (4) the GC content, that should be similar to the GC content of the species, with a deviation higher than 10% indicating some kind of contamination. As in small RNA-Seq the standard read length is 50 or 75 nucleotides (single-end 50 or 75 cycles), but the mature small ncRNAs are <50 nt length, after raw data QC, raw reads should be trimmed (with a command-line tool such as cutadapt, for example) to remove adapters. After trimming, an additional quality check can be the length distribution. Peaks at 0 nt indicate overrepresentation of adapters, peaks at 22 nt high abundance of miRNAs, and peaks at 50 nt longer RNAs have been sequenced.

After confirming that the trimmed reads have the expected size distribution, they should be annotated and/or quantified. In small RNA-Seq, mapping to the reference genome is not essential for identifying and quantifying the miRNAs sequenced, but might help to identify new ones. Identification of new potential miRNAs can be done by using different computational tools to identifying the flanking regions in order to predict a potential hairpin (e.g., Ambros et al., 2003). Identification/annotation of sequenced miRNAs (and the other small ncRNAs) can be alternatively done by blasting against the known miRNAs (and other small ncRNAs) deposited in public repositories (e.g., miRbase (Kozomara and Griffiths-Jones, 2014), MirGeneDB (Fromm et al., 2020), RefSeq and Rfam databases, or the most global one, the RNA Central web site (https://rnacentral.org/), where up to 44 different resources are integrated (RNAcentral Consortium, 2021)), or the respective annotation file on the Ensembl website. Furthermore, the presence of novel miRNAs can be identified with different software (e.g., MIREAP, mirdeep2, and/or miREvo) considering secondary structure, Dicer enzyme cleavage site, and minimum free energy.

Once reads are identified/annotated as miRNAs, read counts are quantified (aggregating raw counts of mapped/annotated reads) using different programs (e.g., HTSeq-count; Anders et al., 2015) and normalized for accurate differential expression analysis. Different examples of normalization strategies particularly applied in fish studies are reads aligned per kilobase mapped (RPKM; Sun et al., 2017a), transcripts or reads per million (specific miRNA count/total count of clean reads) $\times 10^6$); Yang et al., 2019), the relative frequency of reads (specific number of miRNA reads in a sample/total number of miRNA reads in the sample; Bizuayehu et al., 2012a,b), and/or total counts per library (Fernández et al., 2019), among others. Afterward, different programs can be chosen to determine which miRNAs are differentially expressed such as DESeq2 (Love et al., 2014) or EdgeR (Varet et al., 2016) implemented in commercially designed software (among others CLC Genomics Workbench or GenExplain) using different statistical criteria (e.g., *P*-value, padj., log2(Fold Change) and/or false discovery rate).

Different studies have identified critical points (miRNA isolation, starting material input and/or library preparation, including reverse transcription and amplification, among others) that might induce some bias in the quantification of miRNAs through small RNA-Seq (e.g., Coenen-Stass et al., 2018; Dard-Dascot et al., 2018; Giraldez et al., 2018; Yeri et al., 2018; Wright et al., 2019; Chu and Nabity, 2019). Comparisons of the different commercial RNA isolation kits, library preparation methods, and sequencing platforms for small RNA-Seq should be conducted for identifying and implementing the best approaches to the objective and samples of interest. One of the most recent and complete reports assessing the accuracy and reproducibility of the currently used small RNA-Seq library preparation methods, involving 9 different labs, evaluated the results obtained from 9 different library kits differing in the use of adapters (3 of defined sequence and 6 with degenerate bases; Giraldez et al., 2018). Results bias was found related to protocol- and sequence-specific differences. Among them, a reduced ability to accurately measure adenosine-to-inosine editing in miRNAs was highlighted. Nevertheless, these biases were limited when using adapters with degenerate bases. Most importantly, relative miRNA quantification between samples was accurate and reproducible across laboratories, the protocol bias being potentially much greater than the one of lab to lab (Giraldez et al., 2018).

The percentage of small ncRNAs types found in a sequenced population was different when a diverse set of RNA-Seq library kits were used, mostly due to the different adapter-ligation procedures (Coenen-Stass et al., 2018). Also, the number of miRNA reads mapped to the genome varied considerably. Furthermore, the consistency of results in isomiRs of altered length derived from multiple moderate starting inputs and different library preparation kits was also explored by Wright et al. (2019). Although overall quantifications were similar, results for individual miRNAs were divergent. Moreover, significant differences were reported in accuracy, mapping rates, and consistency. Previous studies have already reported that the concordance of small RNA-Seq results using different library kits was dependent on the RNA input (Dard-Dascot et al., 2018; Yeri et al., 2018), and that the validation of results from RNA-Seq by other technologies such as qPCR, EdgeSeq, and FirePlex was expression level dependent: the higher the expression, the better the reproducibility (Yeri et al., 2018). Recently, a comparison of different samples from biofluids showed how the optimal isolation and library preparation kits might differ between biofluids (Chu and Nabity, 2019). These authors demonstrated that the library preparation method might introduce higher variation in miRNA expression than the miRNA isolation method, library preparation-dependent bias being a significant factor for successful miRNA profiling for biomarker discovery.

Independently of the different limitations of each miRNA expression quantification approach (qPCR, microarray, and small RNA-Seq), generated data are relatively reproducible and comparable. However, a low correlation on the expression levels obtained by qPCR and microarray platforms has been reported, particularly for miRNAs with low expression (Chen et al., 2009). Thus, when using high-throughput technologies for miRNA quantification (microarrays and small RNA-Seq), a kind of reproducibility confirmation using a qPCR approach might be recommended, particularly for differentially expressed miRNAs that are lowly expressed.

3.2 Methodological approaches to localize the expression of miRNA

The in situ hybridization (ISH) technique is the most common method to determine the specific cell type where a miRNA is expressed. An interesting review focused on the use of this technique is here recommended (Urbanek et al., 2015), with detailed descriptions of the critical steps for successful miRNA cell/tissue detection. Again, the implementation of LNA probes increased the sensitivity to detect such small size RNAs and facilitated the description of the tissue-specific expression patterns in whole-mount ISH preparations (Fjose and Zhao, 2010). Nevertheless, although one mismatch strongly reduces the hybridization signal when using LNA probes, it does not fully prevent cross-hybridization with several miRNAs from the same family or cluster, requiring at least two or more different internal nucleotides to avoid unspecific signal (Kapsimali et al., 2007).

Another interesting approach for in situ miRNA detection particularly to be applied in paraffin-embedded, formalin-fixed tissue is the in situ PCR. Specific protocols have been designed for the detection of the precursors (pri- and pre-miRNAs) and mature miRNAs and their correlation with their putative target mRNAs (Nuovo et al., 2009).

3.3 Methodological approaches to predict mRNA targets

To get deeper insights on the mechanisms by which a particular miRNA—for instance, found to be differentially expressed by any of the previous methodological approaches—has a key role in a particular biological process, different computational analyses can be used to in silico predict the potential target mRNAs. Furthermore, it is a wise approach toward prioritizing miRNA-mRNA target interactions for experimental validation.

Most predictions are based on algorithms that consider common features of miRNA-mRNA interactions: seed match, conservation, free energy, and site accessibility among other less common features (reviewed in Peterson et al., 2014; Kern et al., 2020; Riolo et al., 2021). One of the most determining features is the presence of a 7–8-nt "canonical" site in the 3′ UTR (as well as on the 5′ UTR or in the CDS, although less commonly occurring) of the target mRNA to which the seed region (2–8 nt from the 5′ end) of the miRNA might bind (Bartel, 2018). Of note, perfect seed complementarity is neither necessary nor sufficient for miRNA regulation (Betel et al., 2010). In addition, some target ("noncanonical") sites with a mismatch or a G:U wobble in the seed region have been proven to confer a noticeable regulatory effect (Didiano and Hobert, 2016). For an accurate prediction, an energetically favorable hybridization between miRNA and target mRNA should be also required, thus most computing analyses include minimum free energy for reliable prediction (Krüger and Rehmsmeier, 2006). Nevertheless, even when the computing tools consider these four parameters, a large number of predictions are generated, many of which are potentially false. A further refinement of these predictions is that the methods also consider the evolutionary conservation of target sequences, eliminating the poorly conserved candidate sites from consideration (Betel et al., 2010).

A very interesting evaluation of the reliability of bioinformatic predictions of miRNA-mRNA interactions suggested that the biological functionality of miRNA binding sites depends on the dose sensitivity of their host gene (Pinzón et al., 2017). Also, the same study found that some conserved miRNA binding sites seem to be conserved in a miRNA-independent fashion, while being fortuitously complementary to miRNAs. Therefore evolutionary conservation of the miRNA binding sites on mRNAs would not be an indication of a higher probability of this prediction being functionally true. Furthermore, through these binding sites, some mRNAs can modulate miRNA activity by titration, which finally might refine miRNA expression patterns. More recently, Fridrich et al. (2019) showed that the length of base pairing between miRNA and its mRNA target seems to be species/taxa specific, with longer base pairing (17 nt) in less evolved animals (cnidarians) than in bilaterian animals where a restricted sequence (seed region of 7–8 nt) might be functional. This implies that using computational prediction based on a seed region conservation might lead to false positives.

More recently, computational tools implemented further features to improve their predictions (reviewed in Kern et al., 2020). This is the case of considering the flanking AU nucleotide content, synergistic regulatory effects by closely located binding sites, or preferential locations of sites within 3′ UTRs that are predominantly combined with machine learning. Additional factors can be added when putative target sites are detected. The expression level of the potential mRNA target—thanks to the advent of more cost-effective transcriptomic high-throughput technologies—can be additional evidence supporting computational prediction when anticorrelation expression patterns of miRNA-mRNA are found. Moreover, the high number of target sites and seed type distributions in the mRNA provide a higher chance of being true positives. Nevertheless, due to the inherent complexity of miRNA gene targeting and that none of the tools outperforms all others, the combined use of several computational methods has been established as common practice to avoid false positives (Kern et al., 2020).

3.4 Methodological approaches for functional characterization of miRNA-mRNA interactions

Since the computational tools to predict mRNA targets still lack some consistency, every putative target identified must undergo functional validation. An extensive and diverse collection of biotechnological procedures has been conceived (see the recent review of Riolo et al., 2021) but will be succinctly enumerated here.

Riolo et al. (2021) have classified validation approaches into two categories. Direct methods investigate the existence of a miRNA-mRNA interaction studying the base pairing or artificially introducing a specific target site bound by miRNAs (a miRNA responsive element) in vector constructions of reporter genes, and assessing the potential changes at protein levels induced by miRNAs, either measuring protein synthesis or indirectly through light emission signals. In contrast, indirect methods are evaluating the effects derived from an altered miRNA expression on mRNA or protein expression, by using low- or high-throughput (e.g., RNA-Seq and mass spectrometry) technologies. However, results from indirect methods might not be directly and necessarily resultant from miRNA-mRNA interaction as other parallel mechanisms may occur. Independently of the method selected, a logically, time-consuming and cost-effective sequential procedure of demonstrations has been proposed as follows: (1) coexpression of miRNA and predicted mRNA target in the same cell/tissue by qPCR, ISH, microarray, and or RNA-Seq; (2) direct interaction of the miRNA and the mRNA binding site through luciferase assay, cross-linking immunoprecipitation combined with high-throughput sequencing (CLIP-Seq) and/or cross-linking, ligation and sequencing of hybrids (CLASH); (3) altered protein expression of the mRNA target by western blotting, ELISA, pulsed stable isotope labeling by amino acids (pSILAC), and/or general proteomic approaches among others; and (4) modified protein expression of the mRNA has an impact on particular biological functions with evaluations of particular biochemical pathways. Nevertheless, successfully achieving this sequential demonstration is not always an easy task or even possible. For instance, although the miRNA-mRNA interaction might be effectively genuine, and an altered protein expression is induced, the altered pathway might not be finally observed in our biological system since other pathways might take place to restore homeostasis.

Different in vivo and in vitro cell systems have been adopted. The cell systems are more ethical and suitable, while embryos from model species (e.g., zebrafish; reviewed in Pase and Lieschke, 2011) can be also considered as an appropriate high-throughput approach. Transfection/microinjection of miRNAs and/or chemically modified/synthetic nucleic acids (mimicking or antagonizing natural miRNA sequences) into a cell or zebrafish eggs from wild type, mutants, and/or transgenic lines carrying reporter gene constructs (e.g., Green Fluorescent Protein (GFP) and the potentially 3′ UTR binding site), provide unique opportunities to directly assay miRNA-mRNA interaction (e.g., Giraldez et al., 2006; Morton et al., 2008; Chu et al., 2019; Sun et al., 2019a) and the subsequent potential functions during embryogenesis and early larval stages (Fjose and Zhao, 2010).

More elegant technological approaches including transient knockdowns and knockouts with morpholino injections and/or gene editing through CRISPR/Cas9 system may be also applied in in vitro and in vivo systems, although gene editing might be a more time-consuming approach (requiring a transgenerational establishment of the cell or zebrafish line). For instance, specifically designed morpholinos to block the processing of the pre-miRNA or the activity of the mature miRNA have been proven as a time- and cost-effective loss-of-function strategy (Kloosterman et al., 2007).

The recent and continuous development of different biotechnological approaches opens new avenues to increase and improve the functional characterization of miRNA-mRNA interactions. Although some of them can be applied to fish species, the success of CLIP-Seq and CLASH methods depends on the availability of antibodies capable of efficiently precipitating the Ago proteins. This is a major constraint in fish species since only a few mammalian antibodies might have good cross-reactivity with the orthologous fish proteins. Other limitations are related to the artificial contexts where they are done, thus being difficult to translate the results obtained to the real biological conditions (Fridrich et al., 2019). Independently of the potential caveats that some of the validation methods might have, undoubtedly the increased experimental confirmation of the occurring miRNA-mRNA interactions might help not only to get further biological insights on how miRNAs control any biological process but also to improve the reliability of computational prediction tools.

4. miRNAs playing key roles in fish development and physiology

During the last decade, the previously described methodologies have been applied to study the potential role of miRNAs on a broad set of biological processes (e.g., development, growth, metabolism, immunology, muscle growth, and fat deposition) in fish species (Table 1). Although a great research effort for identifying, quantifying, and characterizing miRNAs in fish species has

TABLE 1 List of studies using different methodologies to explore the potential role of miRNAs on a broad set of biological processes in fish species.

Physiological process	Species	RNAseq	Microarray	qPCR	ddPCR	ISH	Functional miRNA-mRNA assay
Fish development							
Wienholds et al. (2005)	*Danio rerio*		x			x	
Giraldez et al. (2005)	*Danio rerio*						Mutants
Giraldez et al. (2006)	*Danio rerio*		x				In vivo target validation
Kloosterman et al. (2007)	*Danio rerio*						miRNA morpholino knockdown
Choi et al. (2008)	*Danio rerio*		x	x			
Eberhart et al. (2008)	*Danio rerio*						Mutants, miRNA morpholino knockdown
Leucht et al. (2008)	*Danio rerio*						miRNA morpholino knockdown
Morton et al. (2008)	*Danio rerio*			x		x	miRNA morpholino knockdown, antagomiRs, and luciferase assay
Ramachandra et al. (2008)	*Oncorhynchus mykiss*			x			
Zeng et al. (2009)	*Danio rerio*			x			miRNA morpholino knockdown and microRNA Mimic
Fu et al. (2011)	*Paralichthys olivaceus*	x					
Chi et al. (2011)	*Hypophthalmichthys molitrix*	x					
Chi et al. (2011)	*Hypophthalmichthys nobilis*	x					
Bizuayehu et al. (2012a)	*Hippoglossus hippoglossus*	x					
Eshel et al. (2014)	*Oreochromis niloticus*	x					

Continued

TABLE 1 List of studies using different methodologies to explore the potential role of miRNAs on a broad set of biological processes in fish species.—cont'd

Physiological process	Species	RNAseq	Microarray	qPCR	ddPCR	ISH	Functional miRNA-mRNA assay
Kaitetzidou et al. (2015)	*Dicentrarchus labrax*	x					
Giusti et al. (2016)	*Oreochromis niloticus*			x			Luciferase assay
Huang et al. (2016)	*Cynoglossus semilaevis*	x					
Robledo et al., 2017	*Scophthalmus maximus*	x					
Gomes et al. (2017)	*Colossoma macropomum*	x					
Huang et al. (2018)	*Megalobrama amblycephala*	x					
Pinhal et al. (2018)	*Oreochromis niloticus*	x					
Roberto et al. (2018)	*Danio rerio*			x		x	In vivo and in vitro microRNA mimic
Sarropoulou et al. (2019)	*Dicentrarchus labrax*	x					
Wang et al. (2019a)	*Paralichthys olivaceus*	x		x			
Reproduction and sex differentiation							
Kedde et al. (2007)	*Danio rerio*						miRNA morpholino knockdown
Ma et al. (2012)	*Oncorhynchus mykiss*	x		x			
Bizuayehu et al. (2012b)	*Gadus morhua*						
Gu et al. (2014)	*Paralichthys olivaceus*	x		x			
Jing et al. (2014)	*Pelteobagrus fulvidraco*	x		x			
Xiao et al. (2014)	*Oreochromis niloticus*	x		x			
Li et al. (2016)	*Oryzias melastigma*	x					

Tao et al. (2016)	*Oreochromis niloticus*	x					
Wang et al. (2016)	*Oreochromis niloticus*	x		x			
Presslauer et al. (2017)	*Danio rerio*	x					
Skaftnesmo et al. (2017)	*Salmo salar*	x		x			
Wang et al. (2017b)	*Cyprinus carpio*	x		x			
Zhao et al. (2017)	*Odontobutis potamophila*	x		x			
Gay et al. (2018)	*Oryzias latipes*	x				x	miRNA CRISPR/Cas9 knock-out
Lan et al. (2019)	*Megalobrama amblycephala*			x			
Riesco et al. (2019)	*Solea senegalensis*			x			
Zayed et al. (2019)	*Danio rerio*	x		x			
Fu et al. (2020)	*Symphysodon aequifasciatus*	x		x			
Li et al. (2020)	*Acanthopagrus latus*	x		x			
Papadaki et al. (2020)	*Diplodus puntazzo*	x					
Yan et al. (2021)	*Takifugu rubripes*	x		x			
Zhao et al. (2021)	*Cynoglossus semilaevis*	x					
Adaptation/response to different stressors and environmental signals							
Brzuzan et al. (2012)	*Coregonus lavaretus*	x		x			
Kitano et al. (2013)	*Gasterosteus aculeatus*	x					
Kure et al. (2013)	*Salmo salar*	x					
Lai et al. (2016)	*Oryzias melastigma*	x					
Tse et al. (2016)	*Oryzias melastigma*	x		x			

Continued

TABLE 1 List of studies using different methodologies to explore the potential role of miRNAs on a broad set of biological processes in fish species.—cont'd

Physiological process	Species	RNAseq	Microarray	qPCR	ddPCR	ISH	Functional miRNA-mRNA assay
Diamante et al. (2017)	*Coryphaena hippurus*	x		x			
Feng et al. (2017)	*Hypophthalmichthys molitrix*			x			
Huang et al. (2017)	*Megalobrama amblycephala*						Luciferase assay
Renaud et al. (2017)	*Danio rerio*	x					
Sun et al. (2017a)	*Megalobrama amblycephala*	x		x			
Sun et al. (2019b)	*Cyprinus carpio*	x		x			
Wang et al. (2019a, b)	*Cyprinus carpio*	x		x			
Xu et al. (2019)	*Sciaenops ocellatus*	x					
Kelley et al. (2021)	*Poecilia mexicana*	x					
Shwe et al. (2020)	*Salmo salar*	x					
Immunocompetence and disease resistance							
Ordas et al. (2013)	*Danio rerio*		x				
Chen et al. (2017)	*Epinephelus coioides*	x		x			
Eslamloo et al. (2018)	*Gadus morhua*	x		x			
Chu et al. (2019)	*Miichthys miiuy*			x			miRNA mimic and siRNA, luciferase assay
Geo et al. (2019b)	*Oreochromis niloticus*	x					
Li et al. (2019)	*Paralichthys olivaceus*	x		x			miRNA mimic, luciferase assay

Ma et al. (2021)	*Lampetra morii*	x		x			Western blotting, miRNA mimic and siRNA, luciferase assay
Guan et al. (2020)	*Paralichthys olivaceus*			x			miRNA mimic and siRNA, luciferase assay
Li et al. (2020)	*Paralichthys olivaceus*						
Li et al. (2021)	*Paralichthys olivaceus*			x			miRNA mimic and siRNA, luciferase assay
Smith et al. (2020)	*Salmo salar*	x		x			
Metabolism and nutrition							
Mennigen et al. (2012)	*Oncorhynchus mykiss*			x			Western blotting
Mennigen et al. (2013)	*Oncorhynchus mykiss*			x			Western blotting
Tang et al. (2013)	*Oreochromis niloticus*			x			
Mennigen et al. (2014)	*Oncorhynchus mykiss*			x			Western blotting
Zhang et al. (2014)	*Megalobrama amblycephala*	x		x			
Bizuayehu et al. (2016)	*Gadus morhua*	x					
Zhang et al. (2016)	*Siganus canaliculatus*			x			miRNA mimic
Li et al. (2017)	*Ctenopharyngodon idella*			x			
Miao et al. (2017)	*Megalobrama amblycephala*	x					
Qiang et al. (2017)	*Oreochromis niloticus*	x		x			
Sun et al. (2017b)	*Cynoglossus semilaevis*	x		x			
Chen et al. (2018)	*Siganus canaliculatus*			x			miRNA mimic and inhibitor, luciferase assay
Tao et al. (2018)	*Oreochromis niloticus*	x					

Continued

TABLE 1 List of studies using different methodologies to explore the potential role of miRNAs on a broad set of biological processes in fish species.—cont'd

Physiological process	Species	RNAseq	Microarray	qPCR	ddPCR	ISH	Functional miRNA-mRNA assay
Qiang et al. (2018)	*Oreochromis niloticus*	x		x			
Zhu et al. (2018)	*Oncorhynchus mykiss*			x			
Chen et al. (2019)	*Pelteobagrus fulvidraco*			x			miRNA mimic and inhibitor, luciferase assay
Chen et al. (2019)	*Siganus canaliculatus*			x			Western blotting, miRNA mimic and inhibitor, siRNA, luciferase assay
Fernández et al. (2019)	*Solea senegalensis*	x		x			
Kostyniuk et al. (2019)	*Oncorhynchus mykiss*	x					
Miao et al. (2018)	*Megalobrama amblycephala*			x			AntagomiR
Sun et al. (2019a)	*Siganus canaliculatus*			x			Western blotting, miRNA mimic and inhibitor, siRNA, luciferase assay
Tang et al. (2019)	*Ctenopharyngodon idella*			x			
Yang et al. (2019)	*Cyprinus carpio*	x					
Zhu et al. (2019)	*Oncorhynchus mykiss*			x			
Gonçalves et al. (2020)	*Oncorhynchus mykiss*	x		x			
Oliveira et al. (2020)	*Solea senegalensis*	x					
Sun et al. (2020)	*Siganus canaliculatus*			x			Western blotting, miRNA mimic and inhibitor, siRNA, luciferase assay
Zhu et al. (2020)	*Oncorhynchus mykiss*			x			

been done (Kaitetzidou et al., 2015; Rasal et al., 2016; Fernández et al., 2017; Herkenhoff et al., 2018; Sarropoulou et al., 2019; Zhou et al., 2019; and here, see later), much more remains to come in the nearest future.

Since their discovery, the potential roles of miRNAs in different biological processes have called the attention of researchers. Considering the Web of Science (WoS) repository, 22,218 articles were published regarding the role of miRNAs in human physiology, 4752 in mice, and 553 in fish (just representing less than 2.5% of the articles in humans; Fig. 4). Most articles have been published in the last decade, probably due to the advent of specific tools, such as methods for synthesis of miRs cDNA and/or development and implementation of OMIC technologies such as small RNA-Seq. In particular, regardless of the species considered, research has been focused on studying miRNAs in tissues, while only recent attention has been paid to circulating miRNAs, either included in whole plasma or specifically packaged in plasma exosomes. In this sense, from the total 553 studies on miRNAs in fish species compiled

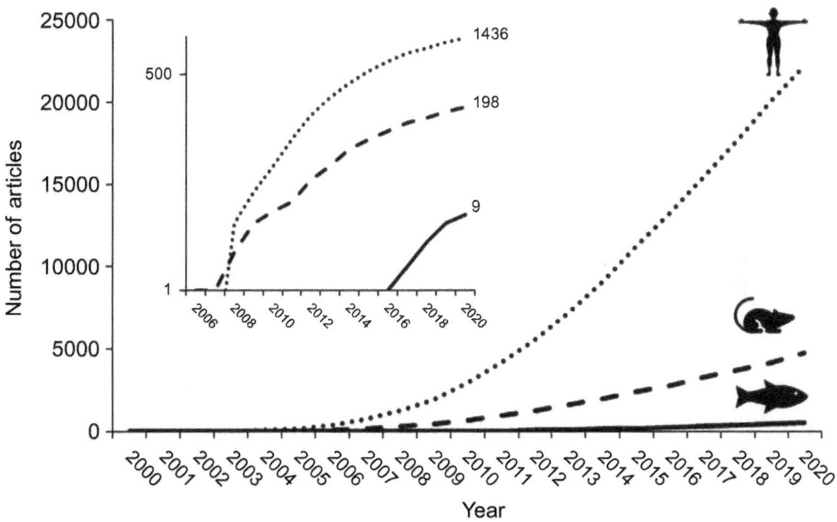

FIG. 4 Annual trend of studies investigating miRNAs in human, mouse, and fish species, compiled in Web of Science (WoS) repository. *Dotted lines*, articles in humans; *dashed lines*, articles in mouse; *solid lines*, articles in fish species. Embedded figure: Studies specifically conducted in circulating miRNAs within WoS repository in logarithmic scale (total number of articles published until 2021 are indicated). In order to collect all literature over noncoding RNAs, and particularly on miRNAs, a keyword bibliographic research was conducted in order to evidence the increased scientific attention on miRNAs during the two last decades. Advanced research was performed with "miRNAs," "noncoding RNAs," and "plasma" terms combined with the OR conjunction within the title, abstract, authors, keywords, and keywords plus fields, using the TS (=topic) tag, and the output was refined with "human," "mouse," or "fish" terms. Search criteria consider only research studies (reviews and research articles) published in English from 2000 to 2020. For studies in fish, a detailed revision of the literature was done to discard studies reporting fluorescence in situ hybridization (FisH) or RNA-FisH techniques.

in WoS repository, only 9 (less than 2% of the total number of articles in fish) were dealing with circulating miRNAs. In this section, we will enumerate the most relevant research studies regarding the role of miRNAs in processes such as development, reproduction and sex differentiation, adaptation to different stressors (including pollutants) and environmental signals, immunocompetence and disease resistance, and/or fish metabolism and nutrition.

4.1 miRNAs in fish development

Since fertilization, miRNAs participate in the control of vertebrate development and homeostasis (Wienholds and Plasterk, 2005). Studies of miRNAs in teleost fish species, as a taxonomic group central in understanding vertebrate evolution, ecologically, economically, and nutritionally important for human beings, and an interesting model species for biomedical research, have been extensively reviewed (see Fjose and Zhao, 2010; Bizuayehu and Babiak, 2014; Bhattacharya et al., 2017). Thus here we will only refer to some of the most representative and recent works.

The first study to identify and explore the expression pattern of miRNAs in fish species was Lim et al. (2003). At that time, 38 miRNAs were detected in zebrafish embryos, revealing their importance in specifying cell differentiation and developmental patterning. The number of miRNAs in zebrafish was later expanded to 115, with its temporal and spatial expression determined through the use of microarrays and ISH (Wienholds et al., 2005). The highly tissue-specific expression of miRNAs during segmentation and later stages, but not early in development, suggested that they might be more involved in differentiation or maintenance of tissue identity rather than in tissue fate establishment. Nevertheless, shortly after these initial studies, definitive proof of the vital role of miRNAs in vertebrate morphogenesis was provided by the defects during the gastrulation, brain formation, somitogenesis, and heart development observed in the maternal and zygotic Dicer mutant zebrafish embryos (Giraldez et al., 2005). Since then, the key role of the miRNAs in early embryogenesis was fully expanded, being involved in processes such as maternal transcript clearance among others (Giraldez et al., 2006). Subsequent studies demonstrated the key roles of particular miRNAs in specific processes during fish development. For instance, members of the miR-200 family were found to be required for olfactory neurogenesis (Choi et al., 2008); miR-375 to be involved in the formation of the insulin-secreting pancreatic islet (Kloosterman et al., 2007); miR-145 directing intestinal maturation (Zeng et al., 2009); miR-138 in heart development, delineating the spatial extent of atrioventricular canal (Morton et al., 2008); miR-9 promoting neurogenesis (Leucht et al., 2008); miR-140 in correcting craniofacial morphogenesis (Eberhart et al., 2008); and/or miR-214 controlling chondrogenesis through chondrocyte differentiation (Roberto et al., 2018). These roles and their underlying mechanisms seem to be evolutionary conserved, as it was demonstrated for miR-2014, probably by targeting the *activating transcription factor 4* (*atf4*; Roberto et al., 2018). In this sense, fish model species such as zebrafish or medaka (*Oryzias latipes*) were essential to get further insights on how miRNAs control development due to both their known biological features (e.g., highly prolific and short lifecycle, among others) and the constant implementation of powerful and novel biotechnological tools (e.g., easy genetic manipulation; Howe et al., 2013; MacRae and Peterson, 2015; Detrich III et al., 2016). However, the existence of novel roles or the lack of conservation within fish and/or vertebrates should not be neglected.

In nonmodel fish species, miRNAs are associated with different phases of development. Fourteen miRNAs were identified in the early embryogenesis (maternal to zygote transition) of rainbow trout (*Oncorhynchus mykiss*) that might guide maternal mRNAs for degradation, a required process for normal embryonic development (Ramachandra et al., 2008). In the embryonic development of Nile tilapia (*Oreochromis niloticus*) miR-10 has been shown to modulate different *hox* genes (Giusti et al., 2016). Furthermore, in the same fish species, miR-4585 was upregulated in male embryos at 2- and 5-days postfertilization (dpf) and decreased at 9 dpf (Eshel et al., 2014). Since miR-4585 seems to target *cyp19a1a* (the major gene for female determination in fish species), the potential significance of this miRNA in sex determination soon after fertilization was evidenced. Bizuayehu et al. (2012a) identified 201 different miRNAs to be expressed along with the Atlantic halibut (*Hippoglossus hippoglossus*) development, particularly on maternal to zygote transition, organogenesis, first feeding and metamorphosis, with some isomiRs (e.g., miR-301c, miR-17a, and miR19b) showing different or even opposing expression patterns. Metamorphosis is a dramatic transformation from a bilaterally symmetrical to an asymmetrical body shape that flatfish species must undergo to acquire the adult morphology (Geffen et al., 2007). This is occurring under the control of thyroid hormones (Campinho, 2019). In Japanese flounder (*Paralichthys olivaceus*), the expression of 140 conserved miRNAs was also characterized through microarray technology, with 66 differentially expressed at two metamorphic stages (17- and 29-days posthatching), reflecting the intricate mechanisms and profound changes involved in this process (Fu et al., 2011). Furthermore, abnormal or incomplete metamorphosis might lead to the appearance of albino fish, a quite important issue in farmed fish since it degrades fish quality perception from consumers and thus, market price. Wang et al. (2019a) showed that induced albinism in Japanese flounder larvae through triiodothyronine hormone exposure was associated with a differential expression of 12 miRNAs. Interestingly, some of the bioinformatically predicted mRNAs targeted by these miRNAs were involved in the melanin pigment production process. Furthermore, RNA-Seq and comparative genomics analyses reported almost 600 miRNAs in Nile tilapia, with a particular arm switching (shift in the ratio of mature 5p and 3p strands that get loaded into the miRISC) of 9 miRNAs in early development, adult stage, and even between males and females (Pinhal et al., 2018). More recently, the different small ncRNAs expressed in 10 different developmental stages (from morula to all fins) of the European seabass (*Dicentrarchus labrax*) were identified, miRNAs being the most of the annotated transcripts. Furthermore, they have the most abundant expression after the earlier stages (morula and 50% epiboly) and at the later stages (first feeding, flexion, and all fins; Sarropoulou et al., 2019). Five miRNAs in particular (miR-155, miR-430a, d1, d2, and miR-458), had a higher abundance only in the morula stage, suggesting a possible relevant role during embryonic development.

All these studies, in addition to those describing the miRNA repertoire in different fish species such as bighead carp (*Hypophthalmichthys nobilis*) and silver carp (*Hypophthalmichthys molitrix*; Chi et al., 2011), half-smooth tongue sole (*Cynoglossus semilaevis*; Huang et al., 2016), turbot (*Scophthalmus maximus*; Robledo et al., 2017), tambaqui (*Colossoma macropomum*; Gomes et al., 2017), or blunt snout bream (*Megalobrama amblycephala*; Huang et al., 2018), clearly widen the complexity of mechanisms regulating fish ontogenetic development and its relationship with miRNA expression.

4.2 miRNAs in reproduction and sex differentiation

In vertebrates, but mostly in mammals, the role of miRNA in sex determination, gametogenesis, and sexual maturation has been revealed (see reviews of Robles et al., 2019; Salas-Huetos et al., 2019).

Particular miRNA-mRNA interactions are crucial for the development of the reproductive system in fish species. Primordial germ cells (PGCs) are originated during early embryogenesis and migrate to the future genital ridges to finally become gonocytes that are posteriorly differentiated to spermatogonia or oogonia (reviewed in Schulz et al., 2010). The interaction of Dead end (Dnd), an RNA-binding protein, with miR-430 is essential for the proper migration, maintenance, and survival of PGCs, where Dnd protects the 3′-UTR-binding sites of *nanos*, *tdrd7*, and *hub* genes from miR-430 (Kedde et al., 2007).

In fish, the miRNome of the female and male reproductive axis (brain, liver, and gonads) from different species has been characterized. Let-7 and miR-21 were the most abundantly expressed in rainbow trout eggs, followed by miR-24, miR-202, miR-148, miR-30, miR-10, miR-146, miR-25, and miR-143 (Ma et al., 2012), which provided a starting point for the better knowledge and understanding of the roles of miRNAs in fish reproduction, controlling egg quality and early embryogenesis in rainbow trout. In Atlantic halibut, the first insight into miRNA involvement in sexual development was provided through the identification of tissue and sexually dimorphic expression of several miRNAs, including miR-29a, miR-34, miR-143, miR-145, miR-202-3p, miR-451, and miR-2188 (Bizuayehu et al., 2012b). While miR-9 was highly expressed in the brain, miR-202 was abundantly expressed in gonads. Moreover, a comparative study of females and males as well as juveniles subjected to a masculinization process (with 17-α-methyltestosterone) identified the differential expression of miR-202-3p in gonads (downregulated in female juveniles when compared to males or masculinized juveniles), suggesting miR-202-3p might be involved in fish sex differentiation and gametogenesis success. Similar results have been obtained in yellow catfish (*Pelteobagrus fulvidraco*) gonad tissues from XX female, XY male, and YY supermale (higher degree of testis maturity) specimens, with low expression of several miR-200 family members in female gonads (Jing et al., 2014). Li et al. (2016) also identified miRNAs expressed in the tissues from the reproductive axis of male and female marine medaka (*Oryzias melastigma*). While miR-109 and miR-124 were highly expressed in the brain, miR-628 and miR-1060 were abundant in gonad tissues, although with higher expression in testes. A qPCR analysis of let-7 family members has shown differential expression in ovary, pituitary, and brain tissues along with the ovary maturation in blunt snout bream (Lan et al., 2019), suggesting functional conservation of this miRNA in fish reproduction. Wider approaches (RNA-Seq) discovered other different miRNAs involved in gonad maturation. In Nile tilapia, Wang et al. (2016) identified 76 differentially expressed miRNAs when comparing the miRNome of testes and ovaries from young specimens. Among them, miR-17-5p and miR-20a (highly expressed in the ovaries) were predicted to target *dmrt1*, which was found to be downregulated and suggested to induce estrogen production. In contrast, in males, lower expression of miR-138, miR-338, miR-200a and miR-456 and miR-138 was consistent with the higher expression of their respective targets (*cyp17a2* and *amh*), suggesting their key role in spermatogonial cell proliferation and spermatogenesis. In addition to let-7, miR-9, and miR-202 miRNA families, other sex-biased miRNAs were also found differentially expressed in gonads from males and females of different fish species such as Japanese flounder (Gu et al., 2014), Nile tilapia (Xiao et al., 2014; Tao et al., 2016), dark sleeper (*Odontobutis potamophila*; Zhao et al., 2017), yellowfin seabream (*Acanthopagrus latus*; Li et al., 2020), discus fish (*Symphysodon aequifasciatus*; Fu et al., 2020), and tiger pufferfish (*Takifugu rubripes*; Yan et al., 2021). According to Bhat et al. (2020), let-7a-5p, miR-143-3p, miR-181-5p, miR-202-5p, and miR-21-5p seemed to be the most important and conserved sex-biased miRNAs when the works performed in zebrafish, Japanese flounder, Nile tilapia, rainbow trout, and marine medaka were considered.

Regarding studies on gonad maturation, among the 8765 miRNAs identified in the miRNome from the gonad/ovary of the common carp (*Cyprinus carpio*), 150, 628, and 431 miRNAs were specifically expressed in the primordial gonad, juvenile, and adult ovary (respectively; Wang et al., 2017b). In particular, miR-6758-3p, miR-3050-5p, and miR-2985-3p were highly expressed in the primordial gonad; miR-3544-5p, miR-6877-3p, and miR-9086-5p in the juvenile ovary; and miR-154-3p, miR-5307-5p, and miR-3958-3p in the adult ovary, enriching the set of miRNAs involved in fish female maturation. When comparing the miRNome of small early vitellogenic (or stage IIIa; unable to undergo oocyte maturation) follicles and oocytes in mid- to late vitellogenic (stage IIIb; able to become mature upon luteinizing hormone stimulation) follicles of zebrafish, 11 miRNAs and 13 miRNAs were significantly up- and down-regulated in the stage IIIb follicular cells, respectively (Zayed et al., 2019). More recently, 10 miRNAs were found to be differentially expressed between mature and immature female sharp snout seabream (*Diplodus puntazzo*) gonads (Papadaki et al., 2020). While miR-125c and miR-24 were upregulated in immature gonads, miR-451, miR-7a, miR-122-1, miR190a, miR129, ENSGACT00000029608, ENSGACT00000029489, and ENSGACT00000029667 were expressed at higher levels in mature gonads. Among them, only miR-451 and miR-190a were commonly found as differentially expressed in both studies, although in an opposite direction. Nevertheless, the mRNA targets predicted in Papadaki et al. (2020) were associated with oocyte development, such as steroidogenesis, proteolysis, and apoptosis, while the ones in Zayed et al. (2019) were related to MAPK signaling and endocytosis pathways.

A comparative analysis of the miRNome from immature, prepubertal, and pubertal Atlantic salmon (*Salmo salar*) testis linked specific miRNAs (miR-101, miR-137, miR-92b, miR-18a, and miR-20a) with testis maturation (Skaftnesmo et al., 2017). Furthermore, these differentially expressed miRNAs were inversely correlated with the expressed targets corresponding with the progesterone signaling and cell cycle pathways. Further complex analysis in zebrafish, including samples from five different developmental stages of male and female gonads, as well as samples of spermatozoa and unfertilized eggs, not only identified sex-biased miRNAs but also correlated with the miRNome of zebrafish gametes (Presslauer et al., 2017). Expression of particular miRNAs was also found to correlate with gamete quality. Ma et al. (2015) found that miR-193b-3p, miR-203c-3p, miR-499-5p, miR-7550-3p, miR-nov-95-5p, and miR-nov-112-5p showed significantly higher expression in aging rainbow trout eggs, this being the first association of the expression of particular miRNAs with egg quality. Regarding sperm quality, a different expression of let-7-d and miR-200a-5p has been found in wild and reared in captivity Senegalese sole individuals, bringing new light on the potential causes of the dysfunctional reproduction of this species when reared in captivity (Riesco et al., 2019). Moreover, the transgenerational inheritance of sex in pseudo males from half-smooth tongue sole has been associated with the lower expression of miR-26a-5p, miR-27b-3p, miR-125b-5p, and miR-199a-5p in pseudo male sperm (Zhao et al., 2021). Nevertheless, although sex-, maturation-, and gamete quality-biased miRNAs were identified in several fish species, and an in silico prediction of miRNA targets provides the first evidence of their role in fish reproduction and sex determination, the functional characterization of both conserved and novel miRNAs is still an important research gap in this topic (Bhat et al., 2020). In this sense, as previously mentioned, the implementation of gene editing technologies and/or other approaches would be invaluable. For example, using medaka as a model, Gay et al. (2018) confirmed the essential role of miR-202 in fish oogenesis. Females from a mutant line (using CRISPR/Cas9 genome editing) of the miR-202 gene do not produce eggs or at least the reduced number of eggs produced cannot be fertilized, leading to no reproductive success.

4.3 miRNAs involved in adaptation/response to different stressors and environmental signals

The molecular mechanisms underlying adaptive evolution have been historically centered on protein-coding sequence variation and/or DNA regulatory motifs (e.g., Kocher, 2004). However, more recent works suggest that miRNA sequences would also evolve (Franchini et al., 2019). Comparing the miRNA and mRNA sequences expressed at 1-day posthatch embryos of 5 cichlid species (*Amphilophus sagittae*, *Amphilophus amarillo*, *Amphilophus citrinellus*, *Amphilophus zaliosus*, and *Amphilophus astorquii*) exhibiting different benthic/limnetic behaviors, the authors found (1) many differentially expressed miRNA/gene pairs as well as (2) novel miRNAs among these species, suggesting a functional linkage among these pairs and novel genomic substrates, respectively, potentially leading to rapid phenotypic adaptation and divergence, even in the presence of gene flow. Similarly, when comparing sympatric species of the Japanese threespine stickleback (*Gasterosteus aculeatus*), the Pacific Ocean and Japan Sea sticklebacks, different expression patterns of small RNAs (miRNAs among others) were reported and considered the potential mechanism underlying differential mRNA expression and transposon activity (Kitano et al., 2013), and probably speciation. In this sense, both the concomitant transgenerational variation of the miRNA sequence and the one of the binding sequences on the mRNA targets should contribute to species evolution and adaptation to changing environments (see also Chapter 10).

Temperature is a key factor in fish development, growth, metabolism, physiology, and a signal triggering fish maturation and reproduction. Furthermore, considering the current context of climate change and global warming, studying how water temperature increase might affect fish is fundamental to perform a better risk assessment of this anthropogenic environmental alteration, as well as to define mitigating or correcting policies (IPCC, 2018). Increased water temperature during embryogenesis altered sex ratios (Ribas et al., 2017; Valdivieso et al., 2020), skeletal development (e.g., Di Santo, 2019), and promoted faster development and increased fish size (Campos et al., 2013a). Such phenotypic outputs seem to be due to an altered DNA methylation profile (Campos et al., 2013b; Valdivieso et al., 2020), among other processes. In particular, Campos et al. (2014) demonstrated that the previously reported greater fish growth under increased water temperature (21 °C versus 15 °C) during embryogenesis of Senegalese sole was associated with a higher expression of some miRNAs (e.g., miR-17a, miR-181-5p, and miR-206) and the downregulation of the miR-181a-3p. More recently, 25 liver miRNAs were found to be differentially expressed in common carp subjected to both cold (5 °C) and heat stress (30 °C) during 18 days (Sun et al., 2019b). Among them, miR-122, miR-30b, miR-15b-5p, miR-20a-5p, miR-1, and miR-7b were found to be associated with the regulation of glycolysis, targeting 12 genes involved in pathways related to pyruvate metabolism, glycolysis/gluconeogenesis, and the citrate cycle. In rainbow trout, acute heat stress (temperature raised from 16 °C to 22.5 °C in 2h and maintained for 48h) altered the expression of several miRs (17 up- and 5 downregulated), some of which were predicted to regulate the metabolism and the immune response (Zhou et al., 2019). This study provides new clues on how producers might reduce the economic losses caused by heat waves, a naturally occurring extreme event in which the frequency and intensity are predicted to be increased with climate change (IPCC, 2018).

Another environmental parameter to which anadromous and catadromous fish species have to deal with along their lifecycle is salinity. For anadromous species such as salmonids, smoltification (adaptation to seawater) is a critical developmental period with strong physiological and biochemical changes. When the Atlantic salmon miRNome of the head kidney before, during smoltification, and post seawater transfer was studied, 71 miRNAs were found to be differentially expressed (Shwe et al., 2020). Furthermore, 42 mRNAs involved in hormone biosynthesis, stress management, immune response, and ion transport, also found differentially expressed in the same tissue, were predicted to be targeted by these miRNAs, suggesting that posttranscriptional regulation of gene expression by miRNAs is important in smoltification and seawater adaption.

Different fish species can be adapted to highly diverging and extreme environments. This is the case of the shortfin molly (*Poecilia mexicana*) that can inhabit springs rich in toxic hydrogen sulfide (H_2S). A large population of mature miRNAs from gills was differentially expressed in *Poecilia mexicana* when compared to locally adapted populations in contrasting habitats (rich versus nonsulfidic springs; Kelley et al., 2021). In silico identification of differentially expressed miRNA-mRNA pairs revealed that some downregulated miRNAs targeted mRNAs involved in sulfide detoxification and cellular homeostasis, while upregulated miRNAs targeted specific mRNAs that act in the mitochondria (the main site of H_2S toxicity and detoxification) of fish inhabiting sulfidic environment. These results indicate that miRNAs may contribute to *Poecilia mexicana* adaptation to sulfidic environments.

Different studies reported miRNAs as a central part of the mechanisms in the toxicological outcomes of different compounds in several tissues and organs (reviewed in Freeman et al., 2013; Ahkin Chin Tai and Freeman, 2020). Liver injury due to microcystin-LR (a toxin produced by cyanobacteria) exposure altered the level of some miRs in different fish species (e.g., silver carp, whitefish (*Coregonus lavaretus*), or zebrafish) (Brzuzan et al., 2012; Li et al., 2013; Feng et al., 2017; Qu et al., 2018). Studies using a qPCR approach for particular miRNAs found that (1) miR-122 and let-7c were upregulated in whitefish liver exposed to microcystin-LR (Brzuzan et al., 2012); (2) let-7b, miR-21, miR-27b, miR-122, miR-125a, miR-143, and miR-148 have a dynamic expression along with time postmicrocystin-LR exposure in liver, spleen, and kidney of silver carp, although all of them showing an upregulation at specific sampling times (Feng et al., 2017); and (3) while miR-21 and miR-27b were upregulated, miR-122 and miR-148 were downregulated in zebrafish liver (Li et al., 2013).

Surprisingly, Qu et al. (2018) did not identify those above-mentioned miRNAs as even expressed in silver carp liver exposed to microcystin-LR when a small RNA-Seq approach was applied. Differences in the concentration of the toxin injected, sampling time, fish species and size should account for these discrepancies.

Other research works using high-throughput approaches identified several miRNAs to be differentially expressed when fish were exposed to different pollutants. For instance, exposure to acidic aluminum-rich water led to an altered expression of 18 miRNAs (14 up- and 4 downregulated) in Atlantic salmon muscle (Kure et al., 2013). Chronic exposure (3 weeks) to 100 nM of bisphenol A, an endocrine disruptor, altered the liver expression of 15 miRNAs which potentially targeted mRNAs involved in oxidative processes, defense response, transmembrane transportation, and cellular parts (Renaud et al., 2017). These results, combined with differential expression analysis of mRNAs in the same tissue, represented a signature associated with nonalcoholic fatty liver disease, oxidative phosphorylation, mitochondrial dysfunction, and cell cycle. Exposure to crude oil causes developmental cardiotoxicity in fish species. Characterization of the miRNome of exposed mahi-mahi (*Coryphaena hippurus*) larvae to oil predicted a disruption of cardiovascular system development due to an overexpression of miR-133, which was correlated with the reduced expression of *potassium voltage-gated channel subfamily H member 2* (*kchn2*; Diamante et al., 2017). Mutations in *kcnh2* have been demonstrated to cause ventricular arrhythmias (Keating and Sanguinetti, 2001), supporting the alteration of miR-133 expression in oil-exposed fish as one of the main mechanisms of abnormal cardiovascular development. In addition, upregulated miRNAs (miR-34a, miR-15, and miR-23) were correlated with *phosphatase and tensin homologue* and *neurogenic locus notch homologue protein 2* gene downregulation, which is involved in abnormal eye development and leads to cataracts and eye degeneration (Diamante et al., 2017). Furthermore, polycyclic aromatic hydrocarbons (PAHs) particularly present in crude oil can alter fish development. Larvae of red drum (*Sciaenops ocellatus*) exposed to different concentrations of specific water accommodated fractions (source oil (6.8, 13.7, and 35.9 μg/L total PAHs) and weathered slick oil (4.7, 8.1, and 18.0 μg/L total PAHs)) showed altered expression of miR-18a, miR-27b, and miR-203a across all exposure concentrations tested (Xu et al., 2019). In this case, targeted mRNAs were predominantly involved in the neuro-cardio system and nervous system developmental processes, consistent with the observed concentration-dependent morphological changes in the brain, iris, eye, and pericardial areas. Other endocrine-disrupting chemicals such as atrazine (widely used in agriculture) were also reported to affect larval development, altering the sex ratio of exposed common carp (Wang et al., 2019b). Depending on the developmental stage where the exposure to atrazine occurred, the effects were observed at early ovary differentiation or in the formation of germ cells, with the sex ratio biased toward females. Indeed, the expression of some miRNAs known to play a key role in gonadal maturation and sex differentiation in fish (let-7, miR-21, and/or miR-143) was found to be altered. Moreover, predicted targets of differentially expressed miRNAs were found to be related to several pathways involved in the reproductive process, namely Wnt signaling, Notch signaling, transforming growth factor-β signaling, p53 signaling, steroid hormone biosynthesis, and estrogen signaling pathway. In particular, atrazine can upregulate aromatase expression through the differentially expressed miRNAs.

Interestingly, less extreme alterations of the environment can also alter the expression of miRNAs. High stocking density is known to be a stressful condition in many farmed fish species. Rainbow trout stocked during 30 days at high density (40 kg/m^3) showed an upregulation of miR-181a-3p/5p, let-7a-5p, miR-10d-5p, and miR-30e-5p/3p, and a downregulation of miR-181b-5p and miR-100-5p in the brain, when compared to those of specimens reared at low density (3 kg/m^3; Gonçalves et al., 2020). This study reveals miRNAs as highly sensitive and suitable biomarkers of fish physiology and well-being, and with great potential to be industrially exploited. Similarly, dissolved oxygen content in rearing systems is another parameter that may affect fish physiology. In blunt snout bream exposed to hypoxia during 3 or 24 h, or to reoxygenation (24 h hypoxia followed by 3 h reoxygenation), 132, 120, and 174 liver miRNAs were differentially expressed when compared to fish under normoxia, most of them associated to oxidative stress (Sun et al., 2017a). Curiously, none of them was the miR-462/731 cluster, which has been shown to exhibit a modified expression as a response to hypoxia in the liver, spleen, gill, muscle, and brain of the same species (Huang et al., 2017). Differences in both experimental approaches (small RNA-Seq versus qPCR), fish size, exposure times, and oxygen levels might explain these discrepancies. Nevertheless, the deleterious effects of hypoxia on fish reproductive fitness through the disruption of the Brain-Pituitary-Gonad axis have been well documented. In marine medaka, among the 558 miRNAs found in the testis, 9 were up- and 5 downregulated by hypoxia (Tse et al., 2016). Bioinformatic analysis identified that many of the potentially targeted mRNAs were closely related to stress response, cell cycle, epigenetic modification, sugar metabolism, and cell motion. Furthermore, the hypoxia epigenetic regulator of transgenerational reproductive impairment (*euchromatic histone-lysine N-methyltransferase* 2) was found to be targeted by the downregulated miR-125-5p, establishing the underlying epigenetic mechanism of the transgenerational inheritance of reproductive dysfunction due to hypoxia on male fish. A similar approach has been conducted in marine medaka females, identifying 33 and 10 miRNAs up- and downregulated under hypoxia, respectively (Lai et al., 2016). A large number of hypoxia-suppressed miRNAs were predicted to target steroidogenic enzymes, including *steroidogenic acute regulatory protein*, *aromatase* and *17-alpha-monooxygenase*, *estrogen receptor* 2, and *androgen receptor*. These results reveal a novel pathway for female reproductive impairment due to hypoxia. Certainly, future studies will enlarge the set of fish responses to different environmental cues involving miRNAs.

4.4 miRNAs involved in fish immunocompetence and disease resistance

Several studies have focused on how miRNAs can promote the immune response of fish against pathogens, but also how they allow bacteria or viruses to escape from the immune system. Indeed, the role that the miRNAs play in fish immunity is suggested to be mediated through targeting pattern recognition receptors and downstream signaling factors. For a detailed description of the miRNAs involved in fish antiviral and antibacterial immunity, and their related pathways, the recent reviews of Wang et al. (2018), Andreassen and Høyheim (2017), and Zhou et al. (2021) are highly recommended for interested readers. Instead, some examples of how relevant pathogens altered the expression of miRNAs in fish tissues will be here briefly presented.

Ordas et al. (2013), applying a miRNA microarray to total RNA isolated from zebrafish embryos infected or not with *Salmonella typhimurium* and adults infected or not with *Mycobacterium marinum*, found the expression of miR-146a and miR-146b to be upregulated in infected fish. Furthermore, the induction of these miRNAs was demonstrated to be through the MyD88-Traf6 pathway. Surprisingly, miR-146 knockdown in zebrafish embryos infected with *Salmonella typhimurium* does not have a major effect on the transcription of pro-inflammatory and transcription factor genes, but altered lipid metabolism. More recently, with the implementation of new sequencing technologies, 51 and 16 differentially expressed miRNAs were found in grouper fin (GF-1) cells when infected or not with red-spotted grouper (*Epinephelus coioides*) nervous necrosis virus (RGNNV) at 3 and 24h postinfection, respectively (Chen et al., 2017). Furthermore, overexpression of four differentially expressed miRNAs (miR-1, miR-30b, miR-150, and miR-184) promoted the replication of RGNNV. Also, how the hepatic miRNome in Nile tilapia was altered by *Streptococcus agalactiae* infection in a longtime course (from 0h to 7day postinfection) was investigated (Gao et al., 2019b). A large number (from 21 to 50) of miRNAs involving apoptotic process, metabolic process, and immune responses were reported to be differentially expressed, evidencing a highly dynamic expression of miRNAs during pathogen infection. More recently, the miRNA cargo packaged in extracellular vesicles (EVs) released from primarily "monocyte-like" and primarily "macrophage-like" head kidney leukocytes from Atlantic salmon was characterized (Smith et al., 2020). A total of 19 differentially enriched miRNAs associated with macrophage differentiation and immune response were found, representing a first step toward the identification of EV biomarkers to monitor immune cell function, fish health, and/or response to disease.

Edwardsiella tarda is a severe fish pathogen with a broad host range. The expression of miRNAs from Japanese flounder kidneys was modulated at different stages of infection (Li et al., 2019). Ninety-six miRNAs differentially expressed were identified and their mRNA targets predicted, which at 12, 24, and 48h postinfection were related to T cell and B cell receptor signaling pathways, among others. In addition, the *sphingosine-1-phosphate receptor 1* (*PoS1PR1*) gene was negatively regulated by miR-182-5p. Furthermore, the overexpression of miR-182-5p promoted apoptosis and inhibited flounder cells' viability, while knockdown enhanced *Edwardsiella tarda* invasion and dissemination in fish tissues. More recently, another three differentially expressed miRNAs (miR-3p-2, miR-novel_547, and miR-novel_642) in Japanese flounder infected by *Edwardsiella tarda* were also functionally characterized (Guan et al., 2020; Li et al., 2020, 2021). The expression of *p53* was negatively regulated by miR-3p-2, promoting autophagy (Guan et al., 2020). MiR-novel_547 repressed the expression of *pten* (*phosphatase and tensin homologue*) and promoted *Edwardsiella tarda* invasion via the AKT/mTOR pathway (Li et al., 2020). Overexpression of miR-novel_642 or interference with the expression of its target, the *structural homologue of murine granulocyte colony-stimulating factor 3* (*pocsf3-1*), also promotes *Edwardsiella tarda* infection, most probably through the activation of autophagy (Li et al., 2021).

Vibrio anguillarum infection is also another relevant pathogen with important outbreaks in different farmed fish species. The expression of miR-21 in miiuy croaker (*Miichthys miiuy*) spleen was upregulated under lipopolysaccharide injection and *Vibrio anguillarum* infection (Chu et al., 2019). In vitro cell systems demonstrated that miR-21 suppressed inflammatory cytokine expression by targeting *IL-1 receptor-associated kinase 4* (*irak4*) gene. Turbot (*Scophthalmus maximus*), an important commercial fish species in Europe and China, also suffers *Vibrio anguillarum* infection. Gao et al. (2019a) studied the expression profiles of miRNAs in the intestine at 1, 4, and 12h postinfection, and 44 miRNAs were found to be differentially expressed, but miR-21 was not included among them. Potentially targeted mRNAs were involved in immune defense/evasion, inflammatory responses, RIG-I signaling pathway, and Toll-like receptor signaling pathway. Korean lamprey (*Lampetra morii*) is a key fish species in evolution (between invertebrates and vertebrates). A generated microRNA expression atlas in leukocytes under *Vibrio anguillarum* infection revealed that miR-4561 potentially regulates its innate immunity via interaction with *lamprey immune protein* (*LIP*), an immune factor expressed in the supraneural body tissue (Ma et al., 2021). Since overexpression of miR-4561 also induced apoptosis in embryonic cells, it was suggested that it might have a role in embryonic development too.

Viral infection has also been reported to induce a miRNA response in Atlantic cod (*Gadus morhua*) macrophages. Treatment with polyriboinosinic polyribocytidylic acid (polyI:C), a viral mimic, induced a differential expression of 4 (miR-731-3p, miR-125b-3-3p, miR-150-3p, and miR-462-3p) and 2 (miR-2188-3p and miR-462-3p) miRNAs at 24 and 72h poststimulation, respectively (Eslamloo et al., 2018). In the same study, other miRNAs known as immune responsive were also found to be differentially expressed.

All these studies open new avenues to better understand the molecular mechanisms of miRNA regulating the host-pathogen interactions in fish species.

Furthermore, their detailed characterization might allow identifying pathogen-specific signatures, allowing their use as reliable biomarkers of fish disease as well as the development of new immune therapies.

4.5 miRs involved in fish metabolism and nutrition

Feeding strategies and behavior in fish species are as much wide as can be expected from the most diverse taxonomic group. Depending on their mode of feeding (filter feeders, parasites, and predators as well as herbivorous and carnivorous) each one has an appropriately adapted digestive system, from a simple short tubular intestine to a more complete digestive system including a foregut (esophagus, stomach, and pylorus), midgut, and hindgut (Egerton et al., 2018). Nutrition and feeding have a great impact on fish growth, reproduction, and health, accounting for the 40%–70% of the fish farm costs. Thus, understanding how food intake, digestion, absorption, and transport of nutrients to the various tissues are determined and promoted is vital for fish farmers from an economic perspective. Here, we will compile the knowledge gained in this issue during the last decades.

Fish, during their early development, pass from a first endogenous feeding phase (acquiring all the nutrients required for normal development and growth from the yolk sac reserves) to an exogenous feeding phase (nutrients are absorbed from the caught living/inert preys/diets), including a mixed feeding mode in between. The first report on how miRNAs might be involved in fish nutrition processes showed that particular miRNAs (as well as their predicted mRNA targets) altered their expression along with this feeding mode transition in rainbow trout (Mennigen et al., 2013). The authors found that the use of external feeds was associated with an increased expression of miRNA-143, concomitant with a decreased expression of the *α/β hydrolasedomain 5* (*abhd5*) gene, a factor related to triglyceride hydrolysis, and suggesting a potential mechanism in the onset of lipid deposition. Similarly, muscle-specific expression of miR-1, miR-133, and *myogenic differentiation 1* (*myod1*) decreased in exogenously feeding alevins, consistent with the intense muscle hypertrophy for promoting muscle growth at such stage. Posteriorly, Bizuayehu et al. (2016) demonstrated how the miRNome was modulated by the type/quality of food at first feeding in Atlantic cod larvae. It is long known how, at first feeding, fish larvae fed natural zooplankton (mostly copepods) outperform fish fed aquaculture live preys (enriched rotifers and *Artemia salina*). The authors found 8 miRNAs (miR-9, miR-19a, miR-130b, miR-146, miR-181a, miR-192, miR-206, and miR-11,240) differentially expressed between these two feeding groups in at least one developmental stage of the six compared. Also, the predicted miRNA targets were found to be involved in metabolic, phototransduction, and signaling pathways. These two nutrigenomic studies highlighted how miRNA expression can be modulated by nutrition and open a new nutrigenomic research line in farmed species.

Different studies have advanced our knowledge on the potential role of miRNAs in fish nutrition and their potential to be used as reliable biomarkers (reviewed in Fernández et al., 2017; Zhou et al., 2021). Glucose and lipid contents have been shown to modulate the expression of several miRNAs. The dynamic expression of miR-122 and miR-33 in the liver of rainbow trout during fasting and refeeding time course was first associated with the insulin pathway (Mennigen et al., 2012). Afterward, the same group showed that injection of miRNA-122 inhibitor resulted in reduced expression of genes involved in hepatic lipid metabolism and decreased lipid availability (Mennigen et al., 2014). This preliminary identification of miRNAs related to glucose metabolism was further extended (by RNA-Seq approach) to 79, 124, and 77 differentially expressed miRNAs in the intestine, liver, and brain (respectively) from blunt snout bream specimens fed high or normal starch dietary content, miR-34a being one of them (Miao et al., 2017). The same authors showed afterward how intraperitoneal injection of a miR-34a inhibitor in blunt snout bream fed a high starch diet induced an altered hepatic expression of genes involved in glucose/lipid metabolic and biosynthetic processes (Miao et al., 2018). In Nile tilapia, another miRNA (miR-1338-5p) was found to modulate growth hormone secretion and glucose utilization (Qiang et al., 2017). More recently, a combined approach of small RNA-Seq and mRNA-Seq to determine the rainbow trout postprandial response to diets rich or devoid in carbohydrates, showed that dietary-induced hyperglycemia resulted in (1) the upregulation of hepatic transcripts involved in the biogenesis of miRNAs (e.g., *drosha*, *xpo5*, and *ago2*) and (2) the altered expression of up to 49 miRNAs (Kostyniuk et al., 2019). Furthermore, a comparative analysis of the hepatic miRNome in glucose tolerant (blunt snout bream) and intolerant (rainbow trout) species to high carbohydrate-rich diets revealed 11 hepatic miRNAs differentially regulated in common. The same research group, through a metaanalysis, suggested an evolutionary conserved (let-7, miRNA-27 family) and rewired (miRNA-30 family, miRNA-152 and miRNA-722) miRNA-metabolic target gene networks in salmonids (Kostyniuk and Mennigen, 2020).

A high-fat diet also induces hepatic regulation of different miRNAs (Zhang et al., 2014; Tang et al., 2019). In blunt snout bream fed high-fat versus normal-fat diets, 12 miRNAs were differentially expressed (Zhang et al., 2014). Among them, miR-30c and miR-30e-3p were upregulated and miR-145 and miR-15a-5p downregulated, having 6 putative lipid metabolism-related genes (*fetuin-B*, *Cyp7a1*, *NADH dehydrogenase (ubiquinone) 1 beta subcomplex subunit 2*, *3-oxoacid CoA transferase 1b*, *stearoyl-CoA desaturase*, and *fatty-acid synthase*) as targets. In Nile tilapia, miR-205-5p was suggested to be an important regulator of hepatic lipid metabolism, controlling the expression of *acetyl-CoA carboxylase* β (*acac*β; Tao et al., 2018). Similarly, miR-122 was also demonstrated to be involved in hepatic lipid metabolism in Nile tilapia, by targeting the *stearoyl-CoA desaturase* gene (Qiang et al., 2018). A downregulation of miR-16, miR-33a, miR-30, and miR-122 was related to increased hepatosomatic index, serum high-density lipoprotein-cholesterol (HDL-C) level, and liver lipid content in grass carp (*Ctenopharyngodon idella*) fed high-fat diet, concomitant with increased expression of *sterol regulatory element-binding transcription factor 1* (*srebp1*), *peroxisome proliferator-activated receptor gamma* (*pparγ*), *liver X receptor alpha* (*lxrα*), and *ATP-binding cassette transporter A1* (*abca1*; Tang et al., 2019). However, it should be noted that the interaction of miRNAs-mRNAs can be also the other way around. For example, the anchor protein repeat (Gankyrin) has been shown to upregulate the expression of miR-16, miR-27b, miR-122, and miR-126 in the zebrafish liver, promoting fat deposition (Her et al., 2011). Specific miRNAs have been reported to regulate the synthesis of long-chain polyunsaturated fatty acids (LC-PUFA). MiR-33 has been reported to control the expression of two enzymes (*Δ4* and *Δ6Δ5 fatty acyl desaturases*) that are central to LC-PUFA biosynthesis, as well as the posttranscriptional regulation of *insulin-induced gene 1* (*Insig1*), promoting LC-PUFA synthesis (Zhang et al., 2016; Sun et al., 2019a). MiR-146a inhibits LC-PUFA synthesis targeting *elongation of very long-chain fatty acids protein 5* gene (Chen et al., 2018), while miR-24 and miR-15/16 cluster increased LC-PUFA production through *insig1* and *pparγ* inhibition, respectively (Chen et al., 2019; Sun et al., 2020).

In fish, glucose and lipid metabolism are closely interconnected, with lipogenesis acting as a glucose sink in most fish species, and particularly in less tolerant fish species such as the salmonids (reviewed in Hemre et al., 2002). In this nutritional interconnection, cholesterol is known to have a central role in lipid deposition, and some miRNAs have been shown to regulate it. MiR-122 was found to be responsible for altered glucose homeostasis in rainbow trout, and its inhibition further resulted in a 20% decrease in plasma cholesterol concentration (Mennigen et al., 2014). Further bodies of evidence were recently provided. In rainbow trout, the expression of miR-1, miR-33a, miR-122, miR-128, and miR-223 was postprandially modulated in the liver, while the expression of miR-122, miR-128, and miR-223 in blood plasma was suggested as a potential biomarker of cholesterol metabolism in rainbow trout (Zhu et al., 2018). In yellow catfish, miR-144 mediates the changes of cholesterol metabolism induced by high-fat diets, which are via the direct regulation of the CCAAT/enhancer-binding protein alpha (C/EBPα) in the liver (Chen et al., 2020).

Other nutritional conditions have been reported to alter the expression profile of miRNAs. Feed additives such as probiotics can modulate the expression profile of miRNAs in rainbow trout brain, helping to cope with stressful conditions (Gonçalves et al., 2020) or improving grass carp growth by skeletal muscle increase (Li et al., 2017). Increasing levels of vitamin E in Nile tilapia diets increased the hepatic expression of miR-16, miR-21, miR-122, miR-125b, miR-146a, miR-155, miR-181a, and miR-223, which was related to the activity of superoxide dismutase (Tang et al., 2013). A comparative analysis of the liver miRNome of common carp fed diets formulated on different protein sources (animal and vegetable) that induce an altered redox status, also reported 5 miRNAs (miR-137-3p, miR-143-3p, miR-146a-5p, miR-21-5p, and miR-125b-5p) to be differentially expressed (Yang et al., 2019). The identification of common miRNA signatures related to redox status in different species and by using different (nutritional) approaches might reinforce the suitability and reliability of using miRNAs as potential biomarkers in fish physiology.

Considering the current shift on fish diet formulations to make aquaculture more sustainable, exploring the expression profile of some of those miRNAs involved in carbohydrate, lipid, and cholesterol metabolism might be useful to understand the potential effects of replacing fishmeal and fish oil with alternative raw materials and/or to define better alternatives to be considered. In this sense, Zhu et al. (2019) already reported how the hepatic expression of miRNAs related to cholesterol metabolism (miR-33a, miR-122, and miR-128) was affected by total fishmeal and fish oil replacement by vegetable sources in rainbow trout. Furthermore, the altered expression of miR-33a by vegetable diets in rainbow trout was further confirmed afterward in Zhu et al. (2020), where the altered expression of miR-33a was also found in blood plasma, further confirming previous findings on the potential use of circulating miRNAs as a less invasive biomarker in fish physiology (Sun et al., 2017b; Fernández et al., 2019; see Section 5).

In general, the research effort of identifying, quantifying, and characterizing miRNAs in fish species (reviewed in Rasal et al., 2016; Fernández et al., 2017; Herkenhoff et al., 2018; Zhou et al., 2021; and here) opens new avenues to find, develop, and implement innovative solutions for old and recent bottlenecks in fish farming, but also to better understand how the miRNA system evolved within multicellular organisms.

5. Recent progress and future perspectives

In the present chapter, a recent overview of the biogenesis and the underlying mechanisms of miRNAs in different biological processes has been presented, particularly in fish species. The increased understanding of how miRNAs play a central role in gene posttranscriptional and DNA methylation regulation was only possible to the continuous development of analytical tools to identify, quantify, locate, predict, and validate the potential mRNA targets in a high-throughput manner during the last two decades. Nevertheless, although the number of studies specifically performed on this topic in fish species has increased, they are still far from representing the relevance that fish have from an ecological, evolutionary, and economic perspective.

Regardless of the great progress on the different approaches applied to gain knowledge on the biological significance of miRNAs in fish biodiversity, development, and physiology, more reliable, high-throughput and cost-effective tools for identifying miRNAs, predicting mRNA targets and their functional characterization are still needed (Forero et al., 2019; Kern et al., 2020). As pointed by Coenen-Stass et al. (2018), small RNA-Seq remains challenging despite considerable improvements. In this sense, better and standardized miRNA isolation methods, library preparation, and normalization procedures should help to reduce bias and identifying common miRNA signatures of biological processes in different fish species. Also, although computational prediction tools allow getting first evidence of potential underlying mechanisms toward prioritizing miRNA candidates for experimental validation, there is still a huge uncertainty whether predicted mRNA targets are just mere false positives. Thus miRNA target predictions should be used with caution (Pinzón et al., 2017). In this sense, both the implementation of machine learning algorithms (see Riolo et al., 2021) and technologies that would allow the discovery of the particular features for an effective miRNA-mRNA interaction in fish species (e.g., Yartseva et al., 2017) such as CLIP-Seq and CLASH combined with CRISPR/Cas9 gene editing, will improve the reliability of their outputs and enhance future miRNA biomarker discovery. For the successful implementation of CLIP-Seq and CLASH methods, one of the limitations when working with fish species is the availability of antibodies capable of efficiently precipitating the Ago proteins. Since most available antibodies have been specifically designed for mammalian species, the procedure depends on a successful generation of custom antibodies (Fridrich et al., 2019), which is not always an easy task. Single-cell cosequencing of miRNAs and mRNAs also offers a great potential to infer the potential miRNA-mRNA interactions at the single cell context and the differentially regulated RISC targetomes (Kern et al., 2020). Furthermore, despite the needed improvements on the powerful high-throughput technologies, simple democratization of programs to design primers for the evaluation of miRNA expression by qPCR, particularly to study isomiRs in nonmodel fish species that are not currently targeted by commercially available methods (Forero et al., 2019), might be also advisable. Overcoming all these limitations would certainly help to increase our knowledge on how miRNA networks evolved in metazoans, and particularly in fish taxa. Most importantly, it will help to exploit the development and implementation of biotechnologies to expand the potential use of miRNAs either as biomarkers and/or therapies, not only to answer some scientific questions, but also to solve important industrial bottlenecks.

Fish is the most abundant and diverse animal taxa, considering the diverse set of ecological niches occupied, the feeding behaviors displayed, as well as their morphological adaptations and reproductive modes. The functional characterization will increase our understanding of how miRNAs are also responsible for phenotypic diversification. A recently conducted comparative analysis of miRNAs in four teleosts and one holostei species (Desvignes et al., 2021) reported poor evolutionary conservation of organ-enriched miRNAs in gonads, supporting the hypothesis that gonad miRNAs evolve with reproductive features in lineage-specific ways. This might explain the different (even opposed) profile of particular miRNAs during gonad maturation (see Section 4.2 regarding contradictory results reported on miR-451 and miR-190a expression by Zayed et al. (2019) and Papadaki et al. (2020)).

The significance of miRNAs studies in fish development and physiology is also beyond the key evolutionary branch that the fish species represents for a better understanding of metazoan evolution. Fish farming is nowadays representing the most promising sustainable and healthy way to cope with the future needs of animal protein sources for human consumption. In this sense, major gains in aquaculture since 2000s (reviewed in Naylor et al. (2021)) have been possible to constant knowledge advances on fish immunity and disease resistance, reproduction control, fish growth, feed digestion efficiency and fish nutrition, among others (e.g., biological knowledge and husbandry practices of farmed species). Many studies reported how miRNAs are involved in all these processes in fish species, but also some of them open new avenues to develop new strategies to cope with the new challenges that wild fisheries and aquaculture are facing (e.g., climate change adaptation and boosting aquafeeds sustainability). In this sense, the recent implementation of biotechnologies already applied in biomedicine is opening new exciting opportunities to improve reproductive performance and aquafeed formulation. The identification and quantification of miRNAs within biofluids are becoming routine monitoring procedures in human health as good, reliable and integrative biomarkers (Viereck and Thum, 2017; Tomasetti et al., 2019), but also recently translated into innovative therapies to treat diseases (Zhang et al., 2019b).

Very recently, the use of high-throughput studies in fish species has shown that nutritional and/or hormonal interventions induce an altered profile of particular circulating miRNAs from the blood plasma (Fernández et al., 2019; Oliveira et al., 2020). Nutritional supplementation of vitamin K (VK) in diets for Senegalese sole broodstock males altered the expression of different isomiRs of the let-7 and miR-146 family members (up- and downregulated, respectively), which was at the same time related with different degrees of sperm DNA fragmentation (Fernández et al., 2019). Bioinformatic analyses of predicted mRNA targets suggested different pathways might be involved in the underlying mechanism, particularly the gonadotropin-releasing hormone (GnRH) receptor pathway being one of them. GnRH acts centrally to mediate sexual behavior, as well as increasing gonadal steroids via the hypothalamic-pituitary-gonadal (HPG) axis to elicit courtship behavior (McGuire et al., 2010). Curiously, Senegalese sole males reared in captivity do not display correctly this courtship behavior. This study unexpectedly suggests an additional and complex regulatory mechanism of the reproductive performance by circulating miRNAs, where optimal dietary levels of VK might help to solve the reproductive dysfunction in Senegalese sole males. Posteriorly, the same research groups further confirmed how circulating miRNAs might be integrative biomarkers of reproductive status in Senegalese sole. Hormonal injection of Kisspeptin, acting upstream of the HPG axis, increased gonadotropins (Fsh and Lh) and testosterone plasma levels (Oliveira et al., 2020). Also, circulating let-7e, miR-199a-3p, and miR-100-5p were found to be differentially expressed in females, while miR-1-3p miRNA was upregulated in kisspeptin-treated males. In this case, in silico prediction of mRNA targets revealed that Kisspeptin treatment might affect paracellular transporters, regulate structural and functional polarity of cells, neural networks, and intracellular trafficking in Senegalese sole females. Both studies opened new research venues to understand how circulating mRNAs are involved in the fish HPG axis. Other studies of miRNAs in biofluids further supported their role in fish reproduction, particularly sperm maturation (Bizuayehu and Babiak, 2020), and/or extended the set of biological processes where differential expression of circulating miRNAs has been observed in fish species such as fish nutrition (Zhu et al., 2018) or response to pollutants (Ma et al., 2018).

In studies of miRNAs from biofluids, there are still some controversies and issues that need to be addressed (e.g., hemolysis; Blondal et al., 2013), as well as some limitations to overcome (e.g., methodological standardization or uncompleted mechanistic understanding of miRNA release and uptake; Iftikhar and Carney, 2016). Nevertheless, they represent an interesting and ethical source of integrative biomarkers with important technical advantages. Biofluids can be easily obtained without the need for animal sacrifice, thus the same individual can be monitored over time, reducing the variability of the evaluated response when different individuals are compared. Also, miRNAs have shown greater stability than mRNAs against different conditions (temperature, pH, etc.). Certainly, in the near future, improvements in biotechnological approaches will pave the way to achieve further exciting discoveries regarding the biological relevance of miRNAs in biofluids.

Acknowledgments

We would like to thank the financial support of the Spanish Ministry of Science Innovation and Universities (MICIU) and the European Regional Development Fund (ERDF) through MET2VI project (Ref. RTI2018-099029-A-I00). F.J.T.S. acknowledges the funding from the Mexican National Council for Science and Technology (CONACYT) for the postdoctoral fellowship No. 2019-000012-01EXTV-00292. E.S. acknowledges the funding from the Ministry of Education and Religious Affairs, under the Call "ARISTEIA I" of the National Strategic Reference Framework 2007–13 (ANnOTATE), cofunded by the European Union and the Hellenic Republic through the European Social Fund, the European Union Seventh Framework Program (FP7 2010–14) under grant agreement no. 265957 (CopeWell), and MedAID (Mediterranean Aquaculture Integrated Development) funded by the European Union in the frame of Horizon 2020, grant agreement number 727315. J.M.O.F. acknowledges the funding from the European Research Council (ERC) under the European Union's Horizon 2020 research and innovation program (grant agreement no. 683210) and from the Research Council of Norway under the Toppforsk program (grant agreement no 250548/F20). I.F. acknowledges the funding from the MICIU and the European Social Fund, "The European Social Fund invests in your future" through the Ramón y Cajal (Ref. RYC2018-025337-I) contract from the Plan Estatal de Investigación Científica y Técnica e Innovación 2017–20. F.J.T.S. and I.F. thanks the support from the network LARVAplus "Estrategias de desarrollo y mejora de la producción de larvas de peces en Iberoamerica" (117RT0521) funded by the Programa Iberoamericano de Ciencia y Tecnologia para el Desarrollo (CYTED).

References

Ahkin Chin Tai, J.K., Freeman, J.L., 2020. Zebrafish as an integrative vertebrate model to identify miRNA mechanisms regulating toxicity. Toxicol. Rep. 7, 559–570. https://doi.org/10.1016/j.toxrep.2020.03.010.

Ambros, V., Bartel, B., Bartel, D.P., Burge, C.B., Carrington, J.C., Chen, X., Dreyfuss, G., Eddy, S.R., Griffiths-Jones, S., Marshall, M., Matzke, M., Ruvkun, G., Tuschl, T., 2003. A uniform system for microRNA annotation. RNA 9, 277–279. https://doi.org/10.1261/rna.2183803.

Anders, S., Pyl, P.T., Huber, W., 2015. HTSeq—a Python framework to work with high-throughput sequencing data. Bioinformatics 31, 166–169. https://doi.org/10.1093/bioinformatics/btu638.

Andreassen, R., Høyheim, B., 2017. miRNAs associated with immune response in teleost fish. Dev. Comp. Immunol. 75, 77–85. https://doi.org/10.1016/j.dci.2017.02.023.

Bartel, D.P., 2018. Metazoan MicroRNAs. Cell 173, 20–51. https://doi.org/10.1016/j.cell.2018.03.006.

Basu, S., Müller, F., Sanges, R., 2013. Examples of sequence conservation analyses capture a subset of mouse long non-coding RNAs sharing homology with fish conserved genomic elements. BMC Genomics 14, S14. https://doi.org/10.1186/1471-2105-14-S7-S14.

Benetti, R., Gonzalo, S., Jaco, I., Muñoz, P., Gonzalez, S., Schoeftner, S., Murchison, E., Andl, T., Chen, T., Klatt, P., Li, E., Serrano, M., Millar, S., Hannon, G., Blasco, M.A., 2008. A mammalian microRNA cluster controls DNA methylation and telomere recombination via Rbl2-dependent regulation of DNA methyltransferases. Nat. Struct. Mol. Biol. 15, 268–279. https://doi.org/10.1038/nsmb.1399.

Berezikov, E., Chung, W.J., Willis, J., Cuppen, E., Lai, E.C., 2007. Mammalian mirtron genes. Mol. Cell 28, 328–336. https://doi.org/10.1016/j.molcel.2007.09.028.

Betel, D., Koppal, A., Agius, P., Sander, C., Leslie, C., 2010. Comprehensive modeling of microRNA targets predicts functional non-conserved and non-canonical sites. Genome Biol. 11. https://doi.org/10.1186/gb-2010-11-8-r90.

Bhat, R.A., Priyam, M., Foysal, M.J., Gupta, S.K., Sundaray, J.K., 2020. Role of sex-biased miRNAs in teleosts—a review. Rev. Aquac. 13, 269–281. https://doi.org/10.1111/raq.12474.

Bhattacharya, M., Sharma, A.R., Sharma, G., Patra, B.C., Nam, J.S., Chakraborty, C., Lee, S.S., 2017. The crucial role and regulations of miRNAs in zebrafish development. Protoplasma 254, 17–31. https://doi.org/10.1007/s00709-015-0931-1.

Bizuayehu, T.T., Babiak, I., 2014. MicroRNA in teleost fish. Genome Biol. Evol. 6, 1911–1937. https://doi.org/10.1093/gbe/evu151.

Bizuayehu, T.T., Babiak, I., 2020. Heterogenic origin of microRNAs in Atlantic salmon (*Salmo salar*) seminal plasma. Int. J. Mol. Sci. 21, e2723, https://doi.org/10.3390/ijms21082723.

Bizuayehu, T.T., Lanes, C.F.C., Furmanek, T., Karlsen, B.O., Fernandes, J.M.O., Johansen, S.D., Babiak, I., 2012a. Differential expression patterns of conserved miRNAs and isomiRs during Atlantic halibut development. BMC Genomics 13, 11. https://doi.org/10.1186/1471-2164-13-11.

Bizuayehu, T.T., Babiak, J., Norberg, B., Fernandes, J.M.O., Johansen, S.D., Babiak, I., 2012b. Sex-biased miRNA expression in Atlantic halibut (*Hippoglossus hippoglossus*) brain and gonads. Sex. Dev. 6, 257–266. https://doi.org/10.1159/000341378.

Bizuayehu, T.T., Furmanek, T., Karlsen, Ø., Van Der Meeren, T., Edvardsen, R.B., Rønnestad, I., Hamre, K., Johansen, S.D., Babiak, I., 2016. First feed affects the expressions of microRNA and their targets in Atlantic cod. Br. J. Nutr. 115, 1145–1154. https://doi.org/10.1017/S0007114516000155.

Blondal, T., Nielsen, S.J., Baker, A., Andreasen, D., Mouritzen, P., Teilum, M.W., Dahlsveen, I.K., 2013. Assessing sample and miRNA profile quality in serum and plasma or other biofluids. Methods 59, S1–S6. https://doi.org/10.1016/j.ymeth.2012.09.015.

Brown, R.A.M., Epis, M.R., Horsham, J.L., Kabir, T.D., Richardson, K.L., Leedman, P.J., 2018. Total RNA extraction from tissues for microRNA and target gene expression analysis: not all kits are created equal. BMC Biotechnol. 18, 16. https://doi.org/10.1186/s12896-018-0421-6.

Brzuzan, P., Woźny, M., Wolińska, L., Piasecka, A., 2012. Expression profiling *in vivo* demonstrates rapid changes in liver microRNA levels of whitefish (*Coregonus lavaretus*) following microcystin-LR exposure. Aquat. Toxicol. 122–123, 188–196. https://doi.org/10.1016/j.aquatox.2012.07.001.

Bustin, S.A., Benes, V., Garson, J.A., Hellemans, J., Huggett, J., Kubista, M., Mueller, R., Nolan, T., Pfaffl, M.W., Shipley, G.L., Vandesompele, J., Wittwer, C.T., 2009. The MIQE guidelines: minimum information for publication of quantitative real- time PCR experiments. Clin. Chem. 55, 611–622. https://doi.org/10.1373/clinchem.2008.112797.

Campinho, M.A., 2019. Teleost metamorphosis: the role of thyroid hormone. Front. Endocrinol. 10, 1–12. https://doi.org/10.3389/fendo.2019.00383.

Campos, C., Valente, L.M., Conceicao, L.E., Engrola, S., Sousa, V., Rocha, E., Fernandes, J.M., 2013a. Incubation temperature induces changes in muscle cellularity and gene expression in Senegalese sole (*Solea senegalensis*). Gene 516, 209–217. https://doi.org/10.1016/j.gene.2012.12.074.

Campos, C., Valente, L.M., Conceicao, L.E., Engrola, S., Fernandes, J.M., 2013b. Temperature affects methylation of the *myogenin* putative promoter, its expression and muscle cellularity in Senegalese sole larvae. Epigenetics 8, 389–397. https://doi.org/10.4161/epi.24178.

Campos, C., Sundaram, A.Y.M., Valente, L.M.P., Conceição, L.E.C., Engrola, S., Fernandes, J.M.O., 2014. Thermal plasticity of the miRNA transcriptome during Senegalese sole development. BMC Genomics 15, 1–13. https://doi.org/10.1186/1471-2164-15-525.

Chen, Y., Gelfond, J.A., Mcmanus, L.M., Shireman, P.K., 2009. Reproducibility of quantitative RT-PCR array in miRNA expression profiling and comparison with microarray analysis. BMC Genomics 10, 407. https://doi.org/10.1186/1471-2164-10-407.

Chen, C.Y., Cheng, Y.Y., Yen, C.Y.T., Hsieh, P.C.H., 2017. Mechanisms of pluripotency maintenance in mouse embryonic stem cells. Cell. Mol. Life Sci. 74, 1805–1817. https://doi.org/10.1007/s00018-016-2438-0.

Chen, C., Zhang, J., Zhang, M., You, C., Liu, Y., Wang, S., Li, Y., 2018. miR-146a is involved in the regulation of vertebrate LC-PUFA biosynthesis by targeting elovl5 as demonstrated in rabbitfish *Siganus canaliculatus*. Gene 676, 306–314. https://doi.org/10.1016/j.gene.2018.08.063.

Chen, C., Wang, S., Zhang, M., Chen, B., You, C., Xie, D., Liu, Y., Monroig, Ó., Tocher, D.R., Waiho, K., Li, Y., 2019. miR-24 is involved in vertebrate LC-PUFA biosynthesis as demonstrated in marine teleost *Siganus canaliculatus*. Biochim. Biophys. Acta Mol. Cell Biol. Lipids 1864, 619–628. https://doi.org/10.1016/j.bbalip.2019.01.010.

Chen, G., Wu, K., Zhao, T., Ling, S., Liu, W., Luo, Z., 2020. MiR-144 mediates high fat-induced changes of cholesterol metabolism via direct regulation of C/EBPα in the liver and isolated hepatocytes of yellow catfish. J. Nutr. 150, 464–474. https://doi.org/10.1093/jn/nxz282.

Chi, W., Tong, C., Gan, X., He, S., 2011. Characterization and comparative profiling of miRNA transcriptomes in bighead carp and silver carp. PLoS One 6, e23549, https://doi.org/10.1371/journal.pone.0023549.

Choi, P.S., Zakhary, L., Choi, W.Y., Caron, S., Alvarez-Saavedra, E., Miska, E.A., McManus, M., Harfe, B., Giraldez, A.J., Horvitz, R.H., Schier, A.F., Dulac, C., 2008. Members of the miRNA-200 family regulate olfactory neurogenesis. Neuron 57, 41–55. https://doi.org/10.1016/j.neuron.2007.11.018.

Chu, C.P., Nabity, M.B., 2019. Comparison of RNA isolation and library preparation methods for small RNA sequencing of canine biofluids. Vet. Clin. Pathol. 48, 310–319. https://doi.org/10.1111/vcp.12743.

Chu, Q., Yan, X., Liu, L., Xu, T., 2019. The inducible microRNA-21 negatively modulates the inflammatory response in teleost fish via targeting IRAK4. Front. Immunol. 10, 1–12. https://doi.org/10.3389/fimmu.2019.01623.

Coenen-Stass, A.M.L., Magen, I., Brooks, T., Ben-Dov, I.Z., Greensmith, L., Hornstein, E., Fratta, P., 2018. Evaluation of methodologies for microRNA biomarker detection by next generation sequencing. RNA Biol. 15, 1133–1145. https://doi.org/10.1080/15476286.2018.1514236.

Collins, L.J., Schönfeld, B., Chen, X.S., 2011. The epigenetics of non-coding RNA. In: Handb. Epigenetics, pp. 49–61, https://doi.org/10.1016/B978-0-12-375709-8.00004-6.

Conesa, A., Madrigal, P., Tarazona, S., Gomez-Cabrero, D., Cervera, A., McPherson, A., Szcześniak, M.W., Gaffney, D.J., Elo, L.L., Zhang, X., Mortazavi, A., 2016. A survey of best practices for RNA-seq data analysis. Genome Biol. 17, 1–19. https://doi.org/10.1186/s13059-016-0881-8.

Crick, F., 1970. Central dogma of molecular biology. Nature 227, 561–563. https://doi.org/10.1038/227561a0.

Dard-Dascot, C., Naquin, D., d'Aubenton-Carafa, Y., Alix, K., Thermes, C., van Dijk, E., 2018. Systematic comparison of small RNA library preparation protocols for next-generation sequencing. BMC Genomics 19, 1–16. https://doi.org/10.1186/s12864-018-4491-6.

Desvignes, T., Sydes, J., Montfort, J., Bobe, J., Postlethwait, J.H., 2021. Evolution after whole-genome duplication: teleost microRNAs. Mol. Biol. Evol. 38, 3308–3331. https://doi.org/10.1093/molbev/msab105.

Detrich III, H., Westerfield, M., Zon, L., 2016. The Zebrafish: Cellular and Developmental Biology, Part B Developmental Biology. Methods in Cell Biology, fourth ed. Academic Press.

Dexheimer, P.J., Cochella, L., 2020. MicroRNAs: from mechanism to organism. Front. Cell Dev. Biol. 8, 1–18. https://doi.org/10.3389/fcell.2020.00409.

Di Santo, V., 2019. Ocean acidification and warming affect skeletal mineralization in a marine fish. Proc. R. Soc. B Biol. Sci. 286, 20182187, https://doi.org/10.1098/rspb.2018.2187.

Diamante, G., Xu, E.G., Chen, S., Mager, E., Grosell, M., Schlenk, D., 2017. Differential expression of microRNAs in embryos and larvae of Mahi-Mahi (*Coryphaena hippurus*) exposed to deepwater horizon oil. Environ. Sci. Technol. Lett. 4, 523–529. https://doi.org/10.1021/acs.estlett.7b00484.

Didiano, D., Hobert, O., 2016. Perfect seed pairing is not a generally reliable predictor for miRNA-target interactions. Nat. Struct. Mol. Biol. 13, 849–851. https://doi.org/10.1038/nsmb1138.

Divisato, G., Passaro, F., Russo, T., Parisi, S., 2020. The key role of microRNAs in self-renewal and differentiation of embryonic stem cells. Int. J. Mol. Sci. 21, 1–21. https://doi.org/10.3390/ijms21176285.

Eberhart, J.K., He, X., Swartz, M.E., Yan, Y.L., Song, H., Boling, T.C., Kunerth, A.K., Walker, M.B., Kimmel, C.B., Postlethwait, J.H., 2008. MicroRNA miR-140 modulates Pdgf signaling during palatogenesis. Nat. Genet. 40, 290–298. https://doi.org/10.1038/ng.82.

Egerton, S., Culloty, S., Whooley, J., Stanton, C., Ross, R.P., 2018. The gut microbiota of marine fish. Front. Microbiol. 9, 1–17. https://doi.org/10.3389/fmicb.2018.00873.

Eshel, O., Shirak, A., Dor, L., Band, M., Zak, T., Markovich-Gordon, M., Chalifa-Caspi, V., Feldmesser, E., Weller, J.I., Seroussi, E., Hulata, G., Ron, M., 2014. Identification of male-specific amh duplication, sexually differentially expressed genes and microRNAs at early embryonic development of Nile tilapia (*Oreochromis niloticus*). BMC Genomics 15. https://doi.org/10.1186/1471-2164-15-774.

Eslamloo, K., Inkpen, S.M., Rise, M.L., Andreassen, R., 2018. Discovery of microRNAs associated with the antiviral immune response of Atlantic cod macrophages. Mol. Immunol. 93, 152–161. https://doi.org/10.1016/j.molimm.2017.11.015.

Fang, Y., Fullwood, M.J., 2016. Roles, functions, and mechanisms of long non-coding RNAs in cancer. Genom. Proteom. Bioinform. 14, 42–54. https://doi.org/10.1016/j.gpb.2015.09.006.

Fatica, A., Bozzoni, I., 2014. Long non-coding RNAs: new players in cell differentiation and development. Nat. Rev. Genet. 15, 7–21. https://doi.org/10.1038/nrg3606.

Feng, Y., Ma, J., Xiang, R., Li, X., 2017. Alterations in microRNA expression in the tissues of silver carp (*Hypophthalmichthys molitrix*) following microcystin-LR exposure. Toxicon 128, 15–22. https://doi.org/10.1016/j.toxicon.2017.01.016.

Fernández, I., Álvarez-González, C.A., Tovar-Ramírez, D., Galaviz, M., 2017. Non-coding RNAs: uncovering their potential relevance in fish nutrition. In: Cruz Suárez, L.E., Ricque Marie, D., Tapia Salazar, M., Nieto López, M.G., Villarreal Cavazos, D.A., Gamboa, J., Rivas Vega, M.E., Miranda, A. (Eds.), Investigación y Desarrollo en Nutrición Acuícola. Avances En Nutrición Acuicola, Nuevo León (México) https://nutricionacuicola.uanl.mx/index.php/acu/article/view/21.

Fernández, I., Fernandes, J.M.O., Roberto, V.P., Kopp, M., Oliveira, C., Riesco, M.F., Dias, J., Cox, C.J., Leonor Cancela, M., Cabrita, E., Gavaia, P., 2019. Circulating small non-coding RNAs provide new insights into vitamin K nutrition and reproductive physiology in teleost fish. Biochim. Biophys. Acta Gen. Subj. 1863, 39–51. https://doi.org/10.1016/j.bbagen.2018.09.017.

Fjose, A., Zhao, X.F., 2010. Exploring microRNA functions in zebrafish. N. Biotechnol. 27, 250–255. https://doi.org/10.1016/j.nbt.2010.02.017.

Forero, D.A., González-Giraldo, Y., Castro-Vega, L.J., Barreto, G.E., 2019. qPCR-based methods for expression analysis of miRNAs. Biotechniques 67, 192–199. https://doi.org/10.2144/btn-2019-0065.

Franchini, P., Xiong, P., Fruciano, C., Schneider, R.F., Woltering, J.M., Hulsey, C.D., Meyer, A., 2019. MicroRNA gene regulation in extremely young and parallel adaptive radiations of crater lake cichlid fish. Mol. Biol. Evol. 36, 2498–2511. https://doi.org/10.1093/molbev/msz168.

Freeman, J.L., Weber, G.J., Sepúlveda, M.S., 2013. Fishing for microRNAs in toxicology. In: microRNAs Toxicol. Med, pp. 49–75, https://doi.org/10.1002/9781118695999.ch4.

Fridrich, A., Hazan, Y., Moran, Y., 2019. Too many false targets for microRNAs: challenges and pitfalls in prediction of miRNA targets and their gene ontology in model and non-model organisms. Bioessays 41, 1–11. https://doi.org/10.1002/bies.201800169.

Fromm, B., Domanska, D., Høye, E., Ovchinnikov, V., Kang, W., Aparicio-Puerta, E., Johansen, M., Flatmark, K., Mathelier, A., Hovig, E., Hackenberg, M., Friedländer, M.R., Peterson, K.J., 2020. MirGeneDB 2.0: the metazoan microRNA complement. Nucleic Acids Res. 48, D132–D141. https://doi.org/10.1093/nar/gkz885.

Fu, Y., Shi, Z., Wu, M., Zhang, J., Jia, L., Chen, X., 2011. Identification and differential expression of microRNAs during metamorphosis of the Japanese flounder (*Paralichthys olivaceus*). PLoS One 6. https://doi.org/10.1371/journal.pone.0022957.

Fu, Y., Xu, Z., Wen, B., Gao, J., Chen, Z., 2020. Gonad-specific transcriptomes reveal differential expression of gene and miRNA between male and female of the discus fish (*Symphysodon aequifasciatus*). Front. Physiol. 11, 1–12. https://doi.org/10.3389/fphys.2020.00754.

Fuso, A., Raia, T., Orticello, M., Lucarelli, M., 2020. The complex interplay between DNA methylation and miRNAs in gene expression regulation. Biochimie 173, 12–16. https://doi.org/10.1016/j.biochi.2020.02.006.

Gallego, R.I., Pai, A.A., Tung, J., Gilad, Y., 2014. RNA-seq: impact of RNA degradation on transcript quantification. BMC Biol. 12, 1–13. https://doi.org/10.1186/1741-7007-12-42.

Gao, C., Cai, X., Fu, Q., Yang, N., Song, L., Su, B., Tan, F., Liu, B., Li, C., 2019a. Dynamics of MiRNA transcriptome in turbot (*Scophthalmus maximus L.*) intestine following *Vibrio anguillarum* infection. Marine Biotechnol. 21, 550–564. https://doi.org/10.1007/s10126-019-09903-z.

Gao, C., Fu, Q., Yang, N., Song, L., Tan, F., Zhu, J., Li, C., 2019b. Identification and expression profiling analysis of microRNAs in Nile tilapia (*Oreochromis niloticus*) in response to *Streptococcus agalactiae* infection. Fish Shellfish Immunol. 87, 333–345. https://doi.org/10.1016/j.fsi.2019.01.018.

Gay, S., Bugeon, J., Bouchareb, A., Henry, L., Delahaye, C., Legeai, F., Montfort, J., Le Cam, A., Siegel, A., Bobe, J., Thermes, V., 2018. MiR-202 controls female fecundity by regulating medaka oogenesis. PLoS Genet. 14, 1–26. https://doi.org/10.1371/journal.pgen.1007593.

Geffen, A.J., van der Veer, H.W., Nash, R.D.M., 2007. The cost of metamorphosis in flatfishes. J. Sea Res. 58, 35–45. https://doi.org/10.1016/j.seares.2007.02.004.

Giraldez, A.J., Cinalli, R.M., Glasner, M.E., Enright, A.J., Thomson, J.M., Baskerville, S., Hammond, S.M., Bartel, D.P., Schier, A.F., 2005. MicroRNAs regulate brain morphogenesis in zebrafish. Science 308, 833–838. https://doi.org/10.1126/science.1109020.

Giraldez, A.J., Mishima, Y., Rihel, J., Grocock, R.J., Van Dongen, S., Inoue, K., Enright, A.J., Schier, A.F., 2006. Zebrafish miR-430 promotes deadenylation and clearance of maternal mRNAs. Science 312, 75–79. https://doi.org/10.1126/science.1122689.

Giraldez, M.D., Spengler, R.M., Etheridge, A., Godoy, P.M., Barczak, A.J., Srinivasan, S., De Hoff, P.L., Tanriverdi, K., Courtright, A., Lu, S., Khoory, J., Rubio, R., Baxter, D., Driedonks, T.A.P., Buermans, H.P.J., Nolte-'T Hoen, E.N.M., Jiang, H., Wang, K., Ghiran, I., Wang, Y.E., Van Keuren-Jensen, K., Freedman, J.E., Woodruff, P.G., Laurent, L.C., Erle, D.J., Galas, D.J., Tewari, M., 2018. Comprehensive multi-center assessment of small RNA-seq methods for quantitative miRNA profiling. Nat. Biotechnol. 36, 746–757. https://doi.org/10.1038/nbt.4183.

Giusti, J., Pinhal, D., Moxon, S., Campos, C.L., Münsterberg, A., Martins, C., 2016. MicroRNA-10 modulates Hox genes expression during Nile tilapia embryonic development. Mech. Dev. 140, 12–18. https://doi.org/10.1016/j.mod.2016.03.002.

Gomes, F., Watanabe, L., Nozawa, S., Oliveira, L., Cardoso, J., Vianez, J., Nunes, M., Schneider, H., Sampaio, I., 2017. Identification and characterization of the expression profile of the microRNAs in the Amazon species *Colossoma macropomum* by next generation sequencing. Genomics 109, 67–74. https://doi.org/10.1016/j.ygeno.2017.02.001.

Gonçalves, A.T., Valenzuela-Muñoz, V., Gallardo-Escárate, C., 2020. Brain microRNAs in rainbow trout are modulated by functional additives and fish density. Aquaculture 519, 734754. https://doi.org/10.1016/j.aquaculture.2019.734754.

Gu, Y., Zhang, L., Chen, X., 2014. Differential expression analysis of *Paralichthys olivaceus* microRNAs in adult ovary and testis by deep sequencing. Gen. Comp. Endocrinol. 204, 181–184. https://doi.org/10.1016/j.ygcen.2014.05.019.

Guan, X.-L., Zhang, B.-C., Sun, L., 2020. Japanese flounder pol-miR-3p-2 suppresses *Edwardsiella tarda* infection by regulation of autophagy via p53. Dev. Comp. Immunol. 103. https://doi.org/10.1016/j.dci.2019.103531.

Ha, M., Kim, V.N., 2014. Regulation of microRNA biogenesis. Nat. Rev. Mol. Cell Biol. 15, 509–524. https://doi.org/10.1038/nrm3838.

Hemre, G.I., Mommsen, T.P., Krogdahl, A., 2002. Carbohydrates in fish nutrition: effects on growth, glucose metabolism and hepatic enzymes. Aquacult. Nutr. 8, 175–194. https://doi.org/10.1046/j.1365-2095.2002.00200.x.

Her, G.M., Hsu, C.C., Hong, J.R., Lai, C.Y., Hsu, M.C., Pang, H.W., Chan, S.K., Pai, W.Y., 2011. Overexpression of gankyrin induces liver steatosis in zebrafish (*Danio rerio*). Biochim. Biophys. Acta Mol. Cell Biol. Lipids 1811, 536–548. https://doi.org/10.1016/j.bbalip.2011.06.011.

Herkenhoff, M.E., Oliveira, A.C., Nachtigall, P.G., Costa, J.M., Campos, V.F., Hilsdorf, A.W.S., Pinhal, D., 2018. Fishing into the microRNA transcriptome. Front. Genet. 9, 88. https://doi.org/10.3389/fgene.2018.00088.

Howe, K., Clark, M.D., Torroja, C.F., et al., 2013. The zebrafish reference genome sequence and its relationship to the human genome. Nature 496, 498–503. https://doi.org/10.1038/nature12111.

Hsieh, C.H., Chen, W.M., Hsieh, Y.S., et al., 2018. A novel multi-gene detection platform for the analysis of miRNA expression. Sci. Rep. 8, 10684. https://doi.org/10.1038/s41598-018-29146-7.

Huang, Y., Yang, Y.B., Sun, X.H., 2016. Genome-wide identification of microRNAs and their target genes in *Cynoglossus semilaevis* using computational approach. Gene Rep. 4, 235–243. https://doi.org/10.1016/j.genrep.2016.07.006.

Huang, C.H., Chen, N., Huang, C.X., Zhang, B., Wu, M., He, L., Liu, H., Tang, R., Wang, W.M., Wang, H.L., 2017. Involvement of the miR-462/731 cluster in hypoxia response in *Megalobrama amblycephala*. Fish Physiol. Biochem. 43, 863–873. https://doi.org/10.1007/s10695-017-0341-8.

Huang, Y., Gong, W., Xiong, J., Gao, X.C., Ren, H.T., 2018. Discovery and characterization of conserved and novel microRNAs from blunt snout bream (*Megalobrama amblycephala*) by deep sequencing. Gene 654, 57–63. https://doi.org/10.1016/j.gene.2018.02.045.

Iftikhar, H., Carney, G.E., 2016. Evidence and potential *in vivo* functions for biofluid miRNAs: from expression profiling to functional testing. Bioessays 38, 367–378. https://doi.org/10.1002/bies.201500130.

IPCC, 2018. Summary for policymakers. In: Masson-Delmotte, V., Zhai, P., Pörtner, H.-O., Roberts, D., Skea, J., Shukla, P.R., Waterfield, T. (Eds.), Global Warming of 1.5°C. An IPCC Special Report on the Impacts of Global Warming of 1.5°C above Pre-Industrial Levels and Related Global Greenhouse Gas Emission Pathways, in the Context of Strengthening the Global Response to the Threat of Climate Change, Sustainable Development, and Efforts to Eradicate Poverty.

Jing, J., Wu, J., Liu, W., Xiong, S., Ma, W., Zhang, J., Wang, W., Gui, J.F., Mei, J., 2014. Sex-biased miRNAs in gonad and their potential roles for testis development in yellow catfish. PLoS One 9, 1–9. https://doi.org/10.1371/journal.pone.0107946.

Kaitetzidou, E., Xiang, J., Antonopoulou, E., Tsigenopoulos, C.S., Sarropoulou, E., 2015. Dynamics of gene expression patterns during early development of the European seabass (*Dicentrarchus labrax*). Physiol. Genomics 47, 158–169. https://doi.org/10.1152/physiolgenomics.00001.2015.

Kapsimali, M., Kloosterman, W.P., de Bruijn, E., Rosa, F., Plasterk, R.H., Wilson, S.W., 2007. MicroRNAs show a wide diversity of expression profiles in the developing and mature central nervous system. Genome Biol. 8, R173. https://doi.org/10.1186/gb-2007-8-8-r173.

Keating, M.T., Sanguinetti, M.C., 2001. Molecular and cellular mechanisms of cardiac arrhythmias. Cell 104, 569–580. https://doi.org/10.1016/S0092-8674(01)00243-4.

Kedde, M., Strasser, M.J., Boldajipour, B., Vrielink, J.A.F.O., Slanchev, K., le Sage, C., Nagel, R., Voorhoeve, P.M., van Duijse, J., Ørom, U.A., Lund, A.H.H., Perrakis, A., Raz, E., Agami, R., 2007. RNA-binding protein Dnd1 inhibits microRNA access to target mRNA. Cell 131, 1273–1286. https://doi.org/10.1016/j.cell.2007.11.034.

Kelley, J.L., Desvignes, T., McGowan, K.L., Perez, M., Rodriguez, L.A., Brown, A.P., Culumber, Z., Tobler, M., 2021. MicroRNA expression variation as a potential molecular mechanism contributing to adaptation to hydrogen sulphide. J. Evol. Biol. https://doi.org/10.1111/jeb.13727.

Kern, F., Backes, C., Hirsch, P., Fehlmann, T., Hart, M., Meese, E., Keller, A., 2020. What's the target: understanding two decades of in silico microRNA-target prediction. Brief. Bioinform. 21, 1999–2010. https://doi.org/10.1093/bib/bbz111.

Kitano, J., Yoshida, K., Suzuki, Y., 2013. RNA sequencing reveals small RNAs differentially expressed between incipient Japanese threespine sticklebacks. BMC Genomics 14, 214. https://doi.org/10.1186/1471-2164-14-214.

Kloosterman, W.P., Lagendijk, A.K., Ketting, R.F., Moulton, J.D., Plasterk, R.H.A., 2007. Targeted inhibition of miRNA maturation with morpholinos reveals a role for miR-375 in pancreatic islet development. PLoS Biol. 5, 1738–1749. https://doi.org/10.1371/journal.pbio.0050203.

Kocher, T.D., 2004. Adaptive evolution and explosive speciation: the cichlid fish model. Nat. Rev. Genet. 5, 288–298. https://doi.org/10.1038/nrg1316.

Kostyniuk, D.J., Mennigen, J.A., 2020. Meta-analysis of differentially-regulated hepatic microRNAs identifies candidate post-transcriptional regulation networks of intermediary metabolism in rainbow trout. Comp. Biochem. Physiol. Part D Genomics Proteomics 36. https://doi.org/10.1016/j.cbd.2020.100750, 100750.

Kostyniuk, D.J., Marandel, L., Jubouri, M., Dias, K., de Souza, R.F., Zhang, D., Martyniuk, C.J., Panserat, S., Mennigen, J.A., 2019. Profiling the rainbow trout hepatic miRNAome under diet-induced hyperglycemia. Physiol. Genomics 51, 411–431. https://doi.org/10.1152/physiolgenomics.00032.2019.

Kozomara, A., Griffiths-Jones, S., 2014. MiRBase: annotating high confidence microRNAs using deep sequencing data. Nucleic Acids Res. 42, 68–73. https://doi.org/10.1093/nar/gkt1181.

Kristensen, L.S., Andersen, M.S., Stagsted, L.V.W., Ebbesen, K.K., Hansen, T.B., Kjems, J., 2019. The biogenesis, biology and characterization of circular RNAs. Nat. Rev. Genet. 20, 675–691. https://doi.org/10.1038/s41576-019-0158-7.

Krüger, J., Rehmsmeier, M., 2006. RNAhybrid: microRNA target prediction easy, fast and flexible. Nucleic Acids Res. 34, W451–W454. https://doi.org/10.1093/nar/gkl243.

Kure, E.H., Sæbø, M., Stangeland, A.M., Hamfjord, J., Hytterød, S., Heggenes, J., Lydersen, E., 2013. Molecular responses to toxicological stressors: profiling microRNAs in wild Atlantic salmon (*Salmo salar*) exposed to acidic aluminum-rich water. Aquat. Toxicol. 138–139, 98–104. https://doi.org/10.1016/j.aquatox.2013.04.004.

Lai, K.P., Li, J.W., Tse, A.C.K., Chan, T.F., Wu, R.S.S., 2016. Hypoxia alters steroidogenesis in female marine medaka through miRNAs regulation. Aquat. Toxicol. 172, 1–8. https://doi.org/10.1016/j.aquatox.2015.12.012.

Lan, T., Chen, Y.L., Gul, Y., Zhao, B.W., Gao, Z.X., 2019. Comparative expression analysis of let-7 microRNAs during ovary development in *Megalobrama amblycephala*. Fish Physiol. Biochem. 45, 1101–1115. https://doi.org/10.1007/s10695-019-00624-7.

Lee, R.C., Feinbaum, R.L., Ambros, V., 1993. The *C. elegans* heterochronic gene lin-4 encodes small RNAs with antisense complementarity to lin-14. Cell 75, 843–854. https://doi.org/10.1016/0092-8674(93)90529-Y.

Leucht, C., Stigloher, C., Wizenmann, A., Klafke, R., Folchert, A., Bally-Cuif, L., 2008. MicroRNA-9 directs late organizer activity of the midbrain-hindbrain boundary. Nat. Neurosci. 11, 641–648. https://doi.org/10.1038/nn.2115.

Li, X., Ma, J., Fang, Q., Li, Y., 2013. Transcription alterations of microRNAs, cytochrome P4501A1 and 3A65, and AhR and PXR in the liver of zebrafish exposed to crude microcystins. Toxicon 73, 17–22. https://doi.org/10.1016/j.toxicon.2013.07.002.

Li, J.W., Lin, X., Tse, A., Cheung, A., Chan, T.F., Kong, R.Y.C., Lai, K.P., Wu, R.S.S., 2016. Discovery and functional characterization of novel miRNAs in the marine medaka *Oryzias melastigma*. Aquat. Toxicol. 175, 106–116. https://doi.org/10.1016/j.aquatox.2016.03.013.

Li, Z.J., Chen, Y.H., Zhang, J.Z., Zhu, X., Zhang, J.S., Chen, D.X., Wang, K.Z., Hu, Y., Chu, W.Y., 2017. Effects of dietary *Bacillus natto* supplementation on growth performance and the growth-related gene/microRNA expression in the skeletal muscle of grass carp *(Ctenopharyngodon idella)*. Aquacult. Nutr. 23, 46–53. https://doi.org/10.1111/anu.12357.

Li, W.r., Hu, Y.H., Jiang, S., Sun, L., 2019. Global profiling and characterization of Japanese flounder *(Paralichthys olivaceus)* kidney microRNAs regulated by *Edwardsiella tarda* infection in a time-dependent fashion. Fish Shellfish Immunol. 93, 766–780. https://doi.org/10.1016/j.fsi.2019.07.078.

Li, S., Lin, G., Fang, W., Gao, D., Huang, J., Xie, J., Lu, J., 2020. Identification and comparison of microRNAs in the gonad of the yellowfin seabream (*Acanthopagrus latus*). Int. J. Mol. Sci. 21, 1–16. https://doi.org/10.3390/ijms21165690.

Li, W., Guan, X., Sun, B., Sun, L., 2021. A novel microRNA of Japanese flounder regulates antimicrobial immunity involving a bacteria-binding CSF3. Front. Immunol. 12, 1–13. https://doi.org/10.3389/fimmu.2021.723401.

Lim, L.P., Glasner, M.E., Yekta, S., Burge, C.B., Bartel, D.P., 2003. Vertebrate microRNA genes. Science 299, 1540. https://doi.org/10.1126/science.1080372.

Love, M.I., Huber, W., Anders, S., 2014. Moderated estimation of fold change and dispersion for RNA-seq data with DESeq2. Genome Biol. 15, 550. https://doi.org/10.1186/s13059-014-0550-8.

Ma, H., Hostuttler, M., Wei, H., Rexroad, C.E., Yao, J., 2012. Characterization of the rainbow trout egg microRNA transcriptome. PLoS One 7, 1–8. https://doi.org/10.1371/journal.pone.0039649.

Ma, H., Weber, G.M., Hostuttler, M.A., Wei, H., Wang, L., Yao, J., 2015. MicroRNA expression profiles from eggs of different qualities associated with post-ovulatory ageing in rainbow trout *(Oncorhynchus mykiss)*. BMC Genomics 16, 1–9. https://doi.org/10.1186/s12864-015-1400-0.

Ma, J., Li, Y., Wu, M., Zhang, C., Che, Y., Li, W., Li, X., 2018. Serum immune responses in common carp *(Cyprinus carpio L.)* to paraquat exposure: the traditional parameters and circulating microRNAs. Fish Shellfish Immunol. 76, 133–142. https://doi.org/10.1016/j.fsi.2018.02.046.

Ma, L., Gou, M., Du, Z., Zhu, T., Li, J., Li, Q.W., Pang, Y., 2021. MicroRNA expression profile in *Lampetra morii* upon *Vibrio anguillarum* infection and miR-4561 characterization targeting lip. Commun. Biol. 4, 1–12. https://doi.org/10.1038/s42003-021-02525-z.

MacRae, C., Peterson, R.T., 2015. Zebrafish as tools for drug discovery. Nat. Rev. Drug Discov. 14, 721–731. https://doi.org/10.1038/nrd4627.

McGuire, N.L., Calisi, R.M., Bentley, G.E., 2010. Seasonality: hormones and behavior. In: Breed, M.D., Moore, J. (Eds.), Encyclopedia of Animal Behavior. vol. 3. Academic Press, Oxford, MA, pp. 108–118, https://doi.org/10.1016/B978-0-08-045337-8.00256-4.

McManus, M.T., Sharp, P.A., 2002. Gene silencing in mammals by small interfering RNAs. Nat. Rev. Genet. 3, 737–747. https://doi.org/10.1038/nrg908.

Mennigen, J.A., Panserat, S., Larquier, M., Plagnes-Juan, E., Medale, F., Seiliez, I., Skiba-Cassy, S., 2012. Postprandial regulation of hepatic microRNAs predicted to target the insulin pathway in rainbow trout. PLoS One 7, e38604, https://doi.org/10.1371/journal.pone.0038604.

Mennigen, J.A., Skiba-Cassy, S., Panserat, S., 2013. Ontogenetic expression of metabolic genes and microRNAs in rainbow trout alevins during the transition from the endogenous to the exogenous feeding period. J. Exp. Biol. 216, 1597–1608. https://doi.org/10.1242/jeb.082248.

Mennigen, J.A., Martyniuk, C.J., Seiliez, I., Panserat, S., Skiba-Cassy, S., 2014. Metabolic consequences of microRNA-122 inhibition in rainbow trout, *Oncorhynchus mykiss*. BMC Genomics 15, 1–20. https://doi.org/10.1186/1471-2164-15-70.

Miao, L.H., Lin, Y., Pan, W.J., Huang, X., Ge, X.P., Ren, M.C., Zhou, Q.L., Liu, B., 2017. Identification of differentially expressed micrornas associate with glucose metabolism in different organs of blunt snout bream *(Megalobrama amblycephala)*. Int. J. Mol. Sci. 18, 1161. https://doi.org/10.3390/ijms18061161.

Miao, L.H., Lin, Y., Huang, X., Pan, W.J., Zhou, Q.L., Liu, B., Ren, M.C., Ge, X.P., Pan, L.K., 2018. *In vivo* analysis of miR-34a regulated glucose metabolism related genes in *Megalobrama amblycephala*. Int. J. Mol. Sci. 19, 2417. https://doi.org/10.3390/ijms19082417.

Miotto, E., Saccenti, E., Lupini, L., Callegari, E., Negrini, M., Ferracin, M., 2014. Quantification of circulating miRNAs by droplet digital PCR: comparison of EvaGreen- and TaqMan-based chemistries. Cancer Epidemiol. Biomarkers Prev. 23, 2638–2642. https://doi.org/10.1158/1055-9965.EPI-14-0503.

Morton, S.U., Scherz, P.J., Cordes, K.R., Ivey, K.N., Stainier, D.Y.R., Srivastava, D., 2008. MicroRNA-138 modulates cardiac patterning during embryonic development. Proc. Natl. Acad. Sci. U. S. A. 105, 17830–17835. https://doi.org/10.1073/pnas.0804673105.

Mutz, K.O., Heilkenbrinker, A., Lönne, M., Walter, J.G., Stahl, F., 2013. Transcriptome analysis using next-generation sequencing. Curr. Opin. Biotechnol. 24, 22–30. https://doi.org/10.1016/j.copbio.2012.09.004.

Naylor, R.L., Hardy, R.W., Buschmann, A.H., Bush, S.R., Cao, L., Klinger, D.H., Little, D.C., Lubchenco, J., Shumway, S.E., Troell, M., 2021. A 20-year retrospective review of global aquaculture. Nature 591, 551–563. https://doi.org/10.1038/s41586-021-03308-6.

Nuovo, G.J., Elton, T.S., Nana-Sinkam, P., Volinia, S., Carlo, M., 2009. A methodology for the combined *in situ* analyses of the precursor and mature forms of microRNAs and correlation with their putative targets. Nat. Protoc. 4, 107–115. https://doi.org/10.1038/nprot.2008.215.A.

Oliveira, C.C.V., Fatsini, E., Fernández, I., Anjos, C., Chauvigné, F., Cerdà, J., Mjelle, R., Fernandes, J.M.O., Cabrita, E., 2020. Kisspeptin influences the reproductive axis and circulating levels of microRNAs in *Senegalese sole*. Int. J. Mol. Sci. 21, 1–21. https://doi.org/10.3390/ijms21239051.

Ordas, A., Kanwal, Z., Lindenberg, V., Rougeot, J., Mink, M., Spaink, H.P., Meijer, A.H., 2013. MicroRNA-146 function in the innate immune transcriptome response of zebrafish embryos to *Salmonella typhimurium* infection. BMC Genomics 14, 1–15. https://doi.org/10.1186/1471-2164-14-696.

Papadaki, M., Kaitetzidou, E., Mylonas, C.C., Sarropoulou, E., 2020. Non-coding RNA expression patterns of two different teleost gonad maturation stages. Marine Biotechnol. 22, 683–695. https://doi.org/10.1007/s10126-020-09991-2.

Pase, L., Lieschke, G.J., 2011. Discerning different in vivo roles of microRNAs by experimental approaches in zebrafish. In: Methods in Cell Biology, third ed. Elsevier Inc., https://doi.org/10.1016/B978-0-12-374814-0.00020-3.

Peterson, S.M., Thompson, J.A., Ufkin, M.L., Sathyanarayana, P., Liaw, L., Congdon, C.B., 2014. Common features of microRNA target prediction tools. Front. Genet. 5, 1–10. https://doi.org/10.3389/fgene.2014.00023.

Pinhal, D., Bovolenta, L.A., Moxon, S., Oliveira, A.C., Nachtigall, P.G., Acencio, M.L., Patton, J.G., Hilsdorf, A.W.S., Lemke, N., Martins, C., 2018. Genome-wide microRNA screening in Nile tilapia reveals pervasive isomiRs' transcription, sex-biased arm switching and increasing complexity of expression throughout development. Sci. Rep. 8, 1–18. https://doi.org/10.1038/s41598-018-26607-x.

Pinzón, N., Li, B., Martinez, L., Sergeeva, A., Presumey, J., Apparailly, F., Seitz, H., 2017. MicroRNA target prediction programs predict many false positives. Genome Res. 27, 234–245. https://doi.org/10.1101/gr.205146.116.

Presslauer, C., Tilahun Bizuayehu, T., Kopp, M., Fernandes, J.M.O., Babiak, I., 2017. Dynamics of miRNA transcriptome during gonadal development of zebrafish. Sci. Rep. 7, 1–16. https://doi.org/10.1038/srep43850.

Pritchard, C.C., Cheng, H.H., Tewari, M., 2012. MicroRNA profiling: approaches and considerations. Nat. Rev. Genet. 13, 358–369. https://doi.org/10.1038/nrg3198.

Qian, X., Ba, Y., Zhuang, Q., Zhong, G., 2014. RNA-seq technology and its application in fish transcriptomics. OMICS 18, 98–110. https://doi.org/10.1089/omi.2013.0110.

Qiang, J., Bao, J.W., Li, H.X., Chen, D.J., He, J., Tao, Y.F., Xu, P., 2017. miR-1338-5p modulates growth hormone secretion and glucose utilization by regulating ghitm in genetically improved farmed tilapia (GIFT, *Oreochromis niloticus*). Front. Physiol. 8, 1–12. https://doi.org/10.3389/fphys.2017.00998.

Qiang, J., Tao, Y.F., Bao, J.W., Chen, D.J., Li, H.X., He, J., Xu, P., 2018. High fat diet-induced miR-122 regulates lipid metabolism and fat deposition in genetically improved farmed tilapia (GIFT, *Oreochromis niloticus*) liver. Front. Physiol. 9, 1–11. https://doi.org/10.3389/fphys.2018.01422.

Qu, X., Hu, M., Shang, Y., Pan, L., Jia, P., Fu, C., Liu, Q., Wang, Y., 2018. Liver transcriptome and miRNA analysis of silver carp *(Hypophthalmichthys molitrix)* intraperitoneally injected with microcystin-LR. Front. Physiol. 9, 1–12. https://doi.org/10.3389/fphys.2018.00381.

Ramachandra, R.K., Salem, M., Gahr, S., Rexroad, C.E., Yao, J., 2008. Cloning and characterization of microRNAs from rainbow trout *(Oncorhynchus mykiss)*: their expression during early embryonic development. BMC Dev. Biol. 8, 1–11. https://doi.org/10.1186/1471-213X-8-41.

Ramakrishnan, V., 2002. Ribosome structure and the mechanism of translation. Cell 108, 557–572. https://doi.org/10.1016/S0092-8674(02)00619-0.

Ramat, A., Simonelig, M., 2021. Functions of PIWI proteins in gene regulation: new arrows added to the piRNA quiver. Trends Genet. 37, 188–200. https://doi.org/10.1016/j.tig.2020.08.011.

Rasal, K.D., Nandanpawar, P.C., Swain, P., Badhe, M.R., Sundaray, J.K., Jayasankar, P., 2016. MicroRNA in aquaculture fishes: a way forward with high-throughput sequencing and a computational approach. Rev. Fish Biol. Fish. 26, 199–212. https://doi.org/10.1007/s11160-016-9421-6.

Rbbani, G., Nedoluzhko, A., Galindo-Villegas, J., Fernandes, J.M.O., 2021. Function of circular RNAs in fish and their potential application as biomarkers. Int. J. Mol. Sci. 22, 7119. https://doi.org/10.3390/ijms22137119.

Renaud, L., da Silveira, W.A., Hazard, E.S., Simpson, J., Falcinelli, S., Chung, D., Carnevali, O., Hardiman, G., 2017. The plasticizer bisphenol a perturbs the hepatic epigenome: a systems level analysis of the miRNome. Genes 8, 1–37. https://doi.org/10.3390/genes8100269.

Ribas, L., Liew, W.C., Díaz, N., Sreenivasan, R., Orbán, L., Piferrer, F., 2017. Heat-induced masculinization in domesticated zebrafish is family-specific & yields a set of different gonadal transcriptomes. Proc. Natl. Acad. Sci. U. S. A. 114, E941–E950. https://doi.org/10.1073/pnas.1609411114.

Riesco, M.F., Valcarce, D.G., Martínez-Vázquez, J.M., Martín, I., Calderón-García, A.Á., Gonzalez-Nuñez, V., Robles, V., 2019. Male reproductive dysfunction in *Solea senegalensis*: new insights into an unsolved question. Reprod. Fertil. Dev. 31, 1104–1115. https://doi.org/10.1071/RD18453.

Riethoven, J.J., 2010. Regulatory regions in DNA: promoters, enhancers, silencers, and insulators. In: Ladunga, I. (Ed.), Computational Biology of Transcription Factor Binding. Methods Mol. Biol. (Methods and Protocols), vol. 674. Humana Press, Totowa, NJ, pp. 33–42, https://doi.org/10.1007/978-1-60761-854-6_3.

Riolo, G., Cantara, S., Marzocchi, C., Ricci, C., 2021. MiRNA targets: from prediction tools to experimental validation. Methods Protoc. 4, 1–20. https://doi.org/10.3390/mps4010001.

RNAcentral Consortium, 2021. RNAcentral 2021: secondary structure integration, improved sequence search and new member databases. Nucleic Acids Res. 49, D212–D220. https://doi.org/10.1093/nar/gkaa921.

Roberto, V.P., Gavaia, P., Nunes, M.J., Rodrigues, E., Cancela, M.L., Tiago, D.M., 2018. Evidences for a new role of miR-214 in chondrogenesis. Sci. Rep. 8, 1–16. https://doi.org/10.1038/s41598-018-21735-w.

Robledo, D., Martin, A.P., Álvarez-Dios, J.A., Bouza, C., Pardo, B.G., Martínez, P., 2017. First characterization and validation of turbot microRNAs. Aquaculture 472, 76–83. https://doi.org/10.1016/j.aquaculture.2016.05.002.

Robles, V., Valcarce, D.G., Riesco, M.F., 2019. Non-coding RNA regulation in reproduction: their potential use as biomarkers. Noncoding RNA Res. 4, 54–62. https://doi.org/10.1016/j.ncrna.2019.04.001.

Salas-Huetos, A., James, E.R., Aston, K.I., Jenkins, T.G., Carrell, D.T., Yeste, M., 2019. The expression of miRNAs in human ovaries, oocytes, extracellular vesicles, and early embryos: a systematic review. Cells 8, 1564. https://doi.org/10.3390/cells8121564.

Sarropoulou, E., Kaitetzidou, E., Papandroulaki, N., Tsalafouta, A., Pavlidis, M., 2019. Inventory of European sea bass (*Dicentrarchus labrax*) sncRNAs vital during early teleost development. Front. Genet. 10, 1–11. https://doi.org/10.3389/fgene.2019.00657.

Sartorelli, V., Lauberth, S.M., 2020. Enhancer RNAs are an important regulatory layer of the epigenome. Nat. Struct. Mol. Biol. 27, 521–528. https://doi.org/10.1038/s41594-020-0446-0.

Schulz, R.W., De França, L.R., Lareyre, J.J., Le Gac, F., Chiarini-Garcia, H., Nobrega, R.H., Miura, T., 2010. Spermatogenesis in fish. Gen. Comp. Endocrinol. 165, 390–411. https://doi.org/10.1016/j.ygcen.2009.02.013.

Schwarzenbach, H., da Silva, A.M., Calin, G., Pantel, K., 2015. Which is the accurate data normalization strategy for microRNA quantification? Clin. Chem. 61, 1333. https://doi.org/10.1373/clinchem.2015.239459.Which.

Sheng, Q., Vickers, K., Zhao, S., Wang, J., Samuels, D.C., Koues, O., Shyr, Y., Guo, Y., 2017. Multi-perspective quality control of Illumina RNA sequencing data analysis. Brief. Funct. Genomics 16, 194–204. https://doi.org/10.1093/bfgp/elw035.

Shinya, M., Kobayashi, K., Masuda, A., Tokumoto, M., Ozaki, Y., Saito, K., Kawasaki, T., Sado, Y., Sakai, N., 2013. Properties of gene knockdown system by vector-based siRNA in zebrafish. Dev. Growth Differ. 55, 755–765. https://doi.org/10.1111/dgd.12091.

Shwe, A., Østbye, T.K., Krasnov, A., Ramberg, S., Andreassen, R., 2020. Characterization of differentially expressed miRNAs and their predicted target transcripts during smoltification and adaptation to seawater in head kidney of Atlantic Salmon. Genes 11, 1059. https://doi.org/10.3390/genes11091059.

Sinkkonen, L., Hugenschmidt, T., Berninger, P., Gaidatzis, D., Mohn, F., Artus-Revel, C.G., Zavolan, M., Svoboda, P., Filipowicz, W., 2008. MicroRNAs control *de novo* DNA methylation through regulation of transcriptional repressors in mouse embryonic stem cells. Nat. Struct. Mol. Biol. 15, 259–267. https://doi.org/10.1038/nsmb.1391.

Skaftnesmo, K.O., Edvardsen, R.B., Furmanek, T., Crespo, D., Andersson, E., Kleppe, L., Taranger, G.L., Bogerd, J., Schulz, R.W., Wargelius, A., 2017. Integrative testis transcriptome analysis reveals differentially expressed miRNAs and their mRNA targets during early puberty in Atlantic salmon. BMC Genomics 18, 1–12. https://doi.org/10.1186/s12864-017-4205-5.

Smith, N.C., Wajnberg, G., Chacko, S., Woldemariam, N.T., Lacroix, J., Crapoulet, N., Ayre, D.C., Lewis, S.M., Rise, M.L., Andreassen, R., Christian, S.L., 2020. Characterization of miRNAs in extracellular vesicles released from Atlantic salmon monocyte-like and macrophage-like cells. Front. Immunol. 11, 1–16. https://doi.org/10.3389/fimmu.2020.587931.

Subtelny, A.O., Eichhorn, S.W., Chen, G.R., Sive, H., Bartel, D.P., 2014. Poly(A)-tail profiling reveals an embryonic switch in translational control. Nature 508, 66–71. https://doi.org/10.1038/nature13007.

Sun, Z., Evans, J., Bhagwate, A., Middha, S., Bockol, M., Yan, H., Kocher, J.P., 2014. CAP-miRSeq: a comprehensive analysis pipeline for microRNA sequencing data. BMC Genomics 15, 423. https://doi.org/10.1186/1471-2164-15-423.

Sun, S., Xuan, F., Ge, X., Zhu, J., Zhang, W., 2017a. Dynamic mRNA and miRNA expression analysis in response to hypoxia and reoxygenation in the blunt snout bream (*Megalobrama amblycephala*). Sci. Rep. 7, 1–11. https://doi.org/10.1038/s41598-017-12537-7.

Sun, Z., Hao, T., Tian, J., 2017b. Identification of exosomes and its signature miRNAs of male and female *Cynoglossus semilaevis*. Sci. Rep. 7, 1–10. https://doi.org/10.1038/s41598-017-00884-4.

Sun, J.J., Zheng, L.G., Chen, C.Y., Zhang, J.Y., You, C.H., Zhang, Q.H., Ma, H.Y., Monroig, Ó., Tocher, D.R., Wang, S.Q., Li, Y.Y., 2019a. MicroRNAs involved in the regulation of LC-PUFA biosynthesis in teleosts: miR-33 enhances LC-PUFA biosynthesis in *Siganus canaliculatus* by targeting *insig1* which in turn upregulates *srebp1*. Marine Biotechnol. 21, 475–487. https://doi.org/10.1007/s10126-019-09895-w.

Sun, J., Zhao, L., Wu, H., Lian, W., Cui, C., Du, Z., Luo, W., Li, M., Yang, S., 2019b. Analysis of miRNA-seq in the liver of common carp (*Cyprinus carpio* L.) in response to different environmental temperatures. Funct. Integr. Genomics 19, 265–280. https://doi.org/10.1007/s10142-018-0643-7.

Sun, J., Chen, C., You, C., Liu, Y., Ma, H., Monroig, Ó., Tocher, D.R., Wang, S., Li, Y., 2020. The miR-15/16 cluster is involved in the regulation of vertebrate LC-PUFA biosynthesis by targeting *pparγ* as demonstrated in rabbitfish *Siganus canaliculatus*. Marine Biotechnol. 22, 475–487. https://doi.org/10.1007/s10126-020-09969-0.

Tang, X.L., Xu, M.J., Li, Z.H., Pan, Q., Fu, J.H., 2013. Effects of vitamin E on expressions of eight microRNAs in the liver of Nile tilapia *(Oreochromis niloticus)*. Fish Shellfish Immunol. 34, 1470–1475. https://doi.org/10.1016/j.fsi.2013.03.353.

Tang, T., Hu, Y., Peng, M., Chu, W., Hu, Y., Zhong, L., 2019. Effects of high-fat diet on growth performance, lipid accumulation and lipid metabolism-related microRNA/gene expression in the liver of grass carp *(Ctenopharyngodon idella)*. Comp. Biochem. Physiol. B Biochem. Mol. Biol. 234, 34–40. https://doi.org/10.1016/j.cbpb.2019.04.006.

Tao, W., Sun, L., Shi, H., Cheng, Y., Jiang, D., Fu, B., Conte, M.A., Gammerdinger, W.J., Kocher, T.D., Wang, D., 2016. Integrated analysis of miRNA and mRNA expression profiles in tilapia gonads at an early stage of sex differentiation. BMC Genomics 17, 1–13. https://doi.org/10.1186/s12864-016-2636-z.

Tao, Y.F., Qiang, J., Bao, J.W., Li, H.X., Yin, G.J., Xu, P., Chen, D.J., 2018. MiR-205-5p negatively regulates hepatic acetyl-CoA carboxylase β mRNA in lipid metabolism of *Oreochromis niloticus*. Gene 660, 1–7. https://doi.org/10.1016/j.gene.2018.03.064.

The dMIQE Group, Huggett, J.F., 2020. The digital MIQE guidelines update: minimum information for publication of quantitative digital PCR experiments for 2020. Clin. Chem. 66, 1012–1029. https://doi.org/10.1093/clinchem/hvaa125.

The ENCODE Project Consortium, 2004. The ENCODE (encyclopedia of DNA elements) project. Science 306, 636–640.

Tomasetti, M., Gaetani, S., Monaco, F., Neuzil, J., Santarelli, L., 2019. Epigenetic regulation of miRNA expression in malignant mesothelioma: miRNAs as biomarkers of early diagnosis and therapy. Front. Oncol. 9, 1–14. https://doi.org/10.3389/fonc.2019.01293.

Treiber, T., Treiber, N., Meister, G., 2019. Regulation of microRNA biogenesis and its crosstalk with other cellular pathways. Nat. Rev. Mol. Cell Biol. 20, 5–20. https://doi.org/10.1038/s41580-018-0059-1.

Tse, A.C.K., Li, J.W., Wang, S.Y., Chan, T.F., Lai, K.P., Wu, R.S.S., 2016. Hypoxia alters testicular functions of marine medaka through microRNAs regulation. Aquat. Toxicol. 180, 266–273. https://doi.org/10.1016/j.aquatox.2016.10.007.

Turchinovich, A., Tonevitsky, A.G., Burwinkel, B., 2016. Extracellular miRNA: a collision of two paradigms. Trends Biochem. Sci. 41, 883–892. https://doi.org/10.1016/j.tibs.2016.08.004.

Urbanek, M.O., Nawrocka, A.U., Krzyzosiak, W.J., 2015. Small RNA detection by *in situ* hybridization methods. Int. J. Mol. Sci. 16, 13259–13286. https://doi.org/10.3390/ijms160613259.

Valdivieso, A., Ribas, L., Monleón-Getino, A., Orbán, L., Piferrer, F., 2020. Exposure of zebrafish to elevated temperature induces sex ratio shifts and alterations in the testicular epigenome of unexposed offspring. Environ. Res. 186. https://doi.org/10.1016/j.envres.2020.109601, 109601.

Varet, H., Brillet-Guéguen, L., Coppée, J.-Y., Dillies, M.-A., 2016. SARTools: a DESeq2- and EdgeR-based R pipeline for comprehensive differential analysis of RNA-Seq data. PLoS One 11. https://doi.org/10.1371/journal.pone.0157022, e0157022.

Viereck, J., Thum, T., 2017. Circulating noncoding RNAs as biomarkers of cardiovascular disease and injury. Circ. Res. 120, 381–399. https://doi.org/10.1161/CIRCRESAHA.116.308434.

Wang, W., Liu, W., Liu, Q., Li, B., An, L., Hao, R., Zhao, J., Liu, S., Song, J., 2016. Coordinated microRNA and messenger RNA expression profiles for understanding sexual dimorphism of gonads and the potential roles of microRNA in the steroidogenesis pathway in Nile *tilapia (Oreochromis niloticus)*. Theriogenology 85, 970–978. https://doi.org/10.1016/j.theriogenology.2015.11.006.

Wang, S., Wu, W., Claret, F.X., 2017a. Mutual regulation of microRNAs and DNA methylation in human cancers. Epigenetics 12, 187–197. https://doi.org/10.1080/15592294.2016.1273308.

Wang, F., Jia, Y., Wang, P., Yang, Q., Du, Q.Y., Chang, Z.J., 2017b. Identification and profiling of *Cyprinus carpio* microRNAs during ovary differentiation by deep sequencing. BMC Genomics 18, 1–17. https://doi.org/10.1186/s12864-017-3701-y.

Wang, M., Jiang, S., Wu, W., Yu, F., Chang, W., Li, P., Wang, K., 2018. Non-coding RNAs function as immune regulators in teleost fish. Front. Immunol. 9, 1–15. https://doi.org/10.3389/fimmu.2018.02801.

Wang, N., Wang, R., Wang, R., Chen, S., 2019a. RNA-seq and microRNA-seq analysis of Japanese flounder (*Paralichthys olivaceus*) larvae treated by thyroid hormones. Fish Physiol. Biochem. 45, 1233–1244. https://doi.org/10.1007/s10695-019-00654-1.

Wang, F., Yang, Q.W., Zhao, W.J., Du, Q.Y., Chang, Z.J., 2019b. Effects of short-time exposure to atrazine on miRNA expression profiles in the gonad of common carp *(Cyprinus carpio)*. BMC Genomics 20, 1–15. https://doi.org/10.1186/s12864-019-5896-6.

Wienholds, E., Plasterk, R.H.A., 2005. MicroRNA function in animal development. FEBS Lett. 579, 5911–5922. https://doi.org/10.1016/j.febslet.2005.07.070.

Wienholds, E., Kloosterman, W.P., Miska, E., Alvarez-Saavedra, E., Berezikov, E., de Bruijn, E., Horvitz, H.R., Kauppinen, S., Plasterk, R.H., 2005. MicroRNA expression in zebrafish embryonic development. Science 309, 310–311. https://doi.org/10.1126/science.1114519.

Wightman, B., Ha, I., Ruvkun, G., 1993. Posttranscriptional regulation of the heterochronic gene lin-14 by lin-4 mediates temporal pattern formation in *C. elegans*. Cell 75, 855–862. https://doi.org/10.1016/0092-8674(93)90530-4.

Wittrup, A., Lieberman, J., 2015. Knocking down disease: a progress report on siRNA therapeutics. Nat. Rev. Genet. 16, 543–552. https://doi.org/10.1038/nrg3978.

Wright, C., Rajpurohit, A., Burke, E.E., Williams, C., Collado-Torres, L., Kimos, M., Brandon, N.J., Cross, A.J., Jaffe, A.E., Weinberger, D.R., Shin, J.H., 2019. Comprehensive assessment of multiple biases in small RNA sequencing reveals significant differences in the performance of widely used methods. BMC Genomics 20, 1–21. https://doi.org/10.1186/s12864-019-5870-3.

Xiao, J., Zhong, H., Zhou, Y., Yu, F., Gao, Y., Luo, Y., Tang, Z., Guo, Z., Guo, E., Gan, X., Zhang, M., Zhang, Y., 2014. Identification and characterization of microRNAs in ovary and testis of Nile tilapia *(Oreochromis niloticus)* by using Solexa sequencing technology. PLoS One 9, e86821, https://doi.org/10.1371/journal.pone.0086821.

Xu, E.G., Khursigara, A.J., Li, S., Esbaugh, A.J., Dasgupta, S., Volz, D.C., Schlenk, D., 2019. MRNA-miRNA-Seq reveals neuro-cardio mechanisms of crude oil toxicity in red drum *(Sciaenops ocellatus)*. Environ. Sci. Technol. 53, 3296–3305. https://doi.org/10.1021/acs.est.9b00150.

Yan, H., Liu, Q., Jiang, J., Shen, X., Zhang, L., Yuan, Z., Wu, Y., Liu, Y., 2021. Identification of sex differentiation-related microRNA and long non-coding RNA in *Takifugu rubripes* gonads. Sci. Rep. 11, 1–13. https://doi.org/10.1038/s41598-021-83891-w.

Yang, S., Luo, J., Long, Y., Du, J., Xu, G.C., Zhao, L., Du, Z., Luo, W., Wang, Y., He, Z., 2019. Mixed diets reduce the oxidative stress of common carp *(Cyprinus carpio)*: based on microRNA sequencing. Front. Physiol. 10. https://doi.org/10.3389/fphys.2019.00631.

Yartseva, V., Takacs, C.M., Vejnar, C.E., Lee, M.T., Giraldez, A.J., 2017. RESA identifies mRNA-regulatory sequences at high resolution. Nat. Methods 14, 201–207. https://doi.org/10.1038/nmeth.4121.

Yeri, A., Courtright, A., Danielson, K., Hutchins, E., Alsop, E., Carlson, E., Hsieh, M., Ziegler, O., Das, A., Shah, R.V., Rozowsky, J., Das, S., Van Keuren-Jensen, K., 2018. Evaluation of commercially available small RNASeq library preparation kits using low input RNA. BMC Genomics 19, 1–15. https://doi.org/10.1186/s12864-018-4726-6.

Yi, M., Chen, F., Luo, M., Cheng, Y., Zhao, H., Cheng, H., Zhou, R., 2014. Rapid evolution of piRNA pathway in the teleost fish: implication for an adaptation to transposon diversity. Genome Biol. Evol. 6, 1393–1407. https://doi.org/10.1093/gbe/evu105.

Zayed, Y., Qi, X., Peng, C., 2019. Identification of novel microRNAs and characterization of microRNA expression profiles in zebrafish ovarian follicular cells. Front. Endocrinol. 10, 1–12. https://doi.org/10.3389/fendo.2019.00518.

Zeng, L., Carter, A.D., Childs, S.J., 2009. MiR-145 directs intestinal maturation in zebrafish. Proc. Natl. Acad. Sci. U. S. A. 106, 17793–17798. https://doi.org/10.1073/pnas.0903693106.

Zhang, D., Lu, K., Dong, Z., Jiang, G., Xu, W., Liu, W., 2014. The effect of exposure to a high-fat diet on microRNA expression in the liver of blunt snout bream *(Megalobrama amblycephala)*. PLoS One 9. https://doi.org/10.1371/journal.pone.0096132, e96132.

Zhang, Q., You, C., Wang, S., Dong, Y., Monroig, Ó., Tocher, D.R., Li, Y., 2016. The miR-33 gene is identified in a marine teleost: a potential role in regulation of LC-PUFA biosynthesis in *Siganus canaliculatus*. Sci. Rep. 6, 1–10. https://doi.org/10.1038/srep32909.

Zhang, P., Wu, W., Chen, Q., Chen, M., 2019a. Non-coding RNAs and their integrated networks. J. Integr. Bioinform. 16, 1–12. https://doi.org/10.1515/jib-2019-0027.

Zhang, L., Liao, Y., Tang, L., 2019b. MicroRNA-34 family: a potential tumor suppressor and therapeutic candidate in cancer. J. Exp. Clin. Cancer Res. 38, 53. https://doi.org/10.1186/s13046-019-1059-5.

Zhao, C., Zhang, G., Yin, S., Li, Z., Wang, Q., Chen, S., Zhou, G., 2017. Integrated analysis of mRNA-seq and miRNA-seq reveals the potential roles of sex-biased miRNA-mRNA pairs in gonad tissue of dark sleeper *(Odontobutis potamophila)*. BMC Genomics 18, 1–15. https://doi.org/10.1186/s12864-017-3995-9.

Zhao, N., Jia, L., He, X., Zhang, B., 2021. Sex bias miRNAs in *Cynoglossus semilaevis* could play a role in transgenerational inheritance. Comp. Biochem. Physiol. Part D Genomics Proteomics 39. https://doi.org/10.1016/j.cbd.2021.100853, 100853.

Zhou, C.Q., Zhou, P., Ren, Y.L., Cao, L.H., Wang, J.L., 2019. Physiological response and miRNA-mRNA interaction analysis in the head kidney of rainbow trout exposed to acute heat stress. J. Therm. Biol. 83, 134–141. https://doi.org/10.1016/j.jtherbio.2019.05.014.

Zhou, W., Xie, Y., Li, Y., Xie, M., Zhang, Z., Yang, Y., Zhou, Z., Duan, M., Ran, C., 2021. Research progress on the regulation of nutrition and immunity by microRNAs in fish. Fish Shellfish Immunol. 113, 1–8. https://doi.org/10.1016/j.fsi.2021.03.011.

Zhu, T., Corraze, G., Plagnes-Juan, E., Skiba-Cassy, S., 2018. Circulating miRNA measurements are reflective of cholesterol-based changes in rainbow trout *(Oncorhynchus mykiss)*. PLoS One 13, 1–22. https://doi.org/10.1371/journal.pone.0206727.

Zhu, T., Corraze, G., Plagnes-Juan, E., Montfort, J., Bobe, J., Quillet, E., Dupont-Nivet, M., Skiba-Cassy, S., 2019. MicroRNAs related to cholesterol metabolism affected by vegetable diet in rainbow trout *(Oncorhynchus mykiss)* from control and selected lines. Aquaculture 498, 132–142. https://doi.org/10.1016/j.aquaculture.2018.08.058.

Zhu, T., Corraze, G., Plagnes-Juan, E., Skiba-Cassy, S., 2020. Cholesterol metabolism regulation mediated by SREBP-2, LXRα and miR-33a in rainbow trout *(Oncorhynchus mykiss)* both *in vivo* and *in vitro*. PLoS One 15, 1–23. https://doi.org/10.1371/journal.pone.0223813.

Chapter 5

Nutritional epigenetics

Kaja H. Skjærven, Anne-Catrin Adam, Saito Takaya, Rune Waagbø, and Marit Espe
Institute of Marine Research, IMR, Bergen, Norway

1. Introduction

Not only the quantity, but also the quality of feed, play key roles in development, maintaining growth, and ensuring a robust health status for fish. Classic nutritional studies show that nutritionally imbalanced diets can cause changes in the metabolic homeostasis, chronic diseases, immune deficiency, increase susceptibility to infectious diseases, and poor growth (Halver and Hardy, 2002; Kiron, 2012; Lall and Lewis-McCrea, 2007; Trichet, 2010; Turchini et al., 2009; Waagbø, 2006, 2008). Nutrition-based studies on aquaculture-related fish species have shown that fish have both species- and stage-specific as well as conditional requirements for macro- and micronutrients, including vitamins and minerals (NRC, 2011; Hemre et al., 2016; Espe et al., 2020; Vera et al., 2017; Waagbø et al., 2020). In the early life stages, fish larvae with an underdeveloped and immature gastrointestinal tract, a higher number of nutrients are classified as essential compared to on-growing stages (Moren et al., 2011). Although it seems to be species-specific feeding strategies, broodstock fish also require increased levels for specific nutrients prior to and during sexual maturation (Izquierdo et al., 2001; Christiansen and Torrissen, 1997; Mæland et al., 2003; Waagbø et al., 1993; Skjaerven et al., 2016, 2018). This means that special care and attention to nutritional needs should be taken at all stages of aquaculture production, so deficiencies are avoided, and fish welfare is secured. Indeed, nutrition is focused as an attractive way to enhance the robustness in farmed fish species through functional feeds (Waagbø et al., 1993; Panserat et al., 2019; Sweetman et al., 2010; Waagbø and Remø, 2020; Andersen et al., 2016).

In recent times, there has been a growing interest in epigenetic gene regulation as a central mechanism that connects the nutrients in the feed with the overall control of activity and signal flow in the various metabolic pathways. The field of nutritional epigenetics, with a particular focus on how metabolites are central to epigenetic regulation, has grown rapidly over the last 10 years, with the number of published articles increasing each year, mainly for mammalian species.

Cellular and Molecular Approaches in Fish Biology. https://doi.org/10.1016/B978-0-12-822273-7.00006-9

New publications show that nutritional epigenetics is also highly relevant for fish and seems particularly important for robust domestication (Best et al., 2018; Hou and Fuiman, 2020), but also in environmentally controlled evolution through a match between the developmental stage and prey availability (Skulason et al., 2019). Understanding how epigenetic regulators respond and direct their signals in response to the chemical composition of the diet are key factors in aquaculture production to maintain healthy growth through generations.

2. Epigenetic regulators

Briefly, the three major mechanisms in epigenetic regulation are DNA methylation, histone tail modifications, and noncoding RNAs (ncRNAs) (Fig. 1). Both DNA methylation and histone modifications play key roles in regulatory control of DNA binding activities without changing the underlying DNA sequence. The ncRNAs, like piRNA, lncRNA, and microRNAs, cover a wide range of biological functions, including epigenetic regulation. For a further detailed information on epigenetic mechanisms, please see also Chapter 3.

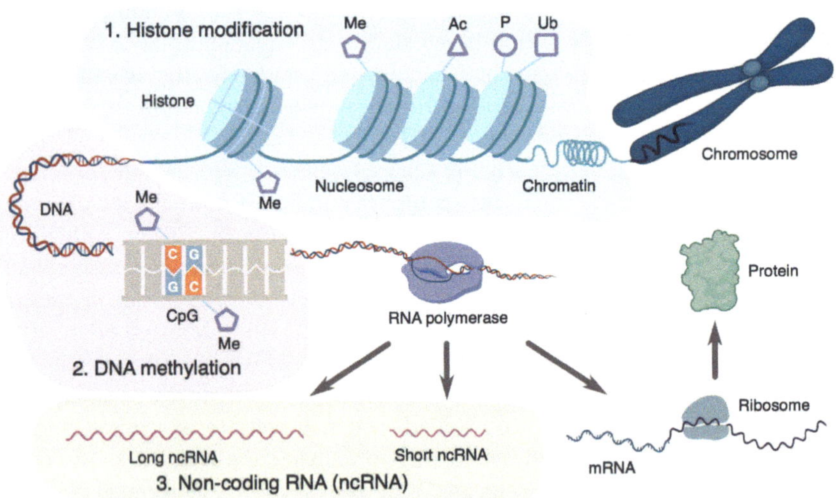

FIG. 1 Three main classes of epigenetic regulators. (1) Histone modifications are posttranslational modifications of eight core histone proteins on their tails, including methylation (Me), acetylation (Ac), phosphorylation (P), and ubiquitination (Ub). Both nucleosomes and chromatins are key structures of DNA packing within chromosomes. Chromatin is a complex of DNA and proteins with nucleosomes as basic repetitive unit. Nucleosome consists of core histone octamer coiled around by DNA. (2) Cytosine methylation at a CpG site is the most common DNA methylation in higher eukaryotes. (3) Noncoding RNAs (ncRNAs) are a class of RNAs that are transcribed by RNA polymerase II and III, but avoid translation by ribosome, resulting in no protein output. ncRNAs are further categorized by their lengths into short ncRNAs (<200 nt) and long ncRNAs (>200 nt). The figure was created using BioRender.com.

Epigenetic regulators share common aspects among vertebrates, but the divergence of functionality is widespread outside vertebrates (Lewis et al., 2020). Prokaryotes have no histones, and most insects and fungi have different types of DNA methylation from vertebrates (Antequera et al., 1984; Bewick et al., 2017). The difference of ncRNA profiles between plants and metazoans is also considerably divergent (Yu et al., 2019).

Nutrition has been traditionally well studied in aquaculture-related fish species and in addition is a major external factor that influences epigenomes. Therefore nutritional epigenetic research in fishes may gain a unique position since they have all three epigenetic regulators similar to those of mammals, the nutritional requirement levels for several species are known and can be controlled at fine-tuned levels, and for some species, their relative short generation time makes it possible for experiments to continue over several generations.

This chapter will specifically focus on DNA methylation and DNA/histone complexes. These epigenetic modifications, located on the genome, operate to control gene expression through the structure of the chromatin (Atlasi and Stunnenberg, 2017) and these structural changes result from the activity of specific enzymes which utilize nutrients directly or as intermediate metabolites in the process (Ducker and Rabinowitz, 2017; Finkelstein et al., 2015; Wang et al., 2018).

2.1 DNA methylation

In brief, DNA methylation is the most examined epigenetic mechanism and is implicated in cell proliferation, determination, and differentiation. Most vertebrates, including mammals, reptiles, and fishes, have their CpG sites (cytosine followed by guanine) extensively methylated (Fig. 1). The DNA can be methylated at cytosines to generate 5-methylcytosine. Cytosine (C) methylation, at CpG sites, is usually symmetric, as a CpG site is methylated when the CpG site of the opposite strand is methylated. Cytosine methylation in higher eukaryotes occurs exclusively in CpGs, but it does not need to be within CpG islands (clusters of high frequent CpG sites). CpG islands in promoter regions tend to have low DNA methylation levels. The definition of a CpG island is simply a CpG-rich region.

DNA methyltransferases are the enzymes that transfer methyl groups to cytosines utilizing *S*-adenosylmethionine (SAM) as the methyl donor. The molecular networks that adjust the SAM availability can thereby modify the genome accessibility, and recent studies have demonstrated that the nutrient availability and quality have a much broader role in mediating the cellular response (Dai et al., 2020). For fish, the cellular concentration of SAM depends on the feed ingredients and the micronutrient concentration in fish diets (Espe et al., 2014; Wischhusen et al., 2020a; Adam et al., 2021). DNA methylation is strongly associated with several histone tail modifications; hence, it can be used as strong markers to indicate gene activities. Moreover, DNA methylation is reversible, which requires several steps from methylated C to unmethylated C via hydroxymethylated C (Kriaucionis and Heintz, 2009). Cytosine hydroxymethylation is increasingly seen as an important epigenetic modification and part of the cells' signaling cascade. DNA methylation and demethylation involves several micronutrients and other metabolites in the signaling cascade (see below, Section 3 and Fig. 3).

DNA methylation plays a key role to establish spatiotemporal patterns of gene expression (De Paoli-Iseppi et al., 2017). DNA methylation status is reset or "washed out" in preblastula stages and during germ cell specification as most methylated cytosine sites turn demethylated (Smallwood and Kelsey, 2012; Lokk et al., 2014). Thereafter, diverse spatiotemporal patterns are gradually established as cytosine sites are progressively methylated (Andersen et al., 2012b).

DNA methyltransferases (DNMTs) are the enzymes that transfer a methyl group to DNA. For vertebrates there are both maintenance (*dnmt1*) and de novo methyltransferases that establish DNA methylation patterns (Jones and Liang, 2009). The main difference is that *dnmt1* is responsible for maintenance of DNA methylation during cell replication, whereas the de novo methyltransferases can methylate new methylation patterns depending on cell type or function (Goll and Bestor, 2005). Teleosts (and other vertebrates) generally have one single-copy gene of *dnmt1* (except that two copies are identified in rainbow trout (*Oncorhynchus mykiss*) and Atlantic salmon (*Salmo salar*) (Liu et al., 2020)), however several paralogs of de novo methyltransferases have arisen from evolutionary genome duplication events (Liu et al., 2020; Campos et al., 2012; Skjaerven et al., 2014). Teleostei have two *dnmt3a*-type genes and three main *dnmt3b*-type genes, but the nomenclature can be confusing probably due to the complexity around the genome duplication events combined with the variations between the species. Several names have been proposed for the teleost de novo methyltransferases. The Ensembl names for *dnmt3a* are *dnmt3aa* (also called *dnmt8*) and *dnmt3ab* (also called *dnmt6*) whereas for the *dnmt3b* genes are called *dnmt3bb.1* (also called *dnmt4*), *dnmt3bb.2* (also called *dnmt3*), *dnmt3bb.3* (also called *dnmt5*), and *dnmt3ba* (also called *dnmt7*). Assessment of the genomic loci of the *dnmt3b*-type genes suggests that the *dnmt3bb.1* and the *dnmt3b* paralogs (*dnmt3bb.2* and *dnmt3ba*) are tandemly duplicated (Skjaerven et al., 2014). The actinopterygian *dnmt3bb.1* and *dnmt3bb.2* are tandem duplicates, but teleosts encode the third gene (*dnmt3ba*) on a separate linkage group. This tandem arrangement is conserved throughout the teleost lineage, but with different species such as zebrafish (*Danio rerio*) encoding extra gene copies of *dnmt3bb.2* called *dnmt3bb.3*, whereas both trout and salmon have extra copies of both the *dnmt3bb.1* and *dnmt3ba* (Liu et al., *2020*). Based upon the locus and phylogenetic clustering, teleost *dnmt3b* orthologs are the product of the teleost-specific whole genome duplication (WGD) event at the root of *dnmt3bb.2* and *dnmt3ba*, as the spotted gar (*Lepisosteus oculatus*) contains both *dnmt3bb.1* and *dnmt3ba*. Similarly, the dual paralogy of the teleost *dnmt3aa* and *dnmt3ab* genes in relation to the single spotted gar *dnmt3a* orthologs is further telling that the extra *dnmt3ab* gene copy is the product of WGD (Skjaerven et al., 2014).

The primary function of promoter vertebrate DNA methylation is silencing (Watt and Molloy, 1988; Boyes and Bird, 1991). Repeated sequences, such as transposons originated from ancient viruses, are extensively methylated as basic defense mechanisms in early development stages (Slotkin and Martienssen, 2007). In gene bodies, CpG sites in exons of highly expressed genes tend to be methylated especially for those remotely located further from the transcription start site (TSS), which presumably assures the prevention of irregular transcription initiation (Maunakea et al., 2010). In enhancers, a single methylated site is enough to alter DNA binding affinity of transcription factors when they have a CpG site in their binding motifs (Zhu et al., 2016).

Conversely, unmethylated CpG sites are useful markers for the identification of active gene regulation. Promoter regions of active genes, especially around TSS, often avoid DNA methylation to bypass the potential disruption of regulatory elements caused by DNA methylation. Also, highly expressed genes, such as house-keeping genes, tend to have unmethylated CpG islands in their promoter regions.

The most widely used methods to detect DNA methylation are bisulfite treatment-based technologies (Chatterjee et al., 2012). Bisulfite treatment converts unmethylated cytosine (C) to uracil (U) while methylated cytosine remains unaffected. Converted uracil is replaced by thymine (T) in the subsequent library preparation processes. Whole genome bisulfite sequencing (WGBS) is genome-wide DNA sequencing for bisulfate treated DNA, whereas reduced representation bisulfite sequencing (RRBS) uses restriction enzymes to target only CpG-rich regions in the genome prior to sequencing (Meissner et al., 2005). Read aligners of bisulfate sequencing need to handle C-T conversion before performing actual alignment. Bismark (Krueger and Andrews, 2011) and Bisulfite Sequence Mapping Program (BSMAP) (Xi and Li, 2009) are two popular and highly cited bioinformatic algorithm-based tools specialized for aligning bisulfate sequencing reads on the genome.

Once reads are aligned, differential methylation calling can be performed to compare two or multiple groups, aiming to identify differentially methylated CpG sites (DMCs) and differentially methylated regions (DMRs). DMCs and DMRs are labeled as either hypermethylated or hypomethylated when the differences of methylation are, respectively, positive and negative based, compared to the control group. Instead of performing parameter estimation of negative binomial (distribution of read counts) and/or beta-binomial (distribution of methylation rates), the logistic regression that models odds ratio of methylation rates can be used to calculate *p*-values of DMCs and DMRs as in methylKit (Akalin et al., 2012).

Apart from WGBS and RRBS, bisulfate treatment has been applied to various technologies, including PCR, microarray, and or pyrosequencing (Tost and Gut, 2007). Moreover, as alternative to bisulfate treatment, MeDIP-seq (methylated

DNA immunoprecipitation sequencing) uses an immunoprecipitation approach with an antibody targeting methylated cytosine sites (Weber et al., 2005). Both WGBS and MeDIP-seq provide genome-wide pictures of the methylation status, whereas RRBS gives limited information represented by only part of the genome. Genome-wide approaches can be useful when no prior knowledge is available regarding the overall methylation patterns for the target species. An evident disadvantage of the whole genome approaches is the high sequencing cost. Furthermore, they tend to generate a large amount of data, which makes the subsequence bioinformatics analysis difficult. RRBS can target cytosine methylation in CpG-rich regions that are most likely linked with a range of important genetic and epigenetic regulations. A potential disadvantage of RRBS is that the patterns and distributions of CpG islands are often species-specific (Cuadrado et al., 2001), therefore it can be difficult to find optimal restriction enzymes to target CpG-rich regions.

2.2 Histone tail modifications

As chromosomes in most eukaryotes contain millions of base pairs, DNA needs to be tightly packed inside the nucleus into small structural units called nucleosomes (Kornberg and Lorch, 1999). DNA can be protected when it is compactly packed; nonetheless, the accessibility of DNA binding proteins is also restricted. Epigenetic regulation provides a dynamic alternation of regional accessibility within chromosomes by remodeling chromatin structures and shifting nucleosome positions.

Chromatin is a complex of DNA and proteins that has nucleosomes as the basic repetitive unit. The nucleosome consists of eight core histone proteins (two each from H2A, H2B, H3, and H4) with DNA coiled around. The histones differ due to molecular weight and amino acid composition but are almost identical in all eukaryotes and specifically conserved in the amino acid profile for H3 and H4. Histones are very rich in the basic amino acids arginine and lysine. N-terminus tails of core histones are under the influence of posttranscriptional modifications (PTMs) like other regular proteins (Rothbart and Strahl, 2014). Each one of the histones can be modified posttranslationally; specific amino acid residues may be enzymatically methylated, acetylated, phosphorylated, and/or ubiquitinated to modify the shape and electric charge of the histones (Mohammed et al., 2020). Most well studied histone tail PTMs are methylation on lysine (K) and arginine (R), and acetylation. Also, methylation can be mono-, di-, or tri-methylation. Different histone modifications depend on various metabolite intermediates such as nicotinamide adenine dinucleotide (NAD^+), which is an essential cofactor for some histone deacetylases; SAM for histone methylation; and β-hydroxybutyrate which inhibits many zinc-dependent histone deacetylases (HDAC) (Fig. 3) (Aon et al., 2016). Protein machinery that interacts with histone tail PTMs can be divided into three categories: erasers, readers, and writers (Gillette and Hill, 2015). Both erasers and writers are enzymes that are responsible for the deletion and addition of PTMs, respectively.

Epigenetic molecular readers recognize PTMs and take part in various biological roles, such as recruitment of transcription factors and other regulatory proteins. Histone marks represent different states of histone tail PTMs that strongly associate with various gene regulatory activities as well as chromatin remodeling and nucleosome positioning (Kouzarides, 2007).

Histone marks are usually denoted as core histone + residue + PTM. For instance, H3K9Me1 represents mono-methylation at the ninth lysine residue (K9) from the N-terminus of histone H3. In promoters, H3K4me3 represents active genes. Specifically, genes that are actively transcribed or ready for transcription tend to have H3K4me3 on their promoters (Sims 3rd et al., 2003). Moreover, genes that have strong gene expression, like housekeeping genes, have enriched H3K4me3 markers. Conversely, H3K27me3 indicates gene silencing when it occurs in promoters (Ferrari et al., 2014). The promoter marked by H3K27me3 is often targeted by polycomb group proteins, which play important roles in gene silencing during development (Aranda et al., 2015).

In enhancers, both H3K4me1 and H3K27ac indicate active genes (Rada-Iglesias, 2018; Chapuy et al., 2013). H3K27ac often occurs at proximal regions around TSS of active genes. Also, H3K27ac-rich regions indicate super-enhancers, where multiple enhancers are simultaneously involved in gene regulation (Chapuy et al., 2013). Some marks appear in both promoters and enhancers, like H3K4me2 that also implies active genes (Creyghton et al., 2010). H3K4me2 can be used as effective screening of active genes in combination with other markers.

In gene bodies, H3K36me3 potentially serves as speed bumps of transcription when appeared in exons (Schwartz et al., 2009). The speed of transcription could go faster in introns that are usually H3K36me3 free and error prone. Some markers appear genome wide. H3K9me3 are repressive markers often found in heterochromatin (closed chromatin). Interestingly, H3K9me3 appears temporarily in repeated sequences, which potentially promote DNA methylation for long-term suppression (Zhao et al., 2016).

Chromatin immunoprecipitation sequencing (ChIP-Seq) is the most widely used method to detect histone modifications (Park, 2009). Prior to immunoprecipitation, DNA binding proteins are crosslinked with DNA, and the linked complexes are fragmented by sonication thereafter. Subsequently, a specific antibody is used to pull down target complexes for sequencing. Mapped reads are stacked and summarized to peaks, followed by peak calling procedures that aim to find significant signals (target histone marks) compared to the background noise. A number of algorithms exist for peak calling, such as model-based analysis of ChIP-Seq (MACS) (Zhang et al., 2008), Quantitative Enrichment of Sequence Tags (QuEST) (Valouev et al., 2008), and Bayesian Change Point (BCP) (Xing et al., 2012).

One of the alternative methods to ChIP-seq is CUT&RUN sequencing (Skene and Henikoff, 2017), which utilizes micrococcal nuclease (MNase) that digests DNA in a sequence-dependent manner. CUT&RUN sequencing skips the sonification step required in ChIP-seq, which potentially promote improved signal/noise ratio than that of ChIP-Seq.

2.3 Bioinformatics workflow of high-throughput sequencing data

A wide range of quantitative and qualitative methods, including microarray, DNA sequencing, microscopic, and mass spectrometry, have been developed to analyze epigenetic regulators. Among them, the second-generation high-throughput sequencing technologies have been successfully applied to numerous studies in epigenetics.

The output data from sequencers are sequence readouts, or simply "reads," that are either single end or paired ends. A single end read represents a region of DNA or RNA in fixed length, whereas a set of two paired end reads represents a region with variable lengths flanked by one read from 5′ and the other read from 3′.

The most widely used tools for read quality control are FastQC (FastQC, 2015) for the library level and MultipleQC (Ewels et al., 2016) for summary of the experiment. Quality controlled "clean" reads are aligned and mapped to the genome or the transcriptome often by using an index-based algorithm, such as bowtie (Langmead and Salzberg, 2012) and Burrows-Wheeler Aligner (BWA) (Li and Durbin, 2009) that are based on suffix array or suffix tree algorithms (Shrestha et al., 2014).

After quality control of sequence reads, workflows of bioinformatic analysis are usually divergent, depending on multiple factors, such as target species, types of experiments, and sequencing technologies. For example, the identification of differentially methylated sites between two groups (treatment vs. control) using RRBS can consist of three bioinformatics modules: preprocessing, differential methylation analysis, and postprocessing (Fig. 2). For nutritional epigenetics studies with fish, the treatment group can be a group of fish that are fed diets with different levels of nutrition (Skjaerven et al., 2018; Adam et al., 2019; Saito et al., 2020; Wischhusen et al., 2020b). The aim of preprocessing of the RRBS workflow is to produce quality-controlled input for differential methylation analysis (Fig. 2). Read trimming and genome mapping are usually RRBS-specific processes due to the usage of restriction enzymes, whereas methylation call and clustering analysis are common processes between RRBS and WGBS. Clustering analysis is for finding overall methylation patterns as well as detecting batch effects and potential outliers.

In differential methylation analysis, most of the processes can be commonly used between RRBS and WGBS. The main purpose of this module is to calculate methylation differences between two groups with *p*-values to identify differentially methylated CpG sites (DMCs) and differentially

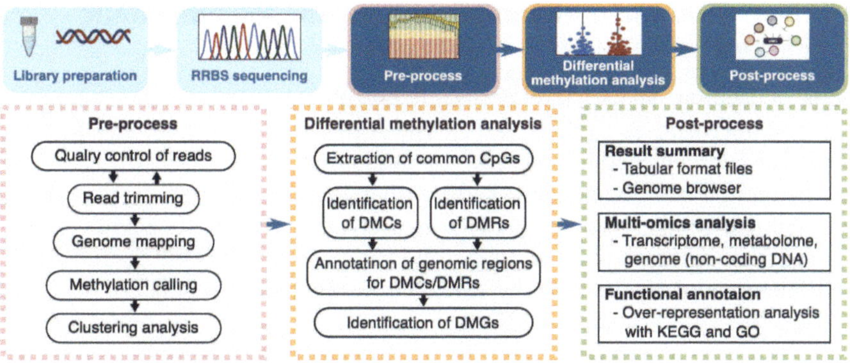

FIG. 2 An example of bioinformatics workflow to analyze differentially methylation sites with RRBS. A workflow diagram shows bioinformatics modules that are necessary to analyze DNA methylation data produced by RRBS sequencing. The flow assumes two groups of DNA samples (treatment vs. control) with multiple biological replicates that are extracted prior to sequencing. The main purpose of the analysis is to find differentially methylated loci/genes affected by the samples in treatment compared to the samples in control. The workflow contains three main modules for bioinformatics analyses: preprocessing, differential methylation analysis, and postprocessing. The processes in preprocessing are mainly responsible for quality control and preparation of the data for the subsequent analysis by aligning reads to the target genome and calculating methylation rates on the mapped CpG sites. In differential methylation analysis, the mapped CpG sites first need to be united to form a set of unique CpG sites that are commonly identified across all the samples from each experimental group. The prime aim of this module is to identify differentially methylated CpG sites (DMCs) and differentially methylation regions (DMRs) with statistical confidence. DMCs and DMRs can be associated with genes to define differentially methylated genes (DMGs) once corresponding genomic regions are determined. In postprocessing, the results are visualized and formatted in a user-friendly manner. Multiple omics data can be combined to investigate potential interactions with different levels of genetic and cellular components. Functional annotation with annotated databases, such as Kyoto Encyclopedia of Genes and Genomes (KEGG) and Gene Ontology (GO), can give biological insights to a set of DMGs.

methylated regions (DMRs). The definition of DMRs can vary (Akalin et al., 2012; Muller et al., 2019), but they are often regions with around 100 nucleotides. It is important to consider multiple testing correction for both DMCs and DMRs, hence *p*-values should be adjusted, or alternatively, *q*-values should be calculated. For subsequent postprocessing, defining differentially methylated genes (DMGs), which are the genes that overlap with at least one DMC or DMR, can be useful.

In postprocessing, the main aim is to represent the result in a user-friendly manner, such as using a genome browser and tabular-format files. Multiomics analysis finds overlaps among different types of omics data from other high-throughput technologies, such as RNA-seq and ChIP-seq, to investigate the genetic and epigenetic interactions at different levels. Functional annotation provides biological insights to DMGs by using curated databases, such as Kyoto Encyclopedia of Genes and Genomes (KEGG) (Kanehisa and Goto, 2000) and Gene Ontology (GO) (Boyle et al., 2004).

3. Nutrients and metabolites influencing the epigenetic regulators

Some aquaculture species have defined requirement levels for each macronutrients and micronutrients (NRC, 2011). The chemical composition, the bioavailability of nutrients, and the presence of antinutrients in the feed will modulate the absorption over the intestine and together alter the homeostasis in both intra- and intercellular spaces. The feed or diet, along with other environmental factors such as temperature or stress, will set a baseline for cellular functions, and interact with the genome by adjusting the epigenetic regulators (Gut and Verdin, 2013). Chromatin-modifying enzymes utilize key epigenetic nutrients and metabolites from intermediary metabolism as cofactors or substrates (Fig. 3). Key epigenetic nutrients and metabolites participate in protein, carbohydrate, lipid, 1C metabolism, and glutathione pathways, all involved in energy balance with tricarboxylic acid (TCA) cycle as the central hub, linking catabolic and anabolic pathways (Fig. 3A). Important key intermediate metabolites are SAM and acetyl-CoA. SAM is produced through the 1C metabolism, while acetyl-CoA is synthesized either through the catabolism from certain amino acids or from pyruvate through glycolysis and β-oxidation of fatty acids. Acetyl-CoA and SAM can be utilized directly as substrates for methyltransferases and acetylases which alter DNA methylation or histone tail modifications (Fig. 3B). Interestingly these pathways also involve α-ketoglutarate (α-KG), NAD^+, flavin adenine dinucleotide (FAD), and adenosine triphosphate (ATP). Thus fluctuations in these metabolites result from dietary changes and may thereby affect the epigenetic regulators.

3.1 Folate and methionine cycles constitute the 1C metabolism

The full list of nutrients that interact and influence the epigenetic regulators is constantly increasing. However, some nutrients have a more direct involvement, as for instance the B-vitamins (Wang et al., 2018). The folate cycle, together with the methionine cycle, constitutes the 1C metabolism (Ducker and Rabinowitz, 2017; Xu and Sinclair, 2015). The 1C cycle has a clear association with the epigenetic methylation reactions through SAM (Fig. 3A). Folate (vitamin B9) converts the methyl group from the amino acid serine (vitamin B6 and B2 dependent) to homocysteine to remethylate it to methionine. Remethylation of methionine depends on vitamin B12 (also named cobalamin), as the cofactor for methionine synthase that operates at the intersection between the folate cycle and methionine cycle. This enzyme catalyzes the conversion of a methyl group from 5-methyl-tetrahydrofolate to homocysteine, generating tetrahydrofolate (THF) and methionine (Ducker and Rabinowitz, 2017).

In the methionine cycle, the enzyme methionine adenosyltransferase (MAT, also known as the *S*-adenosylmethionine synthetase) generates SAM

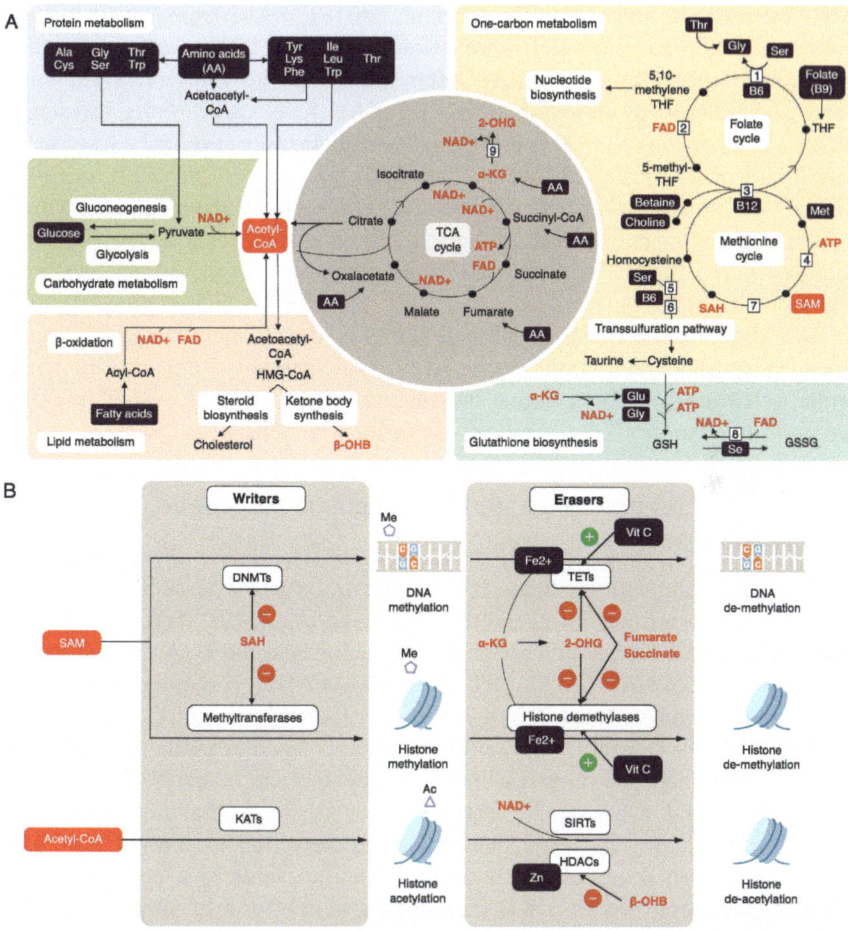

FIG. 3 Key nutrients, pathways, and metabolites in epigenetic regulation. (A) Protein, carbohydrate, lipid, 1C metabolism, and glutathione synthesis and their link to the tricarboxylic acid (TCA) cycle and its intermediate metabolism fundamental to the nutritional influence of epigenetic regulators. (B) Epigenetic writers and erasers mediating methylation or acetylation of DNA or histone tails and their nutritional substrates or cofactors for enzymatic activity. *Dark gray* boxes: key epigenetic nutrients. *Red* boxes/font: key epigenetic metabolites. Numbered enzymes: 1: serine hydroxymethyltransferase, 2: methylenetetrahydrofolate reductase, 3: methionine synthase, 4: methionine adenosyltransferase, 5: cystathionine beta synthase, 6: cystathionine gamma lyase, 7: methyltransferases, 8: glutathione reductase, and 9: malate dehydrogenase. Abbreviations: ala: alanine, cys: cysteine, gly: glycine, glu: glutamate, ser: serine, thr: threonine, trp: tryptophan, tyr: tyrosine, lys: lysine, phe: phenylalanine, ile: isoleucine, leu: leucine, met: methionine, AA: amino acids, CoA: coenzyme A, NAD^+: nicotinamide adenine dinucleotide, FAD: flavin adenine dinucleotide, HMG-CoA: β-hydroxy β-methylglutaryl-CoA, β-OHB: β-hydroxybutyric acid, TCA cycle: tricarboxylic acid cycle, 2-OHG: 2-hydroxyglutaric acid, ATP: adenosine triphosphate, α-KG: α-ketoglutarate, Se: selenium, GSH/GSSG: glutathione (reduced/oxidized), B6: vitamin B6 (pyridoxine), B9: vitamin B9 (folate), B12: vitamin B12 (cobalamin), THF: tetrahydrofolate, SAM: S-adenosylmethionine, SAH: S-adenosylhomocysteine, DNMTs: DNA methyltransferases, TETs: ten eleven translocation enzymes, KATs: lysine acetyltransferases, Vit C: vitamin C, SIRTs: sirtuins, HDACs: histone deacetylases, Fe^{2+}: iron, Zn: zinc.

from methionine and ATP (Wang et al., 2018). SAM transfers a methyl group to biological methylation reactions including DNA methylation and histone tail methylation (Fig. 3). After methyl group transfer, SAM converts to *S*-adenosylhomocysteine (SAH). SAH level can directly influence its own production as SAH inhibits methyltransferases (Fig. 3B). In case of low folate or vitamin B12 levels, homocysteine can be converted to glutathione (the main cellular redox scavenger) via a multistep reaction in which vitamin B6 is an enzymatic cofactor in the transsulfuration pathway and thereby increasing the cellular antioxidant capacity (Fig. 3A) (Ducker and Rabinowitz, 2017).

In addition to the above-mentioned nutrients, the 1C metabolism is also influenced by other dietary components like choline/betaine, serine, glycine, threonine, selenium, and zinc. Signal transduction pathways, like AMPK-mTOR, are also linked to the 1C metabolism via serine availability.

3.2 Acetyl-CoA and α-ketoglutarate regulate epigenetic writers and erasers

Several metabolic intermediates that are central for the cellular respiration and energy expenditure influence the epigenetic regulators. Among them are acetyl-CoA and α-KG, which are both intermediates in the TCA cycle (Fig. 3). The conversion of NAD^+ to NADH, and FAD to $FADH_2$ is required to produce α-KG and ATP through the TCA cycle that is fueled by acetyl-CoA (Fig 3A). The TCA cycle is central for the cellular metabolism and can utilize metabolites from all three macronutrients as lipids (the metabolism of fatty acids through β-oxidation yield acetyl-CoA), carbohydrates (through the metabolism via glycolysis producing pyruvate, which can be converted to acetyl-CoA), and proteins (through deamination of several L-amino acids to α-keto-acids and glutamate that converts to α-KG utilizing the active form of vitamin B6 as a cofactor for aminotransferases).

α-KG serves as cosubstrate and Fe^{2+} as cofactor for both the ten-eleven translocation (TET) enzymes that demethylate DNA and the histone demethylase enzymes that demethylate lysine residues on histone tails (Fig. 3B). α-KG and NADH can be transformed into 2-hydroxyglutarate (2-OHG) and NAD^+ using malate dehydrogenase. TET and histone demethylase activity can be inhibited by 2-hydroxyglutarate, fumarate, or succinate. Vitamin C and Fe^{2+} can also directly affect TET enzymes and histone demethylases through stimulating demethylation of both DNA and histones. Furthermore, histone acetyltransferases (KATs) utilize acetyl-CoA to acetylate lysine residues on histone tails, whereas deacetylation is maintained by zinc-dependent histone deacetylases (HDACs) and diverse NAD^+ dependent sirtuins (SIRTs) that generate nicotinamide. In addition, HDACs can be inhibited by the ketone body β-hydroxybutyrate (β-OHB) synthesized from circulating fatty acids or by acetyl-CoA through β-hydroxy β-methylglutaryl-CoA (also known as HMG-CoA). All together this points toward an active interaction between nutrient adjusted metabolites, the epigenetic regulators and chromatin accessibility.

4. Transfer of cell memory utilizing the epigenetic regulators

The location and type of epigenetic regulator can be modified by the cellular milieu via the nutrients in the feed. The epigenetic changes will alter the control of active metabolic pathways leading to acquired phenotypes. These phenotypes can be transmitted through cell divisions from mother cells to daughter cells (intragenerational), from parental nutrition to offspring (intergenerational), or persist in offspring from the great-grandparents diet (transgenerational) (Fig. 4A). So far, it is still unknown if the feed given to great-grandparents may lead to persistent epigenetic marks transmitted through meiosis in fish; however, recent research has shown that offspring DNA methylation is altered due to parental nutrition (Skjaerven et al., 2018; Adam et al., 2019) or salinity exposure (Heckwolf et al., 2020).

As with other nonmammalian animals, the yolk in fish eggs ensures a nutrient-dependent development as the embryonic development proceeds through the embryonic stages. Firstly, the maternal and paternal haploid germ cells merge to one zygote containing one genome (Lubzens et al., 2010, 2017). Thereafter, the cleavage stage, characterized by rapid synchronous mitotic divisions of egg cytoplasm on top of an enormous yolk, which sequentially reduces the cell size while increasing the cell number. At the blastula stage, the embryo contains about 1 K cells whereby most of them are yet undifferentiated pluripotent cells. Thereafter, the cells start to differentiate alongside coordinated cell and tissue movements with the following zygotic gene activation with active translation during the gastrula stage. Major tissue morphogenesis occurs during the organogenesis stage which eventually results in hatching and the free-swimming larvae stage (Fig. 4B). Throughout this developmental period, the yolk plays a pivotal role influencing the metabolism by nourishing the proliferating and differentiating cells of the growing embryo until hatch, adjusting the interior milieu and contributing to establish the cell- and tissue-specific epigenomes already prior to hatching (Fig. 3B). Interestingly, micronutrients such as folate, vitamin-B6, -C, and -E are transferred from the yolk and into the larval body at different developmental rates in Atlantic halibut (*Hippoglossus hippoglossus*) (Mæland et al., 2003; Rønnestad et al., 1997, 1999).

A few fish nutrition experiments, focusing on DNA methylation (Table 1) and/or histone modifications (Table 2), have provided new perceptions for nutritional epigenetics in fish. They revealed that feed composition or dietary micronutrient levels can adjust epigenetic regulators and alter the metabolic activity in specific pathways in a spatiotemporal pattern and thereby establish persistent phenotypes.

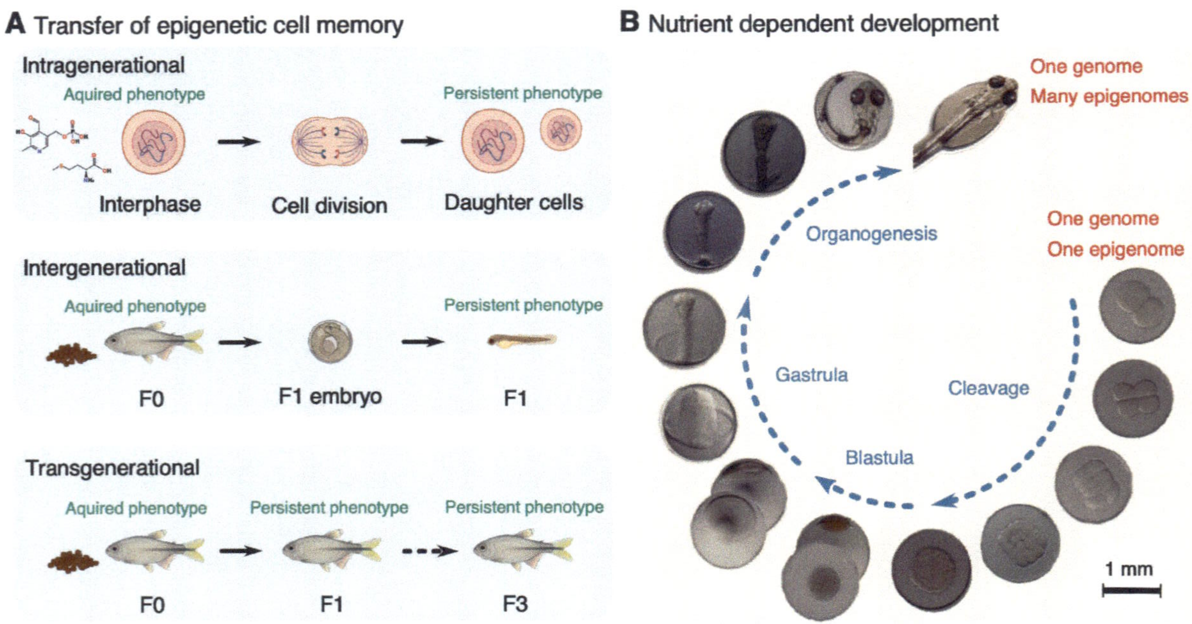

FIG. 4 (A) Transfer of epigenetic cell memory from acquired phenotype to persistent phenotype. Intragenerational inheritance: from mother cell to daughter cells during cell replication. Intergenerational inheritance: from parental diets to offspring. Transgenerational inheritance: from great-grand parental diets to future generations (F3). (B) Nutrient-dependent development due to yolk nutrient content throughout embryonic cleavage, blastula, gastrula, and organogenesis stages. The newly fertilized egg contains one genome, which through developmental cell proliferation and differentiation acquires several cell-specific epigenomes whereby the yolk contributes via availability of central epigenetic nutrients and metabolites (Fig. 3). Photos for illustration, pelagic eggs from Atlantic cod (*Gadus morhua*) with diameter approximately 1.4 mm (Skjaerven et al., 2014).Panel (A) of the figure was created with BioRender.com.

TABLE 1 Studies reporting global and locus-specific DNA methylation changes between nutritional treatments in different fish species.

Authors (year)	Type of study Species/tissue	Nutritional treatment window, stimulus or challenge	Method	Methylation measure	Methylation difference magnitude	Effect and phenotype reported
Skjaerven et al. (2018)	Intergenerational (F1) Zebrafish (*Danio rerio*) liver	Parental life cycle feeding. Parental methionine and B-vitamin supplementation in a plant-based diet	RRBS	2869 DMCs	≥25%	Reduced liver lipid inclusion, mRNA expression, no growth effects
Adam et al. (2019)	Intergenerational (F1) Zebrafish (*D. rerio*) liver	Parental life cycle feeding. Parental arachidonic acid supplementation in a plant-based diet	RRBS	2142 DMCs	≥25%	mRNA expression, no growth effects
Dhanasiri et al. (2020)	Intragenerational Zebrafish (*D. rerio*) intestine	Parental adult feeding. Plant protein-based diets (30% pea, soy, or wheat)	RRBS	1739, 3607 and 3568 DMCs	≥25%	Mild inflammation in mid-intestine (histology), no nutritional analyses
Marandel et al. (2016)	Intragenerational Rainbow trout (*Oncorhynchus mykiss*) liver	Juvenile feeding. Carbohydrate-enriched diet	ELISA, targeted next generation bisulfite sequencing	Global, 2 DMGs	≤10% (global), n/a (DMGs)	Global hypomethylation, hypomethylation of 2 DMGs, hyperglycemic phenotype
Panserat et al. (2017)	Intragenerational Rainbow trout (*O. mykiss*) liver	Larval feeding. Vitamin supplementation at first feeding	ELISA	Global	≤10% (global)	Global hypomethylation, no growth effects
Veron et al. (2018)	Intergenerational (F1) Rainbow trout (*O. mykiss*) fry	Parental adult and fry feeding. Parental and fry methionine deficiency stimulus (combined effect with hypoxia)	Targeted next generation bisulfite sequencing	2 DMGs	≤10%	No growth effects

Continued

TABLE 1 Studies reporting global and locus-specific DNA methylation changes between nutritional treatments in different fish species—cont'd

Authors (year)	Type of study Species/tissue	Nutritional treatment window, stimulus or challenge	Method	Methylation measure	Methylation difference magnitude	Effect and phenotype reported
Song et al. (2019)	Intragenerational Rainbow trout (*O. mykiss*) muscle	Larval feeding. High dietary carbohydrate (low protein) stimulus at first feeding	LUMA	Global	≤10% (global)	Global hypomethylation, glucose metabolism
Wischhusen et al. (2020b)	Intergenerational (F1) Rainbow trout (*O. mykiss*) swim up fry liver	Parental adult feeding, supplementation with two selenium sources in plant-based diets	RRBS	6535 and 6890 DMGs	≥20%	1C metabolism and transsulfuration pathway
Vera et al. (2017)	Intragenerational Atlantic salmon (*Salmo salar*) liver, triploid	Larval and juvenile feeding. Plant-based dietary stimulus at first feeding	HPLC	Global	≤10% (global)	Improved growth and nutrient utilization
Irvine et al. (2019)	Intragenerational Atlantic salmon (*S. salar*) liver	Juvenile feeding. Increased dietary fish oil levels	Sodium bisulfite pyrosequencing	4 DMGs	≤10%	No growth effects, weak methylation difference on single CpGs
Saito et al. (2020)	Intragenerational Atlantic salmon (*S. salar*) liver	Juvenile to adult feeding. Micronutrient supplementation in three (L1 – 3) plant-based diets	RRBS	2521 (L2) and 2555 (L3) DMCs	≥20%	Methylation follows dietary levels in a dose-dependent manner, improved growth
Perera and Yufera (2017)	Gilthead sea bream (*Sparus aurata*) larvae	Larval feeding. SBM-enriched diet at first feeding	ELISA	Global	≤10% (global)	Global hypomethylation, reduced larval growth when feeding SBM beyond 14 days post hatch

Turkmen et al. (2019)	Intergenerational (F1) Gilthead sea bream (*S. aurata*) liver	Parental adult and juvenile feeding. Parental diet enriched with vegetable oil (high α-linolenic acid) and juvenile fish oil challenge diet	Bisulfite pyrosequencing	1 DMG	≤10%	Reduced growth, lipid metabolism
Perera et al. (2020)	Intergenerational (F1) Gilthead sea bream (*S. aurata*) liver	Adult feeding. Parental diet enriched with vegetable oil (high α-linolenic acid) and juvenile challenge diet	Bisulfite pyrosequencing	1 DMG	≤10%	DMG promoter hypermethylation and inverse correlation with expression, reduced liver fat deposition
Cai et al. (2020)	Intragenerational Grass carp (*Ctenopharyngodon idellus*) liver	Adult feeding. Carbohydrate-enriched diet	WGBS	21,542 DMGs	n/a	Decrease in weight gain and food intake
Xu et al. (2014)	Intragenerational Japanese seabass (*Lateolabrax japonicus*) liver	Juvenile feeding. Varying lipid sources and LC-PUFA levels in the diets	BSP	1 DMG	≤10%	Fatty acid metabolism, inverse correlation of promoter methylation and gene expression
Liao et al. (2015)	Intragenerational Large yellow croaker (*Larimichthys crocea*) liver	Adult feeding. Dietary olive and perilla oil levels	Bisulfite-PCR pyrosequencing of mtDNA	3 DMGs	≤10%	Liver lipid content, no growth effects
Liao et al. (2016)	Intragenerational Large yellow croaker (*L. crocea*) liver	Adult feeding. Dietary crude lipid levels	Bisulfite-PCR pyrosequencing of mtDNA	2 DMGs	≤10%	Liver weight, no specific growth rate effects

BSP, bisulfite genomic sequencing PCR; *DMC*, differentially methylated CpG site; *DMG*, differentially methylated gene in a CpG context; *ELISA*, enzyme-linked immunosorbent assay; *HPLC*, high-performance liquid chromatography, LC-PUFA: long chain polyunsaturated fatty acids; *LUMA*, luminometric methylation assay; *mtDNA*, mitochondrial DNA; *RRBS*, reduced representation bisulfite sequencing; *SBM*, soybean meal; *WGBS*, whole genome bisulfite sequencing.

TABLE 2 Studies reporting global histone modification measures and site-specific histone modification changes between nutritional treatments in different fish species.

Authors (year)	Type of study Species/tissue	Nutritional treatment window, stimulus or challenge	Analysis	Type of histone modification	Effect	Phenotype reported
Perera and Yufera (2017)	Intragenerational Gilthead sea bream (*Sparus aurata*) larvae	Larval feeding. SBM-enriched diet at first feeding	Fluorometric assay, ELISA	H3 and H4 acetylation and methylation	Global H3 hyperacetylation, treatment duration dependent	Reduced larval growth when feeding SBM beyond 14 days post hatch
Marandel et al. (2016)	Intragenerational Rainbow trout (*Oncorhynchus mykiss*) liver	Juvenile feeding. Carbohydrate-enriched diet	Western blot	H3K4me3, H3K9me3, H3K9ac and H3K36me3	Global H3K9 hypoacetylation	Hyperglycemic phenotype
Panserat et al. (2017)	Intragenerational Rainbow trout (*O. mykiss*) liver	Larval feeding. Vitamin supplementation at first feeding	Western blot	H3K4me3, H3K9me3, H3ac and H4ac	Global H3K4 hypomethylation, global H3 hypoacetylation	No growth effect
Seite et al. (2019)	Intragenerational Rainbow trout (*O. mykiss*) liver	Fry and juvenile feeding. Methionine deficiency at first feeding and as juvenile challenge	Western blot	H3K4me3, H3K9me3 and H3K36me3	H3K4me3 hypermethylation	No growth effect in fry, increased growth after the challenge
You et al. (2020)	Intragenerational Mandarin fish (*Siniperca chuatsi*) liver	Juvenile feeding. Carbohydrate-enriched diet	Western blot, ChIP assay	H3K4me3	H3K4me3 hypermethylation	Decreased feed intake (anorexia), hyperglycemia
Terova et al. (2016)	Intragenerational European sea bass (*Dicentrarchus labrax*) liver	Juvenile feeding. Sodium butyrate supplementation in a plant-based diet	Western Blot	Histone acetylation	H4K8 Hyperacetylation	No growth effect

ChIP, chromatin immunoprecipitation; *ELISA*, enzyme-linked immunosorbent assay; *SBM*, soybean meal.

4.1 Intragenerational inheritance from cell to cell during growth

For each cell cycle during growth periods, the cell will notice rapid metabolic fluxes and changes in the metabolites and decide for either further proliferation, differentiation, or growth (Gut and Verdin, 2013). The epigenetic pattern of the cell types will be maintained through the enzymatic activity of, for instance, DNA methyltransferases and histone methyltransferases, which respond to intracellular SAM levels (Mentch and Locasale, 2016). For pluripotency versus differentiation of mammalian embryonic stem cells, this sensing of the cell metabolome has shown that the SAM:SAH ratio determined by threonine via glycine and THF (Fig. 3A) can specifically alter the methylation of histones (Gut and Verdin, 2013). However, even postmitotic cells will govern dynamic changes in DNA methylation depending on metabolic stimuli. As such, throughout fish development and life course, the nutrients and metabolites available within each tissue type will establish the foundation for cellular functions through epigenome-diet interaction and intragenerational inheritance (Fig. 4A).

Nutritional programming, an expanding approach in fish nutrition research, has been applied to various fish species, including farmed and model fish such as zebrafish. The early growth period with their remarkable plasticity serves as open sensitive windows to prime fish through nutritional cues to better adapt to diets or environments at later developmental stages.

In intensive aquaculture production, the quantity and quality of fish feed used have direct effects on fish growth and thereby improved profitability for the industry. As such, huge efforts have been made to establish requirement levels in practical diets to avoid either a deficiency or addition of unnecessary high levels of cost-expensive vitamins or proteins (Panserat et al., 2019). Furthermore, to make the sector utilize sustainable feed ingredients, both to reduce the dependency on limited wild fish resources (fishmeal and fish oil) and to increase the volume of available protein and lipid sources for an increasing aquaculture industry, the requirement levels need continuous reconsideration to sustain optimal fish growth and welfare (Hemre et al., 2016; Espe et al., 2020; Taylor et al., 2019). For Atlantic salmon, this transfer from marine- to plant-based diets led to a new evaluation of the micronutrient package. The micronutrient package included 1C nutrients as well as other vitamins and cholesterol and was evaluated throughout the whole life cycle for aquaculture production (www.arraina.eu). More specifically, addition of the micronutrient package in double amounts (L2) or approximately four times of the amounts (L3) recommended in NRC 2011 (L1) increased the growth and welfare but concluded that the L2 level fulfilled the requirements. In terms of DNA methylation, the addition of the micronutrient package revealed 2521 DMCs for L1:L2 and 2555 DMCs for L1:L3 in promoter regions in liver tissues when setting a cutoff of 20% difference in methylation level between the diets compared (Saito et al., 2020). Furthermore, some DMCs show the same response in terms of methylation when fed either the L2 or L3 diet, others reveal higher methylation in L2 than L3, and others lower in L2 than L3. This suggests that the epigenetic changes are independently regulated for each gene promoter, while others are directly regulated in a dose-dependent manner (Saito et al., 2020). The genes epigenetically regulated by the micronutrient package are involved in several cell signaling pathways or regulatory mechanisms including immune functions, phosphorylation, both the folate and methionine cycle in the 1C metabolism, epigenetic writers and erasers mediating methylation or acetylation of DNA or histone tails, transcription factors, and fatty acid synthesis (Saito et al., 2020). These findings show that nutrients given in the feed induced a wave of alterations in DNA methylation level in salmon genome promoters regulating gene accessibility and potential for gene expression. Firstly, several genes are controlled in a dose-dependent manner (L3 > L2 > L1), and secondly, some of the DNA methylation changes affect the liver transcriptome directly like for *acaca*, while others like *cd28* were not transcriptionally altered in the liver. Future studies are needed to determine if they are controlling gene expression in other tissues, developmental stages, or time points.

A stimulus of surplus vitamins or increased amounts of carbohydrates at first feeding changed global DNA methylation, global histone acetylation, and methylation in rainbow trout (Panserat et al., 2017; Marandel et al., 2016; Song et al., 2019). Rainbow trout fed carbohydrate-enriched diets displayed hepatic global DNA hypomethylation and hypoacetylation of histone H3K9 (Marandel et al., 2016). Studies performed in gilthead sea bream (*Sparus aurata*) larvae, mandarin fish (*Siniperca chuatsi*) liver, and rainbow trout liver also reported global histone modification changes due to different larval and juvenile feeding strategies using various plant-based sources or supplementing diets with either vitamins or carbohydrates (Table 2). Varying sources of lipids or dietary fatty acid amounts affected liver lipid metabolism associated with hepatic DNA methylation changes observed in gilthead sea bream (Turkmen et al., 2019), Atlantic salmon (Irvine et al., 2019), and Japanese seabass (*Lateolabrax japonicus*) (Xu et al., 2014).

Methionine is a central metabolite in the 1C metabolism (Fig. 3A) and thereby acts as a potential epigenetic modulator of nutritional programming in fish. Nutritional studies performed in Atlantic salmon, rainbow trout, and zebrafish showed that DNA methylation is sensitive to dietary levels of methionine (supplementation or deficiency), alone or in combination with other important micronutrients such as B-vitamins (Skjaerven et al., 2018; Veron et al., 2018). Together with the studies presented in Tables 1 and 2, the above-mentioned feeding trials demonstrate the contribution of epigenetic regulators in nutritional epigenetics due to feed ingredients given to larval, fry, and/or adult fish. These trials increase the understanding of the importance of nutrients in molecular events and intermediary metabolism for improved health, growth, and reproduction in farmed fish.

4.2 Nutritional intergenerational inheritance from broodstock to offspring

Among vertebrate embryos, it is known that the gene regulatory networks are highly conserved and tightly structured, as these networks control fundamental developmental processes. Karl Ernst von Baer was one of the foremost embryologists in the early 18th century and apart from his discovery of the mammalian egg, he noticed the striking similarities between vertebrate embryos. These common features are regulated by the expression of the same genes or gene networks during development, but also during oogenesis. There is a high number of orthologous genes linked to key processes in cellular energy metabolism, which exhibit similar expression profiles in evolutionary distant vertebrate species (Charlier et al., 2012). However, the zebrafish offspring fine-tune the gene expression profiles to encounter environmental exposures, and parental micronutrient status is one of the environmental signals that results in a massive altered gene expression profile at the organogenesis stage, which is an evolutionary conserved developmental stage among vertebrates (Skjaerven et al., 2016).

As for insects, reptiles, and birds, the egg volume of fish eggs is primarily yolk. The large yolk content is an evolutionary adaptation that enables embryonic development through organogenesis and establishment of most critical physiological organismal processes in the absence of external food resources. The yolk syncytial layer (YSL) will function as the embryo's "intestinal epithelium" and the nutrients enclosed in the yolk will be transferred through YSL and into the perivitelline space and later via the embryonic circulatory system. The yolk contains all the amino acids, proteins, lipids, carbohydrates, calcium, vitamins, and metals that the egg needs to develop from a single cell to a free-swimming larvae ready to nourish itself (Lubzens et al., 2017; Brooks et al., 1997). Lipids and proteins are the most important nutrients to provide energy for the growing embryo (Izquierdo et al., 2001), but their amount and ontogenetic profile depend on the species considered (Finn and Fyhn, 2010). Vitellogenins, the major precursors to the egg yolk proteins, are produced in the liver and secreted as large 350–600 kDa phosphoglycerolipoprotein dimer taken up by receptor-mediated endocytosis in the growing oocyte to provide nutrients for the early embryo. The lipids in the vitellogenins are mainly phospholipids, whereas yolk proteins include lipovitellin (both heavy chain and light chain), phosvitin, β-component, and a C-terminal peptide and other complexes which vary depending on the species (Finn, 2007a, 2007b).

Early development and its physiological processes are mostly regulated via maternally loaded mRNAs, and the paternal and maternal DNA methylation patterns present in the oocyte and sperm haploid pronucleus become largely demethylated (Wang and Bhandari, 2019). After fertilization, fish eggs undergo meroblastic discoidal cleavage whereby the cleavage furrows do not penetrate the yolk all the way to the vegetal pole. The blastomeres develop on top at the animal pole of the egg. Cleavage cells maintain open connections so that 17-kDa metabolites can pass from the yolk but also between the blastomeres. The cleavage stage is characterized by rapid, synchronous cleavage of cells leading to increased cell number as the cell size decreases (Fig. 4B).

At mid-blastula transition, the embryo will start a major transcriptional wave of zygotic gene activation, transcribing its own DNA (Tadros and Lipshitz, 2009). The blastula cells have the lowest levels of DNA methylation as the majority of the blastula cells are pluripotent cells that can become any of the cells in the body. Several promoters are unmethylated and increase their methylation status as the cells start to differentiate during gastrulation (Andersen et al., 2012a, b).

Around the tenth cleavage (1024 cells), the embryonic cells become motile, the cell divisions will lose their synchrony, and the gastrulation starts (Lubzens et al., 2017; Carvalho and Heisenberg, 2010). The first movement is the epiboly of the blastoderm cells, whereby the blastoderm will undergo differentiation to the three germ layers and early lineage commitment. The cells do not appear phenotypically different from its uncommitted state, but there are overt changes in cellular biochemistry followed by functional changes. As the embryonic cells start to transcribe their own DNA, producing their own mRNAs and proteins, then DNA methylation increases to create cell-type specific DNA methylation profiles (Fig 4B). Progenitor cells generally have less methylation than the differentiated cells (Bock et al., 2012). The nuclei that form the YSL can be seen as a ring underlining the blastoderm (Carvalho and Heisenberg, 2010). The YSL nuclei will form a connection with the enveloping layer (the top cell sheath of the blastula). During gastrulation the cells will envelop the yolk content completely at 100% epiboly (Carvalho and Heisenberg, 2010).

During organogenesis, the cells of the three germ layers acquire specialized functions organized into organs. The DNA methylation landscape, its genes and expression patterns follow the specialized cell type.

Embryonic development is also the time for the primordial germ cells to migrate from the position where they are formed toward the site for gonad development where they differentiate into sperm and oocytes (Raz, 2003) that will give rise to the next generation. This means that also for fish, the next generation (F2) is influenced by the parental nutrient status via the yolk content (Skjaerven et al., 2016).

The broodstock diet has tremendous effects on fecundity, fertilization, deformities, hatching rate, and robustness of the larvae. Defining requirement levels for broodstock fish in various species has been problematic due to high running costs of intergenerational broodstock feeding trials. The solution has generally been to make a broodstock feed which is rich in essential fatty acids, vitamins, and minerals (Izquierdo et al., 2001). An enriched broodstock diet is expensive and may not be beneficial, as we know that too high surplus of vitamins might be energetically expensive and thereby reduce growth (Espe et al., 2020). Broodstock diets should also be reevaluated to fulfill nutritional requirements for different production regimes (Skjaerven et al., 2020). The yolk includes a variety of transcripts that accumulate during oocyte growth which have specific functions during oocyte maturation or during embryonic development (Lubzens et al., 2017).

During the embryonic nutrient-dependent development, studies have shown that the expression profile of the genes encoding epigenetic erasers and writers is highly conserved among vertebrates, but also dynamic in terms of environmental adjustments like temperature and nutrients (Campos et al., 2012; Skjaerven et al., 2014; Wang and Bhandari, 2019; Skjaerven et al., 2020). For both medaka (*Oryzias latipes*) and Atlantic cod (*Gadus morhua*), newly fertilized eggs reveal high levels of *dnmt1*, most likely maternally loaded to the eggs, but depleted as the cells divide during cleavage stage. During gastrulation several of the de novo methyltransferases, which methylate previously unmethylated regions, are actively transcribed in a stage and cell type specific controlled manner (Andersen et al., 2012a). However, this group of genes responds to changes in yolk nutrient composition (Skjaerven et al., 2020), and as such, these genes are especially interesting to study in terms of long-term nutritional programming.

Fish nutrition studies applying nutritional programming through maternal and/or larval (offspring) periods have investigated mechanisms particularly through DNA methylation (Table 1), However, to date, no studies have explored histone modifications in an intergenerational perspective. Zebrafish was used to elaborate effects of a parental life cycle plant-based diet low in the micronutrients methionine, folate, choline, vitamin B6 and B12 that gave a fatty-liver-like phenotype, downregulated mRNA levels of lipid related genes, and gene-specific changes in DNA methylation in adult offspring liver (Skjaerven et al., 2018). Another study reported hepatic DNA methylation changes in adult offspring when parental fish were fed on a diet with only one altered nutrient, using an arachidonic acid-enriched diet (Adam et al., 2019). Rainbow trout fed a methionine-deficient diet for 6 months altered DNA methylation in offspring (Veron et al., 2018). Selenium-enriched diets given to rainbow trout broodstock with two different selenium sources produced changes in offspring fry liver DNA methylation at a high number of DMGs in the 1C metabolism that were associated with parental diet (Wischhusen et al., 2020b). The DNA methylation patterns of selected genes expressed in gilthead sea bream liver were changed when α-linoleic acid-rich oils were supplemented in the parental diet (Turkmen et al., 2019; Perera et al., 2020).

4.3 Transgenerational inheritance of meiotically stable marks through generations in fish

Direct environmental changes, exposure or stimuli, can influence the epigenetic status of somatic cells of an individual (F_0), but meiotically stable epigenetic changes to the F_0 germline are necessary to pass on epigenetic status on to the next generations via the progeny's primordial germ cells (F_1 and F_2, Fig. 4A).

If those epigenetic changes become permanently programmed and transferred to the subsequent F_3 generation that is not exposed to the direct environmental change, then this can be characterized as transgenerational epigenetic inheritance. Transgenerational effects can manifest in F_2 generation given the male F_0 is exposed, and in F_3 generation when F_0 females due to the maternal contribution of the egg yolk content that strikes the development of both F_1 and the primordial germ cells constituting the emerging F_2 generation.

In fish, most of the transgenerational epigenetic research has focused on aquatic toxicology due to changing oceans, environmental burdens, and stressors that can show effects of a changing environment over generations (Baker et al., 2014; Bhandari, 2016; Munday, 2014; Carvan et al., 2017; Olsvik et al., 2014). Nutritional alterations such as caloric restriction or high fat diets have epigenetic transgenerational and phenotypic effects across generations in mammals (Pentinat et al., 2010; Dunn and Bale, 2011; Waterland et al., 2008). If nutrition triggers transgenerational epigenetic inheritance across many generations (F_2 or F_3) in fish, it would be an interesting line of research for future advancement in the aquaculture sector in terms of both growth and health.

5. Where do we go with nutritional epigenetics?

For fish, and for domestication of animals in general, several questions remain unsolved for the field of nutritional epigenetics. The field is complex and cross-disciplinary between the fields of reproductive physiology, developmental and molecular biology, and nutrition. Fundamental criteria to consider for nutritional epigenetics studies are the experimental design of feeding experiments (minimum triplicate tanks per group), sample type (choice of organism, tissue, or cells), sample collection, sample analysis, feed formulation, nutritional analyses of feed and tissues, molecular methods, statistics, and bioinformatics methods applied. Awareness must be taken to reveal best comparisons and thresholds to encounter true differences and to reduce false positive assumptions between group epigenetic differences. Because of the cross-disciplinary nature of nutritional epigenetic research, the importance of the reviewers of peer-reviewed papers is crucially valuable to move the field carefully forward for aquaculture.

The intergenerational approach to nutritional epigenetics is of great relevance for aquaculture. Attention toward maturity-dependent nutritional needs for broodstock and the link toward the offspring's phenotypic plasticity for beneficial traits are likely to increase the robustness of the next generation. Modifying feeding strategies, feed ingredients, macro- and micronutrients, minerals, or their amounts to adjust the aquaculture production toward a persistent phenotype based on nutritional epigenetics is a challenge, but also an opportunity for increased robustness and growth of the industry. Domestication of wild fish changes the methylome within just one generation of captivity (Podgorniak et al., 2019). We have seen that the feed given to broodstock will change the mature offspring phenotype, adjust gene expression and DNA methylation

(Skjaerven et al., 2018). Nutritional epigenetics can be viewed as an evolutionary potential for quick adaptation through acquired phenotypes from parental feed or during early life stages, which might act as a domino effect for future growth potential. Optimum cellular nutrient status from first cell divisions might provide a nutritional programming boost in a beneficial metabolic direction and avoid that it becomes a metabolic burden for later life stages or even future generations.

Acknowledgements

This work was supported by the Research Council of Norway through the grant NutrEpi 267787 and by IMRs Nutritional programming project 15469 (Institute of Marine Research, Bergen, Norway).

References

Adam, A.C., Lie, K.K., Whatmore, P., Jakt, L.M., Moren, M., Skjaerven, K.H., 2019. Profiling DNA methylation patterns of zebrafish liver associated with parental high dietary arachidonic acid. PLoS One 14 (8), e0220934.

Adam, A.-C., Saito, T., Espe, M., Whatmore, P., Fernandes, J.M.O., Vikeså, V., Skjaerven, K.H., 2021. Metabolic and molecular signatures of improved growth in Atlantic salmon (*Salmo salar*) fed surplus levels of methionine, folic acid, vitamin B6 and B12 throughout smoltification. Brit. J. Nutr., 1–14. https://doi.org/10.1017/S0007114521002336.

Akalin, A., Kormaksson, M., Li, S., Garrett-Bakelman, F.E., Figueroa, M.E., Melnick, A., Mason, C.E., 2012. MethylKit: a comprehensive R package for the analysis of genome-wide DNA methylation profiles. Genome Biol. 13 (10), R87.

Andersen, I.S., Ostrup, O., Lindeman, L.C., Aanes, H., Reiner, A.H., Mathavan, S., Alestrom, P., Collas, P., 2012a. Epigenetic complexity during the zebrafish mid-blastula transition. Biochem. Biophys. Res. Commun. 417 (4), 1139–1144.

Andersen, I.S., Reiner, A.H., Aanes, H., Alestrom, P., Collas, P., 2012b. Developmental features of DNA methylation during activation of the embryonic zebrafish genome. Genome Biol. 13 (7), R65.

Andersen, S.M., Waagbo, R., Espe, M., 2016. Functional amino acids in fish nutrition, health and welfare. Front. Biosci. (Elite Ed.) 8, 143–169.

Antequera, F., Tamame, M., Villanueva, J.R., Santos, T., 1984. DNA methylation in the fungi. J. Biol. Chem. 259 (13), 8033–8036.

Aon, M.A., Cortassa, S., Juhaszova, M., Sollott, S.J., 2016. Mitochondrial health, the epigenome and healthspan. Clin. Sci. (Lond.) 130 (15), 1285–1305.

Aranda, S., Mas, G., Di Croce, L., 2015. Regulation of gene transcription by Polycomb proteins. Sci. Adv., **1**(11).

Atlasi, Y., Stunnenberg, H.G., 2017. The interplay of epigenetic marks during stem cell differentiation and development. Nat. Rev. Genet. 18 (11), 643–658.

Baker, T.R., Peterson, R.E., Heideman, W., 2014. Using zebrafish as a model system for studying the transgenerational effects of dioxin. Toxicol. Sci. 138 (2), 403–411.

Best, C., Ikert, H., Kostyniuk, D.J., Craig, P.M., Navarro-Martin, L., Marandel, L., Mennigen, J.A., 2018. Epigenetics in teleost fish: from molecular mechanisms to physiological phenotypes. Comp. Biochem. Physiol. B Biochem. Mol. Biol. 224, 210–244.

Bewick, A.J., Vogel, K.J., Moore, A.J., Schmitz, R.J., 2017. Evolution of DNA methylation across insects. Mol. Biol. Evol. 34 (3), 654–665.

Bhandari, R.K., 2016. Medaka as a model for studying environmentally induced epigenetic transgenerational inheritance of phenotypes. Environ Epigenet. 2 (1).

Bock, C., Beerman, I., Lien, W.H., Smith, Z.D., Gu, H., Boyle, P., Gnirke, A., Fuchs, E., Rossi, D.J., Meissner, A., 2012. DNA methylation dynamics during in vivo differentiation of blood and skin stem cells. Mol. Cell 47 (4), 633–647.

Boyes, J., Bird, A., 1991. DNA methylation inhibits transcription indirectly via a methyl-CpG binding protein. Cell 64 (6), 1123–1134.

Boyle, E.I., Weng, S., Gollub, J., Jin, H., Botstein, D., Cherry, J.M., Sherlock, G., 2004. GO::TermFinder—open source software for accessing gene ontology information and finding significantly enriched gene ontology terms associated with a list of genes. Bioinformatics 20 (18), 3710–3715.

Brooks, S., Tyler, C.R., Sumpter, J.P., 1997. Quality in fish: what makes a good egg? Rev. Fish Biol. Fish. 7 (4), 387–416.

Cai, W.J., Liang, X.F., Yuan, X.C., Li, A.X., He, S., 2020. Changes of DNA methylation pattern in metabolic pathways induced by high-carbohydrate diet contribute to hyperglycemia and fat deposition in grass carp (*Ctenopharyngodon idellus*). Front. Endocrinol. 11.

Campos, C., Valente, L.M.P., Fernandes, J.M.O., 2012. Molecular evolution of zebrafish *dnmt3* genes and thermal plasticity of their expression during embryonic development. Gene 500 (1), 93–100.

Carvalho, L., Heisenberg, C.P., 2010. The yolk syncytial layer in early zebrafish development. Trends Cell Biol. 20 (10), 586–592.

Carvan, M.J., Kalluvila, T.A., Klingler, R.H., Larson, J.K., Pickens, M., Mora-Zamorano, F.X., Connaughton, V.P., Sadler-Riggleman, I., Beck, D., Skinner, M.K., 2017. Mercury-induced epigenetic transgenerational inheritance of abnormal neurobehavior is correlated with sperm epimutations in zebrafish. PLoS One 12 (5).

Chapuy, B., McKeown, M.R., Lin, C.Y., Monti, S., Roemer, M.G., Qi, J., Rahl, P.B., Sun, H.H., Yeda, K.T., Doench, J.G., et al., 2013. Discovery and characterization of super-enhancer-associated dependencies in diffuse large B cell lymphoma. Cancer Cell 24 (6), 777–790.

Charlier, C., Montfort, J., Chabrol, O., Brisard, D., Nguyen, T., Le Cam, A., Richard-Parpaillon, L., Moreews, F., Pontarotti, P., Uzbekova, S., et al., 2012. Oocyte-somatic cells interactions, lessons from evolution. BMC Genomics 13, 560.

Chatterjee, A., Stockwell, P.A., Rodger, E.J., Morison, I.M., 2012. Comparison of alignment software for genome-wide bisulphite sequence data. Nucleic Acids Res. 40 (10), e79.

Christiansen, R., Torrissen, O.J., 1997. Effects of dietary astaxanthin supplementation on fertilization and egg survival in Atlantic salmon (*Salmo salar* L). Aquaculture 153 (1–2), 51–62.

Creyghton, M.P., Cheng, A.W., Welstead, G.G., Kooistra, T., Carey, B.W., Steine, E.J., Hanna, J., Lodato, M.A., Frampton, G.M., Sharp, P.A., et al., 2010. Histone H3K27ac separates active from poised enhancers and predicts developmental state. Proc. Natl. Acad. Sci. U. S. A. 107 (50), 21931–21936.

Cuadrado, M., Sacristan, M., Antequera, F., 2001. Species-specific organization of CpG island promoters at mammalian homologous genes. EMBO Rep. 2 (7), 586–592.

Dai, Z., Ramesh, V., Locasale, J.W., 2020. The evolving metabolic landscape of chromatin biology and epigenetics. Nat. Rev. Genet. 21 (12), 737–753.

De Paoli-Iseppi, R., Deagle, B.E., McMahon, C.R., Hindell, M.A., Dickinson, J.L., Jarman, S.N., 2017. Measuring animal age with DNA methylation: from humans to wild animals. Front. Genet. 8, 106.

Dhanasiri, A., Chen, X., Dahle, D., Siriyappagouder, P., Faeste, C.K., Fernandes, J.M.O., 2020. Dietary inclusion of plant ingredients induces epigenetic changes in the intestine of zebrafish. Epigenetics 15 (10), 1035–1051.

Ducker, G.S., Rabinowitz, J.D., 2017. One-carbon metabolism in health and disease. Cell Metab. 25 (1), 27–42.

Dunn, G.A., Bale, T.L., 2011. Maternal high-fat diet effects on third-generation female body size via the paternal lineage. Endocrinology 152 (6), 2228–2236.

Espe, M., Andersen, S.M., Holen, E., Ronnestad, I., Veiseth-Kent, E., Zerrahn, J.E., Aksnes, A., 2014. Methionine deficiency does not increase polyamine turnover through depletion of hepatic S-adenosylmethionine in juvenile Atlantic salmon. Br. J. Nutr. 112 (8), 1274–1285.

Espe, M., Vikesa, V., Thomsen, T.H., Adam, A.C., Saito, T., Skjaerven, K.H., 2020. Atlantic salmon fed a nutrient package of surplus methionine, vitamin B12, folic acid and vitamin B6 improved growth and reduced the relative liver size, but when in excess growth reduced. Aquacult. Nutr. 26 (2), 477–489.

Ewels, P., Magnusson, M., Lundin, S., Kaller, M., 2016. MultiQC: summarize analysis results for multiple tools and samples in a single report. Bioinformatics 32 (19), 3047–3048.

FastQC, 2015. A Quality Control Tool for High Throughput Sequence Data. Available from: https://qubeshub.org/resources/fastqc.

Ferrari, K.J., Scelfo, A., Jammula, S., Cuomo, A., Barozzi, I., Stutzer, A., Fischle, W., Bonaldi, T., Pasini, D., 2014. Polycomb-dependent H3K27me1 and H3K27me2 regulate active transcription and enhancer Fidelity. Mol. Cell 53 (1), 49–62.

Finkelstein, J.L., Layden, A.J., Stover, P.J., 2015. Vitamin B-12 and perinatal health. Adv. Nutr. 6 (5), 552–563.

Finn, R.N., 2007a. The maturational disassembly and differential proteolysis of paralogous vitellogenins in a marine pelagophil teleost: a conserved mechanism of oocyte hydration. Biol. Reprod. 76 (6), 936–948.

Finn, R.N., 2007b. Vertebrate yolk complexes and the functional implications of phosvitins and other subdomains in vitellogenins. Biol. Reprod. 76 (6), 926–935.

Finn, R.N., Fyhn, H.J., 2010. Requirement for amino acids in ontogeny of fish. Aquacult. Res. 41 (5), 684–716.

Gillette, T.G., Hill, J.A., 2015. Readers, writers, and erasers: chromatin as the whiteboard of heart disease. Circ. Res. 116 (7), 1245–1253.

Goll, M.G., Bestor, T.H., 2005. Eukaryotic cytosine methyltransferases. Annu. Rev. Biochem. 74, 481–514.

Gut, P., Verdin, E., 2013. The nexus of chromatin regulation and intermediary metabolism. Nature 502 (7472), 489–498.

Halver, J., Hardy, R., 2002. Fish Nutrition, third ed. Academic Press.

Heckwolf, M.J., Meyer, B.S., Hasler, R., Hoppner, M.P., Eizaguirre, C., Reusch, T.B.H., 2020. Two different epigenetic information channels in wild three-spined sticklebacks are involved in salinity adaptation. Sci. Adv. 6 (12), eaaz1138.

Hemre, G.-I., Lock, E.J., Olsvik, P.A., Hamre, K., Espe, M., Torstensen, B.E., Silva, J., Hansen, A.-C., Waagbø, R., Johansen, J.L., et al., 2016. Atlantic salmon (*Salmo salar*) require increased dietary levels of B-vitamins when fed diets with high inclusion of plant based ingredients. PeerJ 4.

Hou, Z.X., Fuiman, L.A., 2020. Nutritional programming in fishes: insights from mammalian studies. Rev. Fish Biol. Fish. 30 (1), 67–92.

Irvine, N.A., Ruyter, B., Ostbye, T.K., Sonesson, A.K., Lillycrop, K.A., Berge, G., Burdge, G.C., 2019. Dietary fish oil alters DNA methylation of genes involved in polyunsaturated fatty acid biosynthesis in muscle and liver of Atlantic Salmon (*Salmo salar*). Lipids 54 (11 – 12), 725–739.

Izquierdo, M.S., Fernandez-Palacios, H., Tacon, A.G.J., 2001. Effect of broodstock nutrition on reproductive performance of fish. Aquaculture 197 (1–4), 25–42.

Jones, P.A., Liang, G.N., 2009. Rethinking how DNA methylation patterns are maintained. Nat. Rev. Genet. 10 (11), 805–811.

Kanehisa, M., Goto, S., 2000. KEGG: kyoto encyclopedia of genes and genomes. Nucleic Acids Res. 28 (1), 27–30.

Kiron, V., 2012. Fish immune system and its nutritional modulation for preventive health care. Anim. Feed Sci. Technol. 173 (1–2), 111–133.

Kornberg, R.D., Lorch, Y., 1999. Twenty-five years of the nucleosome, fundamental particle of the eukaryote chromosome. Cell 98 (3), 285–294.

Kouzarides, T., 2007. Chromatin modifications and their function. Cell 128 (4), 693–705.

Kriaucionis, S., Heintz, N., 2009. The nuclear DNA Base 5-Hydroxymethylcytosine is present in Purkinje neurons and the brain. Science 324 (5929), 929–930.

Krueger, F., Andrews, S.R., 2011. Bismark: a flexible aligner and methylation caller for bisulfite-Seq applications. Bioinformatics 27 (11), 1571–1572.

Lall, S.P., Lewis-McCrea, L.M., 2007. Role of nutrients in skeletal metabolism and pathology in fish—an overview. Aquaculture 267 (1–4), 3–19.

Langmead, B., Salzberg, S.L., 2012. Fast gapped-read alignment with bowtie 2. Nat. Methods 9 (4), 357–359.

Lewis, S.H., Ross, L., Bain, S.A., Pahita, E., Smith, S.A., Cordaux, R., Miska, E.A., Lenhard, B., Jiggins, F.M., Sarkies, P., 2020. Widespread conservation and lineage-specific diversification of genome-wide DNA methylation patterns across arthropods. PLoS Genet. 16 (6), e1008864.

Li, H., Durbin, R., 2009. Fast and accurate short read alignment with burrows-wheeler transform. Bioinformatics 25 (14), 1754–1760.

Liao, K., Yan, J., Mai, K., Ai, Q., 2015. Dietary olive and Perilla oils affect liver mitochondrial DNA methylation in large yellow croakers. J. Nutr. 145 (11), 2479–2485.

Liao, K., Yan, J., Mai, K., Ai, Q., 2016. Dietary lipid concentration affects liver mitochondrial DNA copy number, gene expression and DNA methylation in large yellow croaker (*Larimichthys crocea*). Comp. Biochem. Physiol. B Biochem. Mol. Biol. 193, 25–32.

Liu, J., Hu, H., Panserat, S., Marandel, L., 2020. Evolutionary history of DNA methylation related genes in chordates: new insights from multiple whole genome duplications. Sci. Rep. 10 (1), 970.

Lokk, K., Modhukur, V., Rajashekar, B., Martens, K., Magi, R., Kolde, R., Koltsina, M., Nilsson, T.K., Vilo, J., Salumets, A., et al., 2014. DNA methylome profiling of human tissues identifies global and tissue-specific methylation patterns. Genome Biol. 15 (4), r54.

Lubzens, E., Bobe, J., Young, G., Sullivan, C.V., 2017. Maternal investment in fish oocytes and eggs: the molecular cargo and its contributions to fertility and early development. Aquaculture 472, 107–143.

Lubzens, E., Young, G., Bobe, J., Cerda, J., 2010. Oogenesis in teleosts: how eggs are formed. Gen. Comp. Endocrinol. 165 (3), 367–389.

Mæland, A., Rønnestad, I., Waagbø, R., 2003. Folate in eggs and developing larvae of Atlantic halibut, *Hippoglossus hippoglossus* L. Aquacult. Nutr. 9 (3), 185–188.

Marandel, L., Lepais, O., Arbenoits, E., Veron, V., Dias, K., Zion, M., Panserat, S., 2016. Remodelling of the hepatic epigenetic landscape of glucose-intolerant rainbow trout (*Oncorhynchus mykiss*) by nutritional status and dietary carbohydrates. Sci. Rep. 6, 32187.

Maunakea, A.K., Nagarajan, R.P., Bilenky, M., Ballinger, T.J., D'Souza, C., Fouse, S.D., Johnson, B.E., Hong, C., Nielsen, C., Zhao, Y., et al., 2010. Conserved role of intragenic DNA methylation in regulating alternative promoters. Nature 466 (7303), 253–257.

Meissner, A., Gnirke, A., Bell, G.W., Ramsahoye, B., Lander, E.S., Jaenisch, R., 2005. Reduced representation bisulfite sequencing for comparative high-resolution DNA methylation analysis. Nucleic Acids Res. 33 (18), 5868–5877.

Mentch, S.J., Locasale, J.W., 2016. One-carbon metabolism and epigenetics: understanding the specificity. Ann. N. Y. Acad. Sci. 1363, 91–98.

Mohammed, S.A., Ambrosini, S., Luscher, T., Paneni, F., Costantino, S., 2020. Epigenetic control of mitochondrial function in the vasculature. Front. Cardiovasc. Med. 7, 28.

Moren, M., Waagbø, R., Hamre, K., 2011. Micronutrients. In: Larval Fish Nutrition Edited by Holt C. John Wiley & Sons Inc, pp. 117–149.

Muller, F., Scherer, M., Assenov, Y., Lutsik, P., Walter, J., Lengauer, T., Bock, C., 2019. RnBeads 2.0: comprehensive analysis of DNA methylation data. Genome Biol. 20 (1), 55.

Munday, P.L., 2014. Transgenerational acclimation of fishes to climate change and ocean acidification. F1000Prime Rep 6, 99.

NRC, 2011. Nutrient Requirements of Fish and Shrimp. The National Academic Press, Washington, D.C.

Olsvik, P.A., Williams, T.D., Tung, H.S., Mirbahai, L., Sanden, M., Skjaerven, K.H., Ellingsen, S., 2014. Impacts of TCDD and MeHg on DNA methylation in zebrafish (*Danio rerio*) across two generations. Comp. Biochem. Physiol. C 165, 17–27.

Panserat, S., Marandel, L., Geurden, I., Veron, V., Dias, K., Plagnes-Juan, E., Pegourie, G., Arbenoits, E., Santigosa, E., Weber, G., et al., 2017. Muscle catabolic capacities and global hepatic epigenome are modified in juvenile rainbow trout fed different vitamin levels at first feeding. Aquaculture 468, 515–523.

Panserat, S., Marandel, L., Seiliez, I., Skiba-Cassy, S., 2019. New insights on intermediary metabolism for a better understanding of nutrition in Teleosts. Annu Rev Anim Biosci 7, 195–220.

Park, P.J., 2009. ChIP-seq: advantages and challenges of a maturing technology. Nat. Rev. Genet. 10 (10), 669–680.

Pentinat, T., Ramon-Krauel, M., Cebria, J., Diaz, R., Jimenez-Chillaron, J.C., 2010. Transgenerational inheritance of glucose intolerance in a mouse model of neonatal overnutrition. Endocrinology 151 (12), 5617–5623.

Perera, E., Turkmen, S., Simo-Mirabet, P., Zamorano, M.J., Xu, H., Naya-Catala, F., Izquierdo, M., Perez-Sanchez, J., 2020. Stearoyl-CoA desaturase (scd1a) is epigenetically regulated by broodstock nutrition in gilthead sea bream (*Sparus aurata*). Epigenetics 15 (5), 536–553.

Perera, E., Yufera, M., 2017. Effects of soybean meal on digestive enzymes activity, expression of inflammation-related genes, and chromatin modifications in marine fish (*Sparus aurata* L.) larvae. Fish Physiol. Biochem. 43 (2), 563–578.

Podgorniak, T., Brockmann, S., Konstantinidis, I., Fernandes, J.M.O., 2019. Differences in the fast muscle methylome provide insight into sex-specific epigenetic regulation of growth in Nile tilapia during early stages of domestication. Epigenetics 14 (8), 818–836.

Rada-Iglesias, A., 2018. Is H3K4me1 at enhancers correlative or causative? Nat. Genet. 50 (1), 4–5.

Raz, E., 2003. Primordial germ-cell development: the zebrafish perspective. Nat. Rev. Genet. 4 (9), 690–700.

Rønnestad, I., Hamre, K., Lie, Ø., Waagbø, R., 1999. Ascorbic acid and alpha-tocopherol levels in larvae of Atlantic halibut before and after exogenous feeding. J. Fish Biol. 55 (4), 720–731.

Rønnestad, I., Lie, Ø., Waagbø, R., 1997. Vitamin B6 in Atlantic halibut, *Hippoglossus hippoglossus*—endogenous utilization and retention in larvae fed natural zooplankton. Aquaculture 157 (3–4), 337–345.

Rothbart, S.B., Strahl, B.D., 2014. Interpreting the language of histone and DNA modifications. Biochim. Biophys. Acta 1839 (8), 627–643.

Saito, T., Whatmore, P., Taylor, J.F., Fernandes, J.M.O., Adam, A.C., Tocher, D.R., Espe, M., Skjaerven, K.H., 2020. Micronutrient supplementation affects transcriptional and epigenetic regulation of lipid metabolism in a dose-dependent manner. Epigenetics, 1–18.

Schwartz, S., Meshorer, E., Ast, G., 2009. Chromatin organization marks exon-intron structure. Nat. Struct. Mol. Biol. 16 (9), 990–U117.

Seite, S., Masagounder, K., Heraud, C., Veron, V., Marandel, L., Panserat, S., Seiliez, I., 2019. Early feeding of rainbow trout (*Oncorhynchus mykiss*) with methionine-deficient diet over a 2 week period: consequences for liver mitochondria in juveniles. J. Exp. Biol. 222 (Pt 18).

Shrestha, A.M., Frith, M.C., Horton, P., 2014. A bioinformatician's guide to the forefront of suffix array construction algorithms. Brief. Bioinform. 15 (2), 138–154.

Sims 3rd, R.J., Nishioka, K., Reinberg, D., 2003. Histone lysine methylation: a signature for chromatin function. Trends Genet. 19 (11), 629–639.

Skene, P.J., Henikoff, S., 2017. An efficient targeted nuclease strategy for high-resolution mapping of DNA binding sites. Elife 6.

Skjaerven, K.H., Hamre, K., Penglase, S., Finn, R.N., Olsvik, P.A., 2014. Thermal stress alters expression of genes involved in one carbon and DNA methylation pathways in Atlantic cod embryos. Comp. Biochem. Physiol. A Mol. Integr. Physiol. 173, 17–27.

Skjaerven, K.H., Jakt, L.M., Dahl, J.A., Espe, M., Aanes, H., Hamre, K., Fernandes, J.M., 2016. Parental vitamin deficiency affects the embryonic gene expression of immune-, lipid transport- and apolipoprotein genes. Sci. Rep. 6, 34535.

Skjaerven, K.H., Jakt, L.M., Fernandes, J.M.O., Dahl, J.A., Adam, A.C., Klughammer, J., Bock, C., Espe, M., 2018. Parental micronutrient deficiency distorts liver DNA methylation and expression of lipid genes associated with a fatty-liver-like phenotype in offspring. Sci. Rep. 8 (1), 3055.

Skjaerven, K.H., Oveland, E., Mommens, M., Samori, E., Saito, T., Adam, A.C., Espe, M., 2020. Out-of-season spawning affects the nutritional status and gene expression in both Atlantic salmon female broodstock and their offspring. Comp. Biochem. Physiol. A Mol. Integr. Physiol. 247.

Skulason, S., Parsons, K.J., Svanback, R., Rasanen, K., Ferguson, M.M., Adams, C.E., Amundsen, P.A., Bartels, P., Bean, C.W., Boughman, J.W., et al., 2019. A way forward with eco evo devo: an extended theory of resource polymorphism with postglacial fishes as model systems. Biol. Rev. Camb. Philos. Soc. 94 (5), 1786–1808.

Slotkin, R.K., Martienssen, R., 2007. Transposable elements and the epigenetic regulation of the genome. Nat. Rev. Genet. 8 (4), 272–285.

Smallwood, S.A., Kelsey, G., 2012. De novo DNA methylation: a germ cell perspective. Trends Genet. 28 (1), 33–42.

Song, Y., Alami-Durante, H., Skiba-Cassy, S., Marandel, L., Panserat, S., 2019. Higher glycolytic capacities in muscle of carnivorous rainbow trout juveniles after high dietary carbohydrate stimulus at first feeding. Nutr. Metab. (Lond.) 16, 77.

Sweetman, J.W., Torrecillas, S., Dimitroglou, A., Rider, S., Davies, S.J., Izquierdo, M.S., 2010. Enhancing the natural defences and barrier protection of aquaculture species. Aquacult. Res. 41 (3), 345–355.

Tadros, W., Lipshitz, H.D., 2009. The maternal-to-zygotic transition: a play in two acts. Development 136 (18), 3033–3042.

Taylor, J.F., Vera, L.M., De Santis, C., Lock, E.J., Espe, M., Skjaerven, K.H., Leeming, D., Del Pozo, J., Mota-Velasco, J., Migaud, H., et al., 2019. The effect of micronutrient supplementation on growth and hepatic metabolism in diploid and triploid Atlantic salmon (*Salmo salar*) parr fed a low marine ingredient diet. Comp. Biochem. Physiol. B Biochem. Mol. Biol. 227, 106–121.

Terova, G., Diaz, N., Rimoldi, S., Ceccotti, C., Gliozheni, E., Piferrer, F., 2016. Effects of sodium butyrate treatment on histone modifications and the expression of genes related to epigenetic regulatory mechanisms and immune response in European Sea bass (*Dicentrarchus labrax*) fed a plant-based diet. PLoS One 11 (7), e0160332.

Tost, J., Gut, I.G., 2007. DNA methylation analysis by pyrosequencing. Nat. Protoc. 2 (9), 2265–2275.

Trichet, V.V., 2010. Nutrition and immunity: an update. Aquacult. Res. 41 (3), 356–372.

Turchini, G.M., Torstensen, B.E., Ng, W.K., 2009. Fish oil replacement in finfish nutrition. Rev. Aquac. 1 (1), 10–57.

Turkmen, S., Perera, E., Zamorano, M.J., Simo-Mirabet, P., Xu, H., Perez-Sanchez, J., Izquierdo, M., 2019. Effects of dietary lipid composition and *fatty acid desaturase 2* expression in broodstock gilthead sea bream on lipid metabolism-related genes and methylation of the *fads2* gene promoter in their offspring. Int. J. Mol. Sci., 20(24).

Valouev, A., Johnson, D.S., Sundquist, A., Medina, C., Anton, E., Batzoglou, S., Myers, R.M., Sidow, A., 2008. Genome-wide analysis of transcription factor binding sites based on ChIP-Seq data. Nat. Methods 5 (9), 829–834.

Vera, L.M., Metochis, C., Taylor, J.F., Clarkson, M., Skjaerven, K.H., Migaud, H., Tocher, D.R., 2017. Early nutritional programming affects liver transcriptome in diploid and triploid Atlantic salmon, *Salmo salar*. BMC Genomics 18 (1), 886.

Veron, V., Marandel, L., Liu, J., Velez, E.J., Lepais, O., Panserat, S., Skiba, S., Seiliez, I., 2018. DNA methylation of the promoter region of *bnip3* and *bnip3l* genes induced by metabolic programming. BMC Genomics 19 (1), 677.

Waagbø, R., 2006. Feeding and disease resistance in fish. In: Mosenthin, R., Zentek, J., Zebrowska, T. (Eds.), Biology of Growing Animal. Elsevier, pp. 387–415.

Waagbø, R., 2008. Reducing production related diseases in farmed fish. In: Lie, Ø. (Ed.), Improving Farmed Fish Quality and Safety. Woodhead Publishing, pp. 363–398.

Waagbø, R., Glette, J., Raa-Nilsen, E., Sandnes, K., 1993. Dietary vitamin C, immunity and disease resistance in Atlantic salmon (*Salmo salar*). Fish Physiol. Biochem. 12 (1), 61–73.

Waagbø, R., Olsvik, P., Remø, S., 2020. Nutritional and metabolic disorders. In: Woo, P., Iwama, G., CAB International (Eds.), Climate Change and Non-infectious Fish Disorders (CCNFD), pp. 102–124.

Waagbø, R., Remø, S., 2020. Functional diets in fish health management. In: Kibenge, F., Powell, M. (Eds.), Aquaculture Health Management. Elsevier, pp. 187–234.

Wang, X., Bhandari, R.K., 2019. DNA methylation dynamics during epigenetic reprogramming of medaka embryo. Epigenetics 14 (6), 611–622.

Wang, Z.J., Long, H., Chang, C., Zhao, M., Lu, Q.J., 2018. Crosstalk between metabolism and epigenetic modifications in autoimmune diseases: a comprehensive overview. Cell. Mol. Life Sci. 75 (18), 3353–3369.

Waterland, R.A., Travisano, M., Tahiliani, K.G., Rached, M.T., Mirza, S., 2008. Methyl donor supplementation prevents transgenerational amplification of obesity. Int. J. Obes. (Lond) 32 (9), 1373–1379.

Watt, F., Molloy, P.L., 1988. Cytosine methylation prevents binding to DNA of a HeLa cell transcription factor required for optimal expression of the adenovirus major late promoter. Genes Dev. 2 (9), 1136–1143.

Weber, M., Davies, J.J., Wittig, D., Oakeley, E.J., Haase, M., Lam, W.L., Schubeler, D., 2005. Chromosome-wide and promoter-specific analyses identify sites of differential DNA methylation in normal and transformed human cells. Nat. Genet. 37 (8), 853–862.

Wischhusen, P., Larroquet, L., Durand, T., Oger, C., Galano, J.M., Rocher, A., Vigor, C., Prabhu, P.A.J., Veron, V., Briens, M., et al., 2020a. Oxidative stress and antioxidant response in rainbow trout fry exposed to acute hypoxia is affected by selenium nutrition of parents and during first exogenous feeding. Free Radic. Biol. Med. 155, 99–113.

Wischhusen, P., Saito, T., Heraud, C., Kaushik, S.J., Fauconneau, B., Antony Jesu Prabhu, P., Fontagne-Dicharry, S., Skjaerven, K.H., 2020b. Parental selenium nutrition affects the one-carbon metabolism and the hepatic DNA methylation pattern of rainbow trout (*Oncorhynchus mykiss*) in the progeny. Life (Basel) 10 (8).

Xi, Y., Li, W., 2009. BSMAP: whole genome bisulfite sequence MAPping program. BMC Bioinf. 10, 232.

Xing, H., Mo, Y., Liao, W., Zhang, M.Q., 2012. Genome-wide localization of protein-DNA binding and histone modification by a Bayesian change-point method with ChIP-seq data. PLoS Comput. Biol. 8 (7), e1002613.

Xu, H., Dong, X., Ai, Q., Mai, K., Xu, W., Zhang, Y., Zuo, R., 2014. Regulation of tissue LC-PUFA contents, Delta6 *fatty acyl desaturase* (*FADS2*) gene expression and the methylation of the putative *FADS2* gene promoter by different dietary fatty acid profiles in Japanese seabass (*Lateolabrax japonicus*). PLoS One 9 (1), e87726.

Xu, J., Sinclair, K.D., 2015. One-carbon metabolism and epigenetic regulation of embryo development. Reprod. Fertil. Dev. 27 (4), 667–676.

You, J.J., Ren, P., He, S., Liang, X.F., Xiao, Q.Q., Zhang, Y.P., 2020. Histone methylation of H3K4 involved in the anorexia of carnivorous mandarin fish (*Siniperca chuatsi*) after feeding on a carbohydrate-rich diet. Front. Endocrinol. (Lausanne) 11, 323.

Yu, Y., Zhang, Y., Chen, X., Chen, Y., 2019. Plant noncoding RNAs: hidden players in development and stress responses. Annu. Rev. Cell Dev. Biol. 35, 407–431.

Zhang, Y., Liu, T., Meyer, C.A., Eeckhoute, J., Johnson, D.S., Bernstein, B.E., Nusbaum, C., Myers, R.M., Brown, M., Li, W., et al., 2008. Model-based analysis of ChIP-Seq (MACS). Genome Biol. 9 (9), R137.

Zhao, Q., Zhang, J.Q., Chen, R.Y., Wang, L.N., Li, B., Cheng, H., Duan, X.Y., Zhu, H.J., Wei, W., Li, J.W., et al., 2016. Dissecting the precise role of H3K9 methylation in crosstalk with DNA maintenance methylation in mammals. Nat. Commun. 7 (1), 12464.

Zhu, H., Wang, G.H., Qian, J., 2016. Transcription factors as readers and effectors of DNA methylation. Nat. Rev. Genet. 17 (9), 551–565.

Chapter 6

Methods for mapping genomic variants associated with production traits in aquaculture species

Clémence Fraslin, Ross D. Houston, and Diego Robledo
The Roslin Institute, University of Edinburgh, Edinburgh, United Kingdom

1. Introduction

Most production traits of relevance to aquaculture are heritable but polygenic in nature, which means that they are controlled by many genetic variants, most of them with a very small effect. These traits (e.g., growth, feed efficiency, fillet quality, disease resistance, or sexual maturation) can be improved via family based breeding programs, which have successfully achieved high rates of genetic gain for several aquaculture species (Gjedrem and Rye, 2018). The aquaculture industry has also been forward looking in the uptake of genomics technologies to support breeding, beginning with the use of molecular genetic markers for parentage assignment in the 1990s (Vandeputte and Haffray, 2014). In parallel to technological progress, genomics has begun to be applied more routinely in aquaculture genetics, for example to discover variants associated with production traits that can be incorporated in selection programs via marker-assisted or genomic selection (Houston et al., 2020; Zenger et al., 2019). However, the discovery of functional variants impacting production traits, frequently termed "causative variants", remains a major challenge, with few success stories to date.

A causative variant refers to variation in the genome sequence between the individuals of a population or a species that is directly responsible for a difference in a trait of interest. This variation in the genomic sequence can be caused by a single nucleotide polymorphism (SNP) or other types of variation such as indels or larger structural variants. The causative variants cause a difference at the molecular level (i.e., changes in gene expression or protein sequence) with a direct impact on the biology of the animal and the trait of interest. Finding the causative variants underlying our traits of interest may lead to an improved

Cellular and Molecular Approaches in Fish Biology. https://doi.org/10.1016/B978-0-12-822273-7.00007-0

rate of genetic gain for aquaculture traits compared to use of linked nonfunctional markers. On one hand, causative variants can be directly incorporated into marker-assisted or genomic selection, potentially leading to improved selection accuracy and therefore genetic gain in a breeding program. On the other hand, causative variants help us understand the functional basis of variation in traits of interest. This basic knowledge enables or facilitates the design of strategies to improve production and welfare of aquaculture animals, targeting the relevant molecular pathways, for example using functional feeds or immunostimulants. Furthermore, a potential future application of the identification of causative variants is genome editing, which can be used to rapidly increase or fix favorable alleles, or to introduce a beneficial variant in a population, and even to transfer the variant across species (Gratacap et al., 2019). The identification of causative variants has been a long-lasting objective of biological research over the past 20 years. In this chapter, past and current methods for mapping causative variants are discussed in detail, along aquaculture examples, while a step-by-step example pipeline for genome-wide association analysis is also provided.

2. Finding causative variants through QTL mapping approaches

Regions of the genome that contribute to genetic variation in traits of interest are known as quantitative trait loci (QTL). Most production traits are explained by many QTLs, each of which explains only a very small proportion of the genetic variation in the trait. Finding these QTLs among the 500 million to 3000 million bases that typically form the genome of an aquaculture species would in theory require sequencing the genomes of hundreds or thousands of animals. Yet, we can detect QTLs (relatively large ones) with just hundreds of genetic markers, and this is possible because the closer two genetic markers are in the genome, the more likely they are to be inherited together from parents to offspring and stay together across generations (Fig. 1). This is a direct consequence of recombination, where meiotic crossovers tend to occur once or twice per chromosome pair, and therefore large stretches of the parental chromosomes are inherited without changes by the next generation. As generations pass, loci in close proximity tend to remain together because the likelihood of a crossover occurring between them is low. Consequently, we will detect a statistical association with a trait of interest for any loci in close proximity to a causative variant, which means there is no need to interrogate the actual causative loci to detect association. Linkage and linkage disequilibrium are two different but closely related concepts to frame the nonrandom association between genetic markers, and each of them is the conceptual basis for a type of QTL detection methods (explained in the next section). Linkage is defined as the physical association between two genetic markers in the same chromosomal segment, inherited directly from a previous generation, while linkage disequilibrium refers

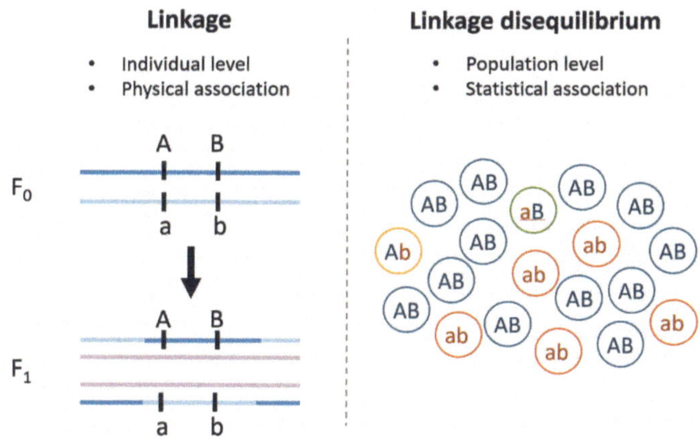

FIG. 1 Illustration of linkage and linkage disequilibrium. Two genetic markers with possible alleles Aa and Bb are used to illustrate linkage and linkage disequilibrium in a diploid organism. In linkage, alleles A and B are located closely in a chromosome, while a and b are on the other copy of the chromosome; this proximity means that it is unlikely that a crossover will occur between the two genetic markers, and therefore A and B will be transmitted together to the offspring F_1 of this individual F_0 (and a and b will also tend to be coinherited). In linkage disequilibrium, alleles A and B are observed together more frequently than expect by chance, and the same is true for alleles a and b; this statistical association observed at the population level is irrespective of the genomic position of the two genetic markers, but is commonly observed for markers in close proximity.

to the statistical association between two allelic variants at the population level, which is frequently also due to their physical proximity in the genome (Fig. 1). Most studies use anonymous genetic markers evenly spread across the genome to associate a genomic region with a trait of interest, exploiting linkage and linkage disequilibrium through different statistical methods.

The basis of QTL analysis is to detect a statistical association between an anonymous genetic marker (with a known position on the genome) and the phenotypic variation in a measured trait (caused by an unknown causative variant). Different approaches can be used to detect QTLs according to the number of markers, number of animals, and population structure.

2.1 Linkage analysis and interval mapping approaches

QTL detection using linkage analysis (LA) is typically performed in experimental designs composed of families and is based on the transmission of large genomic fragments from parents to offspring, taking into account the recombination rate between anonymous genetic markers and the unknown causal mutation. This approach relies on the hypothesis that, if there is a difference in the performance of offspring groups based on the genotype at a given marker, this difference is due to a segregating QTL in linkage with this marker. This approach has been commonly applied to aquaculture species, which benefit from large nuclear families (Fraslin et al., 2020b; You et al., 2020). One main advantage of this approach is that it does not require a high genetic marker density (i.e., ~200 markers can be sufficient), as the segments of the genome tracked from parents to offspring are relatively large. However, there are drawbacks that the resolution of the mapping is low, and that markers discovered in these experimental families do not necessarily have the same association with the trait in the wider population, so a separate validation step is required in independent families and populations.

One of the most common LA approaches is called interval mapping (Lander and Botstein, 1989), which basically divides the genome in evenly spaced intervals (e.g., 1 centimorgan intervals), and tests the hypothesis H0 "no QTL" versus H1 "one QTL". At each interval, a likelihood ratio test (likelihood of HO/likelihood of H1) is performed to assess the presence or absence of a QTL (Elsen et al., 1999; Knott et al., 1996). The main advantage of interval mapping is that it requires both a relatively low number of genetic markers and relatively low number of animals (typically both in the hundreds rather than thousands), and it is a robust QTL detection method (low type I error rate) with the detection power mainly depending on the QTL effect and the population size, and not on marker density (Darvasi et al., 1993). The main caveat of these methods is that the position of the QTL is often imprecise, with large confidence intervals covering often tens of Mb that can even span whole chromosomes in some cases. Since it is performed independently in each family, and large portions of the chromosomes are inherited from parents to offspring, the ability to narrow the QTL depends directly on the recombination rate in the parents. In fish species, recombination rates can vary massively between sexes (Sardell et al., 2018; Singer et al., 2002), which greatly affects the length and density of the constructed linkage maps (Tsai et al., 2016; Wang et al., 2017) and thus the accuracy of QTL localization. In salmonids, female recombination rate is significantly higher (around 5 and up to 16.8 times more than in males of Atlantic salmon (*Salmo salar*) depending on the chromosome), and recombination is relatively evenly spread across the chromosomes, whereas male recombination tends to be restricted to subtelomeric regions (Danzmann et al., 2005). In contrast, a positive aspect is that large full- and half-sib families can easily be obtained in most aquaculture species; this increases the number of crossovers and therefore contributes to narrowing QTL positions (Darvasi and Soller, 1997). Furthermore, the disparate recombination rates between sexes can facilitate a two-stage mapping approach, whereby the low recombination in male parents is used to detect QTL, and the higher recombination in female parents subsequently used to confirm and fine-map the QTL (Hayes et al., 2006; Houston et al., 2008). Prior to the advent of Next Generation Sequencing, interval mapping was widely used in aquaculture to detect QTLs affecting disease resistance and growth-related traits (Table 1).

TABLE 1 Examples of QTL mapping studies using linkage analysis (LA) and genome-wide association studies (GWAS) in various fish species for disease resistance and growth-related traits.

Fish species	Experimental design	Traits	QTL approach	QTL detect.	Publication
Linkage analysis (LA)					
Atlantic salmon (*Salmo salar*)	584 fish (10 FS) ~100 microsatellites loci	Resistance to infectious pancreatic necrosis	Interval mapping	3 QTL	Houston et al. (2008)
Gilthead sea bream (*Sparus aurata*)	409 fish (10 HS) 74 microsatellite loci	Growth-related traits	Interval mapping	4 QTL	Loukovitis et al. (2012)
Turbot (*Scophthalmus maximus*)	758 fish (4 FS) 211 microsatellites loci	Resistance to scuticociliatosis	Interval mapping	5–10 QTL	Rodríguez-Ramilo et al. (2013)
Rainbow trout (*Oncorhynchus mykiss*)	300 DH fish 2130 SNPs	Resistance to bacterial cold-water disease	Interval mapping	15 QTL	Fraslin et al. (2018)
Largemouth bass (*Micropterus salmoides*)	150 progeny and 2 parents (F1 FS family) 6857 SNPs	Growth-related traits	Composite interval mapping	32 QTL	Dong et al. (2019)
Genome-wide association studies (GWAS)					
Atlantic salmon (*Salmo salar*)	4662 fish (118 FS) 116,594 SNPs	Growth-related traits	Bayes C	Polygenic architecture	Yoshida et al. (2017)
Atlantic salmon (*Salmo salar*)	2345 fish (96 FS) 968 SNPs with imputation to ~46K SNPs	Resistance to salmon rickettsial syndrome	GBLUP	Polygenic architecture	Moraleda et al. (2021)

Continued

TABLE 1 Examples of QTL mapping studies using linkage analysis (LA) and genome-wide association studies (GWAS) in various fish species for disease resistance and growth-related traits—cont'd

Fish species	Experimental design	Traits	QTL approach	QTL detect.	Publication
Nile tilapia (*Oreochromis niloticus*)	5866 fish phenotypes (74–89 FS) including 1130 fish genotyped for 43,3127 SNPs	Growth-related traits	wssGBLUP	Polygenic architecture	Yoshida et al. (2019)
Coho salmon (*Oncorhynchus kisutch*)	2606 fish (33 FS) 764 fish genotyped for 9389 SNPs	Resistance to salmon rickettsial syndrome	GBLUP, ssGBLUP, wGBLUP	18 QTL	Barría et al. (2018)
European sea bass (*Dicentrarchus labrax*)	1538 fish (FS) 9195 SNPs	Resistance to viral nervous necrosis	wGBLUP	4 QTL	Palaiokostas et al. (2018a)
Large yellow croaker (*Larimichthys crocea*)	220 fish 27,227 SNPs	Growth related traits	Marker-by-marker: GLM	Up to 44 QTL	Zhou et al. (2019)
Rainbow trout (*Oncorhynchus mykiss*)	706 fish (30 HS) 30,600 SNPs	Resistance to bacterial cold-water disease	MLMA, wssGBLUP	5 QTL	Fraslin et al. (2019)
Rainbow trout (*Oncorhynchus mykiss*)	4987 fish (100 FS) 1044 fish genotyped for 35,397 SNPs and 6308 pedigree records	Resistance to infectious hematopoietic necrosis	ssBMR, wssGBLUP	10 QTL	Vallejo et al. (2019)

DH, doubled haploid; *FS*, full-sib family; *GLM*, general linear model; *HS*, half-sib family; *MLMA*, mixed linear model association; *SNP*, single nucleotide polymorphism; *ssBMR*, single-step Bayesian multiple regression; *ssGBLUP*, single-step genomic best linear unbiased prediction (GBLUP); *wGBLUP*, weighted GBLUP.

2.2 Genome-wide association studies

In the past decade, the development of Genotype-By-Sequencing (GBS) approaches such as restriction site-associated DNA sequencing (RAD-Seq) and the design of SNP arrays in various fish and shellfish species have made possible to obtain > 10 K markers reliably in hundreds of aquaculture species (Houston et al., 2020; Robledo et al., 2017a). Consequently, the research community moved to QTL mapping methods that better exploit high-density genotypes in a large number of families: genome-wide association studies, commonly known as GWAS. In recent years, numerous examples of GWAS to study the genetic architecture of growth and resistance to different diseases have been published in various aquaculture species, with varying degrees of success (Table 1).

GWAS is based on similar principles to linkage mapping, being a hypothesis-free approach to test the statistical association between markers spread along the whole genome and the variation of a target trait. However, GWAS typically uses thousands of markers and relies on the linkage disequilibrium (LD) that exists between anonymous markers (usually SNPs) and the causative variant. GWAS allows the detection of QTLs in a population without prior knowledge on the pedigree or the population structure (relatedness in the population). While increased relatedness among the individuals increases LD in the population and therefore facilitates the detection of QTLs (also likely to be shared by a significant proportion of the animals if relatedness is high), increased LD will also make more difficult to narrow the genomic location of the QTL to a small region. Additionally, in a structured population, the chance of having a spurious association between a SNP and the phenotype due to population stratification is high (Lander and Schork, 1994). Consequently, GWAS typically uses relationship matrices implemented as a random polygenic effect in the model (Meuwissen et al., 2002) to account for the relationship between individuals and reduce the chance of false positives. To do so, GWAS approaches rely on pedigree (A), genomic (G), or a combination of both (H) relationship matrices (depending on what information is available for the population under study). The additive genetic relationship matrix (A) uses only the pedigree information to estimate the relationship between two individuals and is typically applied in linear animal models. The quality of this estimated relationship relies on the quality and depth of the pedigree records. If the pedigree is well known for many generations the estimates are realistic; however, any error in the pedigree will result in a biased estimate, and if the pedigree is only recorded for a few generations (e.g., 1–5) the relationship may be underestimated. The genomic relationship matrix (GRM) G relies on genetic markers (usually SNPs) to estimate the genomic similarity between pairs of individuals, where the number of alleles shared by any two individuals (i.e., how similar their genomes are) is a direct estimate of their relatedness. The relationship between pairs of individuals estimated by the G matrix can be more accurate than the one estimated by the A matrix as genetic markers can capture variation in relatedness within families (important in the large full-sibling families of aquaculture species) and uncover 'hidden' relationships in the pedigrees. Finally, the H matrix is used when both pedigree and genotyping information are available, and those analyses are termed "single-step" (Aguilar et al., 2010).

2.3 GWAS and genome complexity

Many aquaculture species have complex genomes. All teleosts have an additional whole-genome duplication compared to tetrapods, and some groups such as salmonids or sturgeons underwent yet another whole-genome duplication (Glasauer and Neuhauss, 2014). As a consequence, it is common to observe paralogous regions with very high sequence similarity within the genomes of teleost fish species. This feature complicates genome assembly efforts, increasing their fragmentation and the chances of misassemblies (although these risks are minimized by the use of long-read sequencing technologies in the latest genome assemblies (Lu and Luo, 2020)). These misassemblies can have an impact on downstream association analyses, leading for example to the detection of two putative QTLs in paralogous regions when in reality there is a single QTL in one of the two regions. Further, while most fish species are diploid (for species of different ploidy there are specialized GWAS packages, such as GWASpoly; Rosyara et al., 2016), in species with recent whole-genome duplications the reploidization process is still ongoing, and there might be recombination between paralogous regions (Waples et al., 2016), generating aberrant genotypes that can lead to spurious results. A further complication for certain shellfish species is the very high level of heterozygosity and extensive repetitive DNA sequences (Potts et al., 2021). Again for these species the extensive use of long-read sequencing technologies such as Oxford Nanopore and Pacific Biosciences has helped to offset some of these challenges. Nonetheless, stringent quality control is fundamental to reduce the potential impact of genome complexity in GWAS analyses, but in any case the nature of the genome and quality of the genome assembly should be taken into account when interpreting association results.

3. Detailed description of state-of-the-art methodology: Step by step on how to perform GWAS

Here we describe a typical GWAS pipeline for an aquaculture species, which includes format conversion and quality controls on genotyping data and a GWAS with two free, user-friendly and commonly used software: PLINK (Chang et al., 2015) and GCTA (Genome-wide Complex Trait Analysis, Yang et al., 2011). Several other software are available, and these two are used for the purposes of giving an example pipeline for analysis. PLINK is commonly used for data management, quality control, and to recode genotyping files, which can be easily handled and modified to be used as inputs for other data analysis software. GCTA has been developed mainly to perform GWAS to investigate disease resistance traits in humans. This software performs a marker-by-marker GWAS with a correction for population structure using a genomic relationship matrix (GRM, calculated directly from the genetic markers for the GWAS) to help prevent false positive association due to population stratification (Yang et al., 2014).

The basic GWAS approach implemented in GCTA is a maximum linear model association (MLMA) described by Eq. (1):

$$y_i = \mu + \beta_i + a_j x_{ij} + u_i + \varepsilon_i \qquad (1)$$

With y_i being the observed phenotype for the ith individual, μ the overall mean in the population, β_i the fixed effect or covariate, a_j the additive effect of the reference allele of the candidate SNP (j) to be tested as fixed effect for association and x_{ij} the reference allele count (0, 1, or 2) for the SNP j for individual i, u_i is the random polygenic effect of all SNPs for individual i, and ε_i the residual effect for individual i. The vector of residual effects is normally and independently distributed $\varepsilon \sim N(0, \boldsymbol{I}\sigma_e^2)$ with σ_e^2 being the residual variance. The random vector of polygenic effects follows a normal distribution $\boldsymbol{u} \sim N(0, \boldsymbol{G}\sigma_g^2)$ with σ_g^2 being the estimated genetic variance and **G** the GRM calculated using all SNPs.

3.1 Genotype files and format conversion

3.1.1 PLINK inputs files: PED and MAP format

Most genotyping software will be able to export the results in PLINK format, generally as **.ped** and **.map** files (Fig. 2). These files are relatively simple and can be produced manually from other genotype formats.

The **.ped** file contains one row per genotyped individual and (6+2N) columns, N being the number of genotyped markers (SNPs) in the dataset. Columns are separated by a white-space character (tabulation or space). The first 6 columns of the **.ped** file are in this order: Family ID, Fish ID, Sire ID, Dam ID, Sex, Phenotype. If the pedigree is unknown, Family ID can be replaced by the Fish ID (in that case, the first two columns are identical), and then both Sire and Dam IDs should be columns of 0 as all fish will be assumed to be founders. Sex is coded as 1 for male and 2 for female, anything else is treated as unknown. By default, individuals with unknown sex have their phenotype automatically set as missing by PLINK. If some fish have unknown sex but still need to be used in the dataset the "**--allow-no-sex**" option should be used to disable this automatic setting. This file must contain only one phenotype (column 6), which can be either a quantitative trait (use a period to separate decimals) or a qualitative trait, but both types must be recorded as numeric values. Missing values are coded as "-9" by default but the option "**--missing-phenotype** NEW_VALUE" can be used to set a new numeric missing value. From column 7 and onward, the file contains the genotypes for each SNP, with the two alleles of each marker in two different columns. All markers should be biallelic and can be defined by any

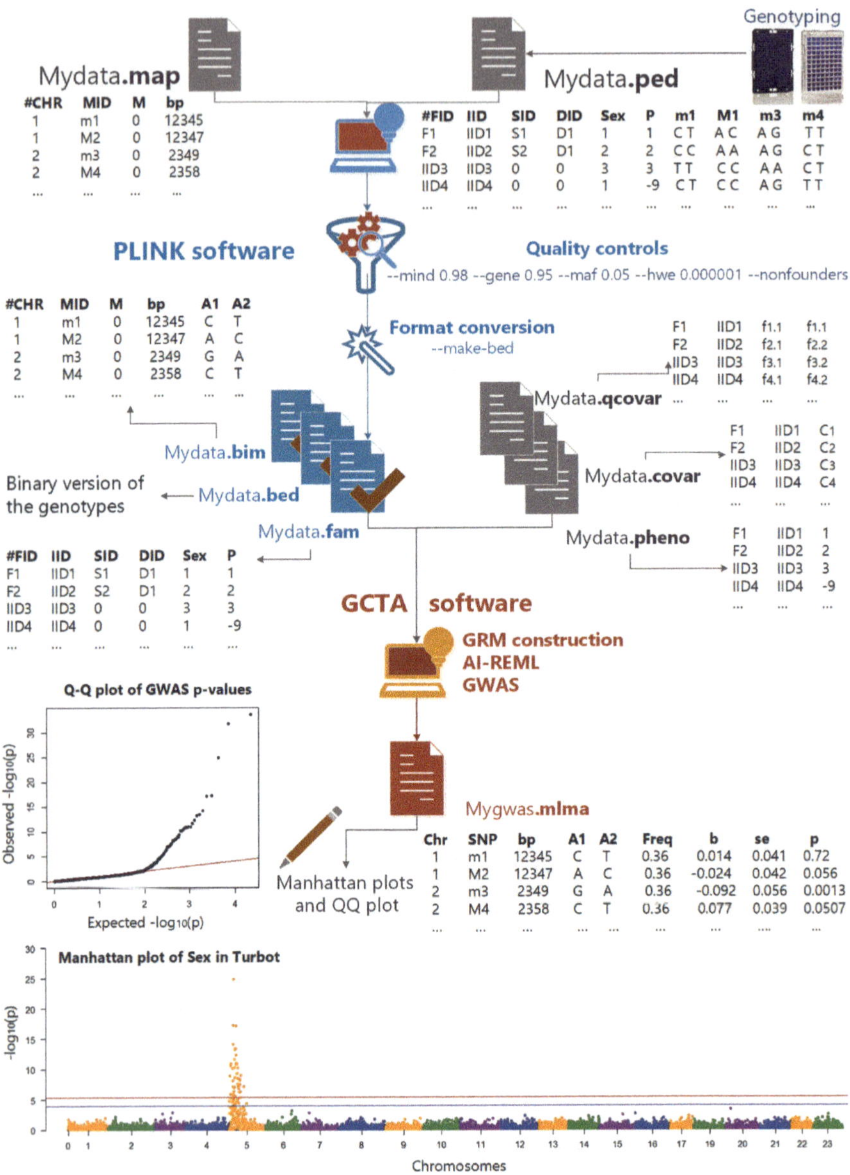

FIG. 2 Pipeline to perform GWAS using PLINK and GCTA. The results shown in the Manhattan plot were obtained from Martínez et al. (2021), and those data can be used as training dataset for the step-by-step pipeline described later. The GWAS for turbot (*Scophthalmus maximus*) sex was performed using the MLMA approach with weight as covariate. In the Manhattan plot, the *red line* is the 5% genome-wide significance threshold and the *blue line* is the 5% chromosome-wide significance threshold (calculated using the Bonferroni correction). The abbreviations in the figure correspond to *CHR*, chromosome; *MID*, marker ID; *M*, Morgan (genetic position); *bp*, base-pair (physical position); *FID*, family ID; *IID*, individual ID; *SID*, sire ID; *DID*, dam ID; *P*, phenotype; *Freq*, frequency of A1; *b*, SNP effect; *se*, standard-error; *p*, *p*-value.

value except 0 which is the default for missing genotypes (can be changed using the option "**--missing-genotype**") (Fig. 2).

The first 6 columns are considered mandatory by default, but if one or more are missing it is possible to use options to indicate it instead of having to modify the file. For the pedigree, the "**--no-fid**" option indicates that there is no Family ID (thus, the first column of the input file is the Fish ID, and in the output files the family ID will be set to be identical to the Fish ID) and the "**--no-parents**" option indicates that there is neither Sire nor Dam ID (they will be set as 0 and all fish will be considered as funders). The "**--no-sex**" option indicates that there are no sex records, and this option must be followed by the "**--allow-no-sex**" option. Finally, the "**--no-pheno**" option indicates that there are no phenotype records.

The .map file contains the genomic position of every genetic marker in the **.ped** file (N rows), and in the same order (which does not necessarily have to be ordered according to their position in the genome). The file has four columns: (1) chromosome name, (2) marker ID, (3) genetic (Morgan), and (4) physical (bp) positions (only one of the two positions is required) (Fig. 2). Chromosome names must be coded from 1 to the actual number of chromosomes (or linkage groups) known in the studied species, and markers that are placed in scaffolds not assigned to any chromosome can be coded as chromosome number 0 (or alternatively removed from the dataset). As PLINK was initially developed for human genetics it assumes that the number of autosomes is 22, but it can be changed using the "**--chr-set**" option.

It is important to avoid errors during data formatting, since this will undoubtedly result in problems downstream. Typical errors include duplicated sample names (if a sample was genotyped in duplicate, rename one of them), or the use of special characters in the IDs (families, fish, chromosomes, or markers). Avoid spaces, tabulations, *, and # in names (# indicates a commented line that will not be read by the software). Use a dot "." to separate decimals (and not a comma). Check carefully that the number of markers is consistent between files, and that markers in the **.map** file are in the same order as in the **.ped** file.

3.1.2 Binary format

PLINK can transform, using the "**--make-bed**" option, the **.ped** and **.map** genotype files into binary files that are used as input for GCTA. The resulting files are a **.bed** file with the genotypes (binary), a **.fam** file with the 6 first columns of the **.ped** file, and a **.bim** file that contains the information in the .map file plus the identity of the two alleles of the SNP.

3.2 Quality controls using the PLINK software

Having a dataset which is filtered to be as free from errors as possible is important to conduct analyses in a GWAS. All quality controls that are usually performed in fish GWAS are explained later, along with their recommended standard thresholds. These thresholds should be modified based on the particularities of the analysis and the characteristics of the dataset, such as population structure. All standard quality controls performed on call rate, minor allele frequency (MAF), and Hardy-Weinberg Equilibrium (HWE) can be performed using a single command line with the PLINK software. The removal of any potentially duplicated loci has to be performed beforehand and can be based on parameters such as excess heterozygosity.

3.2.1 Duplicated samples

Sampling errors might lead to a sample being genotyped twice (same DNA sample in two different tubes, assigned to two different IDs), and those samples have to be detected and removed from the dataset. To do so, the genetic similarity between two samples can be estimated, and two samples with nearly identical genotypes can be considered as duplicates and removed (usually both of them, unless the correct genotype-phenotype pair can be determined or both samples share the same phenotypic value). In PLINK, the "**--genotype**" option is used to estimate IBD (identity by descent) values for each pair of individuals, and the estimates in the output file PI_HAT (proportion IBD) can be used to discard duplicated samples, with a recommended threshold of 0.90.

3.2.2 Individual and genotyping call rate

After discarding all duplicated samples from our dataset, the next step is discarding individuals and markers with low call rates. Individuals can be filtered using the "**--mind** X" command on PLINK, with X being the exclusion threshold. An individual call rate exclusion threshold of 0.02 is commonly used, meaning that all individuals with less than 98% of their SNPs genotyped will be removed from the dataset. This threshold is rather conservative and it can be lowered to 0.05 or even 0.1 depending on the quality of thc genotyping. Likewise, SNPs can be filtered using the "**--geno** Y" option, Y being the SNP exclusion threshold for missing values. The usual exclusion threshold for removing SNPs with a low call rate is 0.05, corresponding to a 95% call rate.

3.2.3 Minor allele frequency

The next quality control step is to discard SNPs with a low minor allele frequency (MAF) as they increase the probability of detecting association when there is none (spurious association). The threshold often used is to exclude SNPs with a MAF below **0.05**. This can be performed using the "**--maf** Z" option, Z being the inclusion threshold, along with the "**--nonfounders**" option to include all individuals in the MAF calculation.

3.2.4 Hardy-Weinberg equilibrium

SNPs that depart from Hardy-Weinberg Equilibrium (HWE) should also be removed, as this deviation can be due to genotyping errors, inbreeding, population structure, or selection pressures, and they can also lead to spurious association. A chi^2 test can be performed to exclude SNPs deviating from HWE with the PLINK command "**--hwe** W" option, being W the inclusion threshold, along with the "**--nonfounders**" option to include all individuals in the HWE calculation. However, SNPs truly associated with the trait of interest can frequently deviate slightly from HWE, thus only strongly deviating SNPs are usually discarded (10^{-5} to 10^{-9}). In fish species, the most commonly used rejection threshold in GWAS is $\boldsymbol{p < 10^{-6}}$.

3.3 GWAS with GCTA

The input files required to run a GWAS in GCTA are the binary files produced by PLINK, a performance file (containing traits of interest), and a file with fixed effect and another one with covariates if there are potential confounding variables that should be considered in the GWAS (e.g., sex or tank effect). Those three last files should have no header and the two first columns should be (1) family ID and (2) fish ID (both matching the .fam file). In the performance file, the third column is the phenotype records with missing values coded as -9 or NA. If the performance file contains records for more than one trait (4 columns or more), by default GCTA performs the GWAS on the first trait (column 3) unless the option "**--mpheno x**" is used to specify that it should be performed on the *x*th trait (4th column for the second trait and so on). Both fixed effect and covariate files can contain multiple columns for quantitative and discrete variables, respectively (Fig. 2).

3.3.1 GRM construction

The first step of the analysis is to build the GRM to correct population structure. The GRM can be obtained either prior to the GWAS, saving the GRM in a file for all future analyses, or, if the dataset is of a manageable size (e.g., a typical medium density SNP array of circa 50K SNPs for a few thousand fish), simultaneously with the GWAS. To build the GRM, the GCTA option is "**--make-grm**" or "**--make-grm-gz**" to save the GRM in a compressed text file to save disk space.

3.3.2 Estimation of genetic parameters

The second step is the use of the "**--reml**" option of GCTA software to estimate genetic parameters as the trait heritability (h^2), the genetic variance explained by all SNPs, as well as the phenotypic and residual variances. These genetic parameters can be used to estimate the proportion of variance explained by a SNP (or a QTL) using the equation described as follows (Eq. 2):

$$V_{gi} = \frac{2p_j q_j g_j^2}{\sigma_g^2} \tag{2}$$

With p_j and q_j the allele frequencies for the jth SNP, g_j the estimated effect of the jth SNP, and σ_g^2 the total genetic variance $\left(\sigma_g^2 = \sum_{j=1}^{n} 2p_j q_j g_j^2\right)$. GCTA performs a restricted maximum likelihood (REML) analysis using average information (AI-REML) by default. This analysis is carried out iteratively until a predefined maximum number of iterations is reached (100 in GCTA, can be changed with the "**--reml-maxit**" option), or until the difference in log-likelihood (measure of the goodness of fit of the current model to the data) between consecutive iterations is sufficiently small (10^{-4} in GCTA). In the latter, we say that the algorithm has reached convergence and the analysis stops. This step can be performed together with the construction of the GRM. Results of the AI-REML analysis can also be found in the log file of the GWAS.

3.3.3 GWAS

The final step of the analysis is to perform the GWAS. Different marker-by-marker GWAS approaches can be implemented in GCTA. The simplest approach is the MLMA analysis, which estimates the effect of each candidate SNP using the previously built GRM to correct for population structure. This GWAS can be performed with the "**--mlma**" option. The LOCO (leave one chromosome out) approach is similar to the MLMA, but it constructs a new GRM without the chromosome carrying the candidate SNP. This approach is a little bit more time consuming than the regular MLMA approach as a new GRM has to be constructed for each chromosome analyzed. However, as the SNP effect is no longer fitted twice in the model (in the GRM as a random effect, and as the target SNP as a fixed effect), this approach is potentially more powerful. The LOCO analysis can be performed using the "**--mlma-loco**" option.

Results are saved in **.mlma** or **.loco.mlma** files. These contain a header and then one row per SNP with 9 columns: chromosome, SNP ID, position (bp), reference allele (which effect has been estimated), the alternative allele, the frequency of the reference allele, SNP effect, standard error, and *p*-value.

3.4 Visualization of results: Manhattan plots

GWAS results are usually visualized in a type of figure called Manhattan plot (Fig. 2), which consists of plotting the SNP position along the genome on the x axis, and the $-\log_{10}$(*p*-value) of each SNP on the y axis. Each SNP is typically plotted as a dot, and each peak of dots is a putative QTL. Several options exist to draw Manhattan plots, generally implemented in different R packages or codes. For users not familiar with R we recommend the use of the QQMAN package (Turner, 2018) since its "**manhattan**" function is simple and the GCTA output can be easily used as input.

3.5 Significance thresholds

The significance of candidate QTL is generally assessed using statistical thresholds, determining genome-wide significance based on *p*-value, and sometimes less-stringent suggestive or chromosome-wide thresholds. When using marker-by-marker approaches, all genotyped SNPs are tested for association with the trait, and therefore the *p*-values must be corrected for multiple testing to limit the number of false positives. The Bonferroni correction threshold is often used to determine the genome-wide and chromosome-wide significance of detected QTL at an α level (typically 5%, *p*-value of 0.05):

$$\text{Genome-wide threshold} = -\log_{10}\left(\alpha / [\text{number of SNP}]\right)$$

$$\text{Chromosome-wide threshold} = -\log_{10}\left(\alpha \Big/ \begin{bmatrix}\text{average number of SNP per} \\ \text{chromosome}\end{bmatrix}\right)$$

The Bonferroni correction is very conservative as it assumes that each association test (for each SNP) is independent of all other tests (all other SNPs), which is untrue due to the LD that exists between SNPs. Due to its conservative nature, the number of false negatives (type II error) is high and real QTLs of small effect may well be declared as nonsignificant. Consequently, SNPs and putative QTLs below the Bonferroni-corrected significance threshold are frequently considered as tentative or suggestive QTLs. In any case, a Manhattan plot at the genome level can provide a broad idea of the trait architecture, and then Manhattan plots at chromosome levels (and with HD panels at QTL levels) can be drawn to better interpret the results. Any outlier SNP that is above the threshold should be carefully considered since it might be a false positive (when that SNP is the only one with a significant *p*-value in a genomic region, i.e., no "peak" is observed). However, when one or two SNP detach from a small peak, they might be true candidate SNPs in close LD with the causative variant. A graphical representation of the deviation of the observed *p* values from the null hypothesis can be represented by a QQ plot: the observed *p* values for each SNP are sorted from largest to smallest and plotted against expected values from a theoretical χ^2-distribution. Major deviation from the theoretical distribution can point to issues in the data such as population structure or other sources of bias.

4. Pipeline

Here we describe a step-by-step pipeline to perform a GWAS. The code is also available in GitHub (https://github.com/Roslin-Aquaculture/GWAS_GCTA), along with the turbot dataset used to generate Fig. 2. The pipeline uses the PLINK software for quality control and formatting of genotyping data, the GCTA software to estimate genetic parameters and perform the GWAS, and the R package "QQMAN" is to produce a Manhattan plot. The three software packages used in this pipeline are freely available and user-friendly. For more detailed information on the options available for each software, you can refer to the PLINK (http://zzz.bwh.harvard.edu/plink/ for version 1.07 or https://www.cog-genomics.org/plink/ for version 1.09) and GCTA websites (http://cnsgenomics.com/software/gcta/). The QQMAN package is available from CRAN (https://cran.r-project.org/web/packages/qqman/vignettes/qqman.html) or GitHub (https://github.com/stephenturner/qqman).

(1) *Optional: Merge datasets*
If using genotyping performed on multiple SNP arrays (chip1 and chip2 here), you should merge PED files from all arrays with the "**--merge**" option:

```
plink --chr-set CHR --file mydata-chip1 --merge mydata-
chip2.ped mydata.map --recode --out mydata
```

Before merging files, you should check if the map files are identical or use two different map files, and verify that alleles are coded equally (i.e., ATGC or 1234). If genotypes are coded differently, the options "**--allele1234**" or "**--alleleACGT**" have to be used before merging the datasets.

(2) *Format conversion*
This first command transforms **.ped** and **.map** data in binary files:

```
plink --ped mydata.ped --map mydata.map --chr-set CHR --
make-bed --out newdata
```

You can add all the following options depending on your dataset "**--allow-no-sex --nonfounders --no-fid --no-parents --no-sex --no-pheno --recode-allele** refile.txt."

The "**--recode-allele**" option is not mandatory but is useful to keep track of the reference alleles and make sure that alleles in future files are consistently coded. Indeed, PLINK recodes alleles according to the minor allele count (by default, the minor allele is A1 and the major allele is A2 in PLINK outputs), and as frequency of alleles might change between populations it is important to force a specific allele to ensure that the results are comparable. "refile.txt" is a two-column file where the first column is the SNP ID and the second column the allele to be used as reference allele (i.e., A1). The reference allele is typically the allele/nucleotide that appears in the reference genome sequence for the species of interest.

Once your files are merged and/or in binary format you can start performing quality control. The command line to load binary files is **--bfile.**

(3) *Removal of duplicated samples*
This command is used to estimate pairwise IBD to detect duplicated samples:

```
plink --bfile newdata --genome --chr-set CHR --out tempdata
```

A file named **tempdata.genome** will be created with the IBD values of each pair of individuals in column 10. To remove duplicated samples, an extra step is necessary to select fish to be removed based on a threshold (IBD_thr=0.90):

```
gawk -v IBD_thr=0.90 ' NR > 1 { if ($10 > IBD_thr)
{ print $1" " $2 " " $3 " " $4 }}' tempdata.genome >
duplicated_ID.txt
gawk '{print $1 " " $2}' duplicated_ID.txt > ID_to_remove.txt
gawk '{print $3 " " $4}' duplicated_ID.txt >> ID_to_re-
move.txt
```

Those three lines of awk are used to create a file that contains the Family_ID and the fish_ID of duplicated samples to be removed from the dataset using the following plink line:

```
plink --bfile newdata --remove ID_to_remove.txt --make-
bed --out noduplicdata
```

(4) *Quality control*
This command is used to perform all other quality controls on the dataset:

```
plink --bfile noduplicdata --chr-set CHR --geno 0.05 --mind
0.05 --hwe 0.000001 --maf 0.05 --nonfounders --make-bed --out
finaldata
```

(5) *GWAS*
The following commands first create the GRM, perform the AI-REML analysis, and run a GWAS with no fixed effect and no covariates using the MLMA approach:

```
gcta64 --bfile finaldata --autosome-num CHR --make-grm --out
GRM_data --thread-num 10
gcta64 --reml --grm GRM_data--pheno TRAIT.pheno --out reml_TRAIT
gcta64 --mlma --bfile finaldata --grm GRM_data --autosome-
num CHR --pheno TRAIT.pheno --out gwas > gwas.log
```

In the **reml_TRAIT.hsq** (output for **--reml**) file you will find estimates of h^2, genetic and phenotypic variances. In the **gwas.log** file you can check that the REML analysis converged (typically AI-REML should converge in less than 20 iterations).

If you want to perform a GWAS using a model with fixed effect and covariates, you can use the following command, this time we used the LOCO approach:

```
gcta64 --mlma-loco --bfile finaldata --grm GRM_data --pheno
TRAIT.pheno --qcovar FixedEffet.qcovar --covar Covariate.
covar --autosome-num CHR --out gwas-fixed-cov --thread-num
10 > gwas-fixed-cov.log
```

(6) *Manhattan plot*

The following R commands use the QQMAN package to plot Manhattan plots from **.mlma** or **.loco.mlma** files:

```
install.package("qqman")
library(qqman)
NCHR=number of chromosomes # you have to set the number
# loading the file
gwasResults = read.table("gwas.mlma", header = TRUE)
# Threshold 1: 5% at the genome-wide level
Thrgen = -log10(0.05/nrow(gwasResults))
# Threshold 2: 5% at the chromosome-wide level
Thrchr = -log10(0.05/(nrow(gwasResults)/NCHR))
manhattan(gwasResults, chr = "Chr", bp="bp", snp ="SNP", p="p", main = "Manhattan
Plot", col = c("blue", "orange", "forestgreen", "purple"), suggestiveline = Thrchr,
genomewideline = Thrgen)
```

You can also use the qq function of the QQMAN package to plot the QQplot (quantile-quantile) of observed versus expected $-\log_{10}(p\text{-values})$:

```
qq(gwasResults$p, main = "Q-Q plot of GWAS p-values")
```

5. The next generation of association mapping methods: Multimarker models

Single-marker models interrogate the association between a trait and a particular SNP at a time, ignoring LD between the markers. While single-marker models are still the most frequently used association test, multimarker models that fit several SNPs simultaneously can potentially be more robust and powerful, especially regarding false positives and rare alleles (Kim et al., 2010; Wang et al., 2007, 2012b).

5.1 GBLUP and single-step GBLUP

The most common multimarker model for GWAS in aquaculture species is the GBLUP (Genomic BLUP), which estimates the effect of all markers simultaneously (assuming every marker has an effect of the same size) using the genomic relationship matrix G (VanRaden, 2008), constructed only with the records of genotyped individuals. A variation is the single-step GBLUP (ssGBLUP), which allows including related animals with phenotypic records but that have not been genotyped, using the H matrix that combine both pedigree (A) and genetic (G) relationship matrices. This model can increase the power and the accuracy of detection of the GWAS (Aguilar et al., 2010; Wang et al., 2012a) and is useful when a significant number of animals have phenotypic records but not genotypes. Both **GBLUP** and **ssGBLUP** models can be described using Eq. (3):

$$y_i = \mu + \beta_i + \sum_{j=1}^{N} g_j x_{ij} + \varepsilon_i \tag{3}$$

Where y_i being the observed phenotype for the ith individual, μ the overall mean in the population, β_i the fixed effect or covariate for the ith individual, g_j the random effect of the SNP j and x_{ij} the genotype of the ith fish for the SNP j (coded 0, 1, or 2 depending on the number of reference alleles), and ε_i the vector of residual effects for the ith fish. The effect of the SNP g_j is defined as $g \sim N(0, GRM\sigma_g^2)$, where σ_g^2 is the estimated genetic variance, and GRM is G in the **GBLUP** approach (the genetic relationship matrix) and H in the **ssGBLUP** approach (relationships based on both pedigree and genotypes).

The GBLUP and ssGBLUP models have been implemented within the *BLUPF90 programs* (Misztal et al., 2015), using the POSTGSF90 options (Aguilar et al., 2014) (http://nce.ads.uga.edu/wiki/doku.php), and they have been widely used in aquaculture genetics studies (Table 1).

5.2 Bayesian stochastic search variable selection approaches

Bayesian approaches are based on the assumption that not all SNPs have the same effect on the trait, which is closer to the reality than the GBLUP models where all SNPs are assumed to have an effect of the same size. The underlying principle of Bayesian approaches is to adapt the GBLUP model to better estimate the genetic variance explained by SNPs with the hypothesis that only a proportion (π) of SNPs have an effect on the trait and each SNP has its own estimated genetic variance. In order to estimate the effect of all SNPs, Bayesian approaches are performed iteratively typically using Markov chain Monte Carlo (MCMC) methods. Different approaches have been proposed: BayesA, BayesB, BayesC, BayesCπ, (Habier et al., 2011), BayesR (Erbe et al., 2012) and Bayesian Lasso (Park and Casella, 2008). The two main algorithms used nowadays are BayesC and BayesCπ, which present a good compromise between computing time and model complexity. Both BayesC and BayesCπ consider, at each iteration, only a proportion π of SNPs that are assumed to have an effect on the trait. Thus, at each iteration, the number of SNP effects to be estimated is typically lower than the number of phenotypic records, unlike with the GBLUP approaches. In BayesC, π is fixed whereas in BayesCπ, π is estimated based on an a priori value given by the user. In both models, at a given iteration, all retained π SNPs have the same variance.

Different software can be used to perform GWAS using Bayesian approaches. GS3 (Legarra et al., 2010) software can be used for Bayesian Lasso or BayesCπ, and BESSiE software (Boerner and Tier, 2016) can be used to perform BayesA, BayesB, BayesCπ, and BayesR.

6. Implementation of QTLs and causative variants into breeding programs

Numerous GWAS have been performed to map QTL affecting commercially important traits in aquaculture species over past two decades (e.g., Table 1). The most impactful one is arguably the discovery of a major QTL for resistance to IPN, where a single locus confers full resistance to the disease (Houston et al., 2008; Moen et al., 2009). This encouraged many other association studies on resistance to diseases in fish and shellfish species, which have revealed that most disease resistance traits have a polygenic architecture (Fraslin et al., 2020b). GWAS studies have also been amply used to understand the genetic basis of sex determination in fish species (e.g., Fraslin et al., 2020a; Palaiokostas et al., 2015; Wilson et al., 2014), assisting in the discovery of the sex determining genes in species with major sex determination QTL (e.g., Lin et al., 2021; Martínez et al., 2021).

The discovery of QTLs for traits of interest allows their implementation into aquaculture breeding programs. Marker-Assisted Selection (MAS) uses markers in LD with the causative variant to indirectly select the animals with favorable alleles. In practice, this is done only for major QTLs that have a large effect over the trait, since the benefit of MAS for QTLs of small effect is negligible compared to standard family based selection. Consequently, despite the high number of QTL mapping studies performed in aquaculture species, only a few major QTLs have been incorporated into commercial breeding programs, with the aforementioned IPNV resistance QTL in Atlantic salmon being a good example that has resulted in a major drop in the incidence of the disease (Houston et al., 2020). However, due to the polygenic nature of most traits, genomic selection (Meuwissen et al., 2001) is typically the most effective route to implementing genetic markers into aquaculture breeding programs. Genomic selection involves the recording of genotype and trait data in a reference population which are then used to predict breeding values for selection candidates for which (typically) only genotype data are available. In the most common form of genomic selection, GBLUP, the relationships between animals are estimated based on the similarity between their genotypes giving the same weight to every genetic marker. Nonetheless, the possibility of increasing the accuracy of genomic selection through the prioritization of QTLs or causative variants is a vibrant area of research, and different methods for the incorporation of this information into breeding programs are likely to arise in the coming years.

7. Funneling candidate variants toward causality

Despite the evolution of QTL mapping methods, the discovery of causative variants remains a cumbersome task with a relatively low success rate. In fact, up to date, only two putative causative variants have been found in aquaculture, both in Atlantic salmon: (1) two missense mutations in the *vgll3* gene control sexual maturation age (Ayllon et al., 2015), and (2) another missense mutation in *cdh1* that has been suggested to partly explain the major QTL affecting resistance to IPNV (Moen et al., 2015). However, both of these putative causative variants/ QTLs have a major effect, explaining a large percentage of the variation of those traits; no causative variants have been proven for QTLs of purely quantitative traits controlled by a large number of causative variants with small effect. While the progressive availability of larger datasets with high-density genotypes or even whole-genome sequencing data will increase our ability to detect QTLs of small effect, the underlying causative variant will likely remain elusive. The reason is that, while linkage and LD are key to find QTLs, they also link together many variants in the same region, making it almost impossible to distinguish the causative one just through association analyses. Furthermore, the effect of the causative variant may vary substantially according to genetic background (epistasis), and it is plausible that different causative variants underlie the same QTL in different families/populations. Therefore other strategies are necessary to distinguish the mutations underlying a QTL from linked markers without any direct effect on the trait.

Current strategies rely on adding different layers of information to progressively narrow the number of candidate variants in a region to a number (<10) that allows testing their causality directly, perhaps through genome editing. A first, simple layer of information would be the annotation of the candidate variants based on the annotation of the genome, classifying the variants in intergenic, intronic, synonymous, etc. (e.g., Palaiokostas et al., 2018b). A major caveat of this approach is that current aquaculture genome assemblies are poorly annotated, missing most noncoding genes and regulatory regions. Consequently, while SNP annotation might help prioritize some candidates with putative functional relevance (i.e., missense SNP or stop mutations), it should not be used to discard mutations that are apparently not functional according to the current annotation (i.e., intergenic SNPs could fall in as-yet-unidentified regulatory elements). As more complete genome annotations become available from ongoing global initiatives such as FAASG (Functional Annotation of All Salmonid Genomes; Macqueen et al., 2017), the prioritization value of SNP annotations will increase.

Understanding the biology underlying differential phenotypes through functional genomics experiments, for example between fast- and slow-growing or between disease resistant and susceptible animals, can also assist in the prioritization of candidate variants (Robledo et al., 2017b, 2018). The rationale is that this information can be used to prioritize genetic markers falling in genes that are related to the biological processes responsible for the phenotypic differences. A more direct approach would be to prioritize candidate variants that fall in genes showing differential expression between disparate phenotypes (or in close proximity to those genes; i.e., Robledo et al., 2019, 2020); if there is no candidate mutation affecting a protein or noncoding gene sequence, then the causative variant must act through the regulation of the expression of the gene in the QTL region. This is directly connected to a third layer of information, which is to directly associate candidate variants with molecular phenotypes such as expression or methylation (i.e., eQTL or metQTL), which can provide direct evidence of the functionality of a genetic marker. While these fields are relatively underdeveloped in aquaculture species, community efforts aimed to generate these resources are likely to arise in the coming years, similarly to the long-standing Genotype-Tissue Expression (GTEx) project in human research (GTEx Consortium, 2013).

While adding different layers of information can help establish a "ranking of probability of causation," conclusive evidence of the connection between genetic markers and traits can only be obtained through direct manipulation of the genome, currently mostly performed through Clustered Regularly Interspaced Short Palindromic Repeats (CRISPR)-Cas mediated genome editing (Edvardsen et al., 2014; Hwang et al., 2013). In many cases, particularly for disease resistance traits, candidate variants can be tested first in cell lines, which can be easier, cheaper, and faster. However, this in vitro editing relies on the ability to obtain relevant cellular phenotypes, which can be troublesome for many complex phenotypes such as feed efficiency or resistance to multicellular parasites such as sea lice. Whether an in vitro step is possible or not, the final step of the causality funnel is genome editing in vivo, which is the only step that will actually confirm the functionality of the putative causative variants, i.e., that they have a real effect on the trait in the animal. However, genome editing is still in its relative infancy in aquaculture, and while the field is developing rapidly there is a current lack of resources (e.g., appropriate cell lines) and methods (low efficiency of in vivo editing). Tackling these limitations to realize the full potential of genome editing in the different species will underpin the successful discovery of causative variants in aquaculture in the coming years.

8. Conclusions

The aquaculture industry has swiftly incorporated molecular tools into their breeding programs to speed up genetic improvement of quantitative traits. The rate of genetic gain of these breeding programs can be further increased through the incorporation of markers linked to QTLs or, even better, the causative variants underlying these QTLs. However, the number of QTL-linked markers and causative variants available in aquaculture species is limited, and to date success stories are based on (i) marker-assisted selection for major-effect QTL, or (ii) use of all markers to predict breeding values using genomic selection.

Here, we have described and discussed historical and current methodologies to detect association between genomic regions and traits of interest, along with a detailed step-by-step workflow using state-of-the-art software and methods. While these methods enable the efficient detection of QTLs, and increased power achieved through larger genotyping efforts will lead to the identification of important numbers of QTLs of increasingly small effect in the near future, causative variants are likely to remain elusive due to the LD between numerous genetic markers in the QTL region, and in aquaculture further complicated due to the complex nature of the genomes of many species. Genome annotation, functional genomics, and association of genetic variants with molecular traits can narrow the number of candidate mutations in a QTL region; however, final proof of the association between a genetic marker and a phenotype can only be achieved through functional tests using techniques such as genome editing. As such, the development of CRISPR-Cas technologies in aquaculture species is expected to underpin the identification of causative variants in the coming years. These causative variants may be applied in breeding programs to increase selection accuracy and therefore genetic gain, or potentially via direct genome editing of broodstock or multiplier animals (Jin et al., 2021). In both cases, identification and use of causative variants for production traits will ultimately improve the sustainability of the industry and global food security.

References

Aguilar, I., Misztal, I., Johnson, D.L., Legarra, A., Tsuruta, S., Lawlor, T.J., 2010. Hot topic: a unified approach to utilize phenotypic, full pedigree, and genomic information for genetic evaluation of Holstein final score1. J. Dairy Sci. 93, 743–752. https://doi.org/10.3168/jds.2009-2730.

Aguilar, I., Misztal, I., Tsuruta, S., Legarra, A., Wang, H., 2014. PREGSF90—POSTGSF90: computational tools for the implementation of single-step genomic selection and genome-wide association with ungenotyped individuals in BLUPF90 programs. In: Proceedings of 10th World Congress on Genetics Applied to Livestock Production. Presented at the 10 World Congress on Genetics Applied to Livestock Production (WCGALP, Vancouver, Canada).

Ayllon, F., Kjærner-Semb, E., Furmanek, T., Wennevik, V., Solberg, M.F., Dahle, G., Taranger, G.L., Glover, K.A., Almén, M.S., Rubin, C.J., Edvardsen, R.B., Wargelius, A., 2015. The vgll3 locus controls age at maturity in wild and domesticated Atlantic Salmon (*Salmo salar L.*) males. PLoS Genet. 11. https://doi.org/10.1371/journal.pgen.1005628, e1005628.

Barría, A., Christensen, K.A., Yoshida, G.M., Correa, K., Jedlicki, A., Lhorente, J.P., Davidson, W.S., Yáñez, J.M., 2018. Genomic predictions and genome-wide association study of resistance against Piscirickettsia salmonis in Coho Salmon (*Oncorhynchus kisutch*) using ddRAD sequencing. G3 (Bethesda) 8, 1183–1194. https://doi.org/10.1534/g3.118.200053.

Boerner, V., Tier, B., 2016. BESSiE: a software for linear model BLUP and Bayesian MCMC analysis of large-scale genomic data. Genet. Sel. Evol. 48, 63. https://doi.org/10.1186/s12711-016-0241-x.

Chang, C.C., Chow, C.C., Tellier, L.C., Vattikuti, S., Purcell, S.M., Lee, J.J., 2015. Second-generation PLINK: rising to the challenge of larger and richer datasets. GigaScience 4, 7. https://doi.org/10.1186/s13742-015-0047-8.

Danzmann, R.G., Cairney, M., Davidson, W.S., Ferguson, M.M., Gharbi, K., Guyomard, R., Holm, L.-E., Leder, E., Okamoto, N., Ozaki, A., Rexroad, C.E., Sakamoto, T., Taggart, J.B., Woram, R.A., 2005. A comparative analysis of the rainbow trout genome with 2 other species of fish (Arctic charr and Atlantic salmon) within the tetraploid derivative Salmonidae family (subfamily: Salmoninae). Genome 48, 1037–1051. https://doi.org/10.1139/g05-067.

Darvasi, A., Soller, M., 1997. A simple method to calculate resolving power and confidence interval of QTL map location. Behav. Genet. 27, 125–132. https://doi.org/10.1023/A:1025685324830.

Darvasi, A., Weinreb, A., Minke, V., Weller, J.I., Soller, M., 1993. Detecting marker-QTL linkage and estimating QTL gene effect and map location using a saturated genetic map. Genetics 134, 943–951.

Dong, C., Jiang, P., Zhang, J., Li, X., Li, S., Bai, J., Fan, J., Xu, P., 2019. High-density linkage map and mapping for sex and growth-related traits of largemouth bass (*Micropterus salmoides*). Front. Genet. 10. https://doi.org/10.3389/fgene.2019.00960, 960.

Edvardsen, R.B., Leininger, S., Kleppe, L., Skaftnesmo, K.O., Wargelius, A., 2014. Targeted mutagenesis in Atlantic Salmon (*Salmo salar L.*) using the CRISPR/Cas9 system induces complete knockout individuals in the F0 generation. PLoS One 9. https://doi.org/10.1371/journal.pone.0108622, e108622.

Elsen, J.-M., Mangin, B., Goffinet, B., Boichard, D., Le Roy, P., 1999. Alternative models for QTL detection in livestock. I. General introduction. Genet. Sel. Evol. 31, 213. https://doi.org/10.1186/1297-9686-31-3-213.

Erbe, M., Hayes, B.J., Matukumalli, L.K., Goswami, S., Bowman, P.J., Reich, C.M., Mason, B.A., Goddard, M.E., 2012. Improving accuracy of genomic predictions within and between dairy cattle breeds with imputed high-density single nucleotide polymorphism panels. J. Dairy Sci. 95, 4114–4129. https://doi.org/10.3168/jds.2011-5019.

Fraslin, C., Dechamp, N., Bernard, M., Krieg, F., Hervet, C., Guyomard, R., Esquerré, D., Barbieri, J., Kuchly, C., Duchaud, E., Boudinot, P., Rochat, T., Bernardet, J.-F., Quillet, E., 2018. Quantitative trait loci for resistance to *Flavobacterium psychrophilum* in rainbow trout: effect of the mode of infection and evidence of epistatic interactions. Genet. Sel. Evol. 50, 60. https://doi.org/10.1186/s12711-018-0431-9.

Fraslin, C., Brard-Fudulea, S., D'Ambrosio, J., Bestin, A., Charles, M., Haffray, P., Quillet, E., Phocas, F., 2019. Rainbow trout resistance to bacterial cold water disease: two new quantitative trait loci identified after a natural disease outbreak on a French farm. Anim. Genet. 50, 293–297. https://doi.org/10.1111/age.12777.

Fraslin, C., Phocas, F., Bestin, A., Charles, M., Bernard, M., Krieg, F., Dechamp, N., Ciobotaru, C., Hozé, C., Petitprez, F., Milhes, M., Lluch, J., Bouchez, O., Poncet, C., Hocdé, P., Haffray, P., Guiguen, Y., Quillet, E., 2020a. Genetic determinism of spontaneous masculinisation in XX female rainbow trout: new insights using medium throughput genotyping and whole-genome sequencing. Sci. Rep. 10, 17693. https://doi.org/10.1038/s41598-020-74757-8.

Fraslin, C., Quillet, E., Rochat, T., Dechamp, N., Bernardet, J.-F., Collet, B., Lallias, D., Boudinot, P., 2020b. Combining multiple approaches and models to dissect the genetic architecture of resistance to infections in fish. Front. Genet. 11. https://doi.org/10.3389/fgene.2020.00677.

Gjedrem, T., Rye, M., 2018. Selection response in fish and shellfish: a review. Rev. Aquac. 10, 168–179. https://doi.org/10.1111/raq.12154.

Glasauer, S.M.K., Neuhauss, S.C.F., 2014. Whole-genome duplication in teleost fishes and its evolutionary consequences. Mol. Genet. Genom. 289, 1045–1060. https://doi.org/10.1007/s00438-014-0889-2.

Gratacap, R.L., Wargelius, A., Edvardsen, R.B., Houston, R.D., 2019. Potential of genome editing to improve aquaculture breeding and production. Trends Genet. 35, 672–684. https://doi.org/10.1016/j.tig.2019.06.006.

GTEx Consortium, 2013. The genotype-tissue expression (GTEx) project. Nat. Genet. 45, 580–585. https://doi.org/10.1038/ng.2653.

Habier, D., Fernando, R.L., Kizilkaya, K., Garrick, D.J., 2011. Extension of the bayesian alphabet for genomic selection. BMC Bioinf. 12, 186. https://doi.org/10.1186/1471-2105-12-186.

Hayes, B.J., Gjuvsland, A., Omholt, S., 2006. Power of QTL mapping experiments in commercial Atlantic salmon populations, exploiting linkage and linkage disequilibrium and effect of limited recombination in males. Heredity 97, 19–26. https://doi.org/10.1038/sj.hdy.6800827.

Houston, R.D., Haley, C.S., Hamilton, A., Guy, D.R., Tinch, A.E., Taggart, J.B., McAndrew, B.J., Bishop, S.C., 2008. Major quantitative trait loci affect resistance to infectious pancreatic necrosis in Atlantic salmon (*Salmo salar*). Genetics 178, 1109–1115. https://doi.org/10.1534/genetics.107.082974.

Houston, R.D., Bean, T.P., Macqueen, D.J., Gundappa, M.K., Jin, Y.H., Jenkins, T.L., Selly, S.L.C., Martin, S.A.M., Stevens, J.R., Santos, E.M., Davie, A., Robledo, D., 2020. Harnessing genomics to fast-track genetic improvement in aquaculture. Nat. Rev. Genet., 1–21. https://doi.org/10.1038/s41576-020-0227-y.

Hwang, W.Y., Fu, Y., Reyon, D., Maeder, M.L., Tsai, S.Q., Sander, J.D., Peterson, R.T., Yeh, J.-R.J., Joung, J.K., 2013. Efficient genome editing in zebrafish using a CRISPR-Cas system. Nat. Biotechnol. 31, 227–229. https://doi.org/10.1038/nbt.2501.

Jin, Y.H., Robledo, D., Hickey, J.M., McGrew, M.J., Houston, R.D., 2021. Surrogate broodstock to enhance biotechnology research and applications in aquaculture. Biotechnol. Adv. 49, 107756. https://doi.org/10.1016/j.biotechadv.2021.107756.

Kim, S., Morris, N.J., Won, S., Elston, R.C., 2010. Single-marker and two-marker association tests for unphased case-control genotype data, with a power comparison. Genet. Epidemiol. 34, 67–77. https://doi.org/10.1002/gepi.20436.

Knott, S.A., Elsen, J.M., Haley, C.S., 1996. Methods for multiple-marker mapping of quantitative trait loci in half-sib populations. Theor. Appl. Genet. 93, 71–80. https://doi.org/10.1007/BF00225729.

Lander, E.S., Botstein, D., 1989. Mapping mendelian factors underlying quantitative traits using RFLP linkage maps. Genetics 121, 185–199.

Lander, E.S., Schork, N.J., 1994. Genetic dissection of complex traits. Science 265, 2037–2048. https://doi.org/10.1126/science.8091226.

Legarra, A., Ricard, A., Filangi, O., 2010. GS3–Genomic Selection, Gibbs Sampling, Gauss Seidel (and BayesCp). INRA, Paris, France. Available at: http://genoweb.toulouse.inra.fr/~alegarra.

Lin, H., Zhou, Z., Zhao, J., Zhou, T., Bai, H., Ke, Q., Pu, F., Zheng, W., Xu, P., 2021. Genome-wide association study identifies genomic loci of sex determination and Gonadosomatic index traits in large yellow croaker (*Larimichthys crocea*). Marine Biotechnol. 23, 127–139. https://doi.org/10.1007/s10126-020-10007-2.

Loukovitis, D., Sarropoulou, E., Batargias, C., Apostolidis, A.P., Kotoulas, G., Tsigenopoulos, C.S., Chatziplis, D., 2012. Quantitative trait loci for body growth and sex determination in the hermaphrodite teleost fish *Sparus aurata L.* Anim. Genet. 43, 753–759. https://doi.org/10.1111/j.1365-2052.2012.02346.x.

Lu, G., Luo, M., 2020. Genomes of major fishes in world fisheries and aquaculture: status, application and perspective. Aquac. Fish. 5, 163–173. https://doi.org/10.1016/j.aaf.2020.05.004.

Macqueen, D.J., Primmer, C.R., Houston, R.D., Nowak, B.F., Bernatchez, L., Bergseth, S., Davidson, W.S., Gallardo-Escárate, C., Goldammer, T., Guiguen, Y., Iturra, P., Kijas, J.W., Koop, B.F., Lien, S., Maass, A., Martin, S.A.M., McGinnity, P., Montecino, M., Naish, K.A., Nichols, K.M., Ólafsson, K., Omholt, S.W., Palti, Y., Plastow, G.S., Rexroad, C.E., Rise, M.L., Ritchie, R.J., Sandve, S.R., Schulte, P.M., Tello, A., Vidal, R., Vik, J.O., Wargelius, A., Yáñez, J.M., 2017. Functional annotation of all salmonid genomes (FAASG): an international initiative supporting future salmonid research, conservation and aquaculture. BMC Genomics 18, 484. https://doi.org/10.1186/s12864-017-3862-8.

Martínez, P., Robledo, D., Taboada, X., Blanco, A., Moser, M., Maroso, F., Hermida, M., Gómez-Tato, A., Álvarez-Blázquez, B., Cabaleiro, S., Piferrer, F., Bouza, C., Lien, S., Viñas, A.M., 2021. A genome-wide association study, supported by a new chromosome-level genome assembly, suggests sox2 as a main driver of the undifferentiatiated ZZ/ZW sex determination of turbot (*Scophthalmus maximus*). Genomics 113, 1705–1718. https://doi.org/10.1016/j.ygeno.2021.04.007.

Meuwissen, T.H., Hayes, B.J., Goddard, M.E., 2001. Prediction of total genetic value using genome-wide dense marker maps. Genetics 157, 1819–1829.

Meuwissen, T.H.E., Karlsen, A., Lien, S., Olsaker, I., Goddard, M.E., 2002. Fine mapping of a quantitative trait locus for twinning rate using combined linkage and linkage disequilibrium mapping. Genetics 161, 373–379.

Misztal, I., Tsuruta, S., Lourenco, D., Aguilar, I., Legarra, A., Vitezica, Z., 2015. Manual for BLUPF90 Family of Programs.

Moen, T., Baranski, M., Sonesson, A.K., Kjøglum, S., 2009. Confirmation and fine-mapping of a major QTL for resistance to infectious pancreatic necrosis in Atlantic salmon (*Salmo salar*): population-level associations between markers and trait. BMC Genomics 10, 368. https://doi.org/10.1186/1471-2164-10-368.

Moen, T., Torgersen, J., Santi, N., Davidson, W.S., Baranski, M., Ødegård, J., Kjøglum, S., Velle, B., Kent, M., Lubieniecki, K.P., Isdal, E., Lien, S., 2015. Epithelial cadherin determines resistance to infectious pancreatic necrosis virus in Atlantic Salmon. Genetics 200, 1313–1326. https://doi.org/10.1534/genetics.115.175406.

Moraleda, C.P., Robledo, D., Gutiérrez, A.P., Del-Pozo, J., Yáñez, J.M., Houston, R.D., 2021. Investigating mechanisms underlying genetic resistance to Salmon Rickettsial Syndrome in Atlantic salmon using RNA sequencing. BMC Genomics 22, 156. https://doi.org/10.1186/s12864-021-07443-2.

Palaiokostas, C., Bekaert, M., Khan, M.G., Taggart, J.B., Gharbi, K., McAndrew, B.J., Penman, D.J., 2015. A novel sex-determining QTL in Nile tilapia (*Oreochromis niloticus*). BMC Genomics 16, 171. https://doi.org/10.1186/s12864-015-1383-x.

Palaiokostas, C., Cariou, S., Bestin, A., Bruant, J.-S., Haffray, P., Morin, T., Cabon, J., Allal, F., Vandeputte, M., Houston, R.D., 2018a. Genome-wide association and genomic prediction of resistance to viral nervous necrosis in European sea bass (*Dicentrarchus labrax*) using RAD sequencing. GSE Genet. Sel. Evol. 50, 30. https://doi.org/10.1186/s12711-018-0401-2.

Palaiokostas, C., Robledo, D., Vesely, T., Prchal, M., Pokorova, D., Piackova, V., Pojezdal, L., Kocour, M., Houston, R.D., 2018b. Mapping and sequencing of a significant quantitative trait locus affecting resistance to koi herpesvirus in common carp. G3 (Bethesda) 8, 3507–3513. https://doi.org/10.1534/g3.118.200593.

Park, T., Casella, G., 2008. The Bayesian lasso. J. Am. Stat. Assoc. 103, 681–686. https://doi.org/10.1198/016214508000000337.

Potts, R.W.A., Gutierrez, A.P., Penaloza, C.S., Regan, T., Bean, T.P., Houston, R.D., 2021. Potential of genomic technologies to improve disease resistance in molluscan aquaculture. Philos. Trans. R. Soc. B Biol. Sci. 376, 20200168. https://doi.org/10.1098/rstb.2020.0168.

Robledo, D., Palaiokostas, C., Bargelloni, L., Martínez, P., Houston, R., 2017a. Applications of genotyping by sequencing in aquaculture breeding and genetics. Rev. Aquac. 10, 670–682. https://doi.org/10.1111/raq.12193.

Robledo, D., Rubiolo, J.A., Cabaleiro, S., Martínez, P., Bouza, C., 2017b. Differential gene expression and SNP association between fast- and slow-growing turbot (*Scophthalmus maximus*). Sci. Rep. 7, 12105. https://doi.org/10.1038/s41598-017-12459-4.

Robledo, D., Gutiérrez, A.P., Barría, A., Yáñez, J.M., Houston, R.D., 2018. Gene expression response to sea lice in Atlantic Salmon skin: RNA sequencing comparison between resistant and susceptible animals. Front. Genet. 9, 287. https://doi.org/10.3389/fgene.2018.00287.

Robledo, D., Gutiérrez, A.P., Barría, A., Lhorente, J.P., Houston, R.D., Yáñez, J.M., 2019. Discovery and functional annotation of quantitative trait loci affecting resistance to sea lice in Atlantic Salmon. Front. Genet. 10. https://doi.org/10.3389/fgene.2019.00056, 56.

Robledo, D., Hamilton, A., Gutiérrez, A.P., Bron, J.E., Houston, R.D., 2020. Characterising the mechanisms underlying genetic resistance to amoebic gill disease in Atlantic salmon using RNA sequencing. BMC Genomics 21, 271. https://doi.org/10.1186/s12864-020-6694-x.

Rodríguez-Ramilo, S.T., Fernández, J., Toro, M.A., Bouza, C., Hermida, M., Fernández, C., Pardo, B.G., Cabaleiro, S., Martínez, P., 2013. Uncovering QTL for resistance and survival time to *Philasterides dicentrarchi* in turbot (*Scophthalmus maximus*). Anim. Genet. 44, 149–157. https://doi.org/10.1111/j.1365-2052.2012.02385.x.

Rosyara, U.R., Jong, W.S.D., Douches, D.S., Endelman, J.B., 2016. Software for genome-wide association studies in autopolyploids and its application to potato. Plant Genome 9. https://doi.org/10.3835/plantgenome2015.08.0073.

Sardell, J.M., Cheng, C., Dagilis, A.J., Ishikawa, A., Kitano, J., Peichel, C.L., Kirkpatrick, M., 2018. Sex differences in recombination in sticklebacks. G3 (Bethesda) 8, 1971–1983. https://doi.org/10.1534/g3.118.200166.

Singer, A., Perlman, H., Yan, Y., Walker, C., Corley-Smith, G., Brandhorst, B., Postlethwait, J., 2002. Sex-specific recombination rates in zebrafish (*Danio rerio*). Genetics 160, 649–657.

Tsai, H.Y., Robledo, D., Lowe, N.R., Bekaert, M., Taggart, J.B., Bron, J.E., Houston, R.D., 2016. Construction and annotation of a High density SNP linkage map of the Atlantic Salmon (*Salmo salar*) genome. G3 (Bethesda) 6, 2173–2179. https://doi.org/10.1534/g3.116.029009.

Turner, S., 2018. Qqman: an R package for visualizing GWAS results using Q-Q and Manhattan plots. J. Open Source Softw. 3, 731. https://doi.org/10.21105/joss.00731.

Vallejo, R.L., Cheng, H., Fragomeni, B.O., Shewbridge, K.L., Gao, G., MacMillan, J.R., Towner, R., Palti, Y., 2019. Genome-wide association analysis and accuracy of genome-enabled breeding value predictions for resistance to infectious hematopoietic necrosis virus in a commercial rainbow trout breeding population. Genet. Sel. Evol. 51, 47. https://doi.org/10.1186/s12711-019-0489-z.

Vandeputte, M., Haffray, P., 2014. Parentage assignment with genomic markers: a major advance for understanding and exploiting genetic variation of quantitative traits in farmed aquatic animals. Front. Genet. 5. https://doi.org/10.3389/fgene.2014.00432, 432.

VanRaden, P.M., 2008. Efficient methods to compute genomic predictions. J. Dairy Sci. 91, 4414–4423. https://doi.org/10.3168/jds.2007-0980.

Wang, T., Zhu, X., Elston, R.C., 2007. Improving power in contrasting linkage-disequilibrium patterns between cases and controls. Am. J. Hum. Genet. 80, 911–920. https://doi.org/10.1086/516794.

Wang, H., Misztal, I., Aguilar, I., Legarra, A., Muir, W.M., 2012a. Genome-wide association mapping including phenotypes from relatives without genotypes. Genet. Res. 94, 73–83. https://doi.org/10.1017/S0016672312000274.

Wang, X., Morris, N.J., Schaid, D.J., Elston, R.C., 2012b. Power of single- vs. multi-marker tests of association. Genet. Epidemiol. 36, 480–487. https://doi.org/10.1002/gepi.21642.

Wang, L., Bai, B., Liu, P., Huang, S.Q., Wan, Z.Y., Chua, E., Ye, B., Yue, G.H., 2017. Construction of high-resolution recombination maps in Asian seabass. BMC Genomics 18. https://doi.org/10.1186/s12864-016-3462-z, 63.

Waples, R.K., Seeb, L.W., Seeb, J.E., 2016. Linkage mapping with paralogs exposes regions of residual tetrasomic inheritance in chum salmon (*Oncorhynchus keta*). Molecular Ecology Resources 16, 17–28. https://doi.org/10.1111/1755-0998.12394.

Wilson, C.A., High, S.K., McCluskey, B.M., Amores, A., Yan, Y., Titus, T.A., Anderson, J.L., Batzel, P., Carvan, M.J., Schartl, M., Postlethwait, J.H., 2014. Wild sex in zebrafish: loss of the natural sex determinant in domesticated strains. Genetics 198, 1291–1308. https://doi.org/10.1534/genetics.114.169284.

Yang, J., Lee, S.H., Goddard, M.E., Visscher, P.M., 2011. GCTA: a tool for genome-wide complex trait analysis. Am. J. Hum. Genet. 88, 76–82. https://doi.org/10.1016/j.ajhg.2010.11.011.

Yang, J., Zaitlen, N.A., Goddard, M.E., Visscher, P.M., Price, A.L., 2014. Advantages and pitfalls in the application of mixed model association methods. Nat. Genet. 46, 100–106. https://doi.org/10.1038/ng.2876.

Yoshida, G.M., Lhorente, J.P., Carvalheiro, R., Yáñez, J.M., 2017. Bayesian genome-wide association analysis for body weight in farmed Atlantic salmon (*Salmo salar L.*). Anim. Genet. 48, 698–703. https://doi.org/10.1111/age.12621.

Yoshida, G.M., Lhorente, J.P., Correa, K., Soto, J., Salas, D., Yáñez, J.M., 2019. Genome-wide association study and cost-efficient genomic predictions for growth and fillet yield in Nile Tilapia (*Oreochromis niloticus*). G3 (Bethesda) 9, 2597–2607. https://doi.org/10.1534/g3.119.400116.

You, X., Shan, X., Shi, Q., 2020. Research advances in the genomics and applications for molecular breeding of aquaculture animals. Aquaculture 526, 735357. https://doi.org/10.1016/j.aquaculture.2020.735357.

Zenger, K.R., Khatkar, M.S., Jones, D.B., Khalilisamani, N., Jerry, D.R., Raadsma, H.W., 2019. Genomic selection in aquaculture: application, limitations and opportunities with special reference to marine shrimp and pearl oysters. Front. Genet. 9. https://doi.org/10.3389/fgene.2018.00693, 693.

Zhou, Z., Han, K., Wu, Y., Bai, H., Ke, Q., Pu, F., Wang, Y., Xu, P., 2019. Genome-wide association study of growth and body-shape-related traits in large yellow croaker (*Larimichthys crocea*) using ddRAD sequencing. Mar. Biotechnol. (N.Y.) 21, 655–670. https://doi.org/10.1007/s10126-019-09910-0.

Chapter 7

Double-labeling immunofluorescence: A simple and effective tool to study pituitary regulation in teleost fish

Paula G. Vissio[a,b], Daniela I. Pérez Sirkin[a,b], and María P. Di Yorio[a,b]
[a]*Department of Biodiversity and Experimental Biology, Faculty of Exact and Natural Sciences, University of Buenos Aires, Buenos Aires, Argentina,* [b]*Institute of Biodiversity and Experimental and Applied Biology (IBBEA), CONICET-University of Buenos Aires, Buenos Aires, Argentina*

1. The importance of gaining knowledge about the neuroendocrine system

Neuroendocrinology is the field of biology that studies the interaction between the endocrine and the nervous system. External (temperature, photoperiod, social interactions, and food availability, among others) and internal (for example, metabolic state and stress) information are integrated within the brain and transduced into neural and neuroendocrine signals that control the different organs, leading to the maintenance of homeostasis in vertebrates. The hypothalamic–pituitary axis is the key system that rules all processes in an organism. Thus, the study of this axis is fundamental to address how different environmental factors modify neuropeptide levels and how these neuropeptides interact between them to regulate metabolic functions.

Since fish development, growth, reproduction, and behavior are especially susceptible to environmental conditions (such as photoperiod, temperature, water quality, crowding, and food quality), neuroendocrine studies could provide important clues on how they impact fish physiology, as well as information to improve aquaculture rearing practices by developing new techniques and strategies for better fish welfare. A classic example of the potential application of neuroendocrine studies in aquaculture is the case of gonadotropin-releasing hormone (Gnrh), the main factor that triggers reproduction. From the first report that demonstrated that Gnrh stimulates gonadotropin secretion (Breton et al., 1972), to the continuous characterization of Gnrh isoforms in fish (Sherwood et al., 1983; Tostivint, 2011; Roch et al., 2014; Muñoz-Cueto et al., 2020) and

Cellular and Molecular Approaches in Fish Biology. https://doi.org/10.1016/B978-0-12-822273-7.00003-3

their regulation by environmental conditions (Maitra and Hasan, 2016; reviewed in Volkoff, 2016), the study of the Gnrh system was fundamental to adequate protocols that substantially improved ovulation, spermiation, and/or spawning performance in species in which spontaneous reproduction was difficult in captivity (Muñoz-Cueto, 2009).

Besides possible aquaculture applications, it is worth to investigate the neuroendocrine system from the viewpoint of comparative endocrinology, to get a picture of the complex evolution of this system through the vertebrate's history. In this sense, fish neuroanatomy makes them exceptional models for studying the function and evolution of this system (Golan et al., 2015), as they conserve the major factors that regulate the same functions as in tetrapods.

2. Neuroendocrine system

The vertebrate's pituitary gland (or hypophysis) constitutes the functional link between the nervous and the endocrine system. It is comprised of two lobes that are distinguished by their embryonic origin, cellular composition, and function: the neurohypophysis (NH) and the adenohypophysis (ADH). Although the entire pituitary derives from the ectoderm, it has a dual origin: an evagination of the ventral hypothalamus (neural ectoderm) gives rise to the NH, while the ADH derives from the adenohypophyseal placode (epidermic ectoderm) that then invaginates and subsequently separates from the stomodeum (Baker and Bronner-Fraser, 2001). Hence, the NH represents the neurohemal part and the ADH the glandular part. The NH is divided into *pars nervosa* and median eminence, whereas the ADH is subdivided into three regions: *pars distalis* (PD), *pars tuberalis* (PT), and *pars intermedia* (PI), the former being the largest of the three regions and the site where the most pituitary hormone-producing cells are localized. The PT, present only in tetrapods, is mainly related to photoperiod and reproductive processes (Shinomiya et al., 2014). The PI is absent in hagfishes and birds, and it is reduced or even absent in mammals (Norris and Carr, 2013). In vertebrates, three classical anatomical models of pituitary-central regulation are described: portal system via the median eminence (tetrapods), direct innervation (teleosts), and diffusion (agnathans) (Sower, 2015).

Particularly in teleost fish, NH consists of an infundibular stem suspended from the ventral region of the hypothalamus. This stem contains an extension of the third ventricle called the infundibular recess, and in its distal part it interdigitates, like tree branches, in the three regions of the ADH: the most anterior named rostral *pars distalis* (RPD), the middle portion named proximal *pars distalis* (PPD), and the most caudal, the PI. Interdigitations in PI are deeper and more elaborate than in RPD and PPD (Ball and Baker, 1969). The whole PI and NH are known as the neurointermediate lobe. The anterior NH is considered homologous to the middle eminence and innervates the RPD and the PPD (Peter et al., 1990). Interestingly, in teleosts, although with variations in some species, different cell types in the ADH are usually localized inspecific regions, like countries on a map. The cells that produce prolactin (Prl) and adrenocorticotropin (Acth), lactotropes and adrenocorticotropes, respectively, are mainly distributed in the RPD. In the PPD, most of the somatotropes, the cells that produce growth hormone (Gh), are bordering the NH, whereas thyrotropes, the cells that produce thyrotropin hormone, are localized in the PPD and RPD. In the case of fish, two types of gonadotropes can be distinguished in the middle portion of the PPD, one that produces luteinizing hormone (Lh) and the other, producing follicle-stimulating hormone (Fsh), whereas, in tetrapods, one single cell type secretes both gonadotropins (Levavi-Sivan et al., 2010). Finally, the PI is the region where melanotropes, the cells that express α-melanophore-stimulating hormone (Msh), and somatolactotropes, the cells that produce somatolactin (Sl), are localized (Fig. 1A). Other nonendocrine pituitary cell type, poorly investigated in fish, is the stellate cells, similar to the mammalian folliculostellate cells. These cells were mainly studied in Nile tilapia (*Oreochromis niloticus*) and are functionally interconnected by gap junction into a continuous network in the PPD and the PI, suggesting that this arrangement would allow that certain signals are conveyed to distant pituitary regions (Golan et al., 2016).

The pituitary innervation originates from neurons mainly localized in the preoptic and hypothalamic area. Nevertheless, it has been reported that other neurons from extra hypothalamic regions such as the olfactory bulbs, the ventral telencephalon, the suprachiasmatic nucleus, the thalamus, or the mesencephalic *tegmentum* may also innervate it (Johnston and Maler, 1992; Anglade et al., 1993; Zohar et al., 2010; Muñoz-Cueto et al., 2020). While it is largely assumed that teleost pituitary cells are directly innervated, different types of associations between nerve terminals and pituitary cells could be observed: (1) direct synaptic contact between the neurosecretory terminals and the endocrine cells; (2) presence of a single or double basal membrane interposed between the neurosecretory ends and the endocrine cells (reviewed in Cerdá-Reverter and Canosa, 2009; Zohar et al., 2010; Trudeau and Somoza, 2020); and (3) a neurovascular mode of delivery, as a new alternative scenario recently proposed in transgenic zebrafish, where a close apposition of gonadotropin-releasing hormone 3 (Gnrh3) fibers and permeable blood vessels at the hypothalamic level has been demonstrated by Golan et al. (2015) (Fig. 1B).

It is important to remark that in nonteleost fish, the hypothalamic neurons release their secretory products into a portal system (Lagios, 1968; Kah and Adrio, 2018), and subsequently delivered to the ADH. Also, a rudimentary portal system was described in few teleosts species such as in the Indian mayor carp (*Labeo rohita*; Jose and Sathyanesan, 1977) and the Indo-Pacific tarpon (*Megalops cyprinoides*; Baskaran and Sathyanesan, 1992).

It is worth noting that the innervation presents a certain degree of plasticity, which means that it could vary during the development or according to particular physiological situations. For example, in the Indian mayor carp and the South American cichlid *Cichlasoma dimerus*, it is possible to observe pituitary innervation of gonadotropin inhibitory hormone (Gnih) during development,

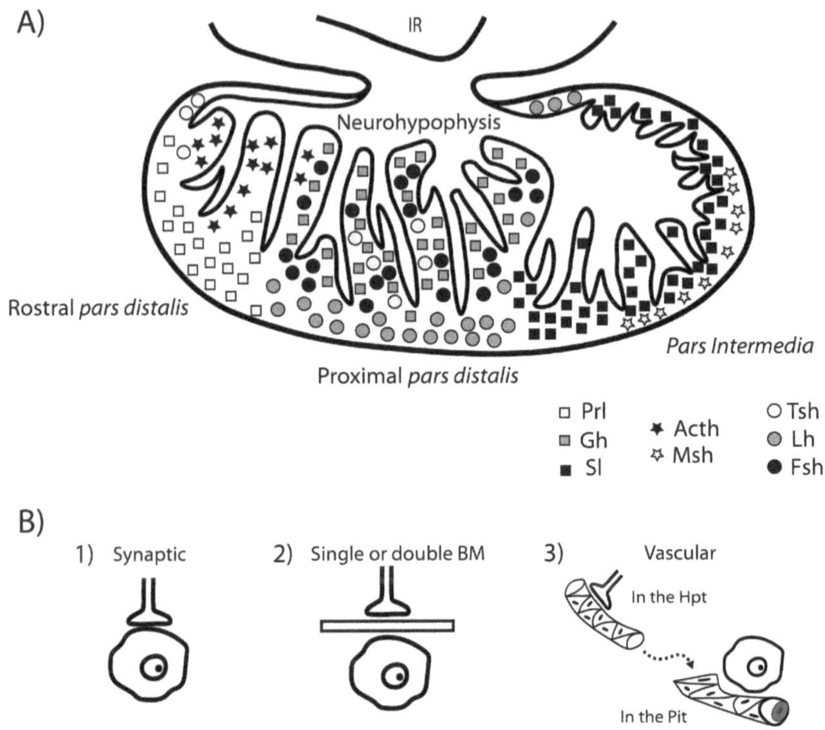

FIG. 1 Scheme of teleost's pituitary and the different types of associations between nerve terminals and pituitary cells. (A) Distribution of adenohypophyseal cell types in a pituitary sagittal section. Cell types are represented by different colored symbols associated with the hormone they produce. (B) Schematic representations of possible associations between the central nervous system terminals and adenohypophyseal cells: (1) Synaptic contact between the nerve terminal and the adenohypophyseal cell. (2) Presence of a single or double basal membrane between the nerve terminal and the adenohypophyseal cell. Neuropeptides diffuse through the extracellular space reaching pituitary cells. (3) At the hypothalamic level, the nerve terminal releases neuropeptides into the vascular system that in turn are delivered to the adenohypophyseal cell (Golan et al., 2015). *White square*: lactotropes, cells producing prolactin (Prl); *gray square*: somatotropes, cells producing growth hormone (Gh); *black square*: somatolactotropes, cells producing somatolactin (Sl); *black star*: adrenocorticotropes, cells producing adrenocorticotropin (Acth); *white star*: melanotropes, cells producing α-melanophore-stimulating hormone (Msh); *white circle*: thyrotropes, cells producing thyrotropin hormone; *gray and black circles*: gonadotropes, cells producing luteinizing hormone (Lh) or follicle stimulating hormone (Fsh), respectively. BM, basal membrane; Hpt, hypothalamus; IR, infundibular recess; Pit, pituitary.

but not in adults, suggesting that the mode of action of Gnih on pituitary cells may differ between life stages (Biswas et al., 2015; Di Yorio et al., 2018). Moreover, sexual stage-dependent plasticity (i.e., number of neurosecretory granules and microvesicles in nerve terminals or variations in the degree of innervation) was also reported in goldfish (*Carassius auratus*; Kaul and Vollrath, 1974; Canosa et al., 2011).

The fact that the different pituitary cells are localized in specific regions, plus the direct innervation, allows using simple strategies that evidence possible associations. In this context, the more cited technique in the literature is the double-labeling immunofluorescence. Besides, the use of genetically modified fish models, such as transgenic zebrafish (*Danio rerio*), medaka (*Oryzias latipes*), and tilapia, with labeled neuropeptides or pituitary cells, represents an excellent system to study neuroendocrine relationships (Trudeau and Somoza, 2020). The use of these novel genetic techniques combined with confocal microscopy allows us to gain knowledge in the hypothalamic–pituitary relationship and will introduce new concepts on fish neuroendocrinology. Although genomic tools are increasingly available for other fish species (Japanese pufferfish (*Takifugu rubripes*), spotted green pufferfish (*Tetraodon nigroviridis*), Atlantic cod (*Gadus morhua*), and little skate (*Leucoraja erinacea*), among others), they are still not representative of all taxa, as teleost fish present a great phenotypic complexity due to their specific 3R genome duplication (Meyer and Van de Peer, 2005). Thus it is important to consider that neuroendocrine studies in nonmodel species might help to build a full picture. In this sense, the strategies that we present in this chapter could provide new insights, especially in nonmodel species.

3. Double-labeling immunofluorescence

In this section, we will present and discuss a simple, cheap, and reliable technique for approaching neuroendocrine studies in fish: the double-labeling immunofluorescence. As the name indicates, this methodology involves the immune recognition of two different peptides with the subsequent detection by fluorochromes. As previously described, although there is a great conservation in hypothalamic nuclei and adenohypophyseal hormones localization in fish, each species could present unique characteristics that are necessary to study. A simple way to start these studies is by performing immunohistochemical techniques in serial histological sections, one for the neuropeptide of interest and the other for the adenohypophyseal hormone. In this way, we could visualize if the innervated region is close to the specific site of the producing cells.

Immunohistochemistry (IHC) is a widely used technique for cellular- or tissue-specific antigen localization by using antigen–antibody recognition. The site of antibody binding can be identified by direct labeling of the first antibody (named direct IHC) or by using a secondary (or additional steps) labeling method, which enables amplifying the signal (named indirect IHC) (Kabiraj et al., 2015). In those cases where a fluorophore is used as a labeling method, the technique is called immunofluorescence (IF). Coons et al. (1941) described for the first time a mammalian tissue constituent using an antibody directly labeled with a fluorescent dye, and more than a decade later, indirect immunofluorescence was introduced by Weller and Coons

(1954) (Sanderson et al., 2018). Nowadays, this technique has improved considerably, being a widely used tool in laboratories worldwide. The IF, as well as the IHC, with the employment of two primary antibodies that recognize different antigens in the same cell culture or histological section, is known as double-labeling IF or double-labeling IHC, respectively. However, before performing the double-labeling techniques, the simple IF/IHC must be fine-tuned separately.

3.1 General immunofluorescence protocols

The first and very important step in IF is the fixation, consisting of stabilization of tissue proteins and preservation of cell morphology and molecules as close as possible to the living stage. Besides, fixation stops cell metabolism, inactivates lysosomal enzymes, and eliminates microorganisms (Martín-Lacave and García-Caballero, 2012). As cell death causes rotting and autolysis, for optimal preservation, fixation should be performed immediately after organ dissection. This is crucial in organs with high blood irrigation, such as brain and pituitary.

The most common fixing procedure used in neuroendocrinology studies is the immersion of the sample in a fixative liquid like additive fixatives, as formaldehyde and paraformaldehyde, both diluted to 4% (*v*/v) in phosphate buffer, or fixative mixtures such as Bouin's liquid (75% saturated picric acid: 25% formaldehyde: 5% glacial acetic acid). Both types provide a good preservation of morphology and fast penetration. It must be considered that formaldehyde presents a health risk given its toxicity and carcinogenicity. Fixation is always a compromise between time, volume, sample size, and temperature. The volume of the fixative must be close to 40 times higher than the sample volume, which at the same time must be small because, instead, fast preservation of the inner part of the sample is not ensured. It is recommended to perform the first fixative hour at room temperature (RT) to accelerate the fixative's diffusion inside the sample and then continue at 4 °C for 18–24 h. To avoid hardening of the organs and loss of antigenicity, the fixation must not exceed 24/48 h when using these fixatives. As will be discussed later, for IF it is recommended to perfuse the samples in order to remove blood cells. However, this is not always possible because of the fish size studied or the laboratory facilities.

Fixation is an irreversible process, and the resulting complexes differ from undenatured proteins. Because of the chemical properties of certain antigens, when formalin fixation and the paraffin processing are performed (Kabiraj et al., 2015), some epitopes are more likely to suffer modifications that generate the nonrecognition of the protein by the antibodies. As will be discussed later, some strategies have been developed to overcome this epitope masking.

After fixation, the tissue samples are usually embedded in paraffin (with a previous dehydration process in a series of increasing ethanol solutions) or frozen, and then sectioned. The protocols commonly used for double-labeling IF for samples embedded in paraffin include the following steps:

1. DEPARAFFINIZATION OF TISSUE SECTIONS by an organic solvent, usually xylene or benzene, for an hour with a change of the solvent in the middle. It is important to make sure to extract all the paraffin from the tissues to allow the posterior antibody diffusion.
2. TISSUE SAMPLE REHYDRATION by the immersion in decreasing graduations of ethanol down to phosphate buffered saline (PBS). The maintenance of the pH buffer at 7.4 for this step and the following ones is essential for the correct tissue preservation.
3. ANTIGEN RETRIEVAL. This step is only recommended in those cases where the antigen is masked by the chemical processes involved in fixation and may require optimization. If it is not necessary, do not perform it because it could damage the tissue. Different strategies are described later.
4. BLOCKING OF NONSPECIFIC BINDING SITES. This is usually done by incubating the sample with bovine serum albumin (BSA), nonfat dry milk, or animal serum. When primary antibodies are raised in rabbit or mouse, nonfat dry milk and BSA are normally used. However, when antibodies are raised in goat, sheep, or horse, the use of these blockers may lead to an increased background. Thus, it is recommended to block with animal serum that is dependent on the host of your primary or secondary antibody. For example, when using antisera derived from goat, block with goat serum.
5. FIRST PRIMARY ANTIBODY INCUBATION. This antibody must recognize and bind with the corresponding antigen (Fig. 2A). The antiserum could be diluted in PBS or the blocking solution previously used, this depends on the antibody specificity. The incubation time, as well as the dilution used, must be set in each case. However, a good beginning is an information reported in the bibliography for IF or ICH performed in the same tissues of closely related species. It is recommended to test different concentrations leaving the time fixed, usually 24h at 4 °C, or less time at room temperature (RT) (be aware of the possible antiserum solution evaporation), or to test a concentration modifying the incubation time or temperature. The lowest concentration, with a clear immunostaining (without a background), should be chosen. These criteria are applied to secondary antiserum or amplification steps. Generally, as the first antibody is not labeled, one of the molecules in the following steps must be coupled with a fluorochrome label:
   5.1 BINDING WITH A SECONDARY ANTIBODY that recognizes the first primary antibody. It could be labeled with fluorochrome or biotinylated (Fig. 2B and C). This antiserum is regularly used at high concentration for 1–2h of incubation at RT.

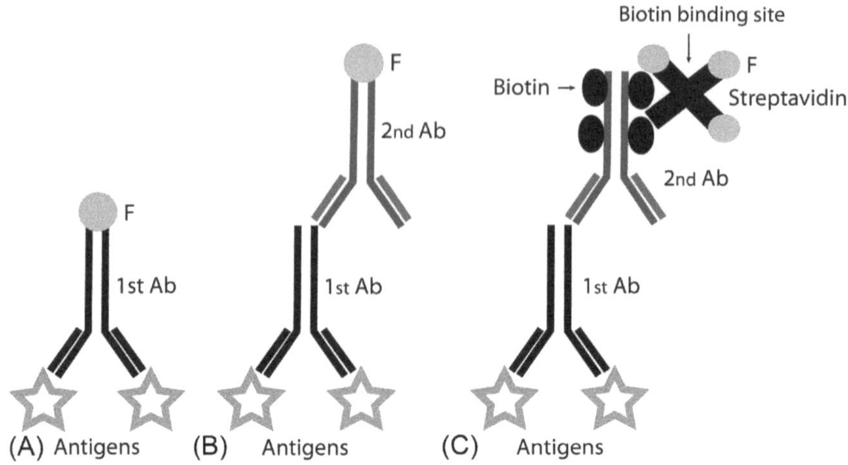

FIG. 2 Schematic representation of possible detection strategies for an immunoassay. (A) Use of a primary antibody coupled with a fluorochrome. (B) Use of a secondary antibody coupled with a fluorochrome that recognizes the primary antibody. (C) Use of streptavidin (amplification molecule) coupled with two or more fluorochromes that recognizes the biotin of the secondary antibody. Ab, antibody; F, fluorochrome.

5.1.1 ADDITION OF ANOTHER AMPLIFICATION STEP. If the secondary antibody is biotinylated, incubation with streptavidin or streptavidin-biotin complex, coupled with a fluorochrome is performed (Fig. 2C). This step helps to improve the detection as it amplifies the signal, but it should be important to avoid it or improve the blockage when tissue has a high concentration of endogenous biotin-containing proteins, lectins or avidin conjugates because of the unwanted nonspecific binding. Generally, this step is performed at high concentration for 1–2 h at RT.

6. Repeat step number 5 for the other protein of interest. In this way, the double-labeling technique is sequential. If the primary antibodies are produced in different animals (see later), they could be incubated simultaneously, so the 5th step is performed for recognition of both antigens.
7. MOUNTING. This step is usually done with a solution of glycerol:PBS (1:1). However, it is recommended to use a commercial antifade mounting medium for preserving fluorescence.

If the tissue is correctly preserved at 4 °C, the double-labeling IF is a transient methodology that lasts a few weeks, because of the loss of fluorescence by quenching and conservation of the sample in a hydrophilic solution.

In Fig. 3, the readers will find some recommendations for double-labeling IF of paraffin-embedded tissue, which could help them to start a new protocol that must be optimized to their species of interest and/or conditions.

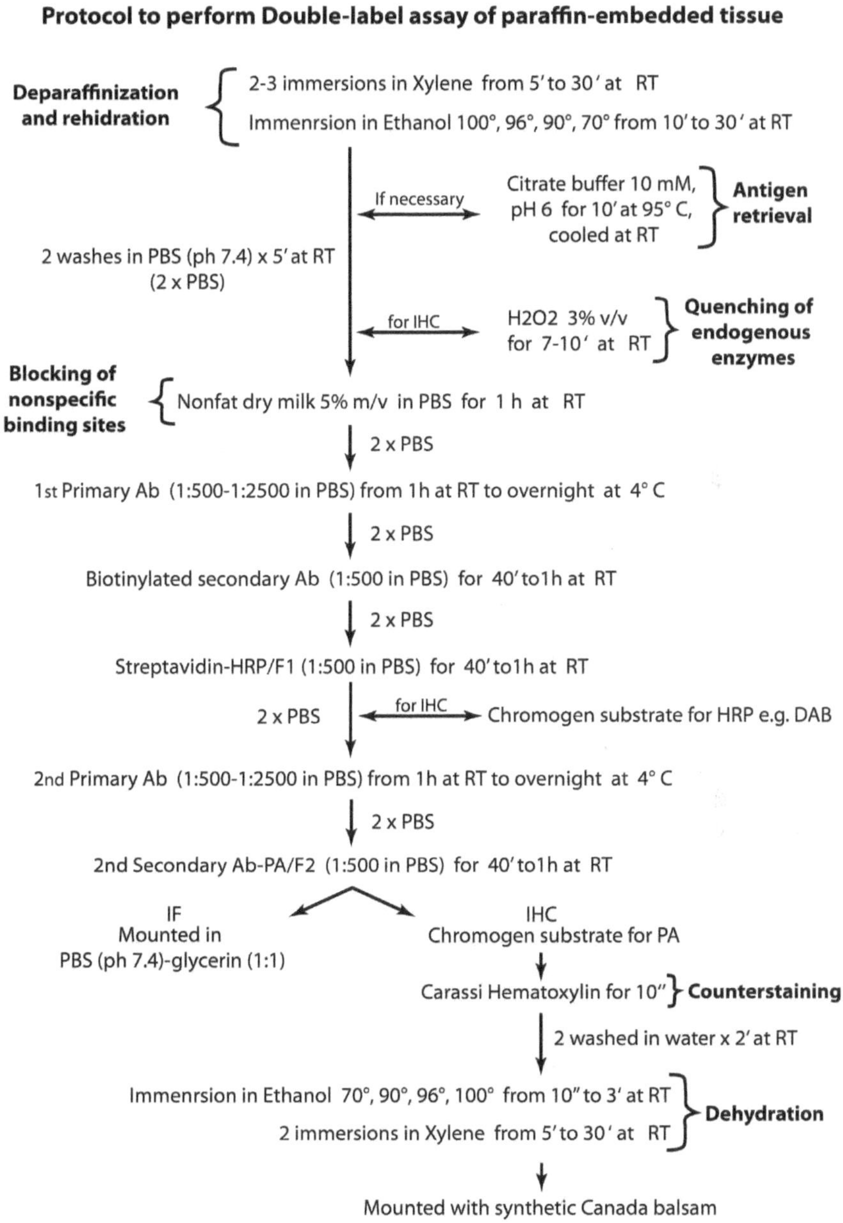

FIG. 3 Schematic protocol to perform double-labeling immunofluorescence (IF) based on Pérez Sirkin et al. (2012), Di Yorio et al. (2015), and Di Yorio et al. (2019a) in paraffin-embedded tissues.

If an IHC is performed, some steps are different from the earlier mentioned protocol. First, before blocking nonspecific binding sites, quenching of endogenous enzymes activity must be done, otherwise they react with detection reagents giving false-positive results. Alkaline phosphatases and peroxidases are examples of endogenous enzymes. This is usually done with H_2O_2 (diluted in PBS or methanol) or free avidin. Second, the detection method is performed by the addition of the specific chromogen substrate for the alkaline phosphatases or peroxidases that are coupled with the secondary antibody or the additional amplification step. The detection method generates a colored precipitate that could be visualized by light microscopy. Later, a counterstaining step is useful because it allows recognizing the context of the tissue where the immunostaining is. Finally, dehydration in an increasing grade of ethanol up to xylene and mounting on a hydrophobic medium must be performed.

Although there is a wide range of variants and problems that could affect these protocols, antigen–antibody recognition and visualization or detection systems are key stages that deserve special attention.

3.2 Antigen–antibody recognition

Antibodies present specificity and high affinity for their antigens and, for that reason, they have been a valuable and indispensable tool in research. The association between an antibody and its antigen is comprised of noncovalent interactions, where amino acid residues on the antigen-binding site, the epitope, and on the antibody-binding site, the paratope, intimately contribute to the energetics needed for the antigen–antibody complex stabilization (Kapingidza et al., 2020).

As mention before, in a double-labeling IF, two specific primary antibodies are used. The secondary antibodies recognize a conserved region of the primary antibody that is specific to the animal in which it has been generated. Thus the different primary antibodies must be raised in different animals (e.g., rabbit, mouse, sheep, and guinea pig, among others) to avoid cross-reactivity. However, a double-labeling technique with two primary antibodies raised in the same animal could be performed when the immune marks are different and easily distinguishable from each other, for example, immunoreactive fibers and immunoreactive cells (Fig. 4) (Pérez Sirkin et al., 2012).

3.2.1 Monoclonal versus polyclonal antibodies

Monoclonal antibodies, derived from a single β-cell clone and produced by the hybridoma technique, provide excellent specificity because the antibody binds to a single epitope. On the other hand, polyclonal antibodies, as they derive from immunoglobulin-rich serum of animals immunized with a purified specific molecule bearing the antigen of interest, have a lower specificity (as they contain antibodies to a wide range of antigens), but a higher affinity and wide reactivity because they are more likely to identify multiple epitopes. So, the use

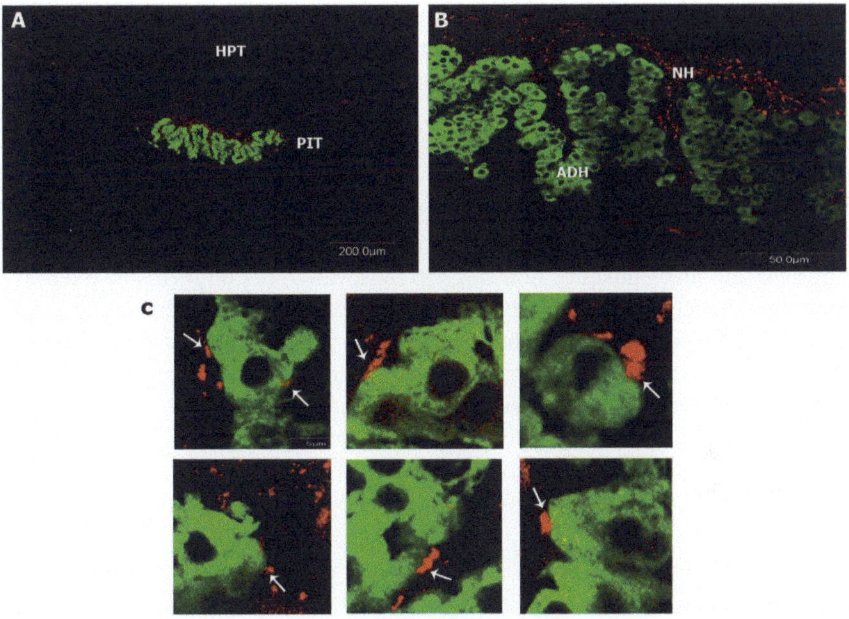

FIG. 4 In this figure, a representative confocal microphotograph of a double-labeling IF showing melanin concentrating hormone (Mch) fibers (*red*) and growth hormone (Gh) cells (*green*) is presented. (A) Coronal sections at the proximal *pars distalis* (PPD) level of *C. dimerus* pituitary. Note that Gh-immunoreactive (-ir) cells are localized in PPD as it was represented in Fig. 1. Scale bar: 200 µm. (B) Higher magnification showing high quantities of Mch-ir fibers in the neurohypophysis at the PPD region of the pituitary (*red*). Mch fibers are in close association with Gh-ir cells which are surrounding the neurohypophysis. Scale bar: 50 µm (C) Detail of Mch-ir fibers and Gh-ir cells contact (*white arrows*). Scale bar: 5 µm. HPT, hypothalamus; PIT, pituitary; NH, neurohypophysis; ADH, adenohypophysis.(Adapted from Pérez Sirkin et al., 2012).

of the latter is usually recommended when no species-specific primary antibodies (heterologous) are applied (Ramos-Vara et al., 2008; Kabiraj et al., 2015).

3.2.2 Homologous versus heterologous antibodies

Homologous antibodies refer to monoclonal or polyclonal antibodies generated against the specific fish species epitope, either obtained by protein purification or synthesis. On the other hand, heterologous antibodies refer to the antibody generated against a specific epitope of other species, such as another fish species or, as it is mostly commercially available, against human or mouse epitope. In nonmodel teleost species, the use of a homologous antibody is less common, so researchers normally use heterologous ones. However, some precautions must be considered. For example, before performing the IHC or IF technique with a heterologous antibody, a western blot control is recommended to corroborate if the antibody recognizes a protein with the expected molecular weight. As mentioned before, one advantage of teleost fish is that pituitary cell types are localized in specific regions of the ADH, so, when testing for the first time a heterologous antibody, the region where the immune mark appears will help to infer its specificity.

3.2.3 The masking effect of fixative

Usually, negative results are obtained when the antigens have been masked by the chemical processes. There are different methodologies for recovering the antigenicity (reviewed in Shi et al., 2011; Ramos-Vara and Miller, 2014). However, the mechanism of how antigen retrieval works on formalin-fixed tissues is not clear (Kabiraj et al., 2015). Among several methods, the ones that imply heating the sections are the most used due to the simplicity of the procedure, and all require a chelating buffer, such as citrate or ethylene-diamine-tetraacetic acid, to keep the tissue moist and stabilize the antigens. Sections can be heated in a microwave oven, pressure cooker, or water bath (between 95 °C and 99 °C), the latter being the one that provides better consistency of the results (Kabiraj et al., 2015).

3.2.4 Negative controls

The specificity of antigen–antibody recognition must be validated for both antibodies used. If primary antibodies are polyclonal, one of the most used methods as a negative control to differentiate specific from nonspecific signal is the replacement of these antibodies by the antiserum preadsorbed with a highly purified antigen (or synthetic peptide). However, it is often difficult for nonmodel fish species to possess the specific antigen. Alternatively, to assess the degree of cross-reactivity of the primary antiserum, a preimmune antiserum could be used. If the primary antibodies are monoclonal, the use of another irrelevant monoclonal antibody is probably the best negative control (Boenisch, 2002).

Besides, it should be considered that additional controls to assess the degree of nonspecific binding by the labeling (secondary) antibody or the detection system must be performed. For example, the omission of the primary or secondary antibody is commonly carried out, replacing them by the antibody diluent (Ramos-Vara et al., 2008).

3.3 Immunofluorescence visualization

For visualizing the results, the selection of the fluorophore dyes will depend on the filters or laser available in the microscope. The most common fluorochromes are FITC (fluorescein isothiocyanate, green dye), rhodamine (red dye) and the synthetics Alexa Fluor. The former presents rapid photobleach, so it is strongly recommended to use an antifade agent. The colors used to present the results could be manipulated. Journals recommend not using green and red colors in the images because it is difficult to interpret the results for color blind people.

3.3.1 Autofluorescence background

Autofluorescence is the major source of nonspecific background throughout the fluorescence spectrum. Sources of autofluorescence include aldehyde fixatives; tissue components with endogenous fluorescence (including extracellular matrix proteins, red blood cells, and macrophages); and lipofuscin, which consists of granules of oxidized proteins and lipids that build up in the lysosomes of aged cells. It is generally recommended by the technical service of the antibody's companies to include an unstained tissue control to determine the level of autofluorescence in the sample. Many treatments have been reported to reduce autofluorescence, including quenching of aldehydes with ammonium sulfate and Tris, bleaching with sodium borohydride, and quenching of autofluorescence with blue or black dyes. Besides, when technically possible, it is highly recommended to perform the fixation by perfusion to remove blood cells to avoid this source of autofluorescence, as previously mentioned.

3.3.2 Fluorescence microscopes

The implementation of IF requires the use of a fluorescence microscope, one of the most widely used imaging methods in biology (Huang et al., 2008). The combination of immunofluorescence with confocal laser-scanning microscopic analysis permits a detailed analysis of larger tissue volumes and at a higher resolution than with conventional fluorescence microscopy (Holmqvist and Ekström, 1997). Besides, the confocal microscope decreases the photobleaching effect because of the use of laser instead of lamps. The double-labeling IF results can be visualized by their examination under both type of microscopes, depending on the biological question being addressed by this method (Joshi and Yu, 2017). However, the confocal system allows observing only the fluorescence that comes from a focal plane and presents a greater resolution than a conventional fluorescence microscope. Resolution is defined as the shortest distance between two distinct points that can still be distinguished as discrete objects. The resolution of optical devices depends on two factors: the wavelength and the numerical aperture of the objective lens (Fouquet et al., 2015). The spatial resolution could be divided into axial resolution (along the z-dimension or optical axis) and lateral resolution (in the x,y plane). Two immunoreactive points could seem to be in contact; however, due to the limited low spatial resolution that this tool presents, about 200–300 nm in the lateral direction and 500–700 nm in the axial direction (Huang et al., 2009), the interpretation could be mistaken. For this reason, confocal microscopy is widely used to moderately enhance spatial resolution (Pawley, 2006). By combining a focused laser for excitation and a pinhole for detection, confocal microscopy can, in principle, have a factor of $\sqrt{2}$ improvement in spatial resolution (Huang et al., 2009). However, as this is still a problem for a colocalization purpose, confocal microscopy with a later computational study of intensity is required. To improve the confocal limitations, several "super-resolution" fluorescence microscopy techniques have been developed to overcome the diffraction barrier. Some examples are the stimulated emission depletion (STED) microscopy, the reversible saturable optically linear fluorescence transitions (RESOLFTs), and the saturated structured-illumination microscopy (SSIM), as well as techniques that are based on the localization of individual fluorescent molecules, such as the stochastic optical reconstruction microscopy (STORM), the photoactivated localization microscopy (PALM), and the fluorescence photoactivation localization microscopy (FPALM) (reviewed in Huang et al., 2009). The description of these techniques is beyond the objectives of the chapter, but it is important to know their limits and how to improve them. Beyond these resolution advantages, one of the most important technical advantages of confocal microscopy and "super-resolution" fluorescence microscopy is decreasing the loss of fluorescence of fluorophore dyes by illuminating discrete points instead of the entire sample at the same time. Other advantages are that a three-dimensional reconstruction can be performed, as well as optical sections.

3.3.3 Immunofluorescence interpretation

As mentioned in Fig. 1, there are different mechanisms by which specific hypothalamic nuclei regulate the adenohypophyseal cells. If direct contacts occurred, specific association dots are expected to be detected when the images obtained from the two fluorophores are merged, as it is observed in Fig. 4. It is important to remember that computational studies of intensity are required due to the spatial resolution. However, if specific association points are not obtained but there is a close association between the two immunomarks, diffusion of the neuropeptide could be considered. For this reason, in the next section, we present methodologies that could be carried out to study the possible regulation over the adenohypophyseal cells.

4. Going deeper on the pituitary hormone regulation

After confirming an association or a close association by double-labeling IF, it is important to go further using other approaches to provide evidences about possible regulation of neuropeptide over pituitary hormone/s. Some techniques that could be useful in this analysis are as follows: (i) in vitro assessments to investigate if the neuropeptide exerts some effect (inhibition/stimulation) over pituitary hormone synthesis and/or release; (ii) immunolabeling, in situ hybridization (ISH), or radiolabeled-ligand binding, combined with immunocytochemistry or IF, to assess if one or more cell types could express receptors for the neuropeptide, which enables them to transduce its signal, meaning a direct regulation (Fig. 5).

4.1 In vitro assays

In these assays, static or perifuse culture of the whole pituitary, or an enriched fraction or multiple populations of the pituitary cells, could be performed in the

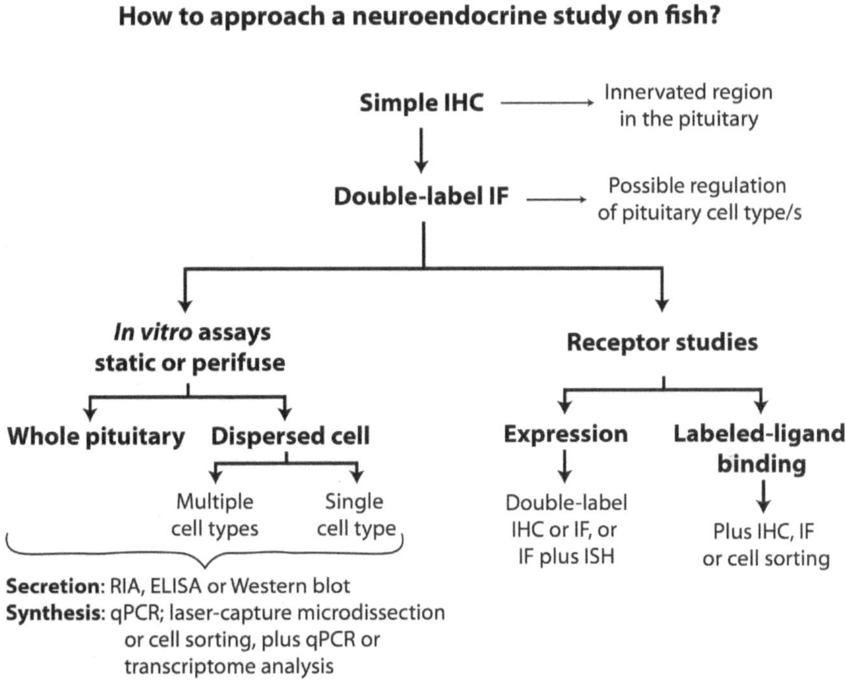

FIG. 5 Outline of the possible steps to follow on a study about a new neuropeptide candidate to regulate pituitary cell function. ELISA, enzyme-linked immunosorbent assay; IF, immunofluorescence; IHC, immunohistochemistry; ISH, in situ hybridization; qPCR, real-time quantitative PCR; RIA, radioimmunoassay.

presence of different concentrations of the neuropeptide. Two fundamental aspects that must be considered in the experimental design are the sex and gonadal stage (especially in females) of the animals used for the experiments since there are sex-specific receptor expression patterns that could vary according to the reproductive cycle as it was reported in the European sea bass (*Dicentrarchus labrax*; González-Martínez et al., 2004), the pejerrey (*Odontesthes bonariensis*; Guilgur et al., 2009), and the orange-spotted grouper (*Epinephelus coioides*; Wang et al., 2015).

For whole pituitary culture, these glands must be rapidly dissected out and then individually placed in multiwell plates or perifusion chambers with culture medium, pH around 7.2–7.4, usually containing fetal bovine serum (FBS) and/or BSA, 4-(2-hydroxyethyl)-1-piperazineethanesulfonic acid (HEPES) buffer, antibiotics, and antimycotics. Before challenging pituitaries against the neuropeptide, a preincubation period is needed for the stabilization of hormone secretion. This and the following incubations are usually carried out at 27–28 °C (Levavi-Sivan and Yaron, 1992; Astola et al., 2004; Cánepa et al., 2008).

In pituitary cell culture, pituitaries are removed from animals and washed with sterile-antibiotic-containing media to remove surface contaminants before processing. Then, they are diced into fragments of approximately $1\,mm^3$ and treated with collagenase type V or trypsin–EDTA solution to digest connective tissue and gently agitated to disperse cells. Note that trypsin-based digestion is faster compared to that of collagenase. It must be considered that big fragments result in a decrease of cell viability, while smaller in an increase of cell damage because of trypsin treatment. The duration and temperature during enzyme incubation are critical in cell yield and its viability. After inhibiting enzymatic digestion, partially digested tissues could be separated by centrifugation. Between the last two steps, most protocols include DNase to prevent cell clumping due to the DNA released from broken cells. Then, cell viability is usually determined with trypan blue dye exclusion and cell number counted. Viability of the recovered cells less than 85% prior to culture is usually not useful for hormone release response studies. Cells are plated in multiwell pates or perifusion chamber/columns with the same culture medium for whole pituitary culture (Levavi-Sivan and Yaron, 1992; Amano et al., 2006; Moussavi et al., 2012; reviewed in Chang and Jobin, 1994).

In line with this, there are some techniques to separate heterogeneous pituitary cell populations into functionally homogeneous groups. Some of them are based on differences in density among cell types, allowing separating each other by gravity or centrifugation over density gradients, consisting of materials such as dextran, albumin, and Percoll. For example, De Leeuw et al. (1984) obtained a better than 95% enriched population of gonadotropes from dispersed pituitary cells of the African catfish (*Clarias lazera*) by using a discontinuous density gradient prepared with a Percoll suspension. Finally, the cell sorting methodology after IF labeling with antibodies against the pituitary hormones of interest could be performed in fish, as it has been applied in mammals (Thorner et al., 1982; Flaws and Suter, 1993).

4.1.1 Whole pituitary vs pituitary dispersed cells

There are advantages and disadvantages to applying one of these methodologies. In whole pituitary cultures, indirect effects could not be ruled out because neuropeptide action may be mediated by the release of remaining factors present in nerve terminals given the impossibility of removing them, since NH interdigitates with the whole ADH. Besides, possible indirect effects through interaction among different pituitary cell types, i.e. a paracrine regulation, can occur. Instead, in cell culture no terminal innervation is present. However, several culture conditions could affect the responsiveness and attachment of the dispersed pituitary cells, including temperature, serum products, cell density, culture surface, as well as the type of medium used (reviewed in Chang and Jobin, 1994). It is important to note that, with trypsin-based digestion, GAP junctions are lost, so a possible effect through the interaction by homotypic cell coupling could be lost as it was reported for gonadotropes in zebrafish and tilapia by Golan et al. (2016). Taken together, enriched cell cultures could be useful to explore specific questions. For example, to study the direct effect of a neuropeptide over a population, an agonist or antagonist of a certain receptor, and/or the signal transduction mechanisms. Instead, multiple cell and pituitary explant cultures, as they are more integrative, they are recommended for more realistic studies because of possible indirect effect given the presence of nerve terminals, or the interaction between different cell types, including stellate cells (Zhou et al., 2004; Golan et al., 2016).

Besides these issues, dispersed cell culture (multiple or homogeneous populations) could be useful to localize the receptors of neuroendocrine factors by labeled-ligand receptor-binding studies (see later) because it reduces nonspecific binding as connective tissue and neuronal tissues have been removed. In this case, it is important to consider that nerve terminals could contain "autoreceptors" or "uptake" mechanisms. Moreover, signal transduction pathways can be accurately determined using these cultures, especially intracellular Ca^{2+} levels that can be measured by monitoring Fura-2 fluorescence in suspensions of dispersed pituitary cells (Johnson et al., 1999).

4.1.2 Static vs perifuse culture

Static incubation is a suitable method for rapid screening that allows testing effects of various neuropeptides or multiple doses of the same neuropeptide over pituitary hormone release/synthesis without generating a large number of samples. Also, a neuropeptide-cumulative effect over the hormone secretion can be observed. It is noteworthy that to detect a cumulative secretion, less sensitive methodologies could be applied, such as western blot, since there is a high concentration of the adenohypophyseal hormone in the culture medium. However, slight differences among treatments could not be detected because of the sensitivity of this technique, and sometimes long-term exposure to certain stimuli could lead to response desensitization and thus, a loss of detectable cumulative response.

Instead, perifusion experiments allow detecting a time-dependent response (the phasic nature of hormone secretion), allowing to infer saturating or refractory responses to neuropeptide action. Moreover, given the episodic nature of neuroendocrine factor release, this procedure can emulate it. However, there are some disadvantages. Perifusion studies generate a large number of samples and they are often of long duration. Also, it needs a more sensitive detection methodology as the hormone released is more diluted.

An appropriate example of the earlier mentioned comparison between static and perifuse pituitary cell cultures has been shown in goldfish by Moussavi et al. (2014). The authors found that in the 12h static culture of dispersed pituitary cells, Gnih stimulated Gh secretion, while in the perifuse culture of these cells, Gnih reduced Gh basal release followed by a rebound increase in Gh release upon Gnih removal. Differences between those results could be explained because the 12h treatment period could produce the down-regulation of Gnih receptors, or their desensitization, leading to a decrease of Gnih effects with the subsequent manifestation of the "rebound increase" of Gh and thus, an overall increase of accumulated Gh in the static culture media.

4.1.3 Neuropeptide action over synthesis and release

The effect of a neuropeptide over synthesis and release can be examined with either of the above-mentioned methodologies. Usually, when those effects are over hormone secretion, the duration of neuropeptide exposure lasts a long time, while the neuropeptide action over the synthesis is within minutes or hours, depending on the culture's characteristics. It is worth mentioning that a direct correlation between the effects over mRNA levels and protein synthesis or release should not be assumed (Liu et al., 2016; Moussavi et al., 2014), neither between protein levels and its release.

Differences in hormone release can be determined by radioimmunoassays (RIA), enzyme-linked immunosorbent assay (ELISA), or western blot using specific antibodies. In this last technique, it is important to use a suitable loading control to perform normalization, since loading the same amount of total protein is not a proper choice. An interesting way to overcome this problem was proposed in Cánepa et al. (2008) where authors normalized against a 148 kDa protein, visualized by Ponceau S in the nitrocellulose membrane, present only in the culture medium in order to avoid possible loading errors in the SDS-PAGE. Another possibility is to validate the use of quantitation of Ponceau S staining before antibody probing as a loading control method (Romero-Calvo et al., 2010; Thacker et al., 2016). Differences in hormone mRNA levels can be measured by quantitative real-time PCR (qPCR) of the whole pituitary or culture cells. Another possibility is to use more specialized techniques such as laser-captured microdissection or cell sorting, followed by qPCR when studying a specific cell/cell-type response. Moreover, pituitary cell dispersed cultures allow to apply laser-captured microdissection and then perform a powerful technique as whole-transcriptome analysis.

4.2 Receptor studies

Neuropeptides exert their actions through binding to their specific receptors. Thus, another way to assess possible direct regulation over hormones is to determine the presence of the respective receptor in the specific pituitary cells.

One possibility to explore this is to perform double-labeling IF in the case you have an antibody against the receptor, previously validated in the species of interest (Ogawa et al., 2016). Another methodology is to combine immunolabeling for pituitary hormones and ISH to detect RNA of the receptor. To perform this methodology, it is essential to have at least a partial sequence (300–900 bp) of the receptor because it is necessary to synthesize sense and antisense riboprobes, and then labeling them with digoxigenin-16-uridine-5′-trifoliate. The hybridization signals are often visualized using a DIG Nucleic Acid Detection

Kit. Every procedure should be performed in a sterile-RNase free environment (Parhar et al., 2002; Biran et al., 2014). Besides, if the studies are performed in pituitary culture, laser-captured microdissection could be used to corroborate the results by analyzing the expression of the receptor by qPCR (Parhar et al., 2005).

Finally, another possibility to detect the presence of receptors in pituitary cells is to use a labeled ligand. Some studies performed in fish species have used radiolabeled ligands (Wong et al., 1993; Stefano et al., 1999). Generally, this assay is performed in pituitary cell cultures, the ligand is labeled with iodine [^{125}I] or tritium [^{3}H], and nonspecific binding is determined by displacement with an excess of unlabeled ligand. After an incubation period, cells are fixed, and then pituitary cell types are identified by IHC. Finally, autoradiograms are performed to detect the radioactive signal. Nowadays, it is possible to synthesize fluorophore-labeled ligand to perform these assays, and then perform immunolabeling for pituitary hormones to analyze cell types that express the receptor (Wong et al., 1993; Stefano et al., 1999).

5. How to approach this kind of neuroendocrine study in fish?

In this section, we will introduce some examples of how, through the different approaches described in the earlier sections, we tried to answer questions regarding the neuroendocrine control of two pituitary hormones. In particular, we will focus on results obtained regarding two lines of research of our laboratory: one about the "enigmatic Sl" so-called by Trudeau and Somoza (2020), and the other about Gh, the hormone involved in somatic growth, differentiation, and reproduction in teleost fish. As the first step for these studies, we have characterized the localization of the different pituitary hormones in *C. dimerus*, our experimental model. By simple IHC using heterologous antisera, we observed that, as it was expected, Sl is expressed in cells of the PI, mainly in those that are close to the NH. Instead Gh is expressed in cells that are surrounding the NH at the PPD (Pandolfi et al., 2001). As previously mentioned, the different hypothalamic neurons that innervate these regions could regulate these pituitary hormones. Considering this, we pursued the following research lines.

5.1 Regulation of somatolactin (Sl) release by melanin concentrating hormone (Mch) and gonadotropin-releasing hormone (Gnrh)

For those who do comparative endocrinology, Sl is an extremely interesting hormone, since it has been only reported in fish and its function has not been clarified yet. Sl, along with Gh and Prl, belongs to a family of cytokine pituitary hormones that are similar in structure, function, and gene organization (Kawauchi and Sower, 2006). It was discovered in Japanese flounder

(*Paralichthys olivaceus*; Ono et al., 1990) and Atlantic cod (Rand-Weaver et al., 1991) and then identified in all Actinopterygii fish examined and in the African lungfish (*Protopterus annectens*; Amemiya et al., 1999; May et al., 1999; Benedet et al., 2008). In teleost, two types of Sl with a highly conserved amino acid sequence are present, Sla and Slb. They are encoded by two genes that are a result of the specific teleost 3R whole-genome duplication (Jaillon et al., 2004; Meyer and Van de Peer, 2005; Ocampo Daza and Larhammar, 2018), but the *slb* gene has been lost in Acanthomorpha (Ocampo Daza and Larhammar, 2018). The neuroendocrine regulation of Sl has been poorly studied, restricted to a few neuropeptides that have been proposed as possible candidates to regulate Sl release (Jiang et al., 2008a,b, 2011; Azuma et al., 2009). Several works suggested that Sl is involved in many processes such as reproduction, stress responses, Ca^{2+} homeostasis, acid–base balance, growth, metabolism, and immune responses (reviewed in Kawauchi et al., 2009). Later, a function related to the regulation of skin pigmentation has been proposed for this hormone in teleosts (Zhu and Thomas, 1996, 1997, 1998; Zhu et al., 1999; Cánepa et al., 2006, 2012; Fukamachi et al., 2004, 2009).

It has been known that Mch is a neurohormone related to the background color adaptation (Baker and Bird, 2002) that directly regulates Msh (Van der Salm et al., 2004), a hormone produced in PI cells that is also related to skin color changes (Fujii and Oshima, 1986). In the first reports, we detected by simple IHC that Mch innervates the three regions of the ADH, especially the PI (Pandolfi et al., 2002, 2003) where Sl-producing cells and melanotropes are localized (Pandolfi et al., 2001) (Fig. 1A). Besides, we also focused on Gnrh possible regulation because Sl could be related to reproduction, and in a previous study performed in pejerrey, a close association between Gnrh fibers and Sl cells had been detected by double IHC (Vissio et al., 1999). Moreover, through studies conducted in primary pituitary cell cultures of pejerrey, we had demonstrated the presence of specific GnRH binding sites in Sl-expressing cells by autoradiography of an iodinated superactive Gnrh agonist ([125I]Gnrh-A) followed by IHC for Sl (Stefano et al., 1999). This result was later confirmed in Nile tilapia using laser-captured microdissection of single digoxigenin-labeled Sl cells coupled with qPCR of endogenous Gnrh receptors (Parhar et al., 2005). Considering this, we studied the immunoreactive Gnrh fibers' distribution in our fish model, finding that they innervated the three regions of the ADH, mainly the PPD (Pandolfi et al., 2002). Based on all these results, we hypothesized that Mch and Gnrh fibers were morphologically and functionally related to Sl cells in *C. dimerus*. To test this, we first performed two double-labeling IFs: one for Mch and Sl, and another for Gnrh and Sl. In this case, the antibodies used were a rabbit anti-*Sparus aurata* Sl (Astola et al., 1996), a rabbit antisynthetic rat MCH and a monoclonal antibody for Gnrh called LRH13 that recognized all the Gnrh variants. Note that in the case of double IF between Sl and Mch, the two primary antibodies were raised against the same animal, but as the immunomark was different and easily distinguishable (cells and fibers) from each other, it did not represent a problem. In all cases, a clear morphological association between Sl cells and both Mch and Gnrh fibers was observed (Cánepa et al., 2008). In detail, high densities of Mch fibers were found reaching Sl cells. On the contrary, a smaller number of Gnrh fibers reached Sl cells. To validate a possible regulation by these neuropeptides, we analyzed the effect of Mch and Gnrh (0.1–10 μM) for 24 h on Sl release by performing in vitro pituitary cultures of *C. dimerus* males and females separately and then analyzed the Sl released into the culture medium by western blot. A sex-dependent response was observed since Mch and Gnrh caused a dose-dependent release of Sl in males' pituitaries, but no differences in females were observed. These results suggest that Sl is involved in diverse physiological processes, including background adaptation and reproduction (Cánepa et al., 2008), under the control of the main neuropeptides involved in those physiological functions. Posterior studies provided more evidence about the role of Sl in background adaptation (Cánepa et al., 2012; Delgadin et al., 2020).

5.2 Regulation of growth hormone (Gh) by melanin concentrating hormone (Mch), neuropeptide Y (Npy), and gonadotropin inhibitory hormone (Gnih)

We are interested in studying the neuroendocrine control of growth and reproduction in fish, two processes that highly depend on external factors, such as food availability, photoperiod, and other environmental conditions. Thus, one of the key-studied hormones is Gh. As it was already said, this hormone is mainly produced by cells located in the PPD, and its neuroendocrine control is multifactorial, with different inhibitors and stimulators (Canosa et al., 2007). In this context, we decided to evaluate the effect of neuropeptides that respond to food availability, as it is the case of Npy, the main orexigenic factor in vertebrates (Kalra et al., 1999; Valassi et al., 2008; Volkoff, 2016; Matsuda et al., 2012); to background color as it is Mch, that is also involved in fish food intake (Kawauchi et al., 1983; Baker et al., 1986; Baker, 1991; Takahashi et al., 2004; Yamanome et al., 2005); and Gnih, that is related to photoperiod signal transduction (reviewed in Di Yorio et al., 2019b). All these conditions are highly relevant in fish physiology, and particularly, in aquaculture. In all these cases, following the line of argument outlined in this chapter, we first performed double-labeling IFs and then static pituitary cultures in *C. dimerus,* evaluating Gh mRNA levels and/or secretion. A clear association was found between Mch and Npy fibers with Gh cells (Pérez Sirkin et al., 2012; Di Yorio et al., 2015). A different scenario was observed with Gnih since during the ontogeny, Gnih innervates the pituitary associated to somatotropes (Di Yorio et al., 2018), but this innervation was not detected in adults (Di Yorio et al., 2016), suggesting that this neuropeptide could differently act in larvae and adults.

In a second step, we developed in vitro cultures of intact adult's pituitary, and we evaluated the effect of these neuropeptides on Gh synthesis and release by qPCR and western blot, respectively. The stimulation with Npy produced no changes in *gh* mRNA levels. However, an increase in Gh release to the culture medium at 2 nM for males and 20 nM for females was observed, suggesting that males' pituitaries were more sensitive to Npy than female ones (Di Yorio et al., 2016). Respect to Mch, 10 μM doses significantly increased both *gh* mRNA levels and Gh release (Pérez Sirkin et al., 2012). In the case of Gnih, although no Gnih innervation is detected in adults, this neuropeptide stimulates Gh release in a dose-dependent manner in pituitary cultures (Di Yorio et al., 2016). These last results suggest that the indirect effects may prevail concerning the regulation of Gh by Gnih, or that a sexual stage-dependent plasticity in the GnIH-pituitary innervation could occur. Unfortunately, because of *C. dimerus* larval size, we cannot obtain pituitaries from this developmental stage to test the Gnih effect during ontogeny.

Based on these results, we moved forward with other questions. For example, in the case of Mch, we decided to manipulate the tanks' background color (white or black) to influence Mch levels and analyze whether *mch* and/or *gh* expression and somatic growth were affected. As expected, fish exposed to a white background presented a greater number of Mch neurons with a larger nuclear area and higher *mch* transcript levels than those fish exposed to a black background. Furthermore, fish reared in a white background showed a greater body weight and total length, suggesting that Mch might regulate somatic growth in *C. dimerus*, in concordance with the in vitro results. No differences were observed in *gh* mRNA levels. However, it is important to consider that Gh has a multifactorial regulation, and therefore in vivo results might not follow the same pattern as in vitro (Pérez Sirkin et al., 2012).

6. Summary

The maintenance of homeostasis in fish is crucial for their welfare, both in natural and captivity conditions. The field of neuroendocrinology has advanced enormously in recent years, providing fundamental knowledge to understand how external and internal signals are integrated within the brain and transduced via the pituitary to regulate different organs and systems. This knowledge allows improving different strategies to overcome some problems that arise in aquaculture. Besides, although some characteristics are exclusive of teleosts, most of the main components and functions of the tetrapods' hypothalamic–pituitary axis are conserved. These characteristics represent a great advantage to carry out neuroendocrinology studies, since it greatly simplifies the techniques used and thus, fish is an excellent model for evolutionary studies.

The use of double-labeling IF provides relevant information about new regulatory factors and pathways that directly control vital functions in an organism. Moreover, the development of groundbreaking technical approaches gives us more detailed information on this possible regulation. This simple tool, combined with in vitro assays, gives us a full scenario of the physiological action of a neuropeptide, and allows us to propose new actors in the regulation of the pituitary function in teleost fish.

Acknowledgments

This study was carried out with the collaboration of Red CYTED LARVAplus (117RT0521) and was supported by Universidad de Buenos Aires (UBA; grant number: 20020160100110BA to PV), Consejo Nacional de Investigaciones Científicas y Técnicas (CONICET; PIP 2014-2016:11220130100501CO to PV), Agencia Nacional de Promoción Científica y Tecnológica (PICT2018-2577 to PV and PICT2018- 02578 to MPDY). We are especially grateful to Dr. Gustavo M. Somoza for his critical reading of this chapter.

References

Amano, M., Moriyama, S., Iigo, M., Kitamura, S., Amiya, N., Yamamori, K., Ukena, K., Tsutsui, K., 2006. Novel fish hypothalamic neuropeptides stimulate the release of gonadotrophins and growth hormone from the pituitary of sockeye salmon. J. Endocrinol. 188 (3), 417–423. https://doi.org/10.1677/joe.1.06494.

Amemiya, Y., Sogabe, Y., Nozaki, M., Takahashi, A., Kawauchi, H., 1999. Somatolactin in the white sturgeon and African lungfish and its evolutionary significance. Gen. Comp. Endocrinol. (2), 181–190. https://doi.org/10.1006/gcen.1998.7250.

Anglade, I., Zandbergen, T., Kah, O., 1993. Origin of the pituitary innervation in the goldfish. Cell Tissue Res. 273, 345–355. https://doi.org/10.1007/bf00312837.

Astola, A., Pendón, C., Ortız, M., Valdivia, M.M., 1996. Cloning and expression of somatolactin, a pituitary hormone related to growth hormone and prolactin from gilthead seabream, *Sparus aurata*. Gen. Comp. Endocrinol. 104 (3), 330–336. https://doi.org/10.1006/gcen.1996.0178.

Astola, A., Calduch-Giner, J.A., Ortiz, M., Pérez-Sánchez, J., Valdivia, M.M., 2004. Genomic structure and functional analysis of promoter region of somatolactin gene of sea bream (*Sparus aurata*). Mar. Biotechnol. 6 (5), 411–418. https://doi.org/10.1007/s10126-004-3210-2.

Azuma, M., Tanaka, M., Nejigaki, Y., Uchiyama, M., Takahashi, A., Shioda, S., Matsuda, K., 2009. Pituitary adenylate cyclase-activating polypeptide induces somatolactin release from cultured goldfish pituitary cells. Peptides 30 (7), 1260–1266. https://doi.org/10.1016/j.peptides.2009.03.011.

Baker, B.I., 1991. Melanin-concentrating hormone: a general vertebrate neuropeptide. Int. Rev. Cytol. 126 (1), 1–47. https://doi.org/10.1016/s0074-7696(08)60681-6.

Baker, B.I., Bird, D.J., 2002. Neuronal organization of the melanin-concentrating hormone system in primitive actinopterygians: evolutionary changes leading to teleosts. J. Comp. Neurol. 442 (2), 99–114. https://doi.org/10.1002/cne.10074.

Baker, C.V.H., Bronner-Fraser, M., 2001. Vertebrate cranial placodes I Embryonic induction. Dev. Biol. 232, 1–61. https://doi.org/10.1006/dbio.2001.0156.

Baker, B.I., Bird, D.J., Buckingham, J.C., 1986. Effects of chronic administration of melanin-concentrating hormone on corticotrophin, melanotrophin, and pigmentation in the trout. Gen. Comp. Endocrinol. 63, 62–69. https://doi.org/10.1016/0016-6480(86)90182-6.

Ball, J.N., Baker, B.I., 1969. The pituitary gland: anatomy and histophysiology. In: Hour, W., Randall, D. (Eds.), Fish Physiology. The Endocrine System, vol.2. Academic press, New York, pp. 1–110.

Baskaran, G., Sathyanesan, A.G., 1992. Tetrapod-like hypothalamo-hypophysial portal system in the teleost *Megalops cyprinoides* (Broussonet). Gen. Comp. Endocrinol. 86, 211–219. https://doi.org/10.1016/0016-6480(92)90104-R.

Benedet, S., Björnsson, B., Taranger, G.L., Andersson, E., 2008. Cloning of somatolactin alpha, beta forms and the somatolactin receptor in Atlantic salmon: seasonal expression profile in pituitary and ovary of maturing female broodstock. Reprod. Biol. Endocrinol. 6, 42. https://doi.org/10.1186/1477-7827-6-42.

Biran, J., Golan, M., Mizrahi, N., Ogawa, S., Parhar, I.S., Levavi-Sivan, B., 2014. LPXRFa, the piscine ortholog of GnIH, and LPXRF receptor positively regulate gonadotropin secretion in tilapia (*Oreochromis niloticus*). Endocrinology 155 (11), 4391–4401. https://doi.org/10.1210/en.2013-2047.

Biswas, S., Jadhao, A.G., Pinelli, C., Palande, N.V., Tsutsui, K., 2015. GnIH and GnRH expressions in the central nervous system and pituitary of Indian major carp, *Labeo rohita* during ontogeny: an immunocytochemical study. Gen. Comp. Endocrinol. 220, 88–92. https://doi.org/10.1016/j.ygcen.2014.06.005.

Boenisch, T., 2002. Métodos inmunohistoquímicos de coloración. Ed. Dako Corp.

Breton, B., Weil, C., Jalabert, B., Billard, R., 1972. Activité reciproque des facteurs hypothalamiques de belier (*Ovis aries*) et de poissons teleosteens sur la secretion in vitro des hormones gonadotropes c-HG et LH respectivement par des hypophyses de carpe et de belier. C. R. Acad. Hebd. Seances. Acad. Sci. D. 274, 2530–2533.

Cánepa, M.M., Pandolfi, M., Maggese, M.C., Vissio, P.G., 2006. Involvement of somatolactin in background adaptation of the cichlid fish *Cichlasoma dimerus*. J. Exp. Zool. 305, 410–419. https://doi.org/10.1002/jez.a.273.

Cánepa, M., Pozzi, A., Astola, A., Maggese, M.C., Vissio, P., 2008. Effect of salmon melanin-concentrating hormone and mammalian gonadotrophin-releasing hormone on somatolactin release in pituitary culture of *Cichlasoma dimerus*. Cell Tissue Res. 333 (1), 49–59. https://doi.org/10.1007/s00441-008-0622-8.

Cánepa, M.M., Zhu, Y., Fossati, M., Stiller, J.W., Vissio, P.G., 2012. Cloning, phylogenetic analysis and expression of somatolactin and its receptor in *Cichlasoma dimerus*: their role in long-term background color acclimation. Gen. Comp. Endocrinol. 176, 52–61. https://doi.org/10.1016/j.ygcen.2011.12.023.

Canosa, L.F., Chang, J.P., Peter, R.E., 2007. Neuroendocrine control of growth hormone in fish. Gen. Comp. Endocrinol. 151 (1), 1–26. https://doi.org/10.1016/j.ygcen.2006.12.010.

Canosa, L.F., Lopez, G.C., Scharrig, E., Lesaux-Farmer, K., Somoza, G.M., Kah, O., Trudeau, V.L., 2011. Forebrain mapping of secretoneurin-like immunoreactivity and its colocalization with isotocin in the preoptic nucleus and pituitary gland of goldfish. J. Comp. Neurol. 519, 3748–3765. https://doi.org/10.1002/cne.22688.

Cerdá-Reverter, J.M., Canosa, L.F., 2009. Neuroendocrine circuits of the fish brain. In: Bernier, N., van Der Kraak, G., Farrell, A., Brauner, C. (Eds.), Fish Neuroendocrinology. vol. 28. Elsevier inc., San Diego, CA, USA, pp. 3–74.

Chang, J.P., Jobin, R.M., 1994. Teleost pituitary cells: isolation, culture and use. Biochem. Mol. Biol. Fish 3, 205–213. https://doi.org/10.1016/B978-0-444-82033-4.50024-6. Elsevier. Vol.

Coons, A.H., Creech, H.J., Jones, R.N., 1941. Immunological properties of an antibody containing a fluorescent group. Proc. Soc. Exp. Biol. Med. 47 (2), 200–202. https://doi.org/10.3181/00379727-47-13084P.

De Leeuw, R., Goos, H.T., Peute, J., Van Pelt, A.M.M., Burzawa-Gérard, E., Van Oordt, P.G.W.J., 1984. Isolation of gonadotrops from the pituitary of the African catfish, *Clarias lazera*. Cell Tissue Res. 236 (3), 669–675. https://doi.org/10.1007/BF00217237.

Delgadin, T.H., Castaneda-Cortes, D.C., Sacks, C., Breccia, A., Fernandino, J.I., Vissio, P.G., 2020. Morphological colour adaptation during development: involvement of growth hormone receptor 1. bioRxiv. https://doi.org/10.1101/2020.06.01.128538.

Di Yorio, M.P., Delgadin, T.H., Sirkin, D.P., Vissio, P.G., 2015. Growth hormone, luteinizing hormone, and follicle-stimulating hormone regulation by neuropeptide Y in both sexes of the cichlid fish, *Cichlasoma dimerus*. Fish Physiol. Biochem. 41 (4), 843–852. https://doi.org/10.1007/s10695-015-0051-z.

Di Yorio, M.P., Pérez Sirkin, D.I., Delgadin, T.H., Shimizu, A., Tsutsui, K., Somoza, G.M., Vissio, P.G., 2016. Gonadotrophin-inhibitory hormone in the cichlid fish *Cichlasoma dimerus*: structure, brain distribution and differential effects on the secretion of gonadotrophins and growth hormone. J. Neuroendocrinol. 28 (5). https://doi.org/10.1111/jne.12377.

Di Yorio, M.P., Sallemi, J.E., Toledo Solís, F.J., Pérez Sirkin, D.I., Delgadin, T.H., Tsutsui, K., Vissio, P.G., 2018. Ontogeny of gonadotropin-inhibitory hormone (GnIH) in the cichlid fish *Cichlasoma dimerus*. J. Endocrinol., e12608. Advanced online publication https://doi.org/10.1111/jne.12608.

Di Yorio, M.P., Pérez-Sirkin, D.I.P., Muñoz-Cueto, J.A., Delgadin, T.H., Tsutsui, K., Somoza, G.M., Vissio, P.G., 2019a. Morphological relationship between GnIH and GnRH neurons in the brain of the neotropical cichlid fish *Cichlasoma dimerus*. Gen. Comp. Endocrinol. 273, 144–151. https://doi.org/10.1016/j.ygcen.2018.06.010.

Di Yorio, M.P., Muñoz-Cueto, J.A., Paullada-Salmerón, J.A., Somoza, G.M., Tsutsui, K., Vissio, P.G., 2019b. The gonadotropin-inhibitory hormone: what we know and what we still have to learn from fish. Front. Endocrinol. 10, 78. https://doi.org/10.3389/fendo.2019.00078.

Flaws, J.A., Suter, D.E., 1993. Preparation of an enriched fraction of luteinizing hormone-secreting cells from a mixed population of rat anterior pituitary cells. Biol. Reprod. 48 (5), 1026–1035. https://doi.org/10.1095/biolreprod48.5.1026.

Fouquet, C., Gilles, J.F., Heck, N., Dos Santos, M., Schwartzmann, R., Cannaya, V., Morel, M.P., Davidson, R.S., Trembleau, A., Bolte, S., 2015. Improving axial resolution in confocal microscopy with new high refractive index mounting media. PLoS One 10 (3). https://doi.org/10.1371/journal.pone.0121096, e0121096.

Fujii, R., Oshima, N., 1986. Control of chromatophore movements in teleost fishes. Zool. Sci. 3 (1), 13–47.

Fukamachi, S., Sugimoto, M., Mitani, H., Shima, A., 2004. Somatolactin selectively regulates proliferation and morphogenesis of neural-crest derived pigment cells in medaka. Proc. Natl. Acad. Sci. U. S. A. 101 (29), 10661–10666. https://doi. https://doi.org/10.1073/pnas.0401278101.

Fukamachi, S., Yada, T., Meyer, A., Kinoshita, M., 2009. Effects of constitutive expression of somatolactin alpha on skin pigmentation in medaka. Gene 442, 81–87. https://doi.org/10.1016/j.gene.2009.04.010.

Golan, M., Zelinger, E., Zohar, Y., Levavi-Sivan, B., 2015. Architecture of GnRH-gonadotrope-vasculature reveals a dual mode of gonadotropin regulation in fish. Endocrinology 156 (11), 4163–4173. https://doi.org/10.1210/en.2015-1150.

Golan, M., Hollander-Cohen, L., Levavi-Sivan, B., 2016. Stellate cell networks in the teleost pituitary. Sci. Rep. 6, 24426. https://doi.org/10.1038/srep24426.

González-Martínez, D., Madigou, T., Mañanos, E., Cerdá-Reverter, J.M., Zanuy, S., Kah, O., Muñoz-Cueto, J.A., 2004. Cloning and expression of gonadotropin-releasing hormone receptor in the brain and pituitary of the European sea bass: an *in situ* hybridization study. Biol. Reprod. 70 (5), 1380–1391. https://doi.org/10.1095/biolreprod.103.022624.

Guilgur, L.G., Strüssmann, C.A., Somoza, G.M., 2009. mRNA expression of GnRH variants and receptors in the brain, pituitary and ovaries of pejerrey (*Odontesthes bonariensis*) in relation to the reproductive status. Fish Physiol. Biochem. 35 (1), 157–166. https://doi.org/10.1007/s10695-008-9215-4.

Holmqvist, B., Ekström, P., 1997. Subcellular localization of neuronal nitric oxide synthase in the brain of a teleost; an immunoelectron and confocal microscopical study. Brain Res. 745 (1–2), 67–82. https://doi.org/10.1016/S0006-8993(96)01128-6.

Huang, B., Wang, W., Bates, M., Zhuang, X., 2008. Three-dimensional super-resolution imaging by stochastic optical reconstruction microscopy. Science 319 (5864), 810–813. https://doi.org/10.1126/science.1153529.

Huang, B., Bates, M., Zhuang, X., 2009. Super-resolution fluorescence microscopy. Annu. Rev. Biochem. 78, 993–1016. https://doi.org/10.1146/annurev.biochem.77.061906.092014.

Jaillon, O., Aury, J.M., Brunet, F., Petit, J.L., Stange-Thomann, N., Mauceli, E., Nicaud, S., 2004. Genome duplication in the teleost fish *Tetraodon nigroviridis* reveals the early vertebrate proto-karyotype. Nature 431 (7011), 946. https://doi.org/10.1038/nature03025.

Jiang, Q., Ko, W.K., Lerner, E.A., Chan, K.M., Wong, A.O., 2008a. Grass carp somatolactin: I. evidence for PACAP induction of somatolactin-α and -β gene expression via activation of pituitary PAC-I receptors. Am. J. Physiol. Endocrinol. Metab. 295 (2), E463–E476. https://doi.org/10.1152/ajpendo.90385.2008.

Jiang, Q., He, M., Wang, X., Wong, A.O., 2008b. Grass carp somatolactin: II. Pharmacological study on postreceptor signaling mechanisms for PACAP-induced somatolactin-α and-β gene expression. Am. J. Physiol. Endocrinol. Metab. 295 (2), E477–E490. https://doi.org/10.1152/ajpendo.90386.2008.

Jiang, Q., Ko, W.K., Wong, A.O., 2011. Insulin-like growth factor as a novel stimulator for somatolactin secretion and synthesis in carp pituitary cells via activation of MAPK cascades. Am. J. Physiol. Endocrinol. Metab. 301 (6), E1208–E1219. https://doi.org/10.1152/ajpendo.00347.2011.

Johnson, J.D., Van Goor, F., Wong, C.J., Goldberg, J.I., Chang, J.P., 1999. Two endogenous gonadotropin-releasing hormones generate dissimilar Ca^{2+} signals in identified goldfish gonadotropes. Gen. Comp. Endocrinol. 116 (2), 178–191. https://doi.org/10.1006/gcen.1999.7349.

Johnston, S.A., Maler, L., 1992. Anatomical organization of the hypophysiotrophic systems in the electric fish, *Apteronotus leptorhynchus*. J. Comp. Neurol. 317 (4), 421–437. https://doi.org/10.1002/cne.903170408.

Jose, T.M., Sathyanesan, A.G., 1977. Hypothalamo-hypophysial neurosecretory and vascular system of the teleost *Labeo rohita* (ham.). Anat. Anz. 142 (4), 424–434.

Joshi, S., Yu, D., 2017. Immunofluorescence. In: Basic Science Methods for Clinical Researchers. Academic Press, pp. 135–150, https://doi.org/10.1016/B978-0-12-803077-6.00008-4.

Kabiraj, A., Gupta, J., Khaitan, T., Bhattacharya, P.T., 2015. Principle and techniques of immunohistochemistry—a review. Int. J. Biol. Med. Res. 6 (3), 5204–5210.

Kah, O., Adrio, F., 2018. Chemical neuroanatomy of the hypothalamo-hypophysealsystem in sturgeons. In: Williot, P., Nonnotte, G., Vizziano-Cantonnet, D., Chebanov, M. (Eds.), The Siberian Sturgeon (Acipenserbaerii, Brandt, 1869). Biology, vol. 1. Springer International Publishing AG, pp. 249–278, https://doi.org/10.1007/978-3-319-61664-3_17.

Kalra, S.P., Dube, M.G., Pu, S., Xu, B., Horvath, T.L., Kalra, P.S., 1999. Interacting appetite-regulating pathways in the hypothalamic regulation of body weight. Endocr. Rev. 20 (1), 68–100. https://doi.org/10.1210/edrv.20.1.0357.

Kapingidza, A.B., Kowal, K., Chruszcz, M., 2020. Antigen–antibody complexes. In: Vertebrate and Invertebrate Respiratory Proteins, Lipoproteins and other Body Fluid Proteins. Springer, Cham, pp. 465–497, https://doi.org/10.1007/978-3-030-41769-7_19.

Kaul, S., Vollrath, L., 1974. The goldfish pituitary. II. Innervation. Cell Tissue Res. 154, 231–249. https://doi.org/10.1007/bf00223166.

Kawauchi, H., Sower, S.A., 2006. The dawn and evolution of hormones in the adenohypophysis. Gen. Comp. Endocrinol. 148, 3–14. https://doi.org/10.1016/j.ygcen.2005.10.011.

Kawauchi, H., Kawazoe, I., Tsubokawa, M., Kishida, M., Baker, B.I., 1983. Characterization of melanin-concentrating hormone in chum salmon pituitaries. Nature 305 (5932), 321–323. https://doi.org/10.1038/305321a0.

Kawauchi, H., Sower, S.A., Moriyama, S., 2009. The neuroendocrine regulation of prolactin and somatolactin secretion in fish. Fish. Physiol. 28, 197–234. https://doi.org/10.1016/S1546-5098(09)28005-8.

Lagios, M.D., 1968. Tetrapod-like organization of the pituitary gland of the polyteriformid fishes, *Calamoichthys calabaricus* and *Polypterus palmas*. Gen. Comp. Endocrinol. 11, 300–315. https://doi.org/10.1016/0016-6480(68)90086-5.

Levavi-Sivan, B., Yaron, Z., 1992. Involvement of cyclic adenosine monophosphate in the stimulation of gonadotropin secretion from the pituitary of the teleost fish, tilapia. Mol. Cell. Endocrinol. 85 (3), 175–182. https://doi.org/10.1016/0303-7207(92)90256-6.

Levavi-Sivan, B., Bogerd, J., Mañanós, E.L., Gómez, A., Lareyre, J.J., 2010. Perspectives on fish gonadotropins and their receptors. Gen. Comp. Endocrinol. 165 (3), 412–437. https://doi.org/10.1016/j.ygcen.2009.07.019.

Liu, Y., Beyer, A., Aebersold, R., 2016. On the dependency of cellular protein levels on mRNA abundance. Cell 165 (3), 535–550. https://doi.org/10.1016/j.cell.2016.03.014.

Maitra, S.K., Hasan, K.N., 2016. The role of melatonin as a hormone and an antioxidant in the control of fish reproduction. Front. Endocrinol. 7, 38. https://doi.org/10.3389/fendo.2016.00038.

Martín-Lacave, I., García-Caballero, T., 2012. Atlas de inmunohistoquímica: caracterización de células, tejidos y órganos normales. Ediciones Díaz de Santos.

Matsuda, K., Sakashita, A., Yokobori, E., Azuma, M., 2012. Neuroendocrine control of feeding behavior and psychomotor activity by neuropeptide Y in fish. Neuropeptides 46 (6), 275–283. https://doi.org/10.1016/j.npep.2012.09.006.

May, D., Alrubaian, J., Patel, S., Dores, R.M., Rand-Weaver, M., 1999. Studies on the GH/SL gene family: cloning of African lungfish (*Protopterus annectens*) growth hormone and somatolactin and toad (*Bufo marinus*) growth hormone. Gen. Comp. Endocrinol. 113 (1), 121–135. https://doi.org/10.1006/gcen.1998.7185.

Meyer, A., Van de Peer, Y., 2005. From 2R to 3R: evidence for a fish-specific genome duplication (FSGD). BioEssays 27 (9), 937–945. https://doi.org/10.1002/bies.20293.

Moussavi, M., Wlasichuk, M., Chang, J.P., Habibi, H.R., 2012. Seasonal effect of GnIH on gonadotrope functions in the pituitary of goldfish. Mol. Cell. Endocrinol. 350 (1), 53–60. https://doi.org/10.1016/j.mce.2011.11.020.

Moussavi, M., Wlasichuk, M., Chang, J.P., Habibi, H.R., 2014. Seasonal effects of GnIH on basal and GnRH-induced goldfish somatotrope functions. J. Endocrinol. 223 (2), 191–202. https://doi.org/10.1530/JOE-14-0441.

Muñoz-Cueto, J.A., 2009. Cerebro y reproducción en peces: bases neurales y neuroendócrinas. In: Carrillo, M. (Ed.), La reproducción en los peces: aspectos básicos y su aplicación en la acuicultura. Fundación OESA, Madrid, pp. 25–75.

Muñoz-Cueto, J.A., Zmora, N., Paullada-Salmerón, J.A., Marvel, M., Mañanos, E., Zohar, Y., 2020. The gonadotropin-releasing hormones: lessons from fish. Gen. Comp. Endocrinol. 291, 113422. https://doi.org/10.1016/j.ygcen.2020.113422.Norris and Carr, 2013Norris, D.O., Carr, J.A., 2013. Vertebrate Endocrinology, Fifth ed. Elsevier Inc., Ed.; Academic Press, Cambridge, MA, USA.

Ocampo Daza, D., Larhammar, D., 2018. Evolution of the growth hormone, prolactin, prolactin 2 and somatolactin family. Gen. Comp. Endocrinol. 264, 94–112. https://doi. https://doi.org/10.1016/j.ygcen.2018.01.007.

Ogawa, S., Sivalingam, M., Biran, J., Golan, M., Anthonysamy, R.S., Levavi-Sivan, B., Parhar, I.S., 2016. Distribution of LPXRFα, a gonadotropin-inhibitory hormone ortholog peptide, and LPXRFα receptor in the brain and pituitary of the tilapia. J. Comp. Neurol. 524 (14), 2753–2775. https://doi.org/10.1002/cne.23990.

Ono, M., Takayama, Y., Rand-Weaver, M., Sakata, S., Yasunaga, T., Noso, T., Kawauchi, H., 1990. cDNA cloning of somatolactin, a pituitary protein related to growth hormone and prolactin. Proc. Natl. Acad. Sci. U. S. A. 87, 4330–4334. https://doi.org/10.1073/pnas.87.11.4330.

Pandolfi, M., Paz, D.A., Maggese, M.C., Meijide, F.J., Vissio, P.G., 2001. Immunocytochemical localization of different cell types in the adenohypophysis of *Cichlasoma dimerus* (Heckel, 1840). Biocell 25 (1), 35–42.

Pandolfi, M., Parhar, I.S., Ravaglia, M.A., Meijide, F.J., Maggese, C.M., Paz, D.A., 2002. Ontogeny and distribution of gonadotropin-releasing hormone (GnRH) neuronal systems in the brain of the cichlid fish *Cichlasoma dimerus*. Anat. Embryol. 205 (4), 271–281. https://doi.org/10.1007/s00429-002-0253-x.

Pandolfi, M., Cánepa, M.M., Ravaglia, M.A., Maggese, M.C., Paz, D.A., Vissio, P.G., 2003. Melanin-concentrating hormone system in the brain and skin of the cichlid fish *Cichlasoma dimerus*: anatomical localization, ontogeny and distribution in comparison to alpha-melanocyte-stimulating hormone-expressing cells. Cell Tissue Res. 311 (1), 61–69. https://doi.org/10.1007/s00441-002-0654-4.

Parhar, I.S., Soga, T., Sakuma, Y., Millar, R.P., 2002. Spatio-temporal expression of gonadotropin-releasing hormone receptor subtypes in gonadotropes, somatotropes and lactotropes in the cichlid fish. J. Neuroendocrinol. 14 (8), 657–665. https://doi.org/10.1046/j.1365-2826.2002.00817.x.

Parhar, I.S., Ogawa, S., Sakuma, Y., 2005. Three GnRH receptor types in laser-captured single cells of the cichlid pituitary display cellular and functional heterogeneity. Proc. Natl. Acad. Sci. U. S. A. 102 (6), 2204–2209. https://doi.org/10.1073/pnas.0409494102.

Pawley, J.B., 2006. Fundamental limits in confocal microscopy. In: Handbook of Biological Confocal Microscopy. Springer, Boston, MA, pp. 20–42, https://doi.org/10.1007/978-0-387-45524-2_2.

Pérez Sirkin, D.I., Cánepa, M.M., Fossati, M., Fernandino, J.I., Delgadin, T., Canosa, L.F., Somoza, G.M., Vissio, P.G., 2012. Melanin concentrating hormone (MCH) is involved in the regulation of growth hormone in *Cichlasoma dimerus* (Cichlidae, Teleostei). Gen. Comp. Endocrinol. 176 (1), 102–111. https://doi.org/10.1016/j.ygcen.2012.01.002.

Peter, R.E., Yu, K.L., Marchant, T.A., Rosenblum, P.M., 1990. Direct neural regulation of the teleost adenohypophysis. J. Exp. Zool. Suppl. 4S, 84–89. https://doi.org/10.1002/jez.1402560415.

Ramos-Vara, J.A., Miller, M.A., 2014. When tissue antigens and antibodies get along: revisiting the technical aspects of immunohistochemistry—the red, brown, and blue technique. Vet. Pathol. 51 (1), 42–87. https://doi.org/10.1177/0300985813505879.

Ramos-Vara, J.A., Kiupel, M., Baszler, T., Bliven, L., Brodersen, B., Chelack, B., West, K., Czub, S., Del Piero, F., Dial, S., Ehrhart, E.J., Graham, T., Manning, L., Paulsen, D., Valli, V.E., 2008. Suggested guidelines for immunohistochemical techniques in veterinary diagnostic laboratories. J. Vet. Diagn. Investig. 20 (4), 393–413. https://doi.org/10.1177/104063870802000401.

Rand-Weaver, M., Noso, T., Muramoto, K., Kawauchi, H., 1991. Isolation and characterization of somatolactin, a new protein related to growth hormone and prolactin from Atlantic cod (*Gadus morhua*) pituitary glands. Biochemist 30, 1509–1515. https://doi.org/10.1021/bi00220a010.

Roch, G.J., Busby, E.R., Sherwood, N.M., 2014. GnRH receptors and peptides: skating backward. Gen. Comp. Endocrinol. 209, 118–134. https://doi.org/10.1016/j.ygcen.2014.07.025.

Romero-Calvo, I., Ocón, B., Martínez-Moya, P., Suárez, M.D., Zarzuelo, A., Martínez-Augustin, O., de Medina, F.S., 2010. Reversible Ponceau staining as a loading control alternative to actin in Western blots. Anal. Biochem. 401 (2), 318–320. https://doi.org/10.1016/j.ab.2010.02.036.

Sanderson, T., Wild, G., Cull, A.M., Marston, J., Zardin, G., 2018. 19 Immunohistochemical and immunofluorescent techniques. In: Bancroft's Theory and Practice of Histological Techniques. E-Book, p. 337.

Sherwood, N., Eiden, L., Brownstein, M., Spiess, J., Rivier, J., Vale, W., 1983. Characterization of a teleost gonadotropin-releasing hormone. Proc. Natl. Acad. Sci. U. S. A. 80 (9), 2794–2798. https://doi.org/10.1073/pnas.80.9.2794.

Shi, S.R., Shi, Y., Taylor, C.R., 2011. Antigen retrieval immunohistochemistry: review and future prospects in research and diagnosis over two decades. J. Histochem. Cytochem. 59 (1), 13–32. https://doi.org/10.1369/jhc.2010.957191.

Shinomiya, A., Shimmura, T., Nishiwaki-Ohkawa, T., Yoshimura, T., 2014. Regulation of seasonal reproduction by hypothalamic activation of thyroid hormone. Front. Endocrinol. 5, 12. https://doi.org/10.3389/fendo.2014.00012.

Sower, S.A., 2015. Breaking dogma on the hypothalamic-pituitary anatomical relations in vertebrates. Endocrinology 156 (11), 3882–3884. https://doi.org/10.1210/en.2015-1778.

Stefano, A.V., Vissio, P.G., Paz, D.A., Somoza, G.M., Maggese, M.C., Barrantes, G.E., 1999. Co-localization of GnRH binding sites with gonadotropin-, somatotropin-, somatolactin-, and prolactin-expressing pituitary cells of the pejerrey, Odontesthes bonariensis, in vitro. Gen. Comp. Endocrinol. 116 (1), 133–139. https://doi.org/10.1006/gcen.1999.7354.

Takahashi, A., Tsuchiya, K., Yamanome, T., Amano, M., Yasuda, A., Yamamori, K., Kawauchi, H., 2004. Possible involvement of melanin-concentrating hormone in food intake in a teleost fish, barfin flounder. Peptides 25, 1613–1622. https://doi.org/10.1016/j.peptides.2004.02.022.

Thacker, J.S., Yeung, D.H., Staines, W.R., Mielke, J.G., 2016. Total protein or high-abundance protein: which offers the best loading control for Western blotting? Anal. Biochem. 496, 76–78. https://doi.org/10.1016/j.ab.2015.11.022.

Thorner, M.O., Borges, J.L., Cronin, M.J., Keefer, D.A., Hellmann, P., Lewis, D., Dabney, L.G., Quesenberry, P.J., 1982. Fluorescence activated cell sorting of functional anterior pituitary cells. Endocrinology 110 (5), 1831–1833. https://doi.org/10.1210/endo-110-5-1831.

Tostivint, H., 2011. Evolution of the gonadotropin-releasing hormone (GnRH) gene family in relation to vertebrate tetraploidizations. Gen. Comp. Endocrinol. 170, 575–581. https://doi.org/10.1016/j.ygcen.2010.11.017.

Trudeau, V.L., Somoza, G.M., 2020. Multimodal hypothalamo-hypophysial communication in the vertebrates. Gen. Comp. Endocrinol. 293, 113475. https://doi.org/10.1016/j.ygcen.2020.113475.

Valassi, E., Scacchi, M., Cavagnini, F., 2008. Neuroendocrine control of food intake. Nutr. Metab. Cardiovasc. Dis. 18 (2), 158–168. https://doi.org/10.1016/j.numecd.2007.06.004.

Van der Salm, A.L., Pavlidis, M., Flik, G., Bonga, S.W., 2004. Differential release of α-melanophore stimulating hormone isoforms by the pituitary gland of red porgy, *Pagrus pagrus*. Gen. Comp. Endocrinol. 135 (1), 126–133. https://doi.org/10.1016/j.ygcen.2003.09.002.

Vissio, P., Stefano, A., Somoza, G.M., Maggese, M.C., Paz, D.A., 1999. Close association of gonadotropin-releasing hormone fibers and gonadotropin, growth hormone, somatolactin and prolactin expressing cells in pejerrey, *Odontesthes bonariensis*. Fish Physiol. Biochem. 21, 121–127. https://doi.org/10.1023/A:1007781606592.

Volkoff, H., 2016. The neuroendocrine regulation of food intake in fish: a review of current knowledge. Front. Neurosci. 10, 540. https://doi.org/10.3389/fnins.2016.00540.

Wang, Q., Qi, X., Guo, Y., Li, S., Zhang, Y., Liu, X., Lin, H., 2015. Molecular identification of GnIH/GnIHR signal and its reproductive function in protogynous hermaphroditic orange-spotted grouper (*Epinephelus coioides*). Gen. Comp. Endocrinol. 216, 9–23. https://doi.org/10.1016/j.ygcen.2015.04.016.

Weller, T.H., Coons, A.H., 1954. Fluorescent antibody studies with agents of varicella and herpes zoster propagated in vitro. Proc. Soc. Exp. Biol. Med. 86 (4), 789–794. https://doi.org/10.3181/00379727-86-21235.

Wong, A.O., Chang, J.P., Peter, R.E., 1993. Characterization of D1 receptors mediating dopamine-stimulated growth hormone release from pituitary cells of the goldfish, *Carassius auratus*. Endocrinology 133 (2), 577–584. https://doi.org/10.1210/en.133.2.577.

Yamanome, T., Amano, M., Takahashi, A., 2005. White background reduces the occurrence of staining, activates melanin-concentrating hormone and promotes somatic growth in barfin flounder. Aquaculture 244, 323–329. https://doi.org/10.1016/j.aquaculture.2004.11.020.

Zhou, H., Wang, X., Ko, W.K., Wong, A.O., 2004. Evidence for a novel intrapituitary autocrine/paracrine feedback loop regulating growth hormone synthesis and secretion in grass carp pituitary cells by functional interactions between gonadotrophs and somatotrophs. Endocrinology 145 (12), 5548–5559. https://doi.org/10.1210/en.2004-0362.

Zhu, Y., Thomas, P., 1996. Elevations of somatolactin in plasma and pituitaries and increased a-MSH cell activity in red drum exposed to black background and decreased illumination. Gen. Comp. Endocrinol. 101, 21–31. https://doi.org/10.1016/s0016-6480(96)90191-4.

Zhu, Y., Thomas, P., 1997. Effects of somatolactin on melanosome aggregation in the melanophores of red drum (*Sciaenops ocellatus*) scales. Gen. Comp. Endocrinol. 105, 127–133. https://doi.org/10.1006/gcen.1996.6810.

Zhu, Y., Thomas, P., 1998. Effects of light on plasma somatolactin levels in red drum (*Sciaenops ocellatus*). Gen. Comp. Endocrinol. 111, 76–82. https://doi.org/10.1006/gcen.1998.7092.

Zhu, Y., Yoshiura, Y., Kikuchi, K., Aida, K., Thomas, P., 1999. Cloning and phylogenetic relationship of red drum somatolactin cDNA and effects of light on pituitary somatolactin mRNA expression. Gen. Comp. Endocrinol. 113 (1), 69–79. https://doi.org/10.1006/gcen.1998.7180.

Zohar, Y., Muñoz-Cueto, J.A., Elizur, A., Kah, O., 2010. Neuroendocrinology of reproduction in teleost fish. Gen. Comp. Endocrinol. 165 (3), 438–455. https://doi.org/10.1016/j.ygcen.2009.04.017.

Chapter 8

Mucosal immunology in fish

Francisco A. Guardiola, Alberto Cuesta, and María Ángeles Esteban
Immunobiology for Aquaculture Group, Department of Cell Biology & Histology, Faculty of Biology, University of Murcia, Murcia, Spain

1. A brief introduction to the immune system in teleost fish

Immunology arose as an academic discipline in its own right out of medical microbiology (Kaufmann and Winau, 2005). The study of immunology started in the last quarter of the nineteenth century, after two major milestones. The first finding was the identification of phagocytic cells (which engulf and destroy invading pathogens) by Metchnikoff, which laid the basis for innate immunity (Kaufmann, 2008). The second discovery was the identification of antibodies (which neutralize microbial toxins) by Behring's and Ehrlich's studies, and this became the basis for acquired immunity (Kaufmann, 2008, 2017). These findings also led to the distinction between cellular and humoral immunity. Before 1950, fish were studied only from the angle of comparative anatomy, physiology, taxonomy, and fish diseases (Van Muiswinkel, 2008). It was mistakenly assumed that fish would have a very simple immune system that would help people understand more complex ones, including the human immune system. However, all of the results that began to be obtained indicated otherwise. That is, as with all the other vertebrates, fish have a very sophisticated immune system, showing specificity and memory (Manning, 1998). In fact, fish are the first group of animals displaying a complete immune system, which attracted the attention of many researchers to deepen and understand the evolution of vertebrate immunity. In this sense, fish lack bone marrow, the main lymphohemopoietic tissue being the head-kidney, which produces all the blood cells. Another interesting issue is the evolution of immunoglobulins, with the teleosts having IgM, IgD, and IgT/IgZ in contrast to the five classes of mammalian species. The great number and variability of fish species and a large evolution history, around 450 million years, generates a diversity of peculiarities in the immune tissues, cells, molecules, regulation, and host-pathogen interactions that bring the interest of immunologist. Furthermore, fish immunity has many particularities not present in mammals, such as the importance of skin mucosal immunity.

Cellular and Molecular Approaches in Fish Biology. https://doi.org/10.1016/B978-0-12-822273-7.00002-1

At present, a considerable body of work exists on the immunology features of several teleost fish species, and they are concentrated on species of commercial importance. The reason for this is that the success of modern aquaculture is based on the proper management of the biology of farmed species. The most important studied fish species are Atlantic salmon (*Salmo salar*), rainbow trout (*Oncorhynchus mykiss*), channel catfish (*Ictalurus punctatus*), carp (*Cyprinus carpio*), tilapia (*Oreochromis mossambicus*), European sea bass (*Dicentrarchus labrax*), gilthead seabream (*Sparus aurata*), Atlantic cod (*Gadus morhua*), turbot (*Scophthalmus maximus*), Atlantic bluefin tuna (*Thunnus thynnus*), and European eel (*Anguilla anguilla*), among others (FAO, 2020).

During the last four decades, aquaculture has developed, diversified, and registered notable technological and scientific advances. However, farmed fish are still confined in restricted spaces, and cultivation has been intensified with a view to increasing productivity per unit volume of water. This overcrowding can lead to impoverished water quality, which would negatively affect the physiological conditions of the specimens in culture; therefore, maintaining the health and welfare of the animals is one of the priorities of the sector. Our research group (before named "Fish Innate Immune System Group" and now named "Immunobiology for Aquaculture") emerged by focusing its attention on the study of the immune system of marine fish of commercial interest in our region (Murcia, Spain) and in the Mediterranean area, mainly gilthead seabream and European sea bass. We focus our research efforts on the immune system because it is key for the survival and maintenance of homeostasis as well as the health of the specimens, which has great interest in both basic and evolutive knowledge and practical applications in aquaculture. The immune system can be defined as the set of organs, tissues, and especially cells and proteins that collaboratively work to defend organisms against any aggression that the body understands as potentially dangerous. It is the so-called sixth sense, since it acts as a detector for the presence of "foreign substances" (antigens) and reacts to them with great sensitivity and specificity (Blalock and Smith, 2007). The evolution of our research line has paralleled the advances in knowledge of fish immunity.

The great interest evolving through hard work in the field of fish immunology for many years has given rise to the fact that, at present, the volume of available data is very high. This makes it almost impossible to resume the available information. With the sole purpose of giving the reader a brief introduction to the subject, some reviews are listed, and they focus on aspects that we consider important. While some of them are general (e.g., Lieschke and Trede, 2009; Biller-Takahashi and Urbinati, 2014), others are focused on specific topics, such as the biology of some immune cells (Scapigliati, 2013; Hodgkinson et al., 2015; Grayfer et al., 2018), different humoral immune parameters (Russell and Lumsden, 2005; Secombes et al., 2011; Ye et al., 2013; Bengtén and Wilson, 2015; Boudinot et al., 2016), innate immunity (Rebl and Goldammer, 2018), adaptive immunity (Secombes et al., 2005; Flajnik, 2018), or mucosal immunity (Salinas et al., 2011; Esteban, 2012; Gómez et al., 2013).

2. Historical perspective and relevance of immunological studies in gilthead seabream

Our study model, the gilthead seabream, is an euryhaline and hermaphroditic protandrous fish species. It is very interesting from the immune response point of view that among circulating or tissue granulocytes, acidophilic granulocytes are the most abundant and phagocytic cells (Esteban et al., 1989; Meseguer et al., 1993), being considered functional equivalents to neutrophils in the rest of vertebrates. In addition, the innate cell-mediated cytotoxic activity is played by a heterogeneous population of lymphocytes, monocyte-macrophages, and acidophilic granulocytes (Cuesta et al., 1999). From the production point of view, gilthead seabream is one of the most important cultured species in the Mediterranean area, and particularly in our region, with high consumer demand and economic value. Therefore studies on its biology, including immunity, have helped to improve the production and productivity by fish farmers thanks to the implementation of solutions to warrant the reproducibility of the production along the years, reduce the stress or mortality by diseases, or perform breeding programs for selection, among others. Therefore basic and evolutionary aspects as well as applied importance led us to deepen the characterization of the gilthead seabream immune response.

The first works on the fish immune system were basically morphological studies that were crucial for the study of cellular immunity. This is also how we started studying the primary organs of the immune system microscopically and the places where their cells are formed, specifically, the head-kidney, which is the main hematopoietic organ (equivalent to the bone marrow of mammals) in fish species. We carried out an exhaustive description of all the hematopoietic series (erythropoiesis, thrombopoiesis, lymphopoiesis, monopoiesis, as well as neutrophilic, acidophilic, and basophilic granulopoiesis), as well as the reticular stroma characteristic of the lymphoid organs and the melano-macrophages, a peculiar type of macrophage-like cells that are capable of accumulating huge amounts of pigments (especially melanin), and that frequently associate in melano-macrophage centers (Esteban et al., 1989; Meseguer et al., 1990, 1991, 1994). We also studied the mature immune cells present in blood and peritoneal exudates from gilthead seabream (López-Ruíz et al., 1992; Meseguer et al., 1993).

After this basic stage, we continued studying the biological function of the main immune cells. We began to fine-tune cell isolation techniques that are typical of cell cultures, and thus achieved the isolation of leucocytes and their study with the use of flow cytometry. This achievement opened up enormous possibilities for us to advance in the study of the physiology of the immune system of gilthead seabream. We studied the phagocytosis mechanism by which large particles are internalized by immune cells in order to be processed and degraded. After this, the resulting materials may be recycled in cellular metabolism or, if it is a microorganism that performs antigenic presentation—a key stage of the innate immune response against pathogens—it may promote the initiation of the adaptive immune response that is mainly carried out by lymphocytes (Esteban et al., 1998). Then we continued with cytotoxicity, a mechanism by which fish cells—known as natural cytotoxic cells (equivalent to natural human killer cells, natural killer or NK cells)—are capable of recognizing and killing tumor cells or cells infected by viruses or bacteria (Cuesta et al., 1999). Eosinophilic granular cells, which are considered functionally equivalent to the mast cells of higher vertebrates, since they can be degranulated experimentally (Noya and Lamas, 1996) are another type of immune cells. At present, we are studying gilthead seabream inflammation, an intricate reaction of the innate immune system that affects all tissues and is frequently involved in different diseases. Usually, this reaction is initiated as a response to a stimulus (e.g., presence of pathogens or irritants, injury, or trauma). Such stimuli trigger a series of cascading reactions, consisting of alerting cells to prepare effective immune responses and initiating wound repair and healing processes to resolve the situation, and to help recover physiological homeostasis (Chen and Nuñez, 2010).

Our research line continued with the characterization of important proteins for the gilthead seabream immune system that are present in blood (humoral immunity), such as those of the complement system (a set of proteins whose direct function is the recognition and destruction of pathogens when they invade an organism), lysozyme (an enzyme capable of causing the rupture of bacterial walls), or immunoglobulin M (or IgM, a plasma protein synthesized by mature B lymphocytes and which was the only type of immunoglobulin described in fish at that time; Ortuño et al., 1998). At this point, after carrying out the main cellular and humoral activities described so far in the same fish, we could already know its immune status regarding innate immunity (main branch of the fish immune system) at the systemic level, that is, at the level of the organism.

Coinciding with the takeoff of polymerase chain reaction (PCR) techniques, we began to study some receptors as well as effector and modulatory molecules involved in the previously listed cell reactions, and we thus began the study of certain cell markers involved in the function of immune cells, as well as the expression of many genes of interest related to varied physiological processes such as growth, cell metabolism, antioxidant status, or intercellular junctions (Cerezuela et al., 2013a; Esteban et al., 2013; Cordero et al., 2016a). These initial studies have led us to recently apply some transcriptomic (RNA-seq) and proteomic analyses, increasing our global knowledge of key immune molecules as well as the cross-relation and interaction between immunity and other systems (Chaves-Pozo et al., 2017, 2019; Cordero et al., 2015, 2016a; Jurado et al., 2015).

3. Fish mucosal immunology

The immune systems in vertebrates, including those present in fish, are made up of many interdependent components. Each component has its own inherent protective value, and it is their final combination that results in a successful immune response. Knowing how to study the immune system of fish, we set out to study the immune system at tissue level. We started by studying intestine immunity, since when supplying dietary additives (with immunostimulant aims, detailed later), we did not know whether we were producing any negative impact on its mucosa. The lymphoid tissue associated with the digestive mucosa was studied and, curiously, it is not as developed as that of mammals. The lymphocytes did not appear organized in Peyer's patches but rather in diffuse accumulations along the intestinal submucosa (Cerezuela et al., 2012a; Lazado and Caipang, 2014).

The study of the intestine increased the interest in other mucosal surfaces in fish since they are the first line of defense against any threat. They act as a physical, chemical, and biological barrier that serves to protect the animal and which also functions as a very active immune tissue. In teleosts, four mucosa-associated lymphoid tissues (MALT) have been established according to their anatomical locations: gut-associated lymphoid tissue (GALT), skin-associated lymphoid tissue (SALT), gill-associated lymphoid tissue (GIALT), and nasopharynx-associated lymphoid tissue (NALT) (Salinas, 2015).

Mucosal immunology is the study of the immune responses that occur at mucosal surfaces. The mucosal surfaces of the fish body are closely associated with the external environment and are particularly vulnerable to infection. They are thin and permeable barriers to the interior of the body due to their physiological activities in food absorption (GALT) and in protection, but also in many other functions, such as communication, sensory perception, locomotion, respiration, ion regulation, and excretion (SALT), in gas exchange (GIALT), and in sensory activities (SALT and NALT). The mucosal immune system must provide protective immunity to pathogens and yet remain tolerant to nonharmful environmental substances. For the last two decades, mucosal immunology has been attracting increasing attention in medicine and veterinary sciences, not only for its critical role in explaining some essential aspects of how immune homeostasis is maintained, but also for its involvement in trying to find vaccines against the many infectious diseases that gain access to the body through mucosal surfaces (Das and Salinas, 2020). All mucosal surfaces in fish consist of an epithelium, which contains intercalated mucous cells (the main are the goblet cells). These mucus cells produce a mucus secretion that pours onto the apical surface of the epithelial cells forming a continuous layer surrounding the fish body. Mucus on the fish and on mucosal surfaces has evolved to have potent mechanisms that trap, immobilize, and inactivate pathogens before they can come into contact with the epithelial surface. Furthermore, the mucus in most fish species is continually secreted and replaced, and this washing-out effect prevents the stable colonization of potential infectious microorganisms and the invasion of parasites. The main molecules present in mucus are high molecular weight and highly glycosylated glycoproteins called mucins. In fact, these mucins are responsible for their adhesive, viscoelastic, and rheological properties. They exert a mechanical barrier by serving as filters for pathogens and preventing pathogen adherence to the underlying tissues (Guardiola et al., 2015a). These mucins form a matrix in which a diverse range of immune molecules can be found. Therefore fish skin (including both mucus and epidermis) acts as a repository of numerous immune components (of both innate and acquired immune system) such as glycoproteins, lysozyme, immunoglobulins (IgM and IgT), complement proteins, lectins, C-reactive protein, flavoenzymes, cytokines, proteolytic enzymes, and antimicrobial peptides (Nigam et al., 2012; Cone, 2009; Shephard, 1994; Salinas, 2015; Valero et al., 2020), which exert inhibitory or lytic activity against different type of pathogens (Subramanian et al., 2008; Whyte, 2007), contributing to fish well-being. For a detailed study of fish mucosal surfaces, there is a very complete book edited by Beck and Peatman (2015) entitled "Mucosal health in aquaculture", highly recommended for further interested readers.

3.1 Intestine mucosal immunology

Some studies were first addressed to identify the structure and immune parameters of the fish intestine. Fish intestine mucosa contains enterocytes, goblet cells, and lamina propria leucocytes (macrophages, granulocytes, lymphocytes, and plasma cells), as well as intraepithelial leucocytes (IELs, mainly B and T lymphocytes; reviewed by Salinas and Parra, 2015). The secreted and external mucus layer protects the intestine from external insults in the digestive tract and consists mainly of mucins, but also contains a wide range and variety of immune molecules [antimicrobial peptides (AMPs), complement, lysozyme, proteases, antiproteases, peroxidase, immunoglobulins, etc.]. Gilthead seabream intestine leucocytes have been isolated, and they contained high numbers of granulocytes with nitro-blue tetrazolium activity (Salinas et al., 2007). In fact, intestine mucus production and mucin-glycosylation patterns were altered upon infection of gilthead seabream with the parasite *Enteromyxum leei* in an effort to washout and reduce the adhesion and colonization of the intestine by this pathogen (Redondo and Alvarez-Pellitero, 2010; Estensoro et al., 2013). Most of the studies focusing on intestine mucosal immunity are limited to routine or classical histological methods due to difficulties in obtaining the intestinal mucus (mainly in small fish, as it was previously commented) as well as the very scarce number of available antibodies to identify leucocyte populations or cytokines.

3.2 Skin mucosal immunology

Fish skin is formed by an external mucus layer, epidermis and dermis. In contrast to mammals, the dermis is entirely formed by live cells along with some mucous cells that produce the mucus. Similar to the intestine, leucocytes, including lymphocytes, plasma cells, macrophages, and granulocytes, are disseminated in the skin. In contrast to the intestine, application of histological procedures to gilthead seabream skin is difficult due to the presence of calcified scales, and this fact makes it difficult to obtain perfect sections in order to study the tegument and mucous layers. However, the isolation of mucus and its study is very simple and effective. Thus gilthead seabream skin extracts or mucus contain many immune factors whose activity or presence can be determined, including proteases, peroxidases, antiproteases, phosphatase alkaline, lysozyme, complement, AMPs, immunoglobulins, or lectins, among others, as well as bactericidal activities (Guardiola et al., 2014a, b). The epidermal thickness is not homogeneous in all fish body. In gilthead seabream, the thickness of the skin above the lateral line is higher than that of below the lateral line, as demonstrated by image analysis using light microscopy, while the expression of immune-related genes is similar in both skin locations, as demonstrated by real-time PCR. In addition, scanning electron microscopy pointed to a greater cell size and area of microridges in the apical part of the dorsal epidermal cells when compared with the epidermal cells from the skin below the lateral line (Cordero et al., 2017a).

3.3 Skin mucosal immunology during wound repair and healing

Fish are continuously exposed to external skin lesions and injuries, which are maximized during intensive culture in farms. Thus lesions, as already mentioned, are an entry portal for pathogens, and the wound healing process is crucial in overcoming this rupture of the protective layer. Confirming the skin asymmetry identified before, a higher ratio of wound healing was observed in the gilthead seabream ventral region, where wounds were closed after 15 days, compared to the dorsal region of skin, where wounds were closed after 21 days (Cordero et al., 2017a). The expression profile also supports this finding since genes related to cell and matrix-cell interactions or cytoskeletons were more up-regulated in the ventral than in the dorsal region (Ceballos-Francisco et al., 2017; Tapia-Paniagua et al., 2018; Chen et al., 2020a, b). In addition, wounds provoke a general decrease in skin mucus immune activities (protease or antiprotease activities) and greatly modifies the expression of immune-related genes [*myeloperoxidase* (*mpo*), *colony stimulating factor 1 receptor* (*csf1r*), *interleukin-1 beta* (*il1b*), *interleukin-8* (*il8*), *tumor necrosis factor alpha* (*tnfa*), and *immunoglobulins M* (*igm*) and *T* (*igt*)] (Table 1) in the skin, favoring the healing process and the refraction to pathogens (Ceballos-Francisco et al., 2017; Tapia-Paniagua et al., 2018; Chen et al., 2020a, b). In addition, a proteomic study of the wound area showed a decreased presence of important immune proteins such as complement component 3 (C3), apolipoprotein A (Apoa1), histones (H2A, H2B, H4), and IgM (Cordero et al., 2017b). Very surprisingly, only the presence of a wound in the skin is enough to alter the intestinal morphology and expression of the immune and cell junction genes that suggest inflammation (Chen et al., 2020b). Therefore loss of skin integrity during a time shift creates

TABLE 1 Summary of the main effects in the mucosal immunity of gilthead seabream (*Sparus aurata*) under different conditions.

Treatment		Intestine tissue	Skin tissue	Mucus	References
Probiotics	*L. delbrueckii lactis* + *B. subtilis*	↑ Acidophilic granulocytes and Ig^+ cells			Salinas et al. (2008)
	B. subtilis	↑ Gene expression of *il8*, *cox2*, *ocl*, and *tf*	↑ Gene expression of *il1b*, *nkefa*, *cox2*, *csfr1*, *tcrb*, *hep*, and *def*	↑ IgM levels ↑ PER and PRO activities	Cerezuela et al. (2013b, 2016) and Esteban et al. (2014)
	S. putrefaciens	↑ Gene expression of *il10*, *tgfb*, *lyz*, *hep*, *sod*, and *fbl* ↓ gene expression of *rbl*	↑ Gene expression of *lyz*	↑ IgM levels ↑ PER and PRO activities ↑ C3, leucocyte elastase inhibitor, NCCRP1, F-type lectin, NKEFB, and LYZ proteins	Cerezuela et al. (2016), Guardiola et al. (2016b), Cordero et al. (2016c), and Chen et al. (2020a),
	K. lactis			↑ Bactericidal activity vs *V. parahaemolyticus*, *V. harveyi*, and *V. anguillarum* ↑ PRO activity ↑ Nitrite levels	Guluarte et al. (2019)
	S. halophilus	↑ Gene expression of *il1b*, *tnfa*, *igm*, *c3*, and *lyz*	↑ Gene expression of *il1b*, *tnfa*, *igm*, *c3*, and *lyz*	↑ IgM levels ↑ PRO, ANTIPRO, PER, and LYZ activities	Reyes-Becerril et al. (2017)
	T. chuii	↑ Gene expression of *il8* and *ocl*			Cerezuela et al. (2013b)
	P. tricornutum	↑ Gene expression of *il8*			Cerezuela et al. (2013b)
Prebiotics	Inulin	↑ Gene expression of *il8*, *actb*, *ocl*, and *tf*			Cerezuela et al. (2013b)

Plants	*Moringa oleifera*		↑ Gene expression of antioxidant, immune, and tight junction genes Restores impact of H_2O_2	↑ PHOS, PER, and LYZ activities ↑ IgM levels Restores impact of H_2O_2	Mansour et al. (2020)
	Phoenix dactylifera	= Gene expression of *sod*, *gr*, and *cat* ↑ Gene expression of *fbl* ↓ Gene expression of *rbl* and *hep*	↑ gene expression of *gr*	↑ IgM levels	Esteban et al. (2014), Cerezuela et al. (2016), and Guardiola et al. (2016b)
	Origanum vulgare			↑ IgM levels ↑ Bactericidal activity vs *P. damselae* = Total antioxidant activity	García-Beltrán et al. (2020)
	Trigonella foenum graecum			↑ PER and PRO activities	Guardiola et al. (2017)
Amino acids	Arginine		↑ Gene expression of *mpo*, *il8*, *tgfb*, *tnfa*, *igf1*, *colxa*, *cat*, and *gr*	↑ IgM levels	Chen et al. (2020b)
	Tryptophan			↑ Bacteriostatic activity vs *V. harveyi*	Azeredo et al. (2017)
Heavy metals	MeHg, Cd, As			↑ FBL levels and glycosylation ↑↓ LYZ, PRO, ANTIPRO PER, ALP, and CP activities	Guardiola et al. (2015c)
	Sediments		↓ Gene expression of *gr*, *il1b* and *cox2*		Ben Hamed et al. (2017)
Microplastics	PVC	↑ Gene expression of *nrf2* ↑ Goblet cell density, villus thickness		↑ IgM levels ↓ PER activity	Espinosa et al. (2017, 2019)
Stress	Crowding, anesthetics, hypoxia			↑ ACH and PRO activities ↓ IgM levels ↑↓ PRO activity	Guardiola et al. (2016a, 2018a)
	Hypoxia	↑ Gene expression of *lyz*, *c3*, *igm*, *hsp70*, *tnfa*, *il6*, *tgfb*, *cox2*, *il1b*, and *il10*	↑ Gene expression of *lyz*, *cox2*, *il1b*, and *il10*	↑ IgM levels	Khansari et al. (2018)

Continued

TABLE 1 **Summary of the main effects in the mucosal immunity of gilthead seabream** (*Sparus aurata*) **under different conditions—cont'd**

Treatment		Intestine tissue	Skin tissue	Mucus	
Wounds	Experimental wounds and ulcers	↑ Gene expression of *mpo*, *csf1r*, *il1b*, and *tnfa*	↑ IgM levels ↑ gene expression of *mpo*, *il8*, *tnfa*, *il1b*, *igm*, and *igt*	↓ C3, APOA1, Histones, and IgM proteins ↑ PER and EST activities ↓ PRO and ANTIPRO activities	Ceballos-Francisco et al. (2017), Cordero et al. (2017b), Tapia-Paniagua et al. (2018) and Chen et al. (2020a, b)
Pathogens and vaccines	*P. damselae*	↓ Gene expression of *nkefa*		↑ IgM and specific IgT levels ↓ ANTIPRO activity	Piazzon et al. (2016) and Azeredo et al. (2017)
	V. parahaemolyticus		↑ Gene expression of *il1b* and *c3*		Reyes-Becerril et al. (2017)
	E. leei	↑ Gene expression of *igm* and *igt*			Piazzon et al. (2016)
	LCDV		↑ Gene expression of *ifn*, *irf3*, *mx*, *mhc2a*, *csfr1*, *tcra*, and *igm*		Cordero et al. (2016b)
	Oral *P. damselae* vaccine	↑ Gene expression of *igm* and *igt*		↑ Specific IgT levels	Piazzon et al. (2016)
	Immersion *V. anguillarum* vaccine	↑ Gene expression of *c3* and *tgfb*	↑ Gene expression of *c3*, *hsp70*, and *cox2*		Khansari et al. (2018)

↑: significant increase; ↑: significant decrease; ↑↓: either increase or decrease; =: nonchanges; vs: versus. **Activities and proteins:** IgM: immunoglobulin M; PER: Peroxidase activity; PRO: Protease activity; C3: Complement component 3; NCCRP1: Nonspecific cytotoxic cell receptor protein 1; NKEFB; Natural killer enhancing factor B; LZY: Lysozyme activity; ANTIPRO: Antiprotease activity; PHOS: Phosphatase activity; ALP: alkaline phosphatase activity; CP: Ceruloplasmin; ACH: hemolytic complement activity; APOA1: Apolipoprotein A; EST: Esterase activity; IgT: immunoglobulin T. **Genes:** *il8 (interleukin-8); cox2 (ciclooxygenasse 2); ocl (occluding); tf (transferrin); il10 (interleukin-10); tgfb (transforming growth factor beta); lyz (lysozyme); hep (hepcidin); sod (superoxide dismutase); fbl (fucose binding lectin); rbl (rhamnose-binding lectin); il1b (interleukin-1 beta); tnfa (tumor necrosis factor alpha); igm (immunoglobulins M); c3 (complement component 3); actb (beta-actin), gr (glutathione reductase); cat (catalase); nrf2 (nuclear factor erythroid 2-related factor 2); hsp70 (heat shock protein 70); il6 (interleukin-6); mpo (myeloperoxidase); csf1r (colony stimulating factor 1 receptor); nkefa (natural killer enhancing factor A), igt (immunoglobulins T); tcrb (T cell receptor beta chain); def (defensin); igf1 (insulin growth factor 1); colxa (collagen Xa), ifn (interferon type I), irf3 (interferon regulatory factor 3); mx (myxovirus resistance protein); mhc2a (major histocompatibility complex II alpha); and tcra (T cell receptor alpha chain).* **Organisms:** *L. delbrueckii lactis (Lactobacillus delbrueckii lactis); B. subtilis (Bacillus subtilis); S. putrefacients (Shewanella putrefaciens); K. lactis (Kluyveromyces lactis); S. halophilus (Sterigmatomyces halophilus); T. chuii (Tetraselmis chuii); P. tricornutum (Phaeodactylum tricornutum); P. damselae (Photobacterium damselae); V. parahaemolyticus (Vibrio parahaemolyticus); E. leei (Enteromyxum leei);* LCDV (lymphocystis disease virus); *V. anguillarum (Vibrio anguillarum); and V. harveyi (Vibrio harveyi).*

a general microenvironment that might favor the entry of pathogens before the wound is healed and the immune system fully restored, though some immune parameters are still present and very active during this period to control the opportunistic pathogens. All these results demonstrate that intricated interconnections exist among different MALT and that very little is known about them.

3.4 Mucosal immunity during infection or vaccination

As the external line of defense, mucosal surfaces are directly exposed to pathogens, and the integrity and immunity of these tissues is crucial to clear or reduce the pathogen burden. Thus, apart from the previously cited alterations in the intestine upon *E. leei* infection (Redondo and Alvarez-Pellitero, 2010; Estensoro et al., 2013), an increment in the gilthead seabream intestinal expression of the *igm* and *igt* genes was observed (Piazzon et al., 2016). Regarding bacterial pathogens, infection with the pathogenic *Photobacterium damselae* subsp. *piscicida*, strain PP3, resulted in increased IgM levels, decreased antiprotease activity in the skin mucus, and reduced gene expression of the *natural killer enhancing factor A* (*nkefa*) in the European sea bass intestine (Azeredo et al., 2017), while the infection with *Vibrio parahaemolyticus* resulted in the up-regulation of *il1b* and *c3* genes in the gilthead seabream skin (Reyes-Becerril et al., 2017). In addition, gilthead seabream vaccination against *P. damselae* subsp. *piscicida* increased the total skin mucous levels of IgM and of specific IgT, and up-regulated the corresponding genes in the intestine (Piazzon et al., 2016). Similar tendencies were observed after the administration by immersion of the *V. anguillarum* vaccine, which induced the intestine, gill, and skin expression of immune genes including *c3*, *transforming growth factor* β (*tgfb*), *heat shock protein 70* (*hsp70*), or *cyclooxygenase 2* (*cox2*) in gilthead seabream (Khansari et al., 2018). Interestingly, the lymphocystis disease virus (LCDV), which produces lesions in the skin, induces a sharp down-regulation of immune genes [*interferon type I* (*ifn*), *interferon regulatory factor 3* (*irf3*), *myxovirus* (influenza) *resistance protein* (*mx*), *major histocompatibility complex II* α (*mhc2a*), *csfr1*, *T cell receptor alpha chain* (*tcra*), and *igm*] in the skin of gilthead seabream (Cordero et al., 2016b). In general, disruption of the physical barrier of immunity is used for pathogens' entry and dissemination into fish tissues, leading to infection. Therefore understanding mucosal immunity and its interaction with pathogens is of major importance from the aquaculture perspective.

3.5 Mucosal immunity during exposure to contaminants

Aquaculture practices and management may lead to immunodepression. The negative impact of pollutants on fish mucosal surfaces has received little attention, even considering the great amount of surface in contact with the toxicant. We have observed that waterborne exposure to heavy metals, such as methylmercury (MeHg), cadmium (Cd), or arsenic (As) alters the protein profile of the gilthead seabream skin mucus, terminal carbohydrate composition, and some of the immune-related parameters that were measured (protease, antiprotease, phosphatase, peroxidase, or bactericidal activities and IgM levels) (Guardiola et al., 2015c). Interestingly, the presence of skin mucus fucose binding lectin (FBL) was increased upon waterborne exposure to As or Cd, as determined by western blotting. In addition, gilthead seabream exposed to heavy metal-contaminated sediments displayed a reduction in the skin expression of genes related to metal detoxification [*metallothionein A* (*mta*)], antioxidants [*glutathione reductase* (*gr*)], cellular stress (*hsp70* and *hsp90*), and immunity (*il1b*), while the expression of the immune-related gene *cox2* was up-regulated and no histological variations were observed in the skin (Ben Hamed et al., 2016, 2017). Recently, plastic debris (micro and nanoplastics) has emerged as novel pollutants. It was demonstrated that feeding gilthead seabream with a diet containing polyvinylchloride (PVC) microplastics resulted in an increase of IgM level and peroxidase activity in the skin mucus (Espinosa et al., 2017).

3.6 Mucosal immunity during stress

Fish farm management and intensive culture provokes stress, which is known to reduce the immune response at chronic stages, though acute stress situations generate an alert state in the animal that also might improve physiological processes, including immunity. Thus acute stress engendered by high densities, anesthesia, or hypoxia produced an increment in the gilthead seabream skin mucus activities of the complement system and in peroxidase, while an increase or reduction of the IgM level and protease activity was recorded (Guardiola et al., 2016a, 2018a; Khansari et al., 2018). In general, short hypoxia exposure induced the expression of important immune genes [*lysozyme* (*lyz*), *cox2*, *il1b*, *il6*, *il10*, or *tgfb*] in both the skin and the intestine (Khansari et al., 2018). Conditions that are considered negative are known to increase susceptibility to disease but, as mentioned before, individual immune parameters might be increased for a short time in order to counterbalance the animal's physiology, though if prolonged they might result in immunity exhaustion, disease, and finally fish death.

4. How to alter or strengthen the fish mucosal immune system

Knowing the immune status of fish, our research group focused on improving or strengthening this immunity. Two great maxims guided this new challenge: we would use natural substances (to contribute to a more safe, sustainable, and eco-friendly aquaculture) and we would administer them in the diet (in order to avoid the use of extra or specialized handling of fish). Thus we began an exciting search for which would be the best immunostimulants for our fish of interest.

After the evaluation of a long list of immunostimulants (vitamins, probiotics, prebiotics, hormones, plants, etc.), depressors (pollutants, stressors, etc.), and pathogens/vaccines, along with their effects on the serum or head-kidney tissue from gilthead seabream and European sea bass, we started to look at the immune parameters in mucosal surfaces, mainly in the intestine and skin. The immune response of mucosal surfaces in fish can be determined at two levels: (1) the mucosa itself by either histochemistry, gene expression, protein profiles, and activities in protein extracts or (2) in the secreted mucus. First attempts aimed to evaluate the intestine structure by histology; histochemistry for goblet cells and its secreted mucus as well as immunohistochemistry, though few studies have focused on the intestine mucus characterization (only for IgM or IgT determinations), mainly due to the interest of this tissue after dietary administration of immunostimulants or vaccines. In the last decade, greater attention has been received by the evaluation of the mucosal immunity of the skin, mainly by the determination of immune factors present in the mucus due to its easy, inexpensive, and noninvasive sampling and thus, contributing to the 3Rs strategy (Replacement, Reduction, and Refinement). Unfortunately, the mucosal immunity at gills still remains scarcely explored.

Mucosal surfaces, mainly the fish skin, represent the first and largest physical and chemical barrier and are vital for health and welfare. Thus skin lesions are a perfect portal for pathogens to enter the fish body. The mucosa integrity can be disrupted by low or inappropriate production of mucus, loss of epithelial integrity (wounds, ulcers, loss of cell- or matrix-cell interactions), nutritional deficiencies, stress, and/or the active penetration by pathogens, among others. Thus, apart from the precise immune cells and molecules, the knowledge of processes involved in mucus production, mucin carbohydrates, cell-cell interactions, cytoskeleton integrity, or antioxidant status is directly related to mucosal integrity, immunity, and health. Therefore we have started to identify the local and mucosal immunity components, and the ways in which they are influenced by pathogens and stressors, to finally identify how to strengthen them and prevent or restore their disruption to avoid their use to the pathogens' profit. The most important findings in the mucosal immunity of the Mediterranean gilthead seabream are summarized in Table 1.

4.1 Dietary improvement of mucosal immunity

It is well known that animal health and welfare, including in fish, are intimately linked to nutrition (Oliva-Teles, 2012). Thus a balanced diet containing the right nutrients, both in kind and in proportion, tends to favor nutritional balance and good health. Apart from the nutritive value, there are many substances that promote or increase immune status, which results in better health and disease resistance: they are called immunostimulants. They have long been known, and classical studies evaluate their impact on immune tissues, mostly those of the head-kidney and in blood. However, the evaluation of MALT has gained attraction since they are the first line of defense. In addition, the search for immunostimulants has moved from the use of pure and difficult-to-obtain or produce immunostimulants to the use of easy, cheap, and eco-friendly alternatives. In this sense, many studies have searched for probiotics, prebiotics, and plants as immunostimulant sources, including the evaluation of these immunostimulants on the mucosal immunity of fish.

4.1.1 Probiotics

Probiotics are usually live (sometimes dead) microorganisms, administered by feeding, that compete with intestinal microbiota, provide nutritive compounds, and improve intestinal health by promoting mucus production, renewal, and immunity (Nayak, 2010; Lazado and Caipang, 2014). Although the list of probiotics used in fish is large, the first studies focused on human lactic bacteria. Thus we first evaluated the use of *Lactobacillus delbrueckii lactis* and *Bacillus subtilis* (Salinas et al., 2005). Afterward, we extensively used a *Shewanella putrefaciens* strain isolated from healthy gilthead seabream skin (Pdp11) with many beneficial effects (Díaz-Rosales et al., 2006) (Table 1). They have been supplemented in the gilthead seabream commercial diet alone or combined with other immunostimulants. Thus bacterial inclusion in the gilthead seabream diet increased the intestinal number of acidophilic granulocytes and Ig^+ cells in the mucosa, suggesting enhanced MALT functioning (Salinas et al., 2008). Later on, dietary supplementation with either *B. subtilis* or Pdp11 showed important alterations in the intestine structure (goblet cells density and composition, high intestinal wall, villus area, IELs, etc.), restricted to the mucosa (Cerezuela et al., 2013a), but this was not linked to greater resistance to the challenge with the pathogenic *P. damselae* (Cerezuela et al., 2012b). These probiotics induce the up-regulation of intestine genes related to immunity [*il8*, *caspase 1* (*casp1*) and *cox2*], cell junctions [*beta-actin* (*actb*), *occludin* (*ocl*)], or iron metabolism [*transferrin* (*tf*)] (Cerezuela et al., 2013b). Nonetheless, dietary feeding with either *B. subtilis* or Pdp11 also affects the skin—they tend to increase the antioxidant potential of the skin mucosal tissues of gilthead seabream—and also in gills, but not in the intestine (Esteban et al., 2014). At gene level, they up-regulate immune-related genes including *il1b*, *lyz*, *mpo*, *tcra*, *hepcidin* (*hep*), and *csf1r1*, among others, in the skin, indicating a higher presence of immune cells and/or their activation (Cerezuela et al., 2016; Cordero et al., 2016c). Regarding the skin mucus, the presence of IgM and the protease and peroxidase activities were also increased by the two probiotics. Very interestingly, gilthead seabream fed Pdp11-supplemented diets showed increased protein levels in skin mucus including C3, leucocyte elastase inhibitor (Lei), nonspecific cytotoxic cell receptor protein 1 (Nccrp1), F-type lectin, natural killer enhancing factor B (NkefB), or lysozyme (Cordero et al., 2016c). Pdp11 also restores the negative impact of the stress caused by crowding in the skin mucus proteome (Cordero et al., 2016c) and alleviates the effects caused by wounds (Chen et al., 2020b). In our hands, the combination of *B. subtilis* or Pdp11 bacteria with other immunostimulants, such as prebiotics, plants, or plant extracts (see later), has not shown any synergistic effect of importance in the mucosal immunity of gilthead seabream.

Other microorganisms—apart from bacteria—were also useful as probiotics for gilthead seabream. For instance, the supplementation of diets with microalgae (*Tetraselmis chuii* or *Phaeodactylum tricornutum*) showed little alteration in the intestine regarding the expression of genes related to immunity (*il8*), cell junctions (*ocl*), the cytoskeleton (*actb*), or iron metabolism (*tf*), which were also increased by inulin supplementation (Cerezuela et al., 2013b). In the case of yeast cells, the dietary inclusion of *Kluyveromyces lactis* in the gilthead seabream diet increased the bactericidal and protease activities in the skin mucus and the nitric oxide production (Guluarte et al., 2019) while *Sterigmatomyces halophilus* increased the IgM level and protease, antiprotease, peroxidase, and lysozyme activities in the skin mucus. It also up-regulated genes related to immunity (*il1b*, *tnfa*, *igm*, *c3*, and *lyz*) in the skin and intestine (Reyes-Becerril et al., 2017). All these studies demonstrated the beneficial effects of probiotics in gilthead seabream mucosal immunity, although precise dietary formulations and administration times need to be optimized, since prolonged administration of immunostimulants fails to increase immune status and consumes a lot of energy.

4.1.2 Prebiotics

Prebiotics are another interesting immunostimulant source for fish (Song et al., 2014). They consist of nondigestible polymers in the food that facilitate intestinal transit and are used by the intestinal microbiota, leading to direct or indirect benefits in fish well-being. They are common and easy to obtain; they are cheap and do not pose environmental issues. Among them, the most used in fish are inulin, fructooligosaccharides (FOS), or mannan oligosaccharides (MOS). Supplementation of the gilthead seabream diet with inulin resulted in an increased expression of intestine genes related to immunity (*il8*), cell junctions (*actb*, *ocl*), or iron metabolism (*tf*), and this was not greatly influenced by the coaddition of *B. subtilis* (Cerezuela et al., 2013b). Supplementation of the European sea bass diet with MOS showed improved intestine morphology, a higher number of lymphocytes and eosinophilic granulocytes, greater production of prostaglandins, and an up-regulated trend of important immune genes [*major histocompatibility complex I* (*mhc1*), *cluster of differentiation 4* (*cd4*), *cd8*, *cox2*, or *tnfa*] in the intestine (Torrecillas et al., 2013, 2015). In addition, in the European sea bass, MOS restored the negative impact produced by the fish oil replacement with soybean oil (Torrecillas et al., 2015) and increased the survival of anally infected specimens with pathogenic *V. anguillarum* (Torrecillas et al., 2012). Care must be taken when prebiotics are administered, since some observations point to intestinal inflammation or enteritis. For this, the correct application with probiotics and a balanced microbiota needs to be considered.

4.1.3 Plants and herbs

Plants, and their extracts (phytogenics), are nowadays extensively investigated, and they are also promising immunostimulants for fish aquaculture (Galina et al., 2009; Awad and Awaad, 2017). Gilthead seabream specimens fed *Moringa oleifera* leaves supplemented diets showed improved skin mucosal immunity, including increases in phosphatase, peroxidase, and lysozyme activities and IgM levels in the skin mucus, as well as the expression of antioxidant [*superoxide dismutase* (*sod*) and *catalase* (*cat*)], encoding tight junction proteins [*ocl* and *zonula occludens 1* (*zo1*)], and immune (*tgfb*, *c3*, and *igm*) genes in both the skin and gills (Mansour et al., 2020). In addition, mucosal surfaces showed better mucosal immunity and lower stress indicators after exposure to H_2O_2 in moringa-fed specimens. In the case of *Origanum vulgare*, an aromatic herb, dietary supplementation improved the skin mucus IgM levels and bactericidal activity in gilthead seabream, though it failed to alter the antioxidant status (García-Beltrán et al., 2020). The inclusion in the diet of *Trigonella foenum graecum* seeds increased the protease activity of skin mucus in gilthead seabream, while coadministration with the probiotics *B. licheniformis*, *L. plantarum*, or *B. subtilis* also increased some immune parameters, including IgM level and bactericidal activity (Guardiola et al., 2017). Date palm fruits or seeds from *Phoenix dactylifera*, administered via the diet, provoked increased levels of IgM in the skin mucus of gilthead seabream and the increased expression of the antioxidant gene *gr*, while in the intestine, it failed to alter the antioxidant system at gene level, up-regulating *fbl* and down-regulating the *rhamnose-binding lectin* (*rbl*) and *hep* immune genes (Esteban et al., 2014; Cerezuela et al., 2016; Guardiola et al., 2016b). Thus plants have also arisen as an eco-friendly alternative for dietary supplementation with promising benefits for fish immunity and well-being, but further efforts are still needed to carefully select and characterize them.

4.1.4 Amino acids

Amino acids are essential nutrients, though some have also been demonstrated to play immunomodulatory roles (Oliva-Teles, 2012; Andersen et al., 2016). Thus gilthead seabream diet supplementation with arginine increased the skin mucus IgM levels and the expression of immune, antioxidant, growth, or skeletal genes [*mpo*, *il8*, *tgfb*, *tnfa*, *insulin growth factor 1* (*igf1*), *collagen Xa* (*colxa*), *cat*, *gr*] in the skin (Chen et al., 2020a). Although methionine failed to show immunostimulatory effects, the addition of tryptophan increased the bacteriostatic activity of the gilthead seabream skin mucus (Azeredo et al., 2017). Interestingly, arginine supplementation facilitated the healing process and immunity upon experimental wounds (Chen et al., 2020a). Much more efforts are needed regarding the use of free amino acids as fish-feed additives.

5. Methodologies to collect and preserve fish mucus

The necessity for permeability of the surface lining of these mucosal surfaces creates obvious vulnerability to infection, and it is not surprising that the vast majority of infectious agents invade the fish body through these routes. A brief description of the most used mucus sample collection methodologies is now described. Special care should be taken in the way of processing and storing the mucus samples before analyzing them.

In the case to work with fish and mucus infection models, a biosecurity level (BSL1) laboratory is enough, but some pathogens would require BSL2 or even BSL3 laboratories. However, for the mucus sample collection is not mandatory to work at BSL1, 2, or 3 but before starting any procedures, the work surface, surrounding areas, and equipment should be disinfected routinely with 70% (v/v) ethanol. All the steps described later must be performed under controlled temperature or below 4 °C in order to minimize the effect of temperature on protein concentration and functions of biologically active molecules. Only healthy fish should be considered for mucus collection, since fish with lesions could lead to contamination of the samples. In all cases, fish handling and management during research must adhere to general and national regulations regarding the bioethical issues of animals in research.

5.1 Intestine mucus

In order to reduce and minimize contamination in the collection of mucus from the intestine, fish should be fasted for at least 12–24 h, although some authors have stipulated longer times (7 days) (Hatten et al., 2001). The methods found in the literature are described later (Fig. 1).

5.1.1 Method 1

Intestinal tissue needs to be carefully dissected from each fish and the midgut and hindgut separated (if applicable) (Hatten et al., 2001). Each intestine part must be opened longitudinally, while being kept on ice, and the lumen surface gently rubbed against a sterile plastic Petri dish. Subsequently, 10 mL of 50 mM Bis-Tris, pH 6.4 and Complete protease inhibitor must be added into Petri dish, transferred to a sterile tube with a plastic or glass Pasteur pipette, and placed on ice. Collected gut mucus must be shaken slowly for 60 min and centrifuged at 2000 × *g* at 4 °C for 60 min. The supernatant can be filtered (0.22 or 0.45 μm pore size) before being stored at − 20 °C or − 80 °C.

Recommendation: The authors use a volume of 10 mL of buffer to dilute the gut mucus from Atlantic salmon of 900–1000 g, which should be adjusted according to the weight and size of the sampled fish. The use of protease inhibitors makes the samples unavailable for the determination of protease or antiprotease activities of the mucus.

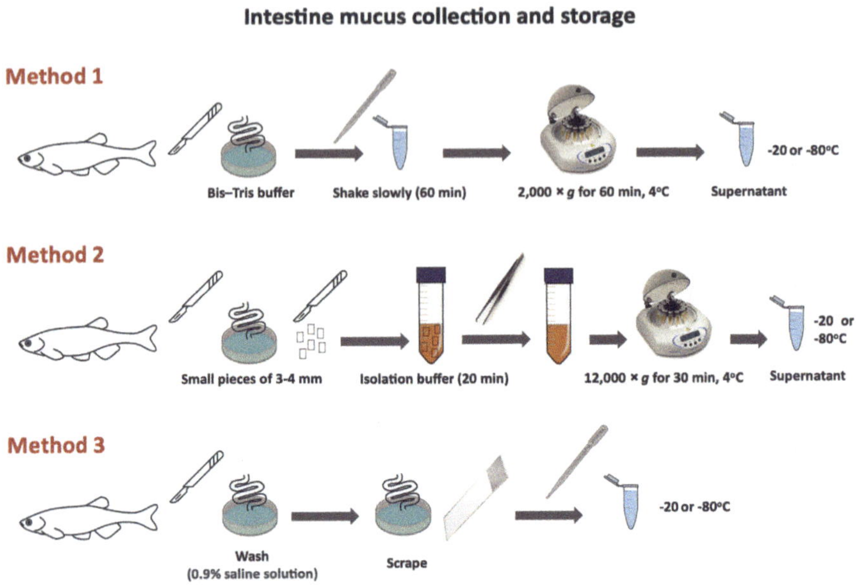

FIG. 1 Methodologies to collect and preserve intestine mucus (Hatten et al., 2001; Schroers et al., 2009; Khan and McGeer, 2013).

5.1.2 Method 2

Intestines must be kept on ice, opened longitudinally, and cut into small pieces of 3–4 mm. Subsequently, secreted luminal mucus must be isolated with isolation buffer (Schroers et al., 2009). Briefly, the intestine pieces are incubated for 20 min in 100 mL of isolation buffer containing protease inhibitors [phosphate buffer saline (PBS) with 1% (w/v) dithiothreitol, 0.03% Amphotericin B, 1% sodium pyruvate, and 0.6% Hepes]. Once the tissue is removed from the isolation buffer, the buffer is centrifuged at 12,000 × *g* for 30 min, and the supernatant collected is frozen at − 20 °C or − 80 °C.

Recommendation: Although the authors do not specify any temperature value, we would recommend to centrifuge at 4 °C.

5.1.3 Method 3

Intestines must be placed in a Petri dish on ice, longitudinally dissected, and washed in 0.9% (w/v) saline solution to remove undigested food (Khan and McGeer, 2013). Afterward, the intestinal mucosal surface must be scraped with a glass coverslip (Glover and Hogstrand, 2002), and this mucus scraping must be transferred into a sterile tube and stored at − 20 °C or − 80 °C.

Recommendation: Glass coverslip should be replaced by a less aggressive material, such as a sterile cell scraper or silicone spatula, since epithelial cells will probably contaminate the mucus.

5.2 Skin mucus

5.2.1 Sample collection

Skin mucus collection must be done before taking any other parameter, fluid or tissue (e.g., weight, length, blood, skin, etc.) from the animals in order to avoid any possible contamination and loss of mucus (Fig. 2). First, fish must be starved for at least 12 to 24h to minimize the contamination of the sample by intestinal excretions. Second, fish should be caught individually from the tank with a scoop net and transferred to the anesthesia bath for a few minutes.

Recommendation: Systematic washing of the scoop net with distilled water between each sampled animal should be mandatory.

Method 1

Once the animal is anesthetized, it must be hooked with toothed dissecting forceps from the mouth and drained for a few seconds to the air in order to remove the excess water. Subsequently, the skin mucus must be collected by gently scraping the dorsolateral surface of animals using a sterile cell scraper or silicone spatula (Guardiola et al., 2018b). This procedure must be done with enough care to avoid contamination with blood, urine-genital, and intestinal excretions. Therefore the cell scarper should evade the area next to the anus or any small wound that could be seen in the body surface. During gentle scraping, the animal must be maintained upside down and held with the toothed dissecting forceps to

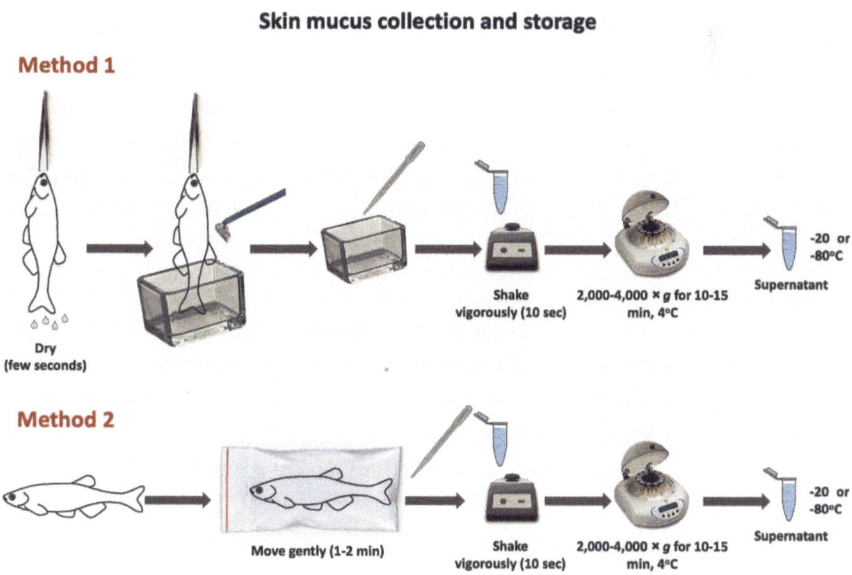

FIG. 2 Methodologies to collect and preserve skin mucus (Guardiola et al., 2018a; Subramanian et al., 2008).

avoid any pressure on the animal's belly with the consequent excretion of urine, feces, eggs, or semen. This methodology should be done on top of a glass jar (e.g. glass slide staining jar) or any other glass container kept on ice (to maintain the collected mucus at low temperature). Subsequently, the collected skin mucus can be transferred to a sterile tube with a plastic or glass Pasteur pipette, and it must be preserved on ice until stored at − 20 °C or − 80 °C.

Recommendations

- Once each sample is collected from each fish, the glass jar must be washed with distilled water and dried in order to avoid contamination between animals.
- Anesthetized fish should not be caught by the operculum with hand or toothed dissecting forceps due to the fact that gills can be damaged, causing a hemorrhage, with probable contamination of the sample.
- In the case of teleost fish characterized by elongated bodies (i.e. order Anguilliformes), it is advisable to replace the toothed dissecting forceps with dry cellulose paper to secure the head of the animal and with a cell scraper or spatula while wearing sterile gloves for wiping the skin of each fish.
- Regarding flatfish, such as Senegalese sole (*Solea senegalensis*), it would be recommended to collect the skin mucus from the ocular and blind sides separately due to the activity levels of some innate immune-related parameters, depending on the place of collection (Escribano et al., 2020). The results reported by these authors seem to suggest that a diverse role exists in the mucosal immune response among the ocular and blind sides, which may be due to the fact that the ocular side would be more exposed to invading pathogens than the blind side.

Method 2

To slough off the skin mucus, individual anesthetized fish must be introduced into a sterile polyethylene bag containing 5 or 10 mL (depending on the species, size, etc.) of 50 or 100 mM NaCl for freshwater and seawater fish species, respectively, according to Subramanian et al. (2008). The fish must be gently moved back and forth inside the polyethylene bag for approximately 1–2 min, in order to obtain as much mucus as possible. Once the fish is removed from the bag, it must be placed on ice, and the skin mucus sample can be transferred into sterile tubes with a Pasteur pipette and stored at − 20 °C or − 80 °C. In a similar approximation, when the plastic bags contain 5 mL of 100 mM ammonium bicarbonate (NH_4HCO_3, pH 7.8) and when the fish are removed from the bag, an additional 5 mL of this buffer is added in the case of Atlantic salmon (according to Ross et al., 2000).

Recommendations

- This method must be carried out individually to avoid injuries on the fins of the animals.

- A drawback of this method could be related to the possible excretion of urine, feces, eggs, or semen, which would contaminate the mucus sample and could invalidate the subsequent analyses.
- Another disadvantage would be the difficulty for the selective collection of mucus samples from a specific area, such as collecting samples from the dorsal or ventral zones (Cordero et al., 2017a) or from around a wound (Ceballos-Francisco et al., 2017).

5.2.2 Sample processing and storage

Regardless of the method used from those described before, the processing and storage of the collected skin mucus samples must be carefully followed. Briefly, skin mucus samples must be vigorously shaken and centrifuged at 2000–4000 × *g* at 4 °C for 10–15 min (Palaksha et al., 2008; Ross et al., 2000; Guardiola et al., 2017) in order to remove scales, insoluble particles, microorganisms, or other foreign material. Subsequently, the supernatant should be aliquoted and stored at − 20 °C or − 80 °C until used (Cordero et al., 2016d). Optionally, the supernatant can be filtrated (0.22 or 0.45 μm pore size) before being aliquoted and stored to ensure better sterility conditions of the samples. This step is advisable when the skin mucus comes from fish challenged with pathogens (bacteria or parasites) mainly by bath exposure (Guardiola et al., 2019; Escribano et al., 2020).

5.3 Gill mucus

Studies about GIALT mucosal immunity in fish are still scarce, but this topic has gained attraction in recent years. However, the isolation of gill mucus is difficult and very limited in quantity and quality (Fig. 3).

5.3.1 Method 1

Once the fish is anesthetized, the gill mucus samples can be obtained using the method described by Costa et al. (2011) in the case of Atlantic salmon. For this, two sterile cotton swabs are used to gently remove the gill mucus of all left and right gill arches, and each swab must be transferred into a tube containing 400 μL of PBS that must be preserved immediately on ice. This procedure must be done with enough care to avoid contamination by blood from the gills. Removal of gill mucus from each swab needs to be achieved by vigorously shaking the tube for 10 s, followed by squeezing each swab against the side of the tube (Vincent et al., 2006). The collected mucus from both tubes must be transferred to a single tube and stored at − 20 °C or − 80 °C until used.

Modifications: PBS can be replaced by another buffer [e.g. Tris-buffered saline (50 mM Tris-HCl, 150 mM NaCl), pH 8.0] (Marcos-López et al., 2017).

Recommendation: In order to minimize the risk of blood contamination, the fish could be euthanized and bled as described in Method 2 with the obvious nonrecovery of the animal.

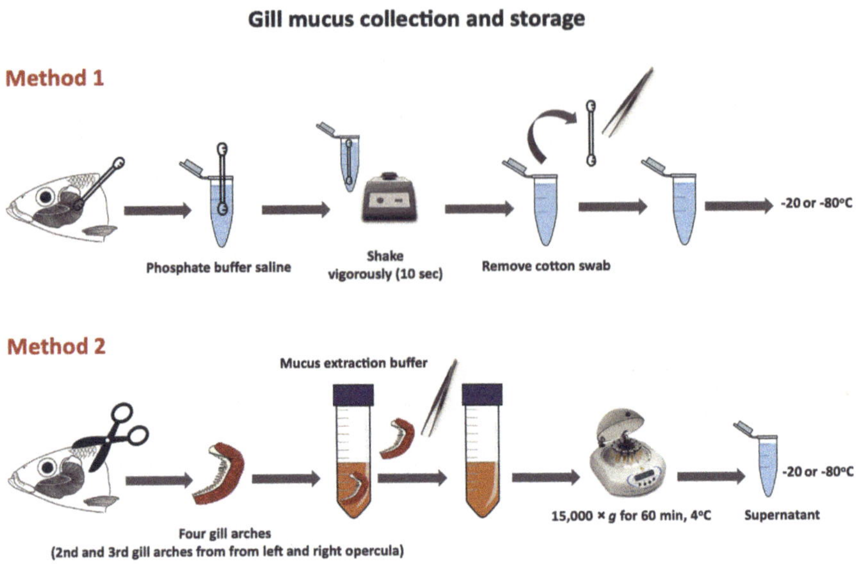

FIG. 3 Methodologies to collect and preserve gill mucus (Costa et al., 2011; Valdenegro-Vega et al., 2014).

5.3.2 Method 2

To apply this methodology, the fish must be euthanized prior to gill mucus collection (Valdenegro-Vega et al., 2014). In this case, the fish must be bled from the caudal vein using syringes or sectioning the peduncle in order to minimize blood contamination in the gill mucus. After that, to remove any remaining blood from the organs, perfusion of the organs via puncture of the bulbous arteriosus needs to be performed with 0.9% (w/v) physiological saline containing 1 IU of heparin mL^{-1} until the gills are white. The operculum must be carefully removed from the fish and four gill arches (2nd and 3rd gill arches from left and right sides) must be placed into tubes containing 20 mL of mucus extraction buffer [2 mM PMSF (Phenylmethylsulfonyl fluoride), 10 mM EDTA, 0.02% (w/v) sodium azide in 0.85% (w/v) saline with 10 μL antiprotease cocktail] and preserved immediately on ice. On the same day, gill mucus samples must be centrifuged at 15,000 × *g* for 60 min at 4 °C, and the collected supernatant frozen at −20 °C or −80 °C before it can be used.

Recommendation: The volume of mucus extraction buffer depends on the species. Our research group has standardized the volume at 5 mL for gill mucus from gilthead seabream (266.35 ± 33.91 g weight and 24.32 ± 0.62 cm length) (unpublished data).

Both methods have advantages and disadvantages. Method 1 is a less invasive methodology since it would not be necessary to sacrifice the animal. Nevertheless, the risk of contamination by serum proteins can be considered high. In the case of Method 2, euthanasia of fish is necessary to collect the whole tissue while the blood contamination would be minimized by the applied procedure. A disadvantage of Method 2 would be the required size of the animals, taking into account that the perfusion procedure is complicated when working with fish around or smaller than 50–60 g.

5.4 General considerations

- Before storing the mucus samples, the protein concentration of each sample should be determined. For instance, the protein concentration can be determined by the dye binding method of Bradford (1976) using bovine serum albumin as the standard.
- Considering the shipment of mucus samples, the samples can be sent fresh (4 °C), frozen (−20 °C or −80 °C), or lyophilized (room temperature). However, according to the results of a study conducted by our research group (Cordero et al., 2016d), the levels of total proteins, sugar residues, and some immune-related enzymes are higher in fresh skin mucus samples than in samples after storage following freezing or lyophilization. In the case of samples stored before determining enzymatic activities, the authors suggest that freezing skin mucus samples is a more suitable methodology than lyophilization. Therefore, to ensure sample integrity, it is recommended that the mucus samples be shipped on dry ice (−80 °C).
- Taking into account that, for specific logistical reasons, the samples must be lyophilized to be preserved or shipped, one should consider the following indications. Once the mucus samples have been centrifuged, the collected supernatant can be lyophilized and stored at room temperature. Subsequently, the mucous powder must be dissolved with the same volume of Milli-Q water (to avoid increasing the content of salts in the sample), vigorously shaken, and centrifuged as described before (2000–4000 × *g* at 4 °C for 10–15 min), and then the undissolved mucus portion should be isolated by centrifugation.
- If the quantity of sample collected is not enough for the analysis raised, the samples can be homogenized with one to six volumes (considering the total protein concentration) of sterile Tris-buffered saline (50 mM Tris-HCl, 150 mM NaCl, pH 8.0) (Guardiola et al., 2014a, b; Mabrok et al., 2016) or a pool of supernatants should be done (not recommended).
- If the total protein concentration is too low for the requirements of any specific analysis, the supernatant of samples can be concentrated by ultrafiltration, for instance, using an Amicon YM3 (3 kDa) column (Palaksha et al., 2008; Schroers et al., 2009). For instance, the total antiprotease activity in the skin mucus of gilthead seabream was undetected unless protein concentrations higher than 2 mg mL^{-1} were used (Guardiola et al., 2014a, b).

5.5 Some examples of downstream analysis with mucus samples

Fish mucus surfaces are dynamic matrices generally composed of structural proteins, enzymes, immune-related proteins, metabolism-related proteins, secondary metabolites, and other molecules. The mucus composition varies among fish species and can be influenced by endogenous (e.g., sex and developmental stage) and exogenous factors [e.g., stress, environmental changes (salinity, water temperature, pH, etc.) and infections] (Esteban, 2012). Under stress conditions (e.g., handling stress, confinement, food deprivation, crowding, and exposure to toxic substances), the fish mucus can change their production and composition compromising fish health and increasing the fish susceptibility to pathogens. Therefore, and considering the important role of the mucus layer in fish health, the sampling and analysis of skin mucus-associated biomarkers have become an issue of interest as noninvasive indicators of fish status. For instance, several studies have investigated the quantification of fish cortisol levels in skin mucus as reliable acute stress biomarkers (Bertotto et al., 2010; Guardiola et al., 2016a; Carbajal et al., 2019; Franco-Martinez et al., 2019). The results provided evidence that skin mucus is a suitable matrix for measuring cortisol levels, acting as good stress indicators in fish exposed to several stress conditions. Similarly, classic indicators associated with the fish stress response, such as the levels of glucose and lactate, have also been detected in skin mucus after physiological challenges (Fernandez-Alacid et al., 2018, 2019). Similarly, the potential of the epidermal mucus as a noninvasive target for evaluating fish antioxidant status has been studied (Guardiola et al., 2018b). More concretely, the hydroxyl radical scavenging, hydrogen peroxide scavenging, ABTS [2,2′-azino-bis (3-ethylbenzothiazoline-6-sulfonic acid] cation scavenging, lipid peroxidation, and linoleic acid autoxidation were identified as good biomarkers for evaluating the antioxidant potential of dietary supplementation (Guardiola et al., 2018b). Similarly, several chemical attractants such as pheromones, sex steroids, prostaglandins, or related metabolites have also been detected in fish skin mucus (reviewed by Reverter et al., 2017) suggesting that it is valuable for reproduction and chemical communication studies.

Recent progress in "omics" technologies such as genomics, transcriptomics, proteomics, and metabolomics, integrated between them or combined with biochemical parameters, could provide the resources and tools needed to make significant progress in our understanding of the role of mucosal immunity and for biomarker discovery. For instance, recent studies published by Saleh et al. (2019) and Xiong et al. (2020) explored the proteome of fish skin mucus in response to pathogens, using label-free LC MS/MS approach, revealing differentially expressed proteins that likely related to a defense strategy in response to microbial infection. Other studies sequenced the DNA (microbiome characterization, microbiomics) of fish skin mucus evidencing their role as natural niches for mucosal pathogen evolution (Carda-Dierguez et al., 2017) and their use to monitor the impacts of pollution through analysis of the bacterial communities (Montenegro et al., 2020). The integration of transcriptomics and proteomics approaches to investigate the regulation of fish mucosal immune response to bacterial, viral, and parasitic infection has also been applied recently (reviewed by Ye et al., 2018). These technologies include high-throughput transcriptomics (e.g., microarray and RNA-Seq approaches) and RT-PCR, and proteomics analysis of gene expression products through shotgun proteomic technology, two-dimensional gel electrophoresis, western blotting, and mass spectrometry. To date, only some studies have developed metabolomic approaches in fish mucus, providing a snapshot of the state of an organism under specific conditions (Reverter et al., 2017). In fact, the results obtained reveal metabolomics as an extremely useful tool for understanding the processes and functions in which fish mucus is involved. In summary, the potential use of skin mucus as an easy, noninvasive, and reliable method to monitor fish welfare has been reported, verified, and validated through new and conventional experimental approaches. New uses as a noninvasive way to analyze fish status will be demonstrated in the short future when other biomolecules might be identified in skin mucus or new techniques applied for other purposes.

6. Main questions still to be addressed

The fish immune system is progressively seen as a highly diffuse organ comprising several organs (head-kidney, spleen, thymus, and MALT) that are interconnected by circulating leucocytes and soluble mediators. Little is still known about how fish immune cells are imprinted by their organ of residence to adjust to special regional needs. Similarly, we do not know the impact that immune cells have on their organs of origin or on their organ(s) of destiny. Future efforts should be directed toward understanding the peculiarities of the immunocytes originated in each of these organs, as well as their functions, both in those and other organs. Studies focused on fish immunology should integrate this crosstalk at all levels (organs, tissues, cells, and molecules).

Moreover, the human microbiome is viewed as an organ vital to health and disease, and it is tightly intertwined with the immune system (Zheng et al., 2020). Similarly, the relationships between fish microbiota and mucosal surfaces should be deeply studied. Fish mucosal surfaces are colonized by commensal microorganisms that live in symbiosis with their host. These bacteria have begun to be studied mostly in terms of presence or abundance. However, very little is known about the many ways in which their presence is beneficial to their host.

Advances in high-throughput multiomics technologies and bioinformatics will allow for the development of studies regarding the fish immune system on multiple levels. They will provide a huge amount of data which will finally result in the building of molecular multinetworks of the immune responses under normal (physiologic) and disease (pathological) conditions, which will be translated into better aquaculture practices.

Further, tools are necessary to evaluate and understand mucosal immunity in fish, such as cell lines or antibodies. This will allow the use of cell cultures of the mucosal surfaces to better understand their role, and it will also allow for control of the integument integrity and mucus production, which would help to reduce the use of fish specimens in experimentation and to adhere to the 3Rs guidelines. In addition, the availability of specific antibodies would help to identify, localize, and quantify the cells and molecules involved in mucosal surfaces, since only routine (hematoxylin-eosin, Mallory's trichrome, Giemsa, etc.) or carbohydrate-semi-specific (periodic acid of Schiff, Alcian blue, or lectin binding) microscopy techniques are available so far. Strikingly, mucus has been mostly used, but the exact role of immune-related molecules within it is not completely understood. Therefore when these molecules are present and if their function is related to protection, which cells produce them and how they are regulated are important questions to be addressed in the near future. Once the tools and methodologies to do so are improved for each fish species, the ways in which they are affected by the fish management during the aquaculture cycle and the different ways to strengthen them are key points of focus in order to use the mucosal surfaces as a way to improve and guarantee fish health and well-being.

Concerning the main protocols for collecting mucus from fish that have been summarized in this chapter, the need to standardize the methodologies is highlighted, as well as evidencing the importance of choosing a mucus sampling protocol regarding the intended purpose of each experiment. In fact, all data generated from fish mucus should be critically reviewed, taking into consideration both the strengths and weaknesses of the selected sampling methodology.

All these approaches will provide not only a better knowledge of fish immunity, but also a far more comprehensive perspective of vertebrate immunology, which will generate new ideas for the prevention and treatment of fish diseases. Through the improvement of immunity of fish, we will improve the performance of growing fish in farms in an eco-friendly and sustainable way.

Acknowledgments

This work was supported by the MINECO cofunded by the European Regional Development Funds (ERDF/FEDER) (grant no. AGL2017-88370-C3-1-R) and *Fundación Seneca de la Región de Murcia* (*Grupo de Excelencia* grant no. 19883/GERM/15).

References

Andersen, S.M., Waagbø, R., Espe, M., 2016. Functional amino acids in fish nutrition, health and welfare. Front. Biosci. 8, 143–169.

Awad, E., Awaad, A., 2017. Role of medicinal plants on growth performance and immune status in fish. Fish Shellfish Immunol. 67, 40–54.

Azeredo, R., Machado, M., Guardiola, F.A., Cerezuela, R., Afonso, A., Peres, H., Oliva-Teles, A., Esteban, M.A., Costas, B., 2017. Local immune response of two mucosal surfaces of the

European seabass, *Dicentrarchus labrax*, fed tryptophan- or methionine-supplemented diets. Fish Shellfish Immunol. 70, 76–86.

Beck, B., Peatman, E. (Eds.), 2015. Mucosal Health in Aquaculture. Elsevier (USA).

Ben Hamed, S., Guardiola, F.A., Martínez, S., Martínez-Sánchez, M.J., Pérez-Sirvent, C., Mars, M., Esteban, M.A., 2016. Exposure of the gilthead seabream (*Sparus aurata*) to sediments contaminated with heavy metals down-regulates the gene expression of stress biomarkers. Toxicol. Rep. 3, 364–372.

Ben Hamed, S., Guardiola, F.A., Cuesta, A., Martínez, S., Martínez-Sánchez, M.J., Pérez-Sirvent, C., Esteban, M.Á., 2017. Head kidney, liver and skin histopathology and gene expression in gilthead seabream (*Sparus aurata* L.) exposed to highly polluted marine sediments from Portman Bay (Spain). Chemosphere 174, 563–571.

Bengtén, E., Wilson, M., 2015. Antibody repertoires in fish. Results Probl. Cell Differ. 57, 193–234.

Bertotto, D., Poltronieri, C., Negrato, E., Majolini, D., Radaelli, G., Simontacchi, C., 2010. Alternative matrices for cortisol measurement in fish. Aquacult. Res. 41, 1261–1267.

Biller-Takahashi, J., Urbinati, E.C., 2014. Fish immunology. The modification and manipulation of the innate immune system: Brazilian studies. An. Acad. Bras. Cienc. 86, 1484–1506.

Blalock, J.E., Smith, E.M., 2007. Conceptual development of the immune system as a sixth sense. Brain Behav. Immun. 21, 23–33.

Boudinot, P., Langevin, C., Secombes, C.J., Levraud, J.P., 2016. The peculiar characteristics of fish type I interferons. Viruses 11, 298.

Bradford, M.M., 1976. A rapid and sensitive method for the quantitation of microgram quantities of protein utilizing the principle of protein-dye binding. Anal. Biochem. 72, 248–254.

Carbajal, A., Reyes-López, F.E., Tallo-Parra, O., Lopez-Bejar, M., Tort, L., 2019. Comparative assessment of cortisol in plasma, skin mucus and scales as a measure of the hypothalamic-pituitary-interrenal axis activity in fish. Aquaculture 506, 410–416.

Carda-Dierguez, M., Ghai, R., Rodriguez-Valera, F., Amaro, C., 2017. Wild eel microbiome reveals that skin mucus of fish could be a natural niche for aquatic mucosal pathogen evolution. Microbiome 5, 162.

Ceballos-Francisco, D., Cordero, H., Guardiola, F.A., Cuesta, A., Esteban, M.Á., 2017. Healing and mucosal immunity in the skin of experimentally wounded gilthead seabream (*Sparus aurata* L.). Fish Shellfish Immunol. 71, 210–219.

Cerezuela, R., Fumanal, M., Tapia-Paniagua, S.T., Meseguer, J., Morinigo, M.Á., Esteban, M.Á., 2012a. Histological alterations and microbial ecology of the intestine in gilthead seabream (*Sparus aurata* L.) fed dietary probiotics and microalgae. Cell Tissue Res. 350, 477–489.

Cerezuela, R., Guardiola, F.A., Meseguer, J., Esteban, M.A., 2012b. Increases in immune parameters by inulin and *Bacillus subtilis* dietary administration to gilthead seabream (*Sparus aurata* L.) did not correlate with disease resistance to *Photobacterium damselae*. Fish Shellfish Immunol. 32, 1032–1040.

Cerezuela, R., Fumanal, M., Tapia-Paniagua, S.T., Meseguer, J., Moriñigo, M.A., Esteban, M.A., 2013a. Changes in intestinal morphology and microbiota caused by dietary administration of inulin and *Bacillus subtilis* in gilthead seabream (*Sparus aurata* L.) specimens. Fish Shellfish Immunol. 34, 1063–1070.

Cerezuela, R., Meseguer, J., Esteban, M.Á., 2013b. Effects of dietary inulin, *Bacillus subtilis* and microalgae on intestinal gene expression in gilthead seabream (*Sparus aurata* L.). Fish Shellfish Immunol. 34, 843–848.

Cerezuela, R., Guardiola, F.A., Cuesta, A., Esteban, M.A., 2016. Enrichment of gilthead seabream (*Sparus aurata* L.) diet with palm fruit extracts and probiotics: effects on skin mucosal immunity. Fish Shellfish Immunol. 49, 100–109.

Chaves-Pozo, E., Valero, Y., Esteve-Codina, A., Gómez-Garrido, J., Dabad, M., Alioto, T., Meseguer, J., Esteban, M.Á., Cuesta, A., 2017. Innate cell-cediated cytotoxic activity of European sea bass leucocytes against nodavirus-infected cells: a functional and RNA-seq study. Sci. Rep. 7, 15396.

Chaves-Pozo, E., Bandín, I., Olveira, J.G., Esteve-Codina, A., Gómez-Garrido, J., Dabad, M., Alioto, T., Esteban, M.A., Cuesta, A., 2019. European sea bass brain DLB-1cell line is susceptible to nodavirus: a transcriptomic study. Fish Shellfish Immunol. 86, 14–24.

Chen, G.Y., Nuñez, G., 2010. Sterile inflammation: sensing and reacting to damage. Nat. Rev. Immunol. 10, 826–837.

Chen, Z., Ceballos-Francisco, D., Guardiola, F.A., Esteban, M.Á., 2020a. Influence of skin wounds on the intestinal inflammatory response and barrier function: protective role of dietary *Shewanella putrefaciens* SpPdp11 administration to gilthead seabream (*Sparus aurata* L.). Fish Shellfish Immunol. 99, 414–423.

Chen, Z., Ceballos-Francisco, D., Guardiola, F.A., Huang, D., Esteban, M.Á., 2020b. Skin wound healing in gilthead seabream (*Sparus aurata* L.) fed diets supplemented with arginine. Fish Shellfish Immunol. 104, 347–358.

Cone, R.A., 2009. Barrier properties of mucus. Adv. Drug Deliv. Rev. 61, 75–85.

Cordero, H., Brinchmann, M.F., Cuesta, A., Meseguer, J., Esteban, M.A., 2015. Skin mucus proteome map of European sea bass (*Dicentrarchus labrax*). Proteomics 15, 4007–4020.

Cordero, H., Guzmán-Villanueva, L.T., Chaves-Pozo, E., Arizcun, M., Ascencio-Valle, F., Cuesta, A., Esteban, M.A., 2016a. Comparative ontogenetic development of two marine teleosts, gilthead seabream and European sea bass: new insights into nutrition and immunity. Dev. Comp. Immunol. 65, 1–17.

Cordero, H., Cuesta, A., Meseguer, J., Esteban, M.A., 2016b. Characterization of the gilthead seabream (*Sparus aurata* L.) immune response under a natural lymphocystis disease virus outbreak. J. Fish Dis. 39, 1467–1476.

Cordero, H., Morcillo, P., Cuesta, A., Brinchmann, M.F., Esteban, M.A., 2016c. Differential proteome profile of skin mucus of gilthead seabream (*Sparus aurata*) after probiotic intake and/or overcrowding stress. J. Proteomics 132, 41–50.

Cordero, H., Cuesta, A., Meseguer, J., Esteban, M.Á., 2016d. Changes in the levels of humoral immune activities after storage of gilthead seabream (*Sparus aurata*) skin mucus. Fish Shellfish Immunol. 58, 500–507.

Cordero, H., Ceballos-Francisco, D., Cuesta, A., Esteban, M.Á., 2017a. Dorso-ventral skin characterization of the farmed fish gilthead seabream (*Sparus aurata*). PLoS One 12, e0180438.

Cordero, H., Brinchmann, M.F., Cuesta, A., Esteban, M.A., 2017b. Chronic wounds alter the proteome profile in skin mucus of farmed gilthead seabream. BMC Genomics 18, 939.

Costa, A.A., Leef, M.J., Bridle, A.R., Carson, J., Nowak, B.F., 2011. Effect of vaccination against yersiniosis on the relative percent survival, bactericidal and lysozyme response of Atlantic salmon, *Salmo salar*. Aquaculture 315, 201–206.

Cuesta, A., Esteban, M.A., Meseguer, J., 1999. Natural cytotoxic activity of gilthead seabream (*Sparus aurata* L.) leucocytes assessment by flow cytometry and microscopy. Vet. Immunol. Immunopathol. 71, 161–171.

Das, P.K., Salinas, I., 2020. Fish nasal immunity: from mucosal vaccines to neuroimmunology. Fish Shellfish Immunol. 104, 165–171.

Díaz-Rosales, P., Salinas, I., Rodríguez, A., Cuesta, A., Chabrillón, M., Balebona, M.C., Moriñigo, M.A., Esteban, M.A., Meseguer, J., 2006. Gilthead seabream (*Sparus aurata* L.) innate immune response after dietary administration of heat-inactivated potential probiotics. Fish Shellfish Immunol. 20, 482–492.

Escribano, M.P., Ramos-Pinto, L., Fernández-Boo, S., Afonso, A., Costas, B., Guardiola, F.A., 2020. Mucosal immune responses in Senegalese sole (*Solea senegalensis*) juveniles after *Tenacibaculum maritimum* challenge: a comparative study between ocular and blind sides. Fish Shellfish Immunol. 104, 92–100.

Espinosa, C., Cuesta, A., Esteban, M.Á., 2017. Effects of dietary polyvinylchloride microparticles on general health, immune status and expression of several genes related to stress in gilthead seabream (*Sparus aurata* L.). Fish Shellfish Immunol. 68, 251–259.

Espinosa, C., Esteban, M.Á., Cuesta, A., 2019. Dietary administration of PVC and PE microplastics produces histological damage, oxidative stress and immunoregulation in European sea bass (*Dicentrarchus labrax* L.). Fish Shellfish Immunol. 95, 574–583.

Esteban, M., 2012. An overview of the immunological defenses in fish skin. ISRN Immunol. 2012, 1–29.

Esteban, M.A., Meseguer, J., Ayala, A.G., Agulleiro, B., 1989. Erythropoiesis and thrombopoiesis in the head-kidney of the sea bass (*Dicentrarchus labrax* L.): an ultrastructural study. Arch. Histol. Cytol. 52, 407–419.

Esteban, M.A., Mulero, V., Muñoz, J., Meseguer, J., 1998. Methodological aspects of assessing phagocytosis of *Vibrio anguillarum* by leucocytes of gilthead seabream (*Sparus aurata* L.) by flow cytometry and electron microscopy. Cell Tissue Res. 293, 133–141.

Esteban, M.A., Chaves-Pozo, E., Arizcun, M., Meseguer, J., Cuesta, A., 2013. Regulation of natural killer enhancing factor (NKEF) genes in teleost fish, gilthead seabream and European sea bass. Mol. Immunol. 55, 275–282.

Esteban, M.A., Cordero, H., Martínez-Tomé, M., Jiménez-Monreal, A.M., Bakhrouf, A., Mahdhi, A., 2014. Effect of dietary supplementation of probiotics and palm fruits extracts on the antioxidant enzyme gene expression in the mucosae of gilthead seabream (*Sparus aurata* L.). Fish Shellfish Immunol. 39, 532–540.

Estensoro, I., Jung-Schroers, V., Álvarez-Pellitero, P., Steinhagen, D., Sitjà-Bobadilla, A., 2013. Effects of *Enteromyxum leei* (Myxozoa) infection on gilthead sea bream (*Sparus aurata*) (Teleostei) intestinal mucus: glycoprotein profile and bacterial adhesion. Parasitol. Res. 112, 567–576.

FAO, 2020. The State of World Fisheries and Aquaculture 2020. Sustainability in Action.

Fernandez-Alacid, L., Sanahuja, I., Ordóñez-Grande, B., Sánchez-Nuño, S., Viscor, G., Gisbert, E., Herrera, M., Ibarz, A., 2018. Skin mucus metabolites in response to physiological challenges: a valuable non-invasive method to study teleost marine species. Sci. Total Environ. 644, 1323–1335.

Fernandez-Alacid, L., Sanahuja, I., Ordonez-Grande, B., Sanchez-Nuno, S., Herrera, M., Ibarz, A., 2019. Skin mucus metabolites and cortisol in meagre fed acute stress-attenuating diets: correlations between plasma and mucus. Aquaculture 499, 185–194.

Flajnik, M.F., 2018. A cold-blooded view of adaptive immunity. Nat. Rev. Immunol. 18, 438–453.

Franco-Martinez, L., Tvarijonaviciute, A., Martinez-Subiela, S., Teles, M., Tort, L., 2019. Chemiluminescent assay as an alternative to radioimmunoassay for the measurement of cortisol in plasma and skin mucus of *Oncorhynchus mykiss*. Ecol. Indic. 98, 634–640.

Galina, J., Yin, G., Ardó, L., Jeney, Z., 2009. The use of immunostimulating herbs in fish. An overview of research. Fish Physiol. Biochem. 35, 669–676.

García-Beltrán, J.M., Silvera, D.G., Ruiz, C.E., Campo, V., Chupani, L., Faggio, C., Esteban, M.Á., 2020. Effects of dietary *Origanum vulgare* on gilthead seabream (*Sparus aurata* L.) immune and antioxidant status. Fish Shellfish Immunol. 99, 452–461.

Glover, C.N., Hogstrand, C., 2002. Amino acid modulation of *in vivo* intestinal zinc absorption in freshwater rainbow trout. J. Exp. Biol. 205, 151–158.

Gómez, D., Sunyer, J.O., Salinas, I., 2013. The mucosal immune system of fish: the evolution of tolerating commensals while fighting pathogens. Fish Shellfish Immunol. 35, 1729–1739.

Grayfer, L., Kerimoglu, B., Yaparla, A., Hodgkinson, J.W., Xie, J., Belosevic, M., 2018. Mechanisms of fish macrophage antimicrobial immunity. Front. Immunol. 9, 1105.

Guardiola, F.A., Cuesta, A., Abellán, E., Meseguer, J., Esteban, M.A., 2014a. Comparative analysis of the humoral immunity of skin mucus from several marine teleost fish. Fish Shellfish Immunol. 40, 24–31.

Guardiola, F.A., Cuesta, A., Arizcun, M., Meseguer, J., Esteban, M.A., 2014b. Comparative skin mucus and serum humoral defence mechanisms in the teleost gilthead seabream (*Sparus aurata*). Fish Shellfish Immunol. 36, 545–551.

Guardiola, F.A., Cuartero, M., Collado-González, M.M., Arizcún, M., Díaz Baños, F.G., Meseguer, J., Cuesta, A., Esteban, M.A., 2015a. Description and comparative study of physico-chemical parameters of the teleost fish skin mucus. Biorheology 52, 247–256.

Guardiola, F.A., de Haro, J.P., Díaz-Baños, F.G., Meseguer, J., Cuesta, A., Esteban, M.Á., 2015b. Terminal carbohydrate composition, IgM level and enzymatic and bacteriostatic activity of European sea bass (*Dicentrarchus labrax*) skin epidermis extracts. Fish Shellfish Immunol. 47, 352–359.

Guardiola, F.A., Dioguardi, M., Parisi, M.G., Trapani, M.R., Meseguer, J., Cuesta, A., Cammarata, M., Esteban, M.A., 2015c. Evaluation of waterborne exposure to heavy metals in innate immune defences present on skin mucus of gilthead seabream (*Sparus aurata*). Fish Shellfish Immunol. 45, 112–123.

Guardiola, F.A., Cuesta, A., Esteban, M., 2016a. Using skin mucus to evaluate stress in gilthead seabream (*Sparus aurata* L.). Fish Shellfish Immunol. 59, 323–330.

Guardiola, F.A., Porcino, C., Cerezuela, R., Cuesta, A., Faggio, C., Esteban, M.A., 2016b. Impact of date palm fruits extracts and probiotic enriched diet on antioxidant status, innate immune response and immune-related gene expression of European seabass (*Dicentrarchus labrax*). Fish Shellfish Immunol. 52, 298–308.

Guardiola, F.A., Bahi, A., Bakhrouf, A., Esteban, M.A., 2017. Effects of dietary supplementation with fenugreek seeds, alone or in combination with probiotics, on gilthead seabream (*Sparus aurata* L.) skin mucosal immunity. Fish Shellfish Immunol. 65, 169–178.

Guardiola, F.A., Saraiva-Fraga, M., Cuesta, A., Esteban, M.A., 2018a. Changes in natural haemolytic complement activity induced by stress in gilthead seabream (*Sparus aurata* L.). Fish Shellfish Immunol. 78, 317–321.

Guardiola, F.A., Bahi, A., Jiménez-Monreal, A.M., Martínez-Tomé, M., Murcia, M.A., Esteban, M.A., 2018b. Dietary administration effects of fenugreek seeds on skin mucosal antioxidant and immunity status of gilthead seabream (*Sparus aurata* L.). Fish Shellfish Immunol. 75, 357–364.

Guardiola, F.A., Mabrok, M., Machado, M., Azeredo, R., Afonso, A., Esteban, M.A., Costas, B., 2019. Mucosal and systemic immune responses in Senegalese sole (*Solea senegalensis* Kaup) bath challenged with *Tenacibaculum maritimum*: a time-course study. Fish Shellfish Immunol. 87, 744–754.

Guluarte, C., Reyes-Becerril, M., Gonzalez-Silvera, D., Cuesta, A., Angulo, C., Esteban, M.Á., 2019. Probiotic properties and fatty acid composition of the yeast *Kluyveromyces lactis* M3. *In vivo* immunomodulatory activities in gilthead seabream (*Sparus aurata*). Fish Shellfish Immunol. 94, 389–397.

Hatten, F., Fredriksen, Å., Hordvik, I., Endresen, C., 2001. Presence of IgM in cutaneous mucus, but not in gut mucus of Atlantic salmon, *Salmo salar*. Serum IgM is rapidly degraded when added to gut mucus. Fish Shellfish Immunol. 11, 257–268.

Hodgkinson, J.W., Grayfer, L., Belosevic, M., 2015. Biology of bony fish macrophages. Biology (Basel) 4, 881–906.

Jurado, J., Fuentes-Almagro, C.A., Guardiola, F.A., Cuesta, A., Esteban, M.Á., Prieto-Álamo, M.J., 2015. Proteomic profile of the skin mucus of farmed gilthead seabream *(Sparus aurata)*. J. Proteomics 120, 21–34.

Kaufmann, S.H.E., 2008. Immunology's foundation: the 100-year anniversary of the Nobel prize to Paul Ehrlich and Elie Metchnikoff. Nat. Immunol. 9, 705–712.

Kaufmann, S.H.E., 2017. Emil von Behring: translational medicine at the dawn of immunology. Nat. Rev. Immunol. 17, 341–343.

Kaufmann, S.H.E., Winau, F., 2005. From bacteriology to immunology: the dualism of specificity. Nat. Immunol. 6, 1063–1066.

Khan, F.R., McGeer, J.C., 2013. Zn-stimulated mucus secretion in the rainbow trout (*Oncorhynchus mykiss*) intestine inhibits Cd accumulation and Cd-induced lipid peroxidation. Aquat. Toxicol. 142–143, 17–25.

Khansari, A.R., Balasch, J.C., Vallejos-Vidal, E., Parra, D., Reyes-López, F.E., Tort, L., 2018. Comparative immune- and stress-related transcript response induced by air exposure and *Vibrio anguillarum* bacterin in rainbow trout (*Oncorhynchus mykiss*) and gilthead seabream (*Sparus aurata*) mucosal surfaces. Front. Immunol. 9, 856.

Lazado, C.C., Caipang, C.M.A., 2014. Mucosal immunity and probiotics in fish. Fish Shellfish Immunol. 39, 78–89.

Lieschke, G.J., Trede, N.S., 2009. Fish immunology. Curr. Biol. 19, R678–R682.

López-Ruíz, A., Esteban, M.A., Meseguer, J., 1992. Blood cells of the gilthead seabream (*Sparus aurata* L.): light and electron microscopic studies. Anat. Rec. 234, 161–171.

Mabrok, M., Machado, M., Serra, C.R., Afonso, A., Valente, L.M.P., Costas, B., 2016. Tenacibaculosis induction in the Senegalese sole (*Solea senegalensis*) and studies of *Tenacibaculum maritimum* survival against host mucus and plasma. J. Fish Dis. 39, 1445–1455.

Manning, M., 1998. Immune defence systems. In: Black, K., Pickering, A. (Eds.), Biology of Farmed Fish. Academic Press, Sheffield, pp. 180–221.

Mansour, A.T., Espinosa, C., García-Beltrán, J.M., Miao, L., Ceballos Francisco, D.C., Alsaqufi, A.S., Esteban, M.Á., 2020. Dietary supplementation of drumstick tree, *Moringa oleifera*, improves mucosal immune response in skin and gills of seabream, *Sparus aurata*, and attenuates the effect of hydrogen peroxide exposure. Fish Physiol. Biochem. 46, 981–996.

Marcos-López, M., Espinosa Ruiz, C., Rodger, H.D., O'Connor, I., MacCarthy, E., Esteban, M.Á., 2017. Local and systemic humoral immune response in farmed Atlantic salmon (*Salmo salar* L.) under a natural amoebic gill disease outbreak. Fish Shellfish Immunol. 66, 207–216.

Meseguer, J., Esteban, M.A., Ayala, A.G., Ruiz, A.L., Agulleiro, B., 1990. Granulopoiesis in the head-kidney of the sea bass (*Dicentrarchus labrax*): an ultrastructural study. Arch. Histol. Cytol. 53, 287–296.

Meseguer, J., Esteban, M.A., Agulleiro, B., 1991. Stromal cells, macrophages and lymphoid cells in the head-kidney of sea bass (*Dicentrarchus labrax* L.). An ultrastructural study. Arch. Histol. Cytol. 54, 299–309.

Meseguer, J., Esteban, M.A., Muñoz, J., López Ruíz, A., 1993. Ultrastructure of the peritoneal exudate cells (PEC) of seawater teleosts, gilthead seabream (*Sparus aurata* L.) and sea bass (*Dicentrarchus labrax* L.). Cell Tissue Res. 273, 301–307.

Meseguer, J., Esteban, M.A., López Ruíz, A., Muñoz, J., 1994. Melano-macrophages of seawater teleosts of sea bass (*Dicentrarchus labrax* L.) and seabream (*Sparus aurata* L.): structural, histochemical and morphofunctional study. Cell Tissue Res. 277, 1–10.

Montenegro, D., Astudillo-García, C., Hickey, T., Lear, G., 2020. A non-invasive method to monitor marine pollution from bacterial DNA present in fish skin mucus. Environ. Pollut. 263, 114438.

Nayak, S.K., 2010. Probiotics and immunity: a fish perspective. Fish Shellfish Immunol. 29, 2–14.

Nigam, A.K., Kumari, U., Mittal, S., Mittal, A.K., 2012. Comparative analysis of innate immune parameters of the skin mucous secretions from certain freshwater teleosts, inhabiting different ecological niches. Fish Physiol. Biochem. 38, 1245–1256.

Noya, M., Lamas, J., 1996. Response of eosinophilic granule cells of gilthead seabream (*Sparus aurata*, Teleostei) to bacteria and bacterial products. Cell Tissue Res. 287, 223–230.

Oliva-Teles, A., 2012. Nutrition and health of aquaculture fish. J. Fish Dis. 35, 83–108.

Ortuño, J., Esteban, M.A., Mulero, V., Meseguer, J., 1998. Methods applied to study the haemolytic, bactericidal, chemoattractant and opsonic activities of seabream (*Sparus aurata*) serum. In: Barnes, A., Davidson, G., Hiney, M., McIntosch, D. (Eds.), Methodology in Fish Diseases Research. MFDR Publications, Albion Press, Aberdeen (U.K.), pp. 97–100.

Palaksha, K.J., Shin, G.W., Kim, Y.R., Jung, T.S., 2008. Evaluation of non-specific immune components from the skin mucus of olive flounder (*Paralichthys olivaceus*). Fish Shellfish Immunol. 24, 479–488.

Piazzon, M.C., Galindo-Villegas, J., Pereiro, P., Estensoro, I., Calduch-Giner, J.A., Gómez-Casado, E., Novoa, B., Mulero, V., Sitjà-Bobadilla, A., Pérez-Sánchez, J., 2016. Differential modulation of IgT and IgM upon parasitic, bacterial, viral, and dietary challenges in a perciform fish. Front. Immunol. 7, 1–16.

Rebl, A., Goldammer, T., 2018. Under control: the innate immunity of fish from the inhibitors' perspective. Fish Shellfish Immunol. 77, 328–349.

Redondo, M.J., Alvarez-Pellitero, P., 2010. Carbohydrate patterns in the digestive tract of *Sparus aurata* L. and *Psetta maxima* (L.) (Teleostei) parasitized by *Enteromyxum leei* and *E. scophthalmi* (Myxozoa). Parasitol. Int. 59, 445–453.

Reverter, M., Sasal, P., Banaigs, B., Lecchini, D., Lecellier, G., Tapissier-Bontemps, N., 2017. Fish mucus metabolome reveals fish life-history traits. Coral Reefs 36, 463–475.

Reyes-Becerril, M., Guluarte, C., Ceballos-Francisco, D., Angulo, C., Esteban, M.Á., 2017. Dietary yeast *Sterigmatomyces halophilus* enhances mucosal immunity of gilthead seabream (*Sparus aurata* L.). Fish Shellfish Immunol. 64, 165–175.

Ross, N.W., Firth, K.J., Wang, A., Burka, J.F., Johnson, S.C., 2000. Changes in hydrolytic enzyme activities of naive Atlantic salmon *Salmo salar* skin mucus due to infection with the salmon louse *Lepeophtheirus salmonis* and cortisol implantation. Dis. Aquat. Organ. 41, 43–51.

Russell, S., Lumsden, J.S., 2005. Function and heterogeneity of fish lectins. Vet. Immunol. Immunopathol. 108, 111–120.

Saleh, M., Kumar, G., Abdel-Baki, A.A.S., Dkhil, M.A., El-Matbouli, M., Al-Quraishy, S., 2019. Quantitative proteomic profiling of immune responses to *Ichthyophthirius multifiliis* in common carp skin mucus. Fish Shellfish Immunol. 84, 834–842.

Salinas, I., 2015. The mucosal immune system of teleost fish. Biology (Basel). 4, 525–539.

Salinas, I., Parra, D., 2015. Fish mucosal immunity: intestine. In: Beck, B., Peatman, E. (Eds.), Mucosal Health in Aquaculture. Academic Press, pp. 135–170.

Salinas, I., Cuesta, A., Esteban, M.A., Meseguer, J., 2005. Dietary administration of *Lactobacillus delbrüeckii* and *Bacillus subtilis*, single or combined, on gilthead seabream cellular innate immune responses. Fish Shellfish Immunol. 19, 67–77.

Salinas, I., Meseguer, J., Esteban, M.Á., 2007. Assessment of different protocols for the isolation and purification of gut associated lymphoid cells from the gilthead seabream (*Sparus aurata* L.). Biol. Proced. Online 9, 43–55.

Salinas, I., Abelli, L., Bertoni, F., Picchietti, S., Roque, A., Furones, D., Cuesta, A., Meseguer, J., Esteban, M.Á., 2008. Monospecies and multispecies probiotic formulations produce different systemic and local immunostimulatory effects in the gilthead seabream (*Sparus aurata* L.). Fish Shellfish Immunol. 25, 114–123.

Salinas, I., Zhang, Y.A., Sunyer, J.O., 2011. Mucosal immunoglobulins and B cells of teleost fish. Dev. Comp. Immunol. 35, 1346–1365.

Scapigliati, G., 2013. Functional aspects of fish lymphocytes. Dev. Comp. Immunol. 41, 200–208.

Schroers, V., van der Marel, M., Neuhaus, H., Steinhagen, D., 2009. Changes of intestinal mucus glycoproteins after peroral application of *Aeromonas hydrophila* to common carp (*Cyprinus carpio*). Aquaculture 288, 184–189.

Secombes, C.J., Bird, S., Zou, J., 2005. Adaptive immunity in teleosts: cellular immunity. Dev. Biol. 121, 25–32.

Secombes, C.J., Wang, T., Bird, S., 2011. The interleukins of fish. Dev. Comp. Immunol. 35, 1336–1345.

Shephard, K.L., 1994. Functions for fish mucus. Rev. Fish Biol. Fish. 4, 401–429.

Song, S.K., Beck, B.R., Kim, D., Park, J., Kim, J., Kim, H.D., Ringø, E., 2014. Prebiotics as immunostimulants in aquaculture: a review. Fish Shellfish Immunol. 40, 40–48.

Subramanian, S., Ross, N.W., MacKinnon, S.L., 2008. Comparison of antimicrobial activity in the epidermal mucus extracts of fish. Comp. Biochem. Physiol. B Biochem. Mol. Biol. 150, 85–92.

Tapia-Paniagua, S.T., Ceballos-Francisco, D., Balebona, M.C., Esteban, M.Á., Moriñigo, M.Á., 2018. Mucus glycosylation, immunity and bacterial microbiota associated to the skin of experimentally ulcered gilthead seabream (*Sparus aurata*). Fish Shellfish Immunol. 75, 381–390.

Torrecillas, S., Makol, A., Caballero, M.J., Montero, D., Dhanasiri, A.K.S., Sweetman, J., Izquierdo, M., 2012. Effects on mortality and stress response in European sea bass, *Dicentrarchus labrax* (L.), fed mannan oligosaccharides (MOS) after *Vibrio anguillarum* exposure. J. Fish Dis. 35, 591–602.

Torrecillas, S., Makol, A., Betancor, M.B., Montero, D., Caballero, M.J., Sweetman, J., Izquierdo, M., 2013. Enhanced intestinal epithelial barrier health status on European sea bass (*Dicentrarchus labrax*) fed mannan oligosaccharides. Fish Shellfish Immunol. 34, 1485–1495.

Torrecillas, S., Montero, D., Caballero, M.J., Pittman, K., Campo, A., Custodio, M., Sweetman, J., Izquierdo, M.S., 2015. Dietary mannan oligosaccharides: counteracting the side effects of soybean oil inclusion on European sea bass (*Dicentrarchus labrax*) gut health? Front. Immunol. 6, 397.

Valdenegro-Vega, V.A., Crosbie, P., Bridle, A., Leef, M., Wilson, R., Nowak, B.F., 2014. Differentially expressed proteins in gill and skin mucus of Atlantic salmon (*Salmo salar*) affected by amoebic gill disease. Fish Shellfish Immunol. 40, 69–77.

Valero, Y., Saraiva-Fraga, M., Costas, B., Guardiola, F.A., 2020. Antimicrobial peptides from fish: beyond the fight against pathogens. Rev. Aquacult. 12, 224–253.

Van Muiswinkel, W.B., 2008. A history of fish immunology and vaccination I. The early days. Fish Shellfish Immunol. 25, 397–408.

Vincent, B.N., Morrison, R.N., Nowak, B.F., 2006. Amoebic gill disease (AGD)-affected Atlantic salmon, *Salmo salar* L., are resistant to subsequent AGD challenge. J. Fish Dis. 29, 549–559.

Whyte, S.K., 2007. The innate immune response of finfish—a review of current knowledge. Fish Shellfish Immunol. 23, 1127–1151.

Xiong, Y., Dan, C., Ren, F., Su, Z., Zhang, Y., Mei, J., 2020. Proteomic profiling of yellow catfish (*Pelteobagrus fulvidraco*) skin mucus identifies differentially-expressed proteins in response to *Edwardsiella ictaluri* infection. Fish Shellfish Immunol. 100, 98–108.

Ye, J., Kaattari, I.M., Ma, C., Kaattari, S., 2013. The teleost humoral immune response. Fish Shellfish Immunol. 35, 1719–1728.

Ye, H., Lin, Q., Luo, H., 2018. Applications of transcriptomics and proteomics in understanding fish immunity. Fish Shellfish Immunol. 77, 319–327.

Zheng, D., Liwinski, T., Elinav, E., 2020. Interaction between microbiota and immunity in health and disease. Cell Res. 30, 492–506.

Chapter 9

Characterization of the neuroendocrine stress status as part of the multiparametric assessment of welfare in fish

Manuel Gesto
Section for Aquaculture, National Institute of Aquatic Resources, Technical University of Denmark, Hirtshals, Denmark

1. Animal welfare

The concept of "animal welfare" refers to the physical and psychological well-being of animals and the focus on its study has been mostly developed during the last three or four decades. Humans have reared and domesticated animals for millennia for a variety of purposes (food, clothing, agricultural work, pets), but the research focused on animals as sentient beings, able to suffer, was only initiated during the second half of the 20th century, probably due to our better understanding about the animals' motivation, cognition, and behavioral complexity (Broom, 2011). Research on animal welfare was initially focused on livestock and laboratory animals, mostly birds and mammals, but has been now extended to cover fish and other vertebrates and even certain invertebrate groups such as cephalopods, crustaceans, and others (Benn et al., 2019; Branson, 2008; Carere and Mather, 2019; Diggles, 2019; Ponte et al., 2019). This has resulted from the increasing body of evidence pointing that those animals have complex cognitive abilities and might be also sentient (Vila Pouca and Brown, 2017). In fish, the increased awareness about fish sentience, together with the extensive current use of fish by humans for food (from fisheries and aquaculture) and other purposes (research, domestic pets, and/or exposition in zoos and aquaria), might explain the growing interest in fish welfare research (Fig. 1). Still, in spite of this recent interest, research studies published on fish welfare represent less than 10% of the studies dealing with animal welfare.

Different definitions of welfare exist (Huntingford and Kadri, 2014) and actually, the exact implications of welfare, and the measures to take to keep the animals in good welfare conditions, will depend on the specific definition of

Cellular and Molecular Approaches in Fish Biology. https://doi.org/10.1016/B978-0-12-822273-7.00001-X

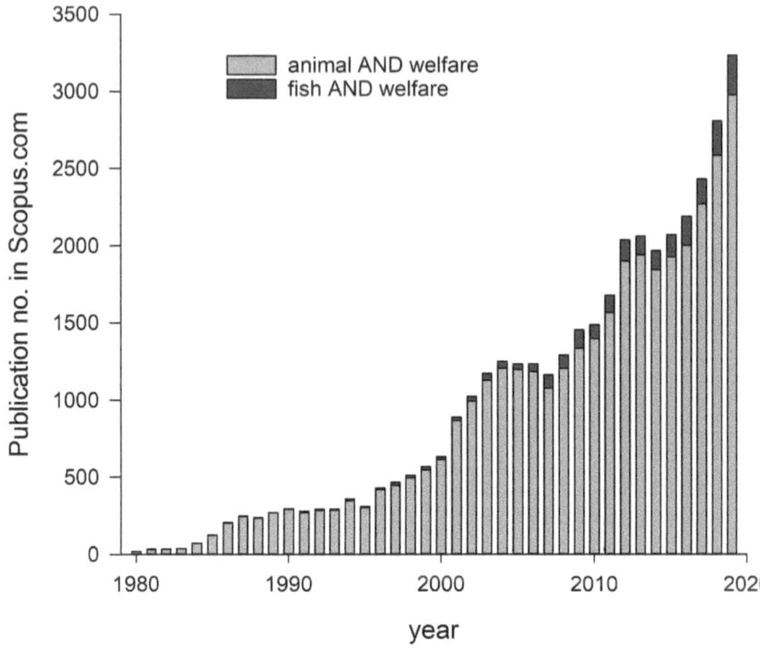

FIG. 1 Number of publications found in Scopus.com by searching the keywords "animal AND welfare." The corresponding share of "fish AND welfare" publications is marked in *dark gray*.

welfare that is adopted. Most definitions of welfare belong to one of three broad categories (Huntingford et al., 2006): (1) the feelings-based definitions consider that good welfare exists only when the animals feel well; (2) the function-based definitions consider that good welfare exists if the animal is in a good physiological state, and all biological functions perform well; and (3) nature-based definitions assume that animals need to be able to live a "natural life," developing the same patterns of behavior that they would have in the wild. These three categories are not mutually exclusive and definitions that combine aspects of the different approaches are common (Sneddon et al., 2016).

No matter the definition considered, the welfare of fish intended for human use is critically important for several reasons (Seibel et al., 2020). In relation to fish farming, welfare is critical for economic reasons since productivity traits such as fish growth are highly dependent on the welfare status of the fish. Besides, avoiding unnecessary animal suffering when capturing, rearing, and slaughtering fish is important according to current ethics standards on the use of animals. Finally, national and international regulations on the use of fish, currently increasing in number and strictness as a result of both ethical standards and available information on fish physiology and behavior, constitute important legal reasons to maintain fish welfare at adequate standards.

1.1 Current knowledge gaps related to fish welfare

While fish welfare is a hot research topic, there are important gaps in our knowledge about the subjective life quality of both captive and wild fish. Furthermore, practical measures, procedures, and strategies to quantify fish welfare are still under development and under debate. Different sets of recommendations or guidelines to promote and evaluate the welfare of common aquaculture species have been recently released. Those include guidelines published by research entities (Noble et al., 2018, 2020) or even by charities such as the RSPCA (Royal Society for the Prevention of Cruelty to Animals) in the United Kingdom, among others (Kristiansen and Bracke, 2020; RSPCA, 2018). The protection of fish by national regulations is uneven (Kristiansen and Bracke, 2020) but increasing, and those kind of guidelines might well end up being the basis of new regulations for fish welfare monitoring under captivity.

Current and future research will help to better answer questions that might seem simple, but have no clear answer such as, for example: Which are the most adequate variables to quantify/evaluate the welfare of wild and farmed fish? What levels of those variables could be considered "good welfare"? What levels should be considered realistic and ethically and legally acceptable in fish farms? How can we compare the welfare state of fish from different sites, farms, and/or environmental conditions, or even from different species? The answers to these questions are in fact very complex due to the highly diverse (inter- and intraspecific) responses of fish to their environment. Just comparing, for example, the welfare of wild versus captive fish is extraordinarily complex: have wild fish a better welfare than farmed fish? Life in captivity can provide very positive conditions for fish performance and survival, when compared to the wild environment, but it comes at the cost of different constraints, including captivity itself, lack of control of the social environment, and reduced environmental diversity, among others. Also, the severity of those constraints varies with the levels of adaptation and, for example, it is believed to decrease with domestication. In general, getting a clear idea of the subjective view of the fish about their environment and their life is extremely complicated, if not impossible, and therefore, efforts to find indirect indicators about their state are currently the basis of fish welfare research.

Among the main difficulties of evaluating fish welfare, finding reliable indicators is one of the most critical. Generally, acute welfare problems might be relatively easy to detect, but things get more difficult when the fish are exposed to prolonged or chronic disturbances of low severity. In those situations, compromised welfare is not always conspicuous. As commented in the following section, stress indicators are often used to evaluate the welfare status in fish, but again, most common stress indicators are not always reliable in conditions of prolonged or chronic stress and/or poor welfare.

1.2 Evaluation of fish welfare

For the evaluation of fish welfare, particularly in the case of farmed fish, most guidelines suggest a double, complementary approach, including the assessment of both the environmental conditions surrounding the fish (do the fish have what they need?) and the actual state of the fish (are the fish doing/feeling ok?). Different environmental or fish-based welfare indicators can be quantified (Stien et al., 2020) and the ones to be used will depend on the purpose and resources available as well as on the fish species. For example, on-farm methods for assessing welfare should be practical and relatively inexpensive, while a more complex and precise approach might be needed and/or possible when conducting research on fish welfare in the laboratory. In this regard, the indicators are generally classified as operational welfare indicators (OWIs, those that can be directly measured on-farm or in the wild) (Martins et al., 2012) and laboratory welfare indicators (LABWIs, those that require that samples are sent to a laboratory) (Noble et al., 2018).

Environmental indicators include variables related to water quality (e.g., temperature, pH, salinity, oxygen content, turbidity) and to the characteristics of the rearing units and their locations (e.g., shape, size, and color of rearing tanks, fish stocking density, lighting conditions, feeding regimes, social environment). Animal-based indicators include variables related to the external appearance (condition indices, external damage in eyes, snout, skin, gills, or fins) and to the physiology, metabolism, and behavior of the fish, individually or collectively (the latter referring to group-based variables such as the quantification of hormones/metabolites in fish rearing water or the evaluation of fish group behavior). Among fish-based indicators, variables related to stress are commonly used, including the levels of the so-called stress hormones, and other parameters related to their induction (e.g., brain serotonergic activity and intermediary hormones such as corticotropin-releasing hormone—CRH or adrenocorticotropic hormone—ACTH) or to their effects on the organism (e.g., increase of plasma glucose and reduction of liver glycogen).

2. The neuroendocrine stress response and its relation to welfare

The concepts of stress and welfare are closely related (Sneddon et al., 2016) and therefore, precise information about the stress status of a fish constitutes important information for assessing its welfare (Iwama, 2007). There exist different definitions of stress (Schreck and Tort, 2016). In general, the term refers to a disturbed state after perception of an external or internal event (a "stressor") and to the physiological alterations that take place in the organism when trying to avoid or overcome the threat. Usually, when fish are exposed to a disturbance, they develop an integrated response to stress that involves a series of alterations at different levels of organization in different tissues (Wendelaar Bonga,

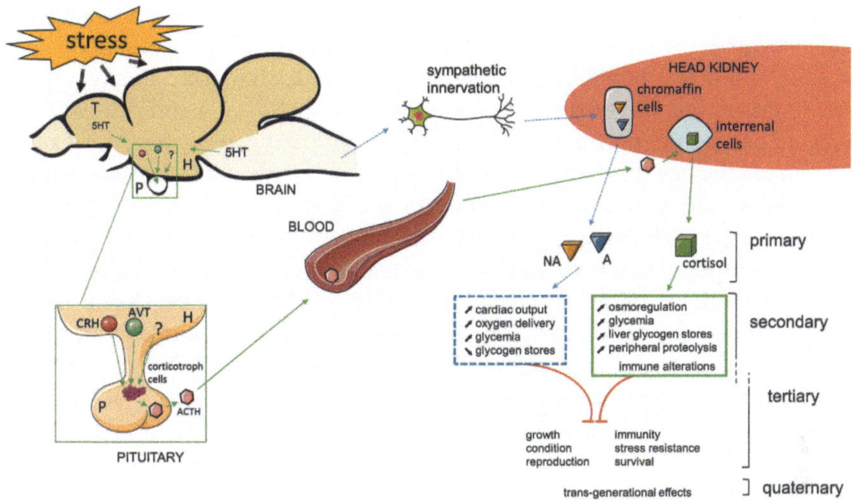

FIG. 2 Main neuroendocrine pathways involved in the fish stress responses and their effects. The two neuroendocrine routes are indicated by *dotted blue* (the brain-sympathetic nervous system-chromaffin cells axis, BSC) and *green solid* lines (the hypothalamus-pituitary-interrenal cells axis, HPI). *5HT*, serotonin; *A*, adrenaline; *ACTH*, adrenocorticotropic hormone; *AVT*, arginine vasotocin; *CRH*, corticotropin-releasing hormone; *H*, hypothalamus; *NA*, noradrenaline; *P*, pituitary; *T*, telencephalon.(Based on Faught, E., Hernandez-Perez, J., Wilson, J.M., Vijayan, M.M., 2020. Stress in response to environmental changes. In: Climate Change and Non-Infectious Fish Disorders, pp. 136–162. https://doi.org/10.1079/9781786393982.0136; Mommsen, T.P., Vijayan, M.M., Moon, T.W., 1999. Cortisol in teleosts: dynamics, mechanisms of action, and metabolic regulation. Rev. Fish Biol. Fish. 9, 211–268. https://doi.org/10.1023/A:1008924418720; Schreck, C.B., Tort, L., 2016. The concept of stress in fish, In: Fish Physiology, pp. 1–34. https://doi.org/10.1016/B978-0-12-802728-8.00001-1; Wendelaar Bonga, S., 1997. The stress response in fish. Physiol. Rev. 77, 591–625.)

1997). The integrated stress response results in effects that have been classified as primary, secondary, or tertiary (Pickering and Pottinger, 1995) (Fig. 2). Also, recently discovered effects of breeders' stress on their progeny, such as altered gene expression or altered stress responses, have been referred to as "quaternary" effects of stress (Faught et al., 2020), which are probably mediated by genetic and epigenetic mechanisms (Faught et al., 2020; Ryu et al., 2018; Vera-Chang et al., 2018). Primary responses include the activation of the stress neuroendocrine axes and the release of stress hormones to circulation. Secondary responses include stress hormone-induced changes in different tissues, generally directed to facilitate a quick evasive or aggressive response and to reinstate homeostasis once the stressor has disappeared. Tertiary responses can become evident if the stressor is too severe or if the exposure to the stressor is persistent. In those cases, the animals might not be able to adequately cope with the stressor, and those tertiary effects might appear, including growth, immune, and reproduction inhibitions, and often a reduced ability to tolerate further stressors (Wendelaar Bonga, 1997).

Describing stress as a disturbance might lead us to think that stress is inherently negative for fish welfare. However, the relationship between stress and welfare is not as straightforward as one might think. Exposure to stressors is part of the normal life of fishes: encounters with a predator, a competitive fight with a conspecific, or entering an area with poor oxygen content, for example, are challenges that fish experience during their lives. Some of these events might constitute relevant opportunities for learning and even increase their capacity to cope with the same or other stressors in the future. In this regard, it has been proposed that both highly stressful and low-challenging environments (hypostimulation) might have negative effects on fish welfare (Korte et al., 2007; Schreck and Tort, 2016; Sneddon et al., 2016). A certain level of exposure to acute stressors that are not severely threatening homeostasis might be considered "eustress" (in contrast to "distress") and be positive (or at least not negative) for the fish in the long term (Schreck, 2010; Schreck and Tort, 2016; Vindas et al., 2016). In general, stressors that are severe, repeated or maintained in time, and unpredictable are more likely to constitute a welfare problem for fish, while punctual exposure to acute stressors, particularly if those are predictable, is more easily tolerated (Balasch and Tort, 2019). Particularly, the duration of the exposure to a stressor is of critical importance for fish welfare. A stressor that is of very low severity when occurring punctually might cause some of the tertiary effects described before when occurring at high frequencies or when prolonged in time. This is of particular relevance in fish under captivity, since the fish are contained in closed environments and cannot escape or avoid being exposed to whatever disturbance they might encounter in their rearing units. Different anatomical, behavioral, and physiological indicators, related to primary, secondary, and tertiary stress effects, are generally used as stress markers (Sopinka et al., 2016) and, taking into consideration the context of those responses, they might be also informative when assessing fish welfare.

2.1 General considerations on the use of stress indicators for welfare evaluation

Different stress variables can be used as indicators of the subjective perception of the fish of its current situation. However, as commented before, that current situation could be temporary and maybe short-lived, so the interpretation of the link between stress and potential negative welfare should be done with caution. In general, primary and secondary stress indicators are good markers for acute stress, but their value as indicators of prolonged or chronic stress is limited (Laberge et al., 2019). In a prolonged stressful situation, the use of tertiary indicators would be more informative when assessing stress and welfare in fish (Sopinka et al., 2016). In the case of indication of acute stress, the context should be taking into consideration with caution when evaluating the welfare situation of the fish. Acute stress might pose a risk for fish welfare or not, depending on the frequency and severity of the stressor and/or the ability of the fish to habituate to it (which is species- and stressor-specific). However, the occurrence of tertiary effects (e.g., reduced growth, higher prevalence of infections and disease, or poor external condition) is in general a clear indicator that the welfare condition of the animals is compromised.

Furthermore, the levels of stress indicators might depend on many different factors (age, strain, population, species, reproductive stage, time of the day, among others), and therefore, the comparison among populations, farms, species, etc. might be problematic. To a large extent, the relevance of using stress indicators as welfare indicators comes from a comparative framework: comparing the actual indicator values to the "expected" values for stress-free animals. These reference values could vary depending on the rearing facilities and environmental conditions and therefore, it is important to have a proper control group of reference to precisely determine if a fish group/population is stressed or not. In other words, absolute values of a stress indicator (e.g., cortisol) are not enough to determine to what extent the fish are stressed. In controlled experiments, it can be relatively easy to establish a proper control group, but this might be complicated when sampling wild fish or fish from a farm to evaluate their welfare. In those cases, historical information about the stress indicators in the same ecosystem or farm/rearing units would be needed to precisely detect any disturbance with respect to conditions considered "basal." In summary, the evaluation of stress in fish is in itself a difficult task. The difficulties associated to this evaluation are beyond the scope of this chapter and interested readers might refer to the chapter published by Sopinka and collaborators (2016), who provided a complete review on the topic of stress evaluation in fish.

3. A proposed method to determine the stress status of fishes as part of the assessment of their welfare

The method proposed here combines indicators at different levels of the stress neuroendocrine cascade to obtain a precise view of the activation of the stress pathways in fish. A schematic summary of the work pipeline for the proposed method is shown in Fig. 3. The method is intended to characterize the fish stress status in the context of stress- or welfare-related research. The method can provide relevant fish-based information about welfare, but it is not intended to be used to characterize fish welfare for practical purposes, for example, for monitoring welfare in a fish farm, due to the relative complexity of sampling and analysis, as well as to their associated costs. As described later, the method requires the sacrifice of the sampled fish. However, the variables included in the proposed method are just a suggestion, and depending on the intention of the stress/welfare assessment or the rationale of the experiments, other alternative and/or complementary variables can be used. Nonlethal sampling is possible if the user is not specifically interested in the brain monoaminergic data, for example: For relatively large fish (> 15–20 g), a nonlethal blood sample would be enough to quantify the blood variables

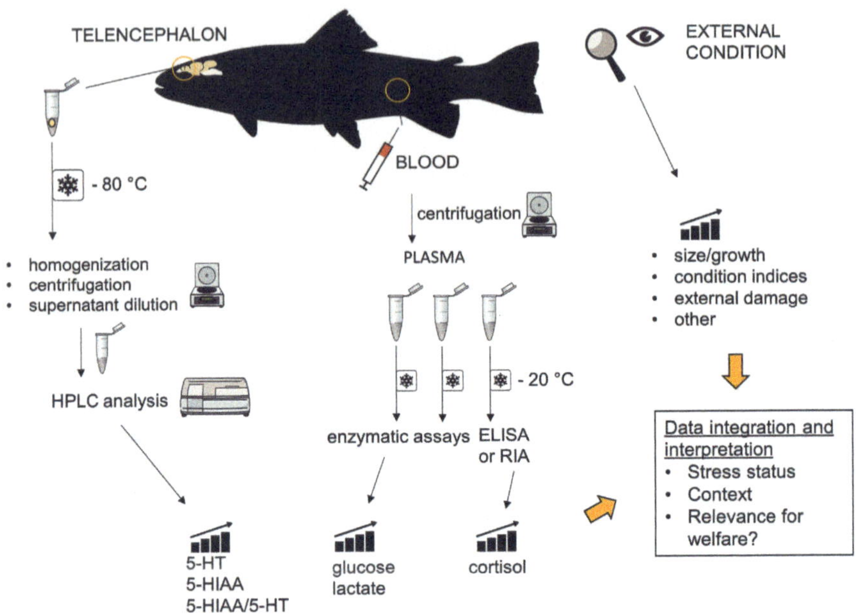

FIG. 3 Summary of the work pipeline for the proposed method to evaluate stress status as part of welfare assessment in fish.

described later. Furthermore, the main stress hormone (cortisol) and other stress-related metabolites can be measured noninvasively, using other tissues (mucus, scales) or even in rearing water (Aerts et al., 2015; Bertotto et al., 2010; Ellis et al., 2004). Each option has its own advantages and drawbacks (Sadoul and Geffroy, 2019) that have to be taken into consideration when interpreting the data for assessing fish stress and/or welfare status. The use of different stress indicators as potential welfare indicators in fish (and in salmonids in particular) has been reviewed before (Noble et al., 2018, 2020; Sopinka et al., 2016). Interested readers can refer to those publications for alternative ways to assess fish stress and welfare.

The variables suggested in this chapter to investigate the neuroendocrine stress status, explained later in more detail, are basically primary and secondary acute stress indicators. Whenever possible, the complementary use of tertiary indicators (e.g., growth, condition factor, and/or external appearance) is highly recommended. The main applications of characterizing the acute stress status of the fish for welfare assessment are as follows:

(1) The evaluation of the stressful potential of specific events, conditions, or protocols in fish maintained in captivity. Experimentally exposing fish to the factor of interest and evaluating their stress status postexposure allows determining the severity of the disturbance for the fish (if any) and whether the factor is to be considered as a threat for fish welfare.

(2) The evaluation of the acute stress status during stress-challenge trials. When welfare conditions are poor, the ability of fish to cope with (further) stressors is often altered (Barton, 2002). Exposing the fish to controlled/standardized acute stressors and evaluating their stress response is informative about the ability of fish to cope, and therefore, about their welfare status (Bonier et al., 2009).

(3) The evaluation of the acute stress indicators in generic fish (i.e., not exposed to specific stressors, or part of a stress/welfare experiment) might provide some information related to welfare. For example, basal cortisol levels can be slightly elevated in chronically stressed fish. However, the responses of acute indicators to prolonged or chronic stress are often inconsistent, due to multiple regulatory factors affecting the release of stress hormones, their receptor levels, etc. (Barton, 2002). Therefore, in chronically stressed fish, acute stress indicators might or might not show increased levels, and the complementary use of the more integrative tertiary indicators, or the use of stress-challenge trials to assess the stress-coping ability of the fish, is recommended.

It is important to highlight that a single-point evaluation of acute stress indicators is usually not enough to fully characterize the fish acute stress response and status, particularly in the case of points (1) and (2) described before. This is because acute stress indicators have a specific time-response profile upon stress exposure, and both the magnitude of the response (how much the indicators increase/decrease upon stress exposure) and the duration of the response (for how long the indicators remain altered—or, in other words, their recovery time) are of relevance when assessing the stress/welfare status. Often, fish in good welfare condition show a response of high magnitude that recovers fast, while fish in poor welfare might show lower peaks of the indicators but extended recovery times (Barton et al., 2005; Gesto et al., 2018b; Jeffrey et al., 2014).

3.1 Proposed variables

The proposed method involves the determination of different acute stress indicators in the fish, including the plasma levels of cortisol, glucose, and lactate and the serotonergic activity in the telencephalon. External condition observations (e.g., growth/size, external tissue damage, and/or condition factor) should be included in the assessment whenever possible (Fig. 3), particularly if the objective of the assessment is just characterizing the general stress/welfare status of the fish individual(s).

The cortisol concentration in the plasma is, most probably, the most commonly used stress indicator in fish. However, its production and release are affected by different endogenous and exogenous factors, so its use as the sole stress indicator is not recommended. Furthermore, using cortisol alone as welfare indicator has even further limitations due to the nonstraightforward link between stress and welfare (Ellis et al., 2012). Therefore, the combined use of cortisol along with other physiological stress indicators is strongly recommended. Among them, brain serotonergic activity provides information about the level of activation of the central nervous system (CNS) upon exposure to a certain stressor. That activation can be viewed as a primary subjective indicator about how the fish perceives the stressor. Acute stress induces a consistent and fast activation of serotonergic transmission in certain areas of the brain (Emerson et al., 2000; Gesto et al., 2013), and the serotonergic system is believed to have a prominent role in the organization and control of the vertebrate stress response (Crawford et al., 2010; Gesto et al., 2013; Jørgensen, 2007). The telencephalon, located in the anterior part of the brain, has a key role in the control of the teleost stress response, and serotonergic activity in this region has been consistently observed to change after stress (Gesto et al., 2013; Vindas et al., 2017). While the response of serotonin (5-hydroxytryptamine or 5-HT) to acute stress is transient and usually recovers after the stressor is no longer present, prolonged and/or repeated stress can also induce adjustments on the serotonergic system in the long term (Backström and Winberg, 2017; Winberg and Thörnqvist, 2016).

Besides primary stress indicators, it is important to have an indication of how the fish physiology and metabolism are actually being affected, so the determination of secondary stress indicators such as the plasma levels of glucose and lactate is recommended. Glucose concentration provides information about its level of mobilization from glycogen stores, or as newly synthetized glucose, to fuel muscle and other tissues to help the fish to overcome the threat generated by the stressor. Glucose levels increase after acute stress due to the action of both catecholamines and cortisol (Wendelaar Bonga, 1997). Lactate levels in plasma often increase after acute stress as an indication of the increased anaerobic metabolism in fish tissues. This is due to certain level of tissue hypoxia resulting from an imbalance between the demand and the delivery of oxygen to the tissues, which is increased when the stressor generates a behavioral response requiring increased exercise/muscle activity, or in particular cases where the stressor involves hypoxia in itself (exposure to air or to hypoxic water) (Milligan and Girard, 1993; Omlin and Weber, 2010). Therefore, contrary to cortisol or glucose, whose response to stress is generally unspecific, the induction of lactate depends on the type of stressor and how it affects fish behavior and oxygen uptake from the environment.

While not described here as part of the proposed method, the determination of plasma osmolality and the plasma concentration of ions such as Mg^{2+}, Na^{+}, and Cl^{-} could be good complementary variables, since both catecholamines and cortisol are known to acutely affect osmoregulation, particularly at the level of the gill (McDonald and Milligan, 1997; Sopinka et al., 2016). In addition, the use of tertiary indicators is advised whenever possible as complementary to any other stress indicator. Tertiary indicators, including external tissue damage (in skin, eyes, gill, and/or fins), growth data, condition indices, energetic variables (glycogen levels in muscle and liver), among others, integrate information about stress and welfare during recent events. Therefore they provide more reliable information about the long-term state of the animal than acute stress indicators (Madaro et al., 2020; Sopinka et al., 2016).

3.2 Technical considerations for experimental design and sampling

3.2.1 *Experimental design*

Stress experiments, or even simple protocols for stress or welfare assessment, require well-planned design and sampling protocols to avoid (1) exposing the fish to uncontrolled factors that could induce stress and (2) disturbing the fish before or during sampling. Having a good control of all potential factors that can induce undesired alterations in the stress indicators during an experiment, or when capturing fish is complicated, particularly for inexperienced workers/researchers. Fish should be allocated to rearing units where the environmental conditions (e.g., noise, light, water quality, and/or stocking density) are not stressful and are equivalent for all rearing units. Readers interested in potential stressors for fish in captivity can refer to published literature on the topic (Ashley, 2007; Biswas et al., 2006; Eissa and Wang, 2016; Smith et al., 2004; Stevens et al., 2017). Factors differentially affecting the different rearing units should be avoided, if possible (e.g., some rearing units might be more exposed to people passing by than others). Furthermore, performing stress experiments in a noisy open area, where many people pass by, is not advisable. On the other hand, isolating fish rearing units in a too-protected environment might make the fish more "stressable" during processes such as tank cleaning or capture for sampling. It is important to have fish under conditions at which it is possible to observe basal levels of the stress indicators in control, nonstressed fish. In this regard, it is of critical importance to provide enough time for the acclimation of fish to new tanks and/or environmental conditions: stress experiments often require reallocating fish from rearing units to experimental units. This reallocation is stressful in itself, and it is generally recommended to wait at least 2 weeks (the time might vary depending on the species) before starting any experimental protocol with the fish (unless the reallocation is considered part of the experimental protocol).

3.2.2 *Sampling design*

Regarding fish capture for sampling, the nature of the response of stress indicators, which often increase fast after stress exposure, requires careful consideration of the way the fish should be captured and the number of fish that will be simultaneously captured. Just as an example: if 10 individuals need to be sampled, it should be considered if all 10 fish should be captured together (taking into account that there are some requirements about the time taken for tissue collection, as commented later) or, alternatively, in batches. If the latter is the case, it might not be adequate capturing fish more than once from the same rearing unit, since the fish remaining in the unit after the capture of the first batch would be stressed to some extent. In general, from the start of the fish capture to the completion of the tissue collection, the process should be as fast as possible and avoiding capturing fish repeatedly from the same units, unless at least several hours have passed in between.

3.2.3 Procedures for tissue collection

As part of the proposed method, both blood and a brain region, the telencephalon, are to be collected. As commented before, the sampling has to be done in a way that does not compromise the samples taken. During sampling, the capture and anesthesia procedure of the fish is in itself stressful and would induce a secondary increase in the stress indicators in a matter of a few minutes. Therefore, both blood and telencephalon samples should be collected fast after capturing and anesthetizing the fish, in order to avoid as much as possible an effect of those processes in the assessed variables. Most usually, a blood sample is collected first, and then, fish is beheaded for the collection of the telencephalon—the fish must be already dead from an anesthetic overdose before this step. The capture of the fish, for example, by netting, should be swift and the fish should be immediately introduced to the anesthesia (most often in the form of an anesthetic bath of MS-222, benzocaine, clove oil, or any other appropriate anesthetic for fish). The use of one or another anesthetic would depend on the species and on the researcher's preferences, but also on the applicable national and international regulations in each case. In the case of salmonids, it is usually better to keep total sampling time (from the start of the capture process until the completion of tissue collection) under 3–4 min. When possible, certain strategies can be used to minimize capture- and sampling-related stress: One would be using high anesthetic doses (relative strength of the dose depends on the anesthetic and on the fish species), reducing the anesthesia time, and therefore, total sampling time. High, lethal doses of anesthetic may be used if the fish is not expected to recover. Another useful strategy is in-tank anesthesia, adding the anesthetic directly to the tank, so the fish do not need to be captured, or even disturbed, before anesthetized, which highly reduces the sampling-associated stress. In-tank anesthesia is normally only used when all the fish in the rearing unit are to be sampled at the same time.

An anticoagulant solution (e.g., ammonium heparin) should be used to rinse the syringes before blood collection, to avoid blood clotting in the syringe or in the collection tubes. It is better to keep the blood on ice after collection and centrifuge the blood (recommended: 10 min at 2500 × g and 4 °C) for obtaining the plasma as soon as possible. Even if cortisol is stable for several hours at room temperature, glucose and lactate levels will be altered if the samples are let for a few hours at room temperature, particularly in the presence of blood cells. In general, both glucose and lactate levels should be stable for at least 30 min in blood and at least 2 h in plasma, when the samples are kept on ice (personal observation). Plasma samples can be aliquoted and stored at −20 °C or below until analysis. The amount of plasma needed for the analyses would depend on the techniques or commercial kits used for the different analyses, but in most of the cases, a total of 60 μL of plasma should be more than enough for assaying all glucose, lactate, and cortisol in duplicate. Some other biological matrices can also be used for the determination of one or more of these variables, such as whole-body homogenates or skin mucus (for cortisol, glucose, and lactate), gill filaments and scales (for cortisol), or fish rearing water (cortisol) (Aerts et al., 2015; Gesto et al., 2015a; Herrera et al., 2020). There are different pros and cons associated to each tissue, and the choice will depend on the objective of the study, the size of the fish, or on whether the fish is expected to survive the sampling or not, among other factors.

The telencephalon should be also sampled as quickly as possible. Once the fish has been killed by an anesthetic overdose, it can be beheaded, and the roof of the skull can then be easily removed with a scalpel, exposing the brain. Then, the two telencephalic lobes can be collected with a pair of fine forceps after vertically severing their anterior and posterior connections with some fine tipped scissors. Most often, the telencephalon can be easily recognized as a paired-lobed region located anterior to the optic tectum, which is generally comprised of two larger lobes. However, brain anatomy varies considerably among fish species and for some, recognizing the different brain areas might not be so straightforward. The chapter by Butler (2000) on the anatomy of fish nervous system might be helpful in this regard for interested readers. Once collected, the tissue should be immediately snap frozen and then stored at −80 °C. 5-HT and its main metabolite 5-hydroxyindoleacetic acid (5-HIAA) is relatively labile. It is very important to be very precise when dissecting out the telencephalon: Monoaminergic cell bodies are grouped forming clusters and therefore, brain tissue is not homogeneous in terms of monoaminergic content. This means that, since the calculated concentration of the compounds will be normalized by tissue mass (see later), small differences in the piece of tissue collected can result in very important differences in the final calculated concentrations per unit of tissue.

3.3 Analytical procedures

Most usually, the plasma levels of glucose and lactate are quantified without the need of any sample preparation. The analysis is usually performed by means of enzymatic assays where, through a series of coupled enzymatic reactions, a colored product is formed in amounts that are directly related to those of the compound of interest in the plasma sample. Using the originally described protocols is possible (see e.g., Bergmeyer, 1974), but diverse analytical kits are now commercially available. Kits are easy to use and assays can be performed with very little previous experience or expertise, just following the kit manual, but they are relatively expensive. The requirements in terms of equipment are relatively low, and are mostly restricted, besides basic laboratory equipment, to a spectrophotometer (using cuvettes or microplates), which is needed to quantify the absorbance at particular wavelength of each reaction tube or microplate well.

Cortisol is most often quantified using immune assays such as RadioImmunoAssay (RIA) or Enzyme-Linked ImmunoSorbent Assay (ELISA). There are different types of immune assays that work in different ways, but all are based on the use of specific antibodies against the analyte (cortisol, in this case). At the end of the assay, a chemical label will be present in the reaction tubes or wells. The type of label will be radioactivity (in the case of RIA) or an enzyme (in the case of ELISA), and the amounts of the label in every case will be directly or inversely linked to the amount of cortisol in the original sample. The intensity of the signal can then be measured with a radioactivity detector in RIA. In the case of ELISA, the amount of the enzyme label can be calculated after adding a reagent that the enzyme is able to transform into a detectable product (most often a colored or luminescent product). Both RIA and ELISA have different advantages and drawbacks. RIA is cheaper, but most usually requires specific allowance and facilities to work with radioactive compounds, depending on the corresponding national regulations. In any case, the development and setup of RIA or ELISA techniques requires specific expertise. However, as commented before for glucose and lactate, there are nowadays many commercially available kits (RIA and ELISA based) that are easy to use, even by inexperienced users. Again, the convenience and easy-to-use characteristics of the kits most often result in a more expensive analysis. Kits of different brands have been used in fish stress experiments, and depending on the kit used, some kind of sample preparation, usually in the form of a preextraction step, might be required. It is worth mentioning that cortisol levels can vary substantially depending on the species, and this can determine the sensitivity required for the analytical kit. Just as an example, salmonids are considered to have intermediate cortisol levels (Barton, 2002), while other species tend to display much higher levels, such as the case of the European seabass (*Dicentrarchus labrax*) (Alfonso et al., 2019). On the opposite side of the spectrum, sturgeons tend to show lower plasma levels of cortisol (Barton, 2002; Hoseini et al., 2016).

The most common and convenient way to analyze brain monoamines, including 5-HT and its main oxidative metabolite 5-HIAA, is by means of HPLC with electrochemical (EC) detection. HPLC equipment is relatively common in physiology or biology laboratories, but the use of EC detectors is not so frequent. There are plenty of HPLC protocols available in the literature for the simultaneous quantification of monoaminergic neurotransmitters, such as 5-HT, dopamine, noradrenaline, and their metabolites, from animal tissue samples (Gesto et al., 2006; Höglund et al., 2000; Jakubovic et al., 1987; Mefford, 1981; Reinhoud et al., 2013). Monoaminergic neurotransmitters have a wide array of functions in the control and regulation of brain processes, including behavior-controlling processes such as the regulation of mood, attention, arousal, motivation, novelty-seeking, aggression, anxiety, or appetite (Abbey-Lee et al., 2018; Messias et al., 2016; Winberg and Thörnqvist, 2016). Furthermore, the activity of these neurotransmitters, particularly the serotonergic activity, provides information about stress signaling in the fish brain (Gesto et al., 2013; Puglisi-Allegra and Andolina, 2015). It is important to highlight that the mere quantification of neurotransmitters from homogenized tissues is not very informative in terms of the neuronal activity, since it does not allow discriminating between active (released) or inactive neurotransmitters, still packed in vesicles in the presynaptic neurons. In order to have an estimation of the rate of release of neurotransmitters into the synapses (i.e., the neurotransmitter activity), the ratios between the main oxidative metabolite and the parent neurotransmitter are often used (Shannon et al., 1986; Winberg and Nilsson, 1993). The rationale behind the use of these ratios is as follows: the neurotransmitters are usually stored in vesicles in the presynaptic neurons. When released, they are partially reintroduced into the presynaptic neuron by specific transporters and are partially catabolized by monoamine oxidase and other enzymes. Therefore the more they are released, the more metabolites will be formed and the ratio will increase. In teleosts, the main metabolite of 5-HT is 5-HIAA (Lepage et al., 2002), and the ratio 5-HIAA/5-HT is used as a proxy of "serotonergic activity." The analysis of monoamines in tissues requires a homogenization step that is usually carried out by using mechanical grinders or ultrasonic disintegrators. Then, after centrifugation (recommended: 10 min at $12000 \times g$ and 4 °C), the diluted supernatant can usually be injected directly into the HPLC. It is important to note that the homogenization step can induce the degradation of labile compounds such as 5-HT and 5-HIAA and therefore, the homogenization protocol should be well standardized (in terms of duration, intensity, temperature, and any other relevant factor). As a rule of thumb, intensity and duration of the homogenization should be kept at the minimum that allows a proper disintegration of the tissue, and the samples should be kept refrigerated at all times. Refreezing should be avoided and samples should be prepared and injected into the HPLC after first thawing. Once analyzed, the amounts of monoamines in the tissue are usually normalized by tissue mass, so the telencephalon should be weighed at some point, either after collection, or during/after thawing (recommended), prior to homogenization.

3.4 Interpretation of the data

Most often, acute stress induces transient increases in the variables described: plasma cortisol, glucose and lactate, and telencephalic 5-HIAA/5-HT ratio. To which extent this increase is related to poor welfare must be interpreted taking into consideration the context of the study or the situation. If the fish is exposed to disturbance for an extended period, some of the proposed variables might remain elevated for some time, sometimes for days or weeks. That is not always the case. As explained before, several indicators can be attenuated with time even if the stress persists. The fish can habituate to the stressor or simply be unable to keep their acute stress pathways activated the same as after acute exposure. As suggested before, tertiary indicators will be needed in those cases to better understand the welfare status of the fish. Those indicators will help to distinguish situations in which the fish are doing fine (i.e., not stressed anymore, maybe indicating habituation), accommodating a load (i.e., affected, but able to cope after physiological adjustments), or just being unable to cope well with their situation (and then showing inhibited growth, higher prevalence of infections and diseases, higher mortalities, among other potential effects). It is important to note that most of the results of experiments focused on stress and welfare open new questions related to the welfare of the fish. In those cases, further research is needed in order for us to be able to predict how the welfare of the fish will be affected by the tested condition. In this regard, for a given potentially disturbing factor, it is fundamental to consider the key role of the exposure timing in relation to welfare: The exposure time and the frequency of exposure to a stressor, together with the stressor severity/intensity, will determine how relevant the stressor is for fish welfare. A factor that is acutely stressful might or might not be a threat to welfare, depending on exposure time and frequency and/or on the fish ability to habituate to the specific stressor. On the contrary, an acute stressor of low severity, or even a factor that is not stressful at all upon acute exposure, might turn into a stressful factor if the exposure is prolonged, repeated, or chronic. Therefore studies demonstrating that a fish is acutely affected (or not) by a given condition or event will require a further research effort to investigate the long-term consequences of the same condition or event, which, in general, cannot be predicted from the acute responses.

3.5 Examples of application of the proposed method

The same or similar approaches to the one suggested here have been applied to assess the stress condition of fish under different circumstances. For example, it has been used to evaluate the stressor potential of different factors, including handling by chasing (Conde-Sieira et al., 2018b; Gesto et al., 2013), alterations of water quality (Gesto et al., 2020), presence of pollutants (Gesto et al., 2006, 2009; Tintos et al., 2007, 2008), and formation of social hierarchies (Xu et al., 2020), among others. It has also been used for basic research on the function and regulation of the teleost stress response (Gesto et al., 2013, 2014, 2015b).

It is worth mentioning that, since stress responses are triggered under specific conditions/stimuli, just assessing stress variables in resting fish is not very informative. Assessing the ability of fish to respond, physiologically and behaviorally, in standardized stress-challenge tests, is usually more relevant. The magnitude and duration of the stress response after a standardized stressor provides a better picture of status of a fish; it provides information about stress reactivity, but also about the metabolic capacity to mount an efficient organismal response (physiological and behavioral) to a threat. The ability of the fish to respond to acute stressors is affected by different factors. In this regard, the suggested protocols have been used on different occasions to assess the effects of different factors on the fish stress response to specific stressors. Among those factors were the diet/nutritional status (Azeredo et al., 2017; Carvalho et al., 2019; Conde-Sieira et al., 2018a; Hundal et al., 2021), early developmental time (Gesto et al., 2017, 2018b), the presence of pollutants (Gesto et al., 2008, 2018a; Melnyk-Lamont et al., 2014), and the presence of stress-relieving compounds (Conde-Sieira et al., 2014; Gesto et al., 2016; López-Patiño et al., 2013), among others. In many occasions, those factors have effects that only become apparent in the "stressed" state (e.g., in the magnitude and/or the duration of the stress response), but have no influence on the basal levels of the stress markers. Finally, the proposed approach has been also used in research about fish personality, which is usually defined based on the animal's stress-coping styles (Alfonso et al., 2019; Gesto, 2019). Those studies contributed to characterize the stress responses of reactive and proactive fish, being the extent of the differences dependent on fish species and on the degree of domestication.

4. Conclusions and future perspectives

The presented technical approach allows for a precise assessment of the immediate stress status of fish, which is relevant as part of the assessment of fish welfare in captive fish, and helps as well in identifying procedures and situations that can threat fish welfare in the long term. This approach is used in fish stress research, since the assessment of welfare in fish farms is generally based on the so-called OWIs, which are required to be simple, inexpensive, and less invasive. Because of those requirements, OWIs are mostly based on either the monitoring of environmental variables (e.g., water quality, currents, stocking density) or in the monitoring of external fish-based variables (e.g., growth, condition, malformations, integument damage) (Noble et al., 2020). Detailed research on fish stress is important to investigate how current and future OWIs correlate with the real psycho- and physiological status of the individual fish. Interestingly, there are ongoing efforts to incorporate the use of stress hormones as OWIs in the form of biosensors that can be used to easily quantify fish cortisol levels. Hopefully, those sensors would be used in the future as diagnostic tools to monitor stress hormones in the water of fish rearing tanks, providing real-time information about the stress status of the fish in the farms (Wu et al., 2017, 2019).

An important drawback in the use of stress variables for welfare monitoring is that there are no clear immediate markers of chronic stress. Chronic stress, resulting from sustained suboptimal conditions during fish rearing, is one of the main determinants of poor welfare in captive fish. Since acute stress variables might be unreliable when assessing chronic stress, its detection usually relies on tertiary effects (such as growth or immune inhibition). However, those effects can also occur without stress, and therefore, an unambiguous indicator of chronic stress is still elusive. Recently, the assessment of cortisol deposited and accumulated in the scales has been developed as a potential indicator of the stress a fish has suffered during its life (or at least during a relatively long period). Though promising, how this deposition of cortisol in the scales (and maybe in other tissues?) is affected by different environmental factors or in different fish species is unknown, and future research should clarify how precise and reliable is this deposition in relation with the stress experienced by the animals during their life.

The recent implementation of metabolomics and other omics approaches might constitute a good opportunity for finding new markers of stress and welfare in fish (Raposo de Magalhães et al., 2020). New indicators would be particularly helpful in relation to the detection and management of sustained suboptimal stress/welfare conditions in captive fish, which would certainly contribute to a more sustainable and ethical fish farming.

References

Abbey-Lee, R.N., Uhrig, E.J., Zidar, J., Favati, A., Almberg, J., Dahlbom, J., Winberg, S., Løvlie, H., 2018. The influence of rearing on behavior, brain monoamines, and gene expression in three-spined sticklebacks. Brain. Behav. Evol. 91 (4), 201–213. https://doi.org/10.1159/000489942.

Aerts, J., Metz, J.R., Ampe, B., Decostere, A., Flik, G., De Saeger, S., 2015. Scales tell a story on the stress history of fish. PLoS One 10. https://doi.org/10.1371/journal.pone.0123411, e0123411.

Alfonso, S., Sadoul, B., Gesto, M., Joassard, L., Chatain, B., Geffroy, B., Bégout, M.L., 2019. Coping styles in European sea bass: the link between boldness, stress response and neurogenesis. Physiol. Behav. 207, 76–85. https://doi.org/10.1016/j.physbeh.2019.04.020.

Ashley, P.J., 2007. Fish welfare: current issues in aquaculture. Appl. Anim. Behav. Sci. 104, 199–235. https://doi.org/10.1016/j.applanim.2006.09.001.

Azeredo, R., Machado, M., Afonso, A., Fierro-Castro, C., Reyes-López, F.E., Tort, L., Gesto, M., Conde-Sieira, M., Míguez, J.M., Soengas, J.L., Kreuz, E., Wuertz, S., Peres, H., Oliva-Teles, A., Costas, B., 2017. Neuroendocrine and immune responses undertake different fates following tryptophan or methionine dietary treatment: tales from a teleost model. Front. Immunol. 8, 1226. https://doi.org/10.3389/fimmu.2017.01226.

Backström, T., Winberg, S., 2017. Serotonin coordinates responses to social stress-what we can learn from fish. Front. Neurosci. 11, 595. https://doi.org/10.3389/fnins.2017.00595.

Balasch, J.C., Tort, L., 2019. Netting the stress responses in fish. Front. Endocrinol. (Lausanne) 10, 62. https://doi.org/10.3389/fendo.2019.00062.

Barton, B.A., 2002. Stress in fishes: a diversity of responses with particular reference to changes in circulating corticosteroids. Integr. Comp. Biol. 42, 517–525. https://doi.org/10.1093/icb/42.3.517.

Barton, B.A., Ribas, L., Acerete, L., Tort, L., 2005. Effects of chronic confinement on physiological responses of juvenile gilthead sea bream, *Sparus aurata* L., to acute handling. Aquac. Res. 36, 172–179. https://doi.org/10.1111/j.1365-2109.2004.01202.x.

Benn, A.L., McLelland, D.J., Whittaker, A.L., 2019. A review of welfare assessment methods in reptiles, and preliminary application of the welfare quality® protocol to the pygmy blue-tongue skink, *Tiliqua adelaidensis*, using animal-based measures. Animals 9, 27. https://doi.org/10.3390/ani9010027.

Bergmeyer, H.U. (Ed.), 1974. Methods of Enzymatic Analysis, second ed. vol. 3. Academic Press.

Bertotto, D., Poltronieri, C., Negrato, E., Majolini, D., Radaelli, G., Simontacchi, C., 2010. Alternative matrices for cortisol measurement in fish. Aquac. Res. 41, 1261–1267. https://doi.org/10.1111/j.1365-2109.2009.02417.x.

Biswas, A.K., Seoka, M., Tanaka, Y., Takii, K., Kumai, H., 2006. Effect of photoperiod manipulation on the growth performance and stress response of juvenile red sea bream (*Pagrus major*). Aquaculture 258, 350–356. https://doi.org/10.1016/j.aquaculture.2006.03.048.

Bonier, F., Martin, P.R., Moore, I.T., Wingfield, J.C., 2009. Do baseline glucocorticoids predict fitness? Trends Ecol. Evol. 24, 634–642. https://doi.org/10.1016/j.tree.2009.04.013.

Branson, E.J., 2008. Fish Welfare. Blackwell Publishing Ltd, Oxford, UK. https://onlinelibrary.wiley.com/doi/book/10.1002/9780470697610.

Broom, D.M., 2011. A history of animal welfare science. Acta Biotheor. 59, 121–137. https://doi.org/10.1007/s10441-011-9123-3.

Butler, A.B., 2000. Chapter 8—Nervous system. In: Ostrander, G.K. (Ed.), The Laboratory Fish. Academic Press, pp. 129–149.

Carere, C., Mather, J., 2019. The Welfare of Invertebrate Animals. Springer Nature, Switzerland AG.

Carvalho, M., Montero, D., Gesto, M., Lencina, A., Lund, I., Izquierdo, M., 2019. The effect of dietary n-3 LC-PUFA on the responses to acute and prolonged stress of meagre (*Argyrosomus regius*, Asso 1801) juveniles. Aquaculture 506, 112–118. https://doi.org/10.1016/j.aquaculture.2019.03.033.

Conde-Sieira, M., Muñoz, J.L.P., López-Patiño, M.A., Gesto, M., Soengas, J.L., Míguez, J.M., 2014. Oral administration of melatonin counteracts several of the effects of chronic stress in rainbow trout. Domest. Anim. Endocrinol. 46, 26–36. https://doi.org/10.1016/j.domaniend.2013.10.001.

Conde-Sieira, M., Gesto, M., Batista, S., Linares, F., Villanueva, J.L.R., Míguez, J.M., Soengas, J.L., Valente, L.M.P., 2018a. Influence of vegetable diets on physiological and immune responses to thermal stress in Senegalese sole (*Solea senegalensis*). PLoS One 13. https://doi.org/10.1371/journal.pone.0194353, e0194353.

Conde-Sieira, M., Valente, L.M.P., Hernández-Pérez, J., Soengas, J.L., Míguez, J.M., Gesto, M., 2018b. Short-term exposure to repeated chasing stress does not induce habituation in Senegalese sole, *Solea senegalensis*. Aquaculture 487, 32–40. https://doi.org/10.1016/j.aquaculture.2018.01.003.

Crawford, L.T.K., Craige, C.P., Beck, S.G., 2010. Increased intrinsic excitability of lateral wing serotonin neurons of the dorsal raphe: a mechanism for selective activation in stress circuits. J. Neurophysiol. 103, 2652–2663. https://doi.org/10.1152/jn.01132.2009.

Diggles, B.K., 2019. Review of some scientific issues related to crustacean welfare. ICES J. Mar. Sci. 76, 66–81. https://doi.org/10.1093/icesjms/fsy058.

Eissa, N., Wang, H.P., 2016. Transcriptional stress responses to environmental and husbandry stressors in aquaculture species. Rev. Aquac. 8, 61–88. https://doi.org/10.1111/raq.12081.

Ellis, T., James, J.D., Stewart, C., Scott, A.P., 2004. A non-invasive stress assay based upon measurement of free cortisol released into the water by rainbow trout. J. Fish Biol. 65, 1233–1252. https://doi.org/10.1111/j.0022-1112.2004.00499.x.

Ellis, T., Yildiz, H.Y., López-Olmeda, J., Spedicato, M.T., Tort, L., Øverli, Ø., Martins, C.I.M., 2012. Cortisol and finfish welfare. Fish Physiol. Biochem. 38, 163–188. https://doi.org/10.1007/s10695-011-9568-y.

Emerson, A.J., Kappenman, D.P., Ronan, P.J., Renner, K.J., Summers, C.H., 2000. Stress induces rapid changes in serotonergic activity: restraint and exertion. Behav. Brain Res. 111, 83–92. https://doi.org/10.1016/S0166-4328(00)00143-1.

Faught, E., Hernandez-Perez, J., Wilson, J.M., Vijayan, M.M., 2020. Stress in response to environmental changes. In: Climate Change and Non-Infectious Fish Disorders, pp. 136–162, https://doi.org/10.1079/9781786393982.0136.

Gesto, M., 2019. Consistent individual competitive ability in rainbow trout as a proxy for coping style and its lack of correlation with cortisol responsiveness upon acute stress. Physiol. Behav. 208, 112576. https://doi.org/10.1016/j.physbeh.2019.112576.

Gesto, M., Tintos, A., Soengas, J.L., Míguez, J.M., 2006. Effects of acute and prolonged naphthalene exposure on brain monoaminergic neurotransmitters in rainbow trout (*Oncorhynchus mykiss*). Comp. Biochem. Physiol., Part C: Toxicol. Pharmacol. 144, 173–183. https://doi.org/10.1016/j.cbpc.2006.08.002.

Gesto, M., Soengas, J.L., Míguez, J.M., 2008. Acute and prolonged stress responses of brain monoaminergic activity and plasma cortisol levels in rainbow trout are modified by PAHs (naphthalene, β-naphthoflavone and benzo(a)pyrene) treatment. Aquat. Toxicol. 86, 341–351. https://doi.org/10.1016/j.aquatox.2007.11.014.

Gesto, M., Tintos, A., Soengas, J.L., Míguez, J.M., 2009. β-Naphthoflavone and benzo(a)pyrene alter dopaminergic, noradrenergic, and serotonergic systems in brain and pituitary of rainbow trout (*Oncorhynchus mykiss*). Ecotoxicol. Environ. Saf. 72, 191–198.

Gesto, M., López-Patiño, M.A., Hernández, J., Soengas, J.L., Míguez, J.M., 2013. The response of brain serotonergic and dopaminergic systems to an acute stressor in rainbow trout: a time course study. J. Exp. Biol. 216, 4435–4442. https://doi.org/10.1242/jeb.091751.

Gesto, M., Soengas, J.L., Rodríguez-Illamola, A., Míguez, J.M., 2014. Arginine vasotocin treatment induces a stress response and exerts a potent anorexigenic effect in rainbow trout, *Oncorhynchus mykiss*. J. Neuroendocrinol. 26, 89–99. https://doi.org/10.1111/jne.12126.

Gesto, M., Hernández, J., López-Patiño, M.A., Soengas, J.L., Míguez, J.M., 2015a. Is gill cortisol concentration a good acute stress indicator in fish? A study in rainbow trout and zebrafish. Comp. Biochem. Physiol. Part A Mol. Integr. Physiol. https://doi.org/10.1016/j.cbpa.2015.06.020.

Gesto, M., López-Patiño, M.A., Hernández, J., Soengas, J.L., Míguez, J.M., 2015b. Gradation of the stress response in rainbow trout exposed to stressors of different severity: the role of brain serotonergic and dopaminergic systems. J. Neuroendocrinol. 27, 131–141. https://doi.org/10.1111/jne.12248.

Gesto, M., Álvarez-Otero, R., Conde-Sieira, M., Otero-Rodiño, C., Usandizaga, S., Soengas, J.L., Míguez, J.M., López-Patiño, M.A., 2016. A simple melatonin treatment protocol attenuates the response to acute stress in the sole *Solea senegalensis*. Aquaculture 452, 272–282. https://doi.org/10.1016/j.aquaculture.2015.11.006.

Gesto, M., Skov, P.V., Jokumsen, A., 2017. Emergence time and skin melanin spot patterns do not correlate with growth performance, social competitive ability or stress response in farmed rainbow trout. Front. Neurosci. 11, 319. https://doi.org/10.3389/fnins.2017.00319.

Gesto, M., Liu, D., Pedersen, L.F., Meinelt, T., Straus, D.L., Jokumsen, A., 2018a. Confirmation that pulse and continuous peracetic acid administration does not disrupt the acute stress response in rainbow trout. Aquaculture 492. https://doi.org/10.1016/j.aquaculture.2018.04.009.

Gesto, M., Madsen, L., Andersen, N.R., Jokumsen, A., 2018b. Differences in stress and disease resilience related to emergence time for first feeding in farmed rainbow trout (*Oncorhynchus mykiss*). J. Exp. Biol. 221. https://doi.org/10.1242/jeb.174623, jeb174623.

Gesto, M., Zupa, W., Alfonso, S., Spedicato, M.T., Lembo, G., Carbonara, P., 2020. Using acoustic telemetry to assess behavioral responses to acute hypoxia and ammonia exposure in farmed rainbow trout of different competitive ability. Appl. Anim. Behav. Sci. 230, 105084. https://doi.org/10.1016/j.applanim.2020.105084.

Herrera, M., Fernández-Alacid, L., Sanahuja, I., Ibarz, A., Salamanca, N., Morales, E., Giráldez, I., 2020. Physiological and metabolic effects of a tryptophan-enriched diet to face up chronic stress in meagre (*Argyrosomus regius*). Aquaculture 522, 735102. https://doi.org/10.1016/j.aquaculture.2020.735102.

Höglund, E., Balm, P.H.M., Winberg, S., 2000. Skin darkening, a potential social signal in subordinate Arctic charr (*Salvelinus alpinus*): the regulatory role of brain monoamines and proopiomelanocortin-derived peptides. J. Exp. Biol. 203, 1711–1721.

Hoseini, S.M., Mirghaed, A.T., Mazandarani, M., Zoheiri, F., 2016. Serum cortisol, glucose, thyroid hormones' and non-specific immune responses of Persian sturgeon, *Acipenser persicus* to exogenous tryptophan and acute stress. Aquaculture 462, 17–23. https://doi.org/10.1016/j.aquaculture.2016.04.031.

Hundal, B.K., Liland, N.S., Rosenlund, G., Höglund, E., Araujo, P., Stubhaug, I., Sissener, N.H., 2021. Increasing the dietary n-6/n-3 ratio alters the hepatic eicosanoid production after acute stress in Atlantic salmon (*Salmo salar*). Aquaculture 534, 736272. https://doi.org/10.1016/j.aquaculture.2020.736272.

Huntingford, F.A., Kadri, S., 2014. Defining, assessing and promoting the welfare of farmed fish. OIE Rev. Sci. Tech. 33, 233–244. https://doi.org/10.20506/rst.33.1.2286.

Huntingford, F.A., Adams, C., Braithwaite, V.A., Kadri, S., Pottinger, T.G., Sandøe, P., Turnbull, J.F., 2006. Current issues in fish welfare. J. Fish Biol. 68, 332–372. https://doi.org/10.1111/j.0022-1112.2006.001046.x.

Iwama, G.K., 2007. The welfare of fish. Dis. Aquat. Org. 75, 155–158. https://doi.org/10.3354/dao075155.

Jakubovic, A., Fu, D., Fibiger, H.C., 1987. Factors affecting the stability and separation of biogenic amines and their metabolites. Simultaneous measurement by HPLC with electrochemical detection. J. Pharmacol. Methods 17, 1–14. https://doi.org/10.1016/0160-5402(87)90032-5.

Jeffrey, J.D., Gollock, M.J., Gilmour, K.M., 2014. Social stress modulates the cortisol response to an acute stressor in rainbow trout (*Oncorhynchus mykiss*). Gen. Comp. Endocrinol. 196, 8–16. https://doi.org/10.1016/j.ygcen.2013.11.010.

Jørgensen, H.S., 2007. Studies on the neuroendocrine role of serotonin. Dan. Med. Bull. 54, 266–288.

Korte, S.M., Olivier, B., Koolhaas, J.M., 2007. A new animal welfare concept based on allostasis. Physiol. Behav. 92, 422–428. https://doi.org/10.1016/j.physbeh.2006.10.018.

Kristiansen, T.S., Bracke, M.B.M., 2020. A brief look into the origins of fish welfare science. In: Kristiansen, T.S., Fernö, A., Pavlidis, M.A., van de Vis, H. (Eds.), The Welfare of Fish. Springer Nature, Switzerland AG, pp. 1–18, https://doi.org/10.1007/978-3-030-41675-1.

Laberge, F., Yin-Liao, I., Bernier, N.J., 2019. Temporal profiles of cortisol accumulation and clearance support scale cortisol content as an indicator of chronic stress in fish. Conserv. Physiol. 7, 1–13. https://doi.org/10.1093/conphys/coz052.

Lepage, O., Tottmar, O., Winberg, S., 2002. Elevated dietary intake of L-tryptophan counteracts the stress-induced elevation of plasma cortisol in rainbow trout (*Oncorhynchus mykiss*). J. Exp. Biol. 205, 3679–3687.

López-Patiño, M.A., Conde-Sieira, M., Gesto, M., Librán-Pérez, Soengas, J.L., Míguez, J.M., 2013. Melatonin partially minimizes the adverse stress effects in Senegalese sole (*Solea senegalensis*). Aquaculture 388–391, 165–172. https://doi.org/10.1016/j.aquaculture.2013.01.023.

Madaro, A., Kristiansen, T.S., Pavlidis, M.A., 2020. How fish cope with stress? In: Kristiansen, T.S., Fernö, A., Pavlidis, M.A., van de Vis, H. (Eds.), The Welfare of Fish. Springer Nature, Switzerland AG, pp. 251–282.

Martins, C.I.M., Galhardo, L., Noble, C., Damsgård, B., Spedicato, M.T., Zupa, W., Beauchaud, M., Kulczykowska, E., Massabuau, J.C., Carter, T., Planellas, S.R., Kristiansen, T., 2012. Behavioural indicators of welfare in farmed fish. Fish Physiol. Biochem. 38, 17–41. https://doi.org/10.1007/s10695-011-9518-8.

McDonald, G., Milligan, L., 1997. Ionic, osmotic and acid-base regulation in stress. In: Iwama, G.K., Pickering, A.D., Sumpter, J.P., Schreck, C.B. (Eds.), Fish Stress and Health in Aquaculture, pp. 119–145.

Mefford, I.N., 1981. Application of high performance liquid chromatography with electrochemical detection to neurochemical analysis: measurement of catecholamines, serotonin and metabolites in rat brain. J. Neurosci. Methods 3, 207–224. https://doi.org/10.1016/0165-0270(81)90056-X.

Melnyk-Lamont, N., Best, C., Gesto, M., Vijayan, M.M., 2014. The antidepressant venlafaxine disrupts brain monoamine levels and neuroendocrine responses to stress in rainbow trout. Environ. Sci. Technol. 48, 13434–13442. https://doi.org/10.1021/es504331n.

Messias, J.P.M., Paula, J.R., Grutter, A.S., Bshary, R., Soares, M.C., 2016. Dopamine disruption increases negotiation for cooperative interactions in a fish. Sci. Rep. 6, 20817. https://doi.org/10.1038/srep20817.

Milligan, C., Girard, S., 1993. Lactate metabolism in rainbow trout. J. Exp. Biol. 193, 175–193.

Noble, C., Gismervik, K., Iversen, M.H., Kolarevic, J., Nilsson, J., Stien, L.H., Turnbull, J.F., 2018. Welfare Indicators for Farmed Atlantic Salmon: Tools for Assessing Fish Welfare. Nofima Marine, Norway, p. 351 (ISBN 978-82-8296-556-9).

Noble, C., Gismervik, K., Iversen, M.H., Kolarevic, J., Nilsson, J., Stien, L.H., Turnbull, J.F., 2020. Welfare Indicators for Farmed Rainbow Trout: Tools for Assessing Fish Welfare. Nofima Marine, Norway, p. 351 (ISBN 978-82-8296-620-7).

Omlin, T., Weber, J.M., 2010. Hypoxia stimulates lactate disposal in rainbow trout. J. Exp. Biol. 213, 3802–3809. https://doi.org/10.1242/jeb.048512.

Pickering, A.D., Pottinger, T.G., 1995. Biochemical effects of stress. In: Hochachka, P.W., Mommsen, T.P. (Eds.), Biochemistry and Molecular Biology of Fishes. vol. 5. Elsevier, pp. 349–379, https://doi.org/10.1016/S1873-0140(06)80043-3.

Ponte, G., Andrews, P., Galligioni, V., Pereira, J., Fiorito, G., 2019. Cephalopod welfare, biological and regulatory aspects: an EU experience. In: Carere, C., Mather, J. (Eds.), The Welfare of Invertebrate Animals. Springer, pp. 209–228, https://doi.org/10.1007/978-3-030-13947-6_9.

Puglisi-Allegra, S., Andolina, D., 2015. Serotonin and stress coping. Behav. Brain Res. 277, 58–67. https://doi.org/10.1016/j.bbr.2014.07.052.

Raposo de Magalhães, C.S.F., Cerqueira, M.A.C., Schrama, D., Moreira, M.J.V., Boonanuntanasarn, S., Rodrigues, P.M.L., 2020. A proteomics and other omics approach in the context of farmed fish welfare and biomarker discovery. Rev. Aquac. https://doi.org/10.1111/raq.12308.

Reinhoud, N.J., Brouwer, H.J., Van Heerwaarden, L.M., Korte-Bouws, G.A.H., 2013. Analysis of glutamate, GABA, noradrenaline, dopamine, serotonin, and metabolites using microbore UHPLC with electrochemical detection. ACS Chem. Neurosci. 4, 888–894. https://doi.org/10.1021/cn400044s.

RSPCA, 2018. RSPCA Welfare Standards for Farmed Rainbow Trout. RSPCA Farm Animals Department. ISBN: 978-0-901098-20-7 https://science.rspca.org.uk/documents/1494935/9042554/RSPCA+Trout+Welfare+Standards+2020.pdf/3f74881f-577b-d4bb-22f0-a9792a298db6?t=1618819287216.

Ryu, T., Veilleux, H.D., Donelson, J.M., Munday, P.L., Ravasi, T., 2018. The epigenetic landscape of transgenerational acclimation to ocean warming. Nat. Clim. Chang. 8, 504–509. https://doi.org/10.1038/s41558-018-0159-0.

Sadoul, B., Geffroy, B., 2019. Measuring cortisol, the major stress hormone in fishes. J. Fish Biol. 94, 540–555. https://doi.org/10.1111/jfb.13904.

Schreck, C.B., 2010. Stress and fish reproduction: the roles of allostasis and hormesis. Gen. Comp. Endocrinol. 165, 549–556. https://doi.org/10.1016/j.ygcen.2009.07.004.

Schreck, C.B., Tort, L., 2016. The concept of stress in fish. In: Fish Physiology, pp. 1–34, https://doi.org/10.1016/B978-0-12-802728-8.00001-1.

Seibel, H., Weirup, L., Schulz, C., 2020. Fish welfare—between regulations, scientific facts and human perception. Food Ethics 5, 4. https://doi.org/10.1007/s41055-019-00063-3.

Shannon, N.J., Gunnet, J.W., Moore, K.E., 1986. A comparison of biochemical indices of 5-hydroxytryptaminergic neuronal activity following electrical stimulation of the dorsal raphe nucleus. J. Neurochem. https://doi.org/10.1111/j.1471-4159.1986.tb00704.x.

Smith, M.E., Kane, A.S., Popper, A.N., 2004. Acoustical stress and hearing sensitivity in fishes: does the linear threshold shift hypothesis hold water? J. Exp. Biol. 207, 3591–3602. https://doi.org/10.1242/jeb.01188.

Sneddon, L.U., Wolfenden, D.C.C., Thomson, J.S., 2016. Stress management and welfare. In: Fish Physiology, pp. 463–539, https://doi.org/10.1016/B978-0-12-802728-8.00012-6.

Sopinka, N.M., Donaldson, M.R., O'Connor, C.M., Suski, C.D., Cooke, S.J., 2016. Stress indicators in fish. In: Schreck, C.B., Tort, L., Farrell, A.P., Brauner, C.J. (Eds.), Fish Physiology, pp. 405–462, https://doi.org/10.1016/B978-0-12-802728-8.00011-4.

Stevens, C.H., Croft, D.P., Paull, G.C., Tyler, C.R., 2017. Stress and welfare in ornamental fishes: what can be learned from aquaculture? J. Fish Biol. 91, 409–428. https://doi.org/10.1111/jfb.13377.

Stien, L.H., Bracke, M., Noble, C., Kristiansen, T.S., 2020. Assessing fish welfare in aquaculture. In: Kristiansen, T.S., Fernö, A., Pavlidis, M.A., van de Vis, H. (Eds.), The Welfare of Fish. Springer Nature, Switzerland, pp. 303–321. ISBN: 978-3-030-41674-4.

Tintos, A., Gesto, M., Míguez, J.M., Soengas, J.L., 2007. Naphthalene treatment alters liver intermediary metabolism and levels of steroid hormones in plasma of rainbow trout (*Oncorhynchus mykiss*). Ecotoxicol. Environ. Saf. 66, 139–147.

Tintos, A., Gesto, M., Míguez, J.M., Soengas, J.L., 2008. β-Naphthoflavone and benzo(a)pyrene treatment affect liver intermediary metabolism and plasma cortisol levels in rainbow trout *Oncorhynchus mykiss*. Ecotoxicol. Environ. Saf. 69, 180–186.

Vera-Chang, M.N., St-Jacques, A.D., Gagné, R., Martyniuk, C.J., Yauk, C.L., Moon, T.W., Trudeau, V.L., 2018. Transgenerational hypocortisolism and behavioral disruption are induced by the antidepressant fluoxetine in male zebrafish *Danio rerio*. Proc. Natl. Acad. Sci. U. S. A. 115, E12435–E12442. https://doi.org/10.1073/pnas.1811695115.

Vila Pouca, C., Brown, C., 2017. Contemporary topics in fish cognition and behaviour. Curr. Opin. Behav. Sci. 16, 46–52. https://doi.org/10.1016/j.cobeha.2017.03.002.

Vindas, M.A., Madaro, A., Fraser, T.W.K., Höglund, E., Olsen, R.E., Øverli, Ø., Kristiansen, T.S., 2016. Coping with a changing environment: the effects of early life stress. R. Soc. Open Sci. https://doi.org/10.1098/rsos.160382.

Vindas, M.A., Gorissen, M., Höglund, E., Flik, G., Tronci, V., Damsgård, B., Thörnqvist, P.-O., Nilsen, T.O., Winberg, S., Øverli, Ø., Ebbesson, L.O.E., 2017. How do individuals cope with stress? Behavioural, physiological and neuronal differences between proactive and reactive coping styles in fish. J. Exp. Biol. https://doi.org/10.1242/jeb.153213.

Wendelaar Bonga, S., 1997. The stress response in fish. Physiol. Rev. 77, 591–625.

Winberg, S., Nilsson, G.E., 1993. Roles of brain monoamine neurotransmitters in agonistic behaviour and stress reactions, with particular reference to fish. Comp. Biochem. Physiol. C Pharmacol. Toxicol. Endocrinol. https://doi.org/10.1016/0742-8413(93)90216-8.

Winberg, S., Thörnqvist, P.O., 2016. Role of brain serotonin in modulating fish behavior. Curr. Zool. 62, 317–323. https://doi.org/10.1093/cz/zow037.

Wu, H., Ohnuki, H., Ota, S., Murata, M., Yoshiura, Y., Endo, H., 2017. New approach for monitoring fish stress: a novel enzyme-functionalized label-free immunosensor system for detecting cortisol levels in fish. Biosens. Bioelectron. 93, 57–64. https://doi.org/10.1016/j.bios.2016.10.001.

Wu, H., Shinoda, R., Murata, M., Matsumoto, H., Ohnuki, H., Endo, H., 2019. Real-time fish stress visualization came true: a novel multi-stage color-switching wireless biosensor system. Biosens. Bioelectron. 130, 360–366. https://doi.org/10.1016/j.bios.2018.09.042.

Xu, X., Zhang, Z., Guo, H., Qin, J., Zhang, X., 2020. Changes in aggressive behavior, cortisol and brain monoamines during the formation of social hierarchy in black rockfish (*Sebastes schlegelii*). Animals 10, 2357. https://doi.org/10.3390/ani10122357.

Chapter 10

Evolution of ray-finned fish genomes: Status and directions with a primer on microRNA characterization

Richard S. Taylor, Rose Ruiz Daniels, Diego Perojil Morata, Manu Kumar Gundappa, and Daniel J. Macqueen

The Roslin Institute and Royal (Dick) School of Veterinary Studies, The University of Edinburgh, Edinburgh, United Kingdom

1. Introduction

In this chapter, "fish" refers to the ray-finned fish clade (Actinopterygii), which split from the lobe-finned fishes (Sarcopterygii, including tetrapods) approximately 450 million years ago (mya). The issue of what defines a "fish" is considered elsewhere (see Helfman et al., 2009), but for clarity, this chapter excludes cartilaginous fishes (chimaeras, sharks, rays, and skates), jawless fishes (lampreys and hagfishes), and nontetrapod sarcopterygian fishes including coelacanths and lungfishes.

The Actinopterygii is noted for its remarkable species richness and phenotypic diversity, concentrated within the extant teleost group (Fig. 1). Fishes are ubiquitous in aquatic habitats, having adapted morphologically and physiologically to most conceivable environments, finding niches in the deepest oceans, underground caves, highest altitudes, and near hydrothermal vents (e.g., Helfman et al., 2009). The diversity among fishes has not only long fascinated biologists interested in adaptation and evolution, but is also heavily exploited by humans, as we rely on many different species from various taxa for food—harvested from the wild or cultivated through aquaculture (Houston et al., 2020).

The term "genome" encompasses the complete genetic material (i.e., the inherited DNA packaged within the nuclei of each cell) that encodes for and regulates the expression of molecules (i.e., proteins and RNAs) that influence phenotypes developed during life, which are selected across generations. We define "epigenome" as the collection of possible variable biochemical states impacting a genome sequence independent of DNA sequence variation, including

Cellular and Molecular Approaches in Fish Biology. https://doi.org/10.1016/B978-0-12-822273-7.00010-0

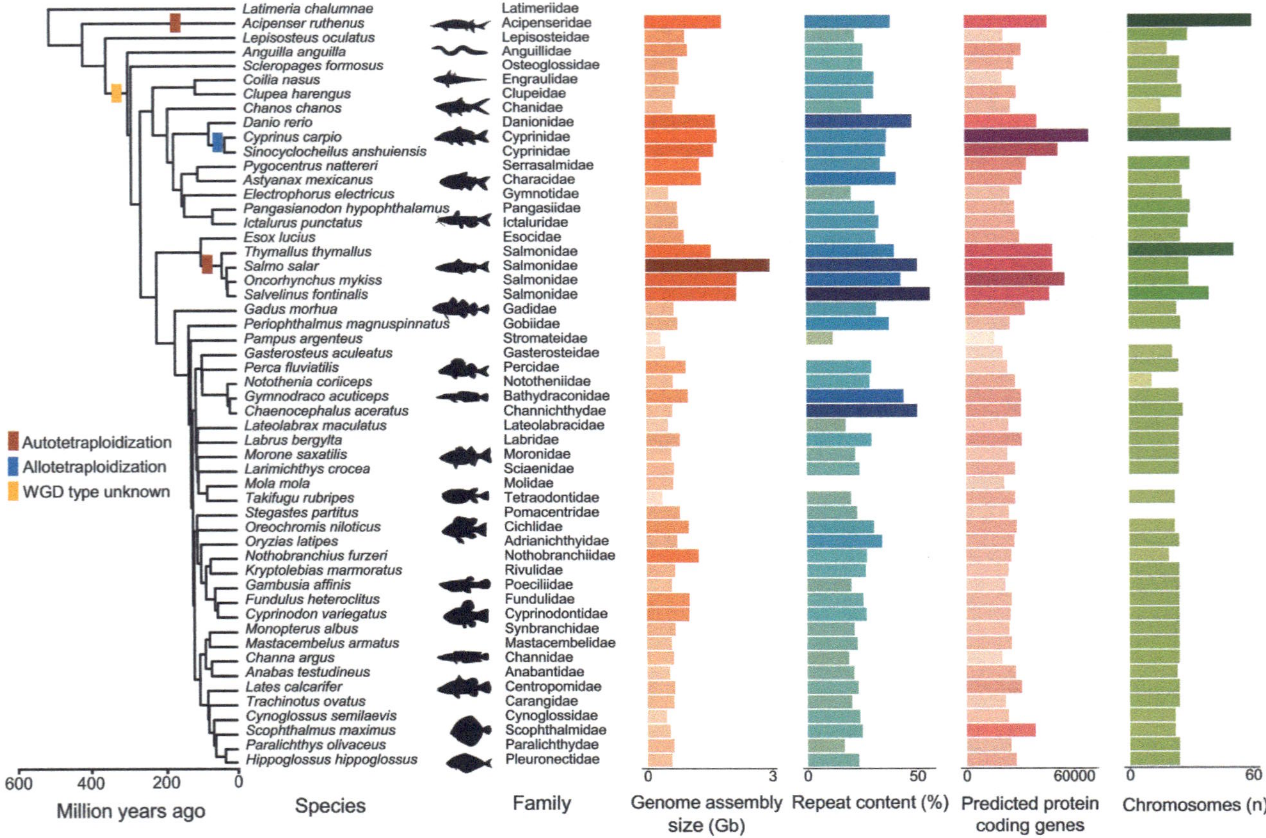

FIG. 1 Genetic diversity of fish genomes highlighted on a time-scaled tree. The phylogenetic time tree was generated from a Newick file provided within Hughes et al. (2018). We retained species to the level of genera within families, where we could identify a reference genome sequence in NCBI, as well as information on genome assembly size, repeat content, predicted number of protein coding genes, and the number of chromosomes (haploid representation). We used ggtree (Yu et al., 2018) to update the original tree by retaining a single representative species in instances where a family had multiple species, leading to 252 tips being removed from the original tree. We used ggplot (Wickham, 2016) to generate the shown barplots (data were missing for some genome features in a small number of species). WGD events in teleost evolution are marked on the tree.

DNA methylation and histone posttranslational modifications, which together modify how DNA is packaged into nucleosomes and available for transcription under different biological conditions (discussed in Section 7.3).

During the last 15 years, our knowledge of fish genomes has advanced dramatically. We have recently moved from a situation where a handful of fish genome sequences were available to having high-quality genome sequences for hundreds of species (e.g., Fig. 1; Section 2), and in some cases numerous individuals *per* species (e.g., Bertolotti et al., 2020). Consequently, fish researchers benefit from an abundance of data to investigate how evolutionary changes in the genome are linked to biological traits of interest.

There are many ways to study the evolution of genomes. This includes comparing different individuals or populations within the same species (i.e., microevolutionary scales). In this chapter we focus mainly on long-term genome evolution (i.e., macroevolutionary scales), approached by comparing different species or distinct regions within the same genomes. The latter is important as fish have experienced multiple rounds of whole genome duplication (WGD) during their evolutionary history, where the entire genetic code was doubled (Fig. 1; Section 3). Such events have allowed the retention of duplicated features, including genes encoding proteins and noncoding RNAs, as well as regulatory sequences controlling transcription, creating new variation and potential for adaptation and change. As expanded later, causative links between WGD and fish diversity have long been proposed, but this is an area of ongoing debate. Importantly, recent advancements in molecular biology are enabling us to annotate the functional molecules and epigenomic features harbored within fish genomes at increasing resolution and breadth, driving forward new levels of understanding.

Therefore the great biological and genetic diversity within fishes makes these animals an exceptional system for studies of genome evolution. The objective of this chapter is to introduce key advances and tools relevant to such investigations, providing a reference for the status and directions in the field. The reader is directed to several excellent reviews for broader or more exhaustive information on many of the themes explored (Roest Crollius and Weissenbach, 2005; Volff, 2005; Glasauer and Neuhauss, 2014; Braasch et al., 2015; Best et al., 2018; Ravi and Venkatesh, 2018). Additionally, we provide a methodological primer, guiding the reader on how to perform a reliable study into the evolution of microRNAs (miRNAs, Section 8).

2. Resources to study fish genome evolution

The second vertebrate genome sequenced and assembled (closely following our own species) was a fish, *Takifugu rubripes*, selected as a study system for its small genome size by vertebrate standards (Aparicio et al., 2002) (Fig. 1). During the 2000s, several other fish species had their genome sequenced and made available to researchers, including the green-spotted pufferfish

(*Tetraodon nigroviridis*) in 2004 (Jaillon et al., 2004), the zebrafish (*Danio rerio*) in 2005 (Howe et al., 2013), and the medaka (*Oryzias latipes*) (Kasahara et al., 2007) and three-spined stickleback (*Gasterosteus aculeatus*) (Jones et al., 2012) in 2006. These consortium efforts provided major up-steps in our understanding of fish genome evolution, for instance compelling evidence of a long-proposed ancestral WGD event in teleosts (Section 3) and major genomic reorganizations that followed in the ancestors to pufferfish, medaka, and zebrafish (Jaillon et al., 2004; Kasahara et al., 2007).

The 2010s saw an increase in the uptake and democratization of next-generation sequencing in fish biology. What followed was an increasing rate of release of genome sequences for many species of commercial, ecological, and evolutionary interest. For instance, 2011 saw the first release of the genomes of Atlantic cod (*Gadus morhua*) (Star et al., 2011) and Atlantic salmon (*Salmo salar*) (Lien et al., 2016). Notably, the same year saw the first release of a genome for a nonteleost member of Actinopterygii, the spotted gar (*Lepisosteus oculatus*), of key evolutionary significance due to its membership to the Holostei (extant gar species and bowfin), the sister group to teleosts (Braasch et al., 2016). Moving into the 2020s, the recent trend has not only been for an increase in the number of genomes for different species (668 different fish species have genome assemblies in the NCBI database as of August 2021; ~90% being deposited since 2016), but also increased representation of species within key families and genera of interest (e.g., salmonids; Fig. 1).

For many species, improved genome assemblies are being released regularly with higher completeness and more accurate annotations of protein coding and noncoding genes (Section 7), supported by technological improvements in sequencing (e.g., long-read sequencing; Amarasinghe et al., 2020) and methods to join assembled genome fragments (contigs) across larger distances into whole chromosomes (e.g., Hi-C), representing the current field standard (e.g., Ghurye and Pop, 2019; Fan et al., 2020). Such advancements, matched by many powerful bioinformatic tools, enable the accurate assembly of fish genomes, even for species showing genetic complexities that challenge this task, e.g., highly similar duplicated regions retained after WGD events in cyprinids (e.g., Chen et al., 2019; Luo et al., 2020), sturgeons (Du et al., 2020), and salmonids (Macqueen et al., 2017) (Section 3). Looking ahead, fishes feature heavily in large cross-national initiatives to sequence vertebrate genomes *en masse* including the Genome 10k effort (Fan et al., 2020), as well as the Earth BioGenome Project, aiming to sequence all eukaryotic species during the next decade (Lewin et al., 2018).

There have also been major advancements in high-throughput functional genomic technologies to study how sequences in the genome are expressed, regulated, and interacting. This includes RNA sequencing (RNA-Seq) methods (Stark et al., 2019), but also a range of approaches providing insights into the epigenomic status of tissues and cells, which are likely to be increasingly used to address evolutionary questions (Section 7). Today's student of fish genome evolution thus benefits from an abundance of data and different data types, and the increasing challenge is how to exploit it meaningfully. Several online frameworks support comparative and evolutionary analyses of fish genomes. The Ensembl genome browser (Yates et al., 2020) offers a powerful set of resources and tools to compare the genomes of different animal species, with the latest release including >80 fish species. Genomicus is another online set of tools supporting studies of fish genome evolution, allowing visualization and analysis of genes and chromosomes in a positional context (Nguyen et al., 2018).

3. Overview of fish genome evolution

Fish genomes have diversified during evolution to an extent rivaling the phenotypic and ecological diversity mentioned earlier, leading to substantial variation in genome size, gene number, and chromosome number (Fig. 1). It is not our intention to review all aspects of fish genomic diversity in detail, rather we outline some of the most important features to consider in evolutionary studies.

3.1 WGD events—Introduction

WGD events are a pervasive feature of fish genome evolution. WGD is a class of gene duplication, the main way new genes are created from existing ones via several mechanisms (reviewed by Zhang, 2003). Duplication events create two gene copies that initially share the same function. While duplicated genes often quickly return to single copy during evolution, they can typically diverge as independent loci, allowing new or specialized functional roles and expression patterns to evolve (Section 7.4). Alternatively, selection can act to maintain ancestral functions or expression after retention of both duplicated copies. Gene duplication may affect sections of genes, entire genes, chromosomal regions containing many genes, or all chromosomes (and hence all genes in concert) after WGD, leading to higher ploidy levels. Duplicated genes can reside in proximal locations on the same chromosome (i.e., tandem gene duplicates) or on different chromosomes after WGD and potentially gene retrotransposition events (Zhang, 2003). Duplicated genes within the same genome are termed "paralogues," in contrast to "orthologues," which are the same genes found in the genome of different species (i.e., genes that have been separated during speciation events) (Koonin, 2005). "Ohnologue" is often used to refer specifically to paralogues retained from WGD events (Wolfe, 2000), honoring the contribution of Susumu Ohno in this area (discussed later). Bioinformatic tools exist to support the inference of such homology relationships among large gene sets, including "Orthofinder" (Emms and Kelly, 2019) and related tools (Kapli et al., 2020), in addition to annotations available through the Ensembl genome browser.

While many duplicated genes are initially polymorphic within a species (i.e., as structural variants, e.g., Bertolotti et al., 2020), across macroevolutionary scales, gene duplication is a key driver for the expansion and functional diversification of gene families across species. Examples include opsin genes, which have diversified during the adaptation of fish visual systems to different environments (e.g., Lin et al., 2017; Liu et al., 2019), globin genes involved in oxygen transport and oxygen-based metabolic reactions (Storz et al., 2013; Gallagher and Macqueen, 2017), and immune gene families involved in defense to pathogens (e.g., Stein et al., 2007).

Returning to WGD, these events have impacted fish genome evolution on multiple evolutionary timescales (Fig. 1). It is important to remember that every ancient WGD event initially arose in one or a small number of individuals within a species, and much like DNA mutations, may change in frequency leading to populations with different ploidy levels, e.g., as seen in some sturgeons (Vasil'ev, 2009). In species where all members share an ancestral WGD event, the initial genome doubling was fixed during evolution, either replacing the ancestral ploidy state, or potentially by creating a new lineage.

3.2 Types of WGD event—Auto vs. Allo

WGD comes in two main "flavors"—the doubling of the same genome within a species or of two genomes following or concurrent to hybridization of different species (Fig. 2). These two main routes to WGD, autopolyploidization and allopolyploidization, respectively, classically lead to either four identical chromosome sets that can initially pair randomly during meiosis (Fig. 2A), or two distinct sets of chromosomes that pair independently during meiosis (Fig. 2B) (also see Otto, 2007). Studies of multiple taxa including fishes have shown that major changes in the genome inevitably follow both types of WGD, implying selection for a more diploid state. A suite of processes are involved, together termed rediploidization (Wolfe, 2001), involving a range of physical and temporal scales. We will return to rediploidization in Section 4.

The term "paleopolyploid" is widely used to refer to lineages that have completed, or are in an advanced stage of rediploidization process, i.e., they underwent an WGD event in the past, but there may be few obvious traces of the originally duplicated chromosome sets—in such cases, the retention of ohnologues may be one of the main remaining clues. Many fish genomes, including most shown in Fig. 1, despite being diploid in their complement of chromosomes and inheritance patterns, are paleopolyploid due to ancient WGD events at the stem of vertebrate or teleost evolution (Sections 3.3 and 3.4).

3.3 WGDs with deep vertebrate origins

There exists a long history of research into WGD events that occurred at the stem of the vertebrate clade, which students of fish genome evolution should delve into deeply; we cannot cover all this literature here. Briefly, Ohno and colleagues were the first to argue that one or more WGDs occurred at an early

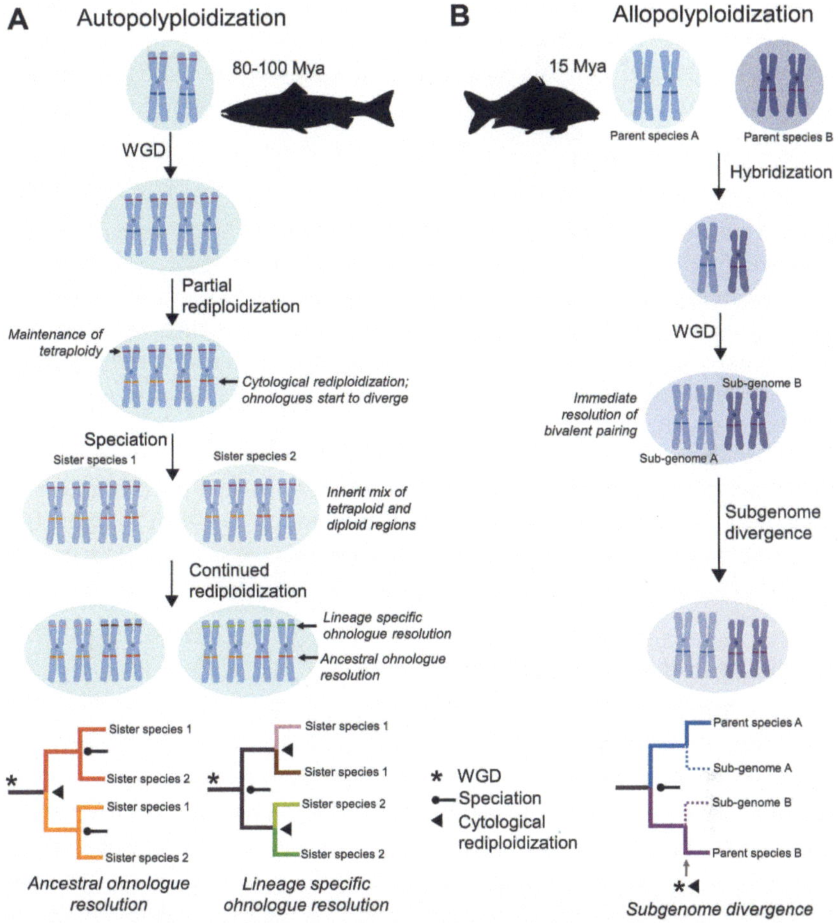

FIG. 2 Cytological rediploidization after WGD and its phylogenetic outcomes. (A) Rediploidization following autopolyploidization, involving spontaneous doubling of the same genome, exemplified by the salmonid Ss4R WGD. We emphasize different rates of rediploidization in different regions of the genome, meaning some loci may remain tetraploid for long periods of evolution. We have depicted the evolutionary scenario observed in salmonids, where speciation occurred in advance of cytological rediploidization in large genomic regions. Subsequently, lineage-specific rediploidization occurred, leading to lineage-specific ohnologue resolution (LORe). The phylogenetic expectation of LORe for ohnologue pairs in sister species is contrasted against genomic regions where rediploidization was ancestral. (B) Rediploidization following classical allopolyploidization, involving genome doubling following the hybridization of two species, exemplified by the WGD event shared by common carp and goldfish. Here, cytological rediploidization is resolved immediately, leading to two subgenomes on different evolutionary paths from the outset. In a phylogenetic analysis, the two subgenomes will branch more closely to their parent species of origin (or its lineage) than one another. As mentioned in the text, after segmental allopolyploidization, the pattern of evolution shown in part A is possible.

"fish" stage of vertebrate evolution, promoting evolutionary innovations impossible without large-scale gene duplication (Ohno et al., 1968; Ohno, 1970). This hypothesis has gained increasing support through the years as more genomic evidence has accumulated (key references include Garcia-Fernández and Holland, 1994; Dehal and Boore, 2005; Putnam et al., 2008).

Though specific details about the timing and mechanisms involved are the subject of continued debate, it is widely accepted that at least one WGD round occurred in the common vertebrate ancestor (Smith et al., 2018), with strong evidence for an additional round either in the common vertebrate ancestor (Sacerdot et al., 2018) or in the ancestor to jawed fishes (i.e., including cartilaginous and bony fishes) after the divergence from jawless fish (Simakov et al., 2020). The two WGD events in jawed vertebrate evolution are often referred to as "2R" (two WGD rounds) and are shared by all Actinopterygii members. These events have left an indelible imprint in all vertebrate genomes through the retention of ohnologues with functions biased toward development, transcriptional regulation, and intracellular signaling (e.g., Putnam et al., 2008), allowing the evolution of novel complexity in the regulation of gene expression when compared to chordates lacking 2R (Marlétaz et al., 2018).

3.4 WGD in the teleost ancestor

Additional WGD events within fish evolution are well established. A highly studied WGD event occurred in the common ancestor to teleosts ~ 300–350 mya (Jaillon et al., 2004; see reviews by Volff, 2005; and Glasauer and Neuhauss, 2014) (Fig. 1). As this event, often called Ts3R (i.e., "teleost-specific 3rd WGD round"), was followed by the apparent diversification of teleosts, it is often speculated to be causatively linked to species radiation and the evolution of phenotypic novelty. The underlying rationale here is logical, considering the known retention of duplicated molecular pathways that have evolved new functions and expression, perhaps offering increased scope for adaptation (e.g., Section 7.4; Glasauer and Neuhauss, 2014). Certainly, there are good examples of Ts3R ohnologues that have contributed to teleost-specific innovations, e.g., in cardiac function (Moriyama et al., 2016). Further, as for 2R, there is clear evidence that Ts3R ohnologues are (i) biased in their retention toward functions that imply molecular complexity and (ii) have frequently evolved novel/specialized expression patterns (e.g., Kassahn et al., 2009).

However, when viewed through a nongenomic perspective, it has been shown that (i) while species diversification rates increased immediately after Ts3R, more substantial species radiations in teleost evolution were detached from Ts3R by many tens of millions of years (Santini et al., 2009; Sallan, 2014; Davesne et al., 2021) and (ii) immediately following Ts3R, phenotypic diversification occurred at a similar rate in teleosts and their sister group Holostei (lacking Ts3R) (Clarke et al., 2016). Such studies remind us that ideas about WGD causing immediate bursts of evolutionary diversification must be considered with due caution. On the other hand, such studies do not rule out that the increased variation provided by WGD events leads to increased adaptive potential that need not be realized until ecological opportunities arise, prompting diversification tens of millions of years later (e.g., Macqueen and Johnston, 2014; Robertson et al., 2017), though such ideas have been questioned (Laurent et al., 2017). Scientific debate about the role of WGD events in the evolutionary diversification of fishes and other organisms will rightly continue long into the future.

3.5 Further WGDs in fish evolution

Several independent WGD events occurred in actinopterygians outside teleosts, and indeed a range of WGD events are found nested within the teleost clade (Fig. 1). A recent study presented a high-quality genome for the sterlet (*Acipenser ruthenus*), a representative of the sturgeons (Acipenseriformes; also including paddlefishes), with strong evidence for an autopolyploidization event ~180 mya (discussed further in Section 4) (Du et al., 2020) (Fig. 1). Additional independent WGD events are suspected in sturgeon (Ludwig et al., 2001) and paddlefish (Cheng et al., 2021) evolution. As pointed out by others (Sallan, 2014), the Acipenseriformes are extremely species poor and correspondingly less diverse compared to teleosts, despite their history of repeated WGD events, again arguing for caution when linking WGD to the evolution of biological diversity.

Another autopolyploidization event occurred in the ancestor to the salmonid family (Salmonidae), a group of high societal value and scientific interest (Houston and Macqueen, 2019). This event, often termed Ss4R (i.e., "salmonid-specific 4th WGD round"), occurred ~80–100 mya (Berthelot et al., 2014; Macqueen and Johnston, 2014) (Fig. 1). Despite being recognized for well over 50 years (Ohno et al., 1968; Allendorf and Thorgaard, 1984), the recent characterization of the Atlantic salmon genome (Lien et al., 2016) provided new insights into Ss4R and its impact on genetic and biological diversity. One of the main findings concerned how the rediploidization process occurred following autopolyploidization, which is further explored in Section 4, and we will return to Ss4R in several additional sections.

Allopolyploidization is a common mode of WGD in cyprinid fishes (Cypriniformes, Cyprinidae) with strong evidence for both recent (i.e., population level) (e.g., Luo et al., 2014) and ancient events including a concentration of independent WGDs ~15 mya (Li and Guo, 2020) (e.g., Fig. 1). These WGDs were proposed to have promoted adaptation and species diversity, and were possibly stimulated by climate change (Li and Guo, 2020). An allopolyploidization that occurred around this time in the ancestor to common carp (*Cyprinus carpio*) and domestic goldfish (*Carassius auratus*) has been recently studied in a series of detailed genomic analyses (i.e., Chen et al., 2019; Xu et al., 2019; Luo et al., 2020). These studies collectively provide insights into the ancestral lineages that hybridized before WGD, in addition to rediploidization (Section 4) and gene expression evolution after allopolyploidization. An additional suspected allopolyploidization occurred in the common ancestor of suckers (Catostomidae) (Ferris, 1984), another Cypriniformes family. A new genome assembly for a representative catostomid species indicated this WGD event occurred ~25 mya (Krabbenhoft et al., 2021).

4. Evolutionary consequences of rediploidization—No "one size fits all"

Until recently, there was insufficient data to establish the extent of variability in the rediploidization process across different WGD events. This is an important topic considering the large impact of WGD events on fish genome evolution. Recent analysis of the genomes of fish species with independent autopolyploidization and allopolyploidization events is creating valuable insights in this area.

4.1 Rediploidization begins at the chromosomal level

Rediploidization is intricately linked to the pairing of duplicated chromosome sets during meiosis. This differs most depending on whether a WGD event resulted from autopolyploidization or allopolyploidization, with consequences for ongoing genome evolution. After autopolyploidization, this level of rediploidization involves an evolutionary transition from a tetraploid mode of pairing (with random pairing of four chromosome sets) to a preferred bivalent pairing of two pairs of chromosome sets (Fig. 2A). Studies of the Atlantic salmon genome suggest that genetic changes, particularly large chromosomal rearrangements such as inversions (i.e., the flipping of DNA on different strands, which suppresses recombination), were drivers for this process after Ss4R (Lien et al., 2016). The important outcome is a switch from a tetraploid mode of inheritance where four alleles are maintained by recombination, to a diploid mode of inheritance, marking the "birth" of ohnologues that can genetically diverge from each other as two nonrecombining loci on distinct chromosomes (Fig. 2A).

From a functional perspective, because major evolutionary changes in ohnologue function and expression are expected to require significant genetic divergence (Section 7.4), the importance of this mode of cytological rediploidization after autopolyploidization is emphasized. The same transition does not occur following classical allopolyploidization events, when the two hybridizing species have distinct genomes that immediately pair preferentially as two sets during meiosis (Fig. 2B). It is likely that WGD in such cases resolves the lack of sufficient homology between the two hybridizing species genomes, allowing viable meiosis and sexual reproduction (Ohno, 1970). However, it should be noted that when two hybridizing species are very closely related, and chromosomes or chromosomal regions are insufficiently divergent to form preferential pairings, an autopolyploid mode of cytological rediploidization is possible (Otto, 2007). This is often termed segmental allopolyploidization.

After classic allopolyploidization events, the term "subgenome" (e.g., subgenome A and B) is often used to refer to the two genomes in the hybridizing "parent" species (Fig. 2B). Following the ancestral carp/goldfish (i.e., 15 mya) and catostomid (i.e., 25 mya) WGDs, the two subgenomes can be distinguished along entire chromosomes in genomic analyses (e.g., Xu et al., 2019; Kon et al., 2020; Luo et al., 2020; Krabbenhoft et al., 2021). Further, a potential ancestral parent lineage for one of the subgenomes was identified for the carp/goldfish WGD event (Xu et al., 2019). Due to the typically close genetic relatedness of hybrids, the two subgenomes start out with a similar complement of genes, and previously orthologous genes (i.e., existing before hybridization) become ohnologues when present within the same nucleus/organism (Fig. 2B). The allopolyploid rediploidization process concerns the genetic/functional divergence and loss of ohnologues and other features between the two subgenomes. It has been observed that one of the subgenomes becomes more dominant in terms of gene expression (e.g., reported in carp; Xu et al., 2019) or evolves at a faster rate (e.g., reported in goldfish; Kon et al., 2020), though this is not always the case (Krabbenhoft et al., 2021). A recent analysis demonstrated that both subgenomes contributed to phenotypes selected during the domestication of goldfish, indicating their combined ongoing importance (Kon et al., 2020).

Subgenomes are not part of autopolyploid rediploidization. Analysis of the Atlantic salmon genome revealed that salmonid genomes comprise 98 pairs of duplicated blocks, ranging in size from a few million DNA bases to entire chromosome arms. Each duplicated block previously underwent the transition described before (Fig. 2A), likely as larger regions, with subsequent rearrangements occurring throughout the genome, mixing up the ancestral organization (Lien et al., 2016). A similar pattern was observed during analysis of the sterlet genome (i.e., that underwent an independent autopolyploidization), with the genome being on a related pathway of rediploidization, though perhaps at a slower evolutionary rate (Du et al., 2020).

Fascinatingly, when studying the level of genetic divergence between duplicated blocks retained from Ss4R, it is clear that cytological rediploidization did not occur at an even temporal rate across the genome (Lien et al., 2016), and this has interesting consequences for long-term evolution (Section 4.2). Indeed, several duplicated blocks, representing around 25% of the genome, underwent cytological rediploidization long after the WGD event. These duplicated blocks are identified by their lower sequence divergence ($<5\%$) compared to the $>10\%–15\%$ divergence level observed across a larger fraction of the genome that underwent cytological rediploidization earlier, and thus have had more time to diverge (Lien et al., 2016). Variability in rediploidization rate across the genome after Ss4R has long been recognized (Allendorf and Thorgaard, 1984), as some regions (as much as 20% of all the DNA), particularly near the tips of chromosomes (telomeres), show evidence of tetraploid meiotic pairing and inheritance in some salmonids (Allendorf et al., 2015), meaning cytological rediploidization is still not resolved 80–100 million years after WGD.

4.2 Lineage-specific outcomes of rediploidization

The importance of a variable rate of cytological rediploidization in different regions of the genome comes into sharp focus when considering the potential for lineage-specific outcomes in sister species sharing the same WGD event. Robertson et al. (2017) described a process called lineage-specific ohnologue resolution (LORe), where the evolutionary transition from tetraploid to diploid meiotic pairing occurs *after* lineages sharing the same WGD event are permanently separated by speciation (Fig. 2A). Consequently, cytological rediploidization can occur independently in different species, leading to independent ohnologue genetic and functional divergence. LORe is highly evident in salmonid genomes and has been classified globally using phylogenetic methods (see Section 6.2). About one-quarter of salmonid genomes experienced LORe in different salmonid subfamilies, which split off ~50 mya (i.e., tens of millions of years post Ss4R). Thus comparing an Atlantic salmon (or any member of subfamily Salmonidae) to any grayling species (subfamily Thymallinae), thousands of ohnologues commonly derived from Ss4R have diverged independently, including at the functional levels, as explored in later sections.

LORe is likely to have occurred after autopolyploidization WGD events other than Ss4R, in addition to segmental allopolyploidization events (Robertson et al., 2017), though this is yet to be proven unequivocally. However, there certainly exists evidence consistent with LORe following Ts3R (see Martin and Holland, 2014; Robertson et al., 2017; Rozenfeld et al., 2019), and the recently published sterlet and paddlefish genomes offer additional fish WGDs to examine for this evolutionary process.

5. Transposable elements as agents of fish genome evolution

Another facet of fish genome biology to explore briefly are transposable elements (TEs). TEs are mobile sequences representing a range of different classes with the ability to move locations and proliferate within and between genomes via several mechanisms. TEs are ubiquitous in all eukaryotic genomes, and their proliferation leads to repeat sequences typically of high copy number, which represent one of the major explanatory factors for differences in genome size among fish species (Fig. 1). They represent agents for mutation, structural changes and rearrangements in genomes, and rarely promote the evolution of genetic novelties and innovations. For recent reviews about TE biology and evolution, see Sotero-Caio et al. (2017) and Bourque et al. (2018).

5.1 TEs and their roles in fish evolution

TE insertions create cleavages and mutations throughout genomes that are rarely beneficial, meaning the eventual location of TE repeats is not random, partly due to selection against deleterious effects (Bourque et al., 2018). As a result, an evolutionary arms race between TEs and genomes has been ongoing since or before the dawn of complex life, which may explain the evolution of epigenetic silencing, one of the main tools genomes use to limit TE proliferation (McDonald, 1998). While most TEs have been long silenced by the host genome, they leave a legacy of repeat sequences of different ages. Such repeats promote rearrangements in chromosomes beyond the impact of TE insertions, promoting structural mutations such as inversions, which may play an important role following WGD events. For example, Ss4R was followed by a major proliferation of TEs (de Boer et al., 2007) that likely promoted rearrangements driving cytological rediploidization (Lien et al., 2016). A similar burst of TE proliferation occurred after the sterlet WGD (Du et al., 2020). Considering these two independent autopolyploidizations, it can be argued that WGD events promote TE expansion due to selection for rediploidization or the dysregulation of TE repression mechanisms (Lien et al., 2016). However, a similar signal was not detected following Ts3R, which may have eroded due to the passage of a much longer period of time, or because the same expansion never occurred (Chalopin and Volff, 2017).

The proportion of repeat sequences derived from TEs is highly variable across fish genomes (Fig. 1) ranging 10-fold from ~6% to ~60% of the total genome size (e.g., Chalopin and Volff, 2017). The mechanisms driving such diversity are not well understood, but include an important horizontal component, with TE transmission across species highly evident. A recent analysis showed that >97% of all horizontal TE transfers in the vertebrate lineage involved Actinopterygii species, indicating a propensity for this type of TE accumulation in fishes (Zhang et al., 2020), which may be facilitated by pathogens such as viruses (Gilbert and Cordaux, 2017).

Conflicts between TEs and their host are reduced by the coevolution of strategies that modify the cost of TE propagation, promoting the emergence of TE-related functions. One example is the accumulation of repetitive TEs into telomeres to limit telomere degradation (Cosby et al., 2019). Another is that TEs are an important source for the biogenesis of noncoding RNAs (Section 7.2), some of which may acquire functions selected during evolution (Chuong et al., 2017). As TEs can insert anywhere in the genome, including coding and regulatory regions of genes, they may rarely become directly "coopted" due to beneficial changes in gene function and/or expression (Chuong et al., 2017; Cosby et al., 2019). A nice example comes from the cooption of two distinct TEs that introduced novel transcriptional regulation involved in sex determination in medaka (Herpin et al., 2010; Schartl et al., 2018). At more recent evolutionary scales, a recent study in cichlids reported changes in visual sensitivity between closely related species that was explained by TE insertions that altered the expression of opsin genes (Carleton et al., 2020), following previous work that associated TEs with the rapid diversification of cichlid fishes (Brawand et al., 2014). Thus TEs have clearly shaped fish genomic diversity on multiple levels, and many exciting findings about their evolutionary significance are anticipated in the near future, greatly assisted by the latest generation of genome assemblies, which have a more accurate representation of TE sequences.

6. Fish phylogenomics

Phylogenomics is a modern framework for evolutionary investigations using phylogenetic methods and genomic datasets. A phylogenetic tree is a way to interpret and visualize the evolutionary descent of biological characters, allowing us to ask which are ancestral or derived alongside other possible scenarios. Phylogenetic analysis is widely used to reconstruct the evolution of homologous genetic characters, e.g., DNA sequences and molecules encoded by DNA (e.g., amino acid sequences), which in turn allows hypotheses on species relationships or gene family evolution to be explored. The different steps used in phylogenetic analysis and the various methods available (including their strengths and weaknesses) sit outside the scope of this section but are briefly explored later (Section 8). While large genomic datasets often improve the quality of phylogenetic analysis, even an infinite number of markers may fail to resolve some evolutionary problems and hypotheses (e.g., species relationships), and the choice of genetic marker, data type, and analysis method is considered paramount. The reader is encouraged to read around these issues in comprehensive recent reviews (see Kapli et al., 2020). Our main aim here is to highlight some of the insights that phylogenomic studies are generating on the evolution of fish species and genomes.

6.1 Phylogenomics to infer species relationships

Phylogenetic analyses of increasingly large sequence character datasets have allowed the evolutionary relationships of many fish species to be defined with higher resolution in recent years, rewriting classic hypotheses on several relationships and the timing of species divergence (e.g., Betancur-R et al., 2017). This represents important progress and is necessary to frame many findings in evolutionary genomics in a broader biological context. The typical design of such studies aims to generate a large volume of orthologous sequence data in a set of species of interest, capturing the vertical descent of genetic variation to the exclusion of paralogous genes, which can be technically challenging (Kapli et al., 2020).

An example of a recent phylogenomic study reported the evolutionary relationships of species representing over 90% of known ray-finned fish orders using a genomic and transcriptomic dataset of > 1000 orthologous exon sequences (translated regions of mRNAs left after splicing) filtered for confounding paralogous sequences (Hughes et al., 2018). Another recent analysis used a targeted sequencing approach to obtain > 1000 orthologous ultraconserved DNA elements sampled in 118 species from the extremely diverse spiny-rayed fish clade (Acanthopterygii), which account for more than half of all ray-finned fish species (Alfaro et al., 2018). A related study used a similar dataset at a finer scale of taxonomic resolution to infer relationships and divergence times among a spiny-rayed fish clade called Pelagaria, including tunas and related pelagic species (Friedman et al., 2019). Such studies have resolved or confirmed species relationship hypotheses that have proven historically equivocal or challenging. By adding temporal constraints into the analysis, it becomes possible to test competing hypotheses about the tempo and drivers of species diversification, and potentially their link to genomic features such as WGD (e.g., Santini et al., 2009; Macqueen and Johnston, 2014; Li and Guo, 2020). The future of this field will involve further expansions in data volume across an increasing number of taxa. At some point in the coming decades it seems likely that the evolutionary relationships and diversification mode of most living fish groups will be defined to the level of genera and potentially species (Alfaro, 2018).

6.2 Phylogenomics for investigations of WGD events

Phylogenetic analysis can also be used to understand genome evolution, above and beyond its proven applications for resolving evolutionary histories within gene families of interest (e.g., Storz et al., 2013; Gallagher and Macqueen, 2017; Lin et al., 2017; Section 8). One application is for inferring and dating large-scale patterns of gene duplication in genomes, which may provide insights into WGD events. Robertson et al. (2017) provide an example of such an approach, in a study where hundreds of phylogenetic trees were generated, cosampling the diversity of diverse salmonid species alongside the evolutionary history of Ss4R ohnologues spread across the genome. Here, the combined use of phylogenetics and comparative genomics provided evidence for LORe, revealing that large, duplicated blocks in salmonid genomes shared unique rediploidization histories, several of which were lineage specific (Robertson et al., 2017) following the process outlined in Fig. 2A.

Phylogenetic approaches incorporating ohnologue data sampled across the genome have similarly been used to infer the timing of Ss4R and rediploidization across different duplicated blocks of the Atlantic salmon genome (Lien et al., 2016) and as cumulative evidence supporting allopolyploidization in carp (Xu et al., 2019). Other studies have used phylogenetic trees including high-confidence ohnologues from multiple teleosts to reveal temporal patterns of ohnologue loss following Ts3R (Inoue et al., 2015). Such studies highlight the fact that WGD events create phylogenetic and positional genomic expectations that can be coexploited, and a recent study provided an approach to combine such signals for more accurate phylogenetic analysis (Parey et al., 2020).

7. Evolution of "functional" regions in fish genomes

Fish genomes harbor many sequence elements with ascribed biological functions, which are of interest when investigating the molecular and evolutionary basis of traits. Before proceeding, we must first acknowledge the potential criticism surrounding what defines "function" in the context of genome biology, and the importance of considering this in an evolutionarily grounded framework. It is well known that not all sequences that are expressed, show biochemical modifications, or are subject to measurable regulation, are performing "functions" that influence phenotypes under selection. For example, a lot of detectable transcriptional activity seems to represent "noise" in the system (e.g., Pertea et al., 2018), including a low level of expression of repetitive sequences and pseudogenes (that have lost protein-coding potential through debilitating mutations) lacking any obvious roles. In other words, a substantial portion of DNA expressed may reflect an evolutionary by-product of functional activities that are necessary for organismal survival. Clearly it becomes important to consider whether measurable genomic activities under investigation are visible to selection, which is best inferred through the lens of evolution, for instance in terms of conservation in DNA sequence or epigenomic state.

It is not our objective to further explore what comprises biological function in a philosophical sense and critical reviews addressing such issues are available elsewhere (Graur et al., 2013; Palazzo and Lee, 2015; Doolittle and Brunet, 2017; Linquist et al., 2020). Needless to say, a cautious approach to inferences of genomic function is advocated. In the rest of this section we provide an overview of the main sequence elements and layers of genomic regulation with *potential* to influence trait variation, and scope to be visible to natural selection and hence evolutionary constraint.

7.1 Protein coding regions

Genome research has long focused predominantly on protein coding genes, making up a minor fraction (usually 1%–2%) of the total DNA in vertebrate genomes. Coding genes are transcribed as precursor mRNAs in the nucleus, where they undergo splicing to remove introns, leaving a mature mRNA, which is exported out of the nucleus and translated into a string of amino acids that in turn folds into a protein. Many sequences and epigenomic layers of regulation within genomes (discussed later) are known to influence the quantity, as well as temporal and spatial distribution of proteins within cells, tissues, and the bloodstream. The small fraction of the genome dedicated to coding regions highlights that genome size is largely decoupled from potential for producing proteins. The number of coding genes per fish species ranges over ~2-fold magnitude whereas genome size varies ~10-fold (e.g., Fig. 1). A key explanator for variation in coding gene number is ohnologue retention after WGD, with salmonids, sturgeons, and cyprinids clearly having an up-step in gene number compared to related species lacking the respective WGD events (Fig. 1).

The functions of coding genes are evidenced at multiple levels, including the maintenance of an intact open reading frame encoding a protein that shows detectable evolutionary conservation and expression regulation in response to stimuli that matches with functional expectations (e.g., a gene coding an antiviral protein is expected to be transcriptionally activated in response to viral infection). The conservation and robust expression of the majority of coding genes aids their identification during the computational annotation of genomes.

Coding genes are widely utilized in evolutionary and functional genomics, including to understand adaptation to different environments. For example, using transcriptome approaches that capture all the expressed genes, signatures of adaptation were reported in coding genes linked to high altitude in cyprinid fish (e.g., Tong et al., 2017). A study reporting the genome sequence of a snailfish (Liparidae) living at depths greater than 6 km in the Mariana trench offered evidence for adaptation in coding genes that may enable such a hadal existence (Wang et al., 2019). Coding genes are also widely utilized to explore patterns of evolution after WGD, including changes in selection pressure and gene expression linked to genome-wide functional redundancy (Section 7.4).

7.2 Noncoding regions

Identifying "function" within the majority noncoding fraction of genomes is one of the great challenges of contemporary genome biology. It has long been evident that much of the variation in genome size is due to sequences evolving under little or no selective constraint (e.g., most repeat sequences), often termed "junk DNA." Nonetheless, it is also important to note the surprise following the initial sequencing of the human genome concerning the small number of protein coding genes (20,440 in the most current Ensembl 104 annotation of the GRCh38.p13 assembly, released in 2019), compared to the 100,000 or more anticipated among the scientific community (Graur et al., 2013). It was realized that humans (and the same is true for fishes) do not have a particularly greater number of coding genes than simpler animals, however that is judged phenotypically. For example, the nematode *Caenorhabditis elegans* has 20,191 coding genes predicted in its genome (according to Ensembl 104 annotation of the WBcel235 assembly, released in 2013). This reinforced the long proposed idea that there must exist more complex regulation of how the coding fraction of vertebrate genomes is expressed. Indeed, comparison of vertebrate genome regulation to closely related chordate invertebrates empirically supports this notion (Marlétaz et al., 2018).

It is now apparent that noncoding regions are of great importance in the biology and evolution of multicellular organisms, including conserved regulatory elements that modulate the transcription of coding genes (Wray, 2007; Douglas and Hill, 2014), as well as genes that express RNA with regulatory roles. A small component of noncoding genomes, often termed conserved noncoding elements ("CNE", or sometimes ultraconserved CNE), shows unusually high conservation across vertebrate taxa. These CNEs tend to be located near conserved developmental genes, with many likely acting as enhancers, i.e., short DNA regions located nearby genes that bind to transcription factor proteins to increase transcription rates under certain conditions/stimuli (e.g., Stephen et al., 2008). It was also recently shown that the extreme conservation of some CNEs is linked to their role in the higher order structure and regulation of chromatin (Harmston et al., 2017). Interestingly, CNEs retained from sharks to humans are evolving at a faster rate in teleost fishes, with losses of CNEs perhaps linked to functional redundancy among duplicated regions retained following Ts3R (Lee et al., 2011).

The noncoding portion of genomes expresses many RNAs, which have been characterized to a variable extent in fishes. Several have fundamental biochemical functions within all cells, including ribosomal RNA and transfer RNA. We do not consider such well-defined molecules here, and instead introduce more recently discovered noncoding RNAs that may play regulatory roles. Noncoding RNA comprises two main classes, small RNAs (sRNAs) (Chapman and Carrington, 2007) and long noncoding RNAs (lncRNAs) (Ulitsky, 2016). sRNAs have regulatory and silencing functions and include, among others, the miRNAs, which operate at the posttranscriptional level to regulate mRNA translation and stability (miRNAs are further explored in Section 8 and in Chapter 4).

An interesting class of less-studied sRNAs are piwi-interacting RNAs (piRNAs), which have similar roles across the animal kingdom, including fishes (Houwing et al., 2007), including a primary role of TE silencing and maintenance of genome integrity during germ cell development, but also mRNA decay, among others (reviewed in Ramat and Simonelig, 2021). The mechanisms by which piRNAs are born and then silence TEs are being defined across taxa, and it is clear that the broader set of molecules comprising the Piwi-piRNA system is evolving rapidly, including in fishes (Yi et al., 2014) perhaps due to an arms race between the host genome and TE "parasite" (reviewed in Parhad and Theurkauf (2019)). It will be interesting in the near future to investigate whether the burst of TE proliferation following some WGD events (e.g., in salmonids and sturgeon) is linked to the Piwi-piRNA system and if so, how this played out during the rediploidization process.

lncRNAs are usually defined as transcribed RNAs of > 200 nucleotides (nt) in length, which lack coding potential, despite otherwise sharing several features with mRNAs (Ulitsky, 2016). A vast number of lncRNAs have been discovered in humans, transcribed from at least as many unique loci as coding genes (e.g., Pertea et al., 2018). While a range of possible roles are assigned to lncRNAs (e.g., Lee et al., 2019), this class of noncoding RNA exemplifies the challenge in assigning functions to noncoding regions, with a large and rapidly growing literature addressing this problem, characterized by substantial debate. Indeed, while a small number of lncRNAs have defined functions studied across taxa, it has been challenging to demonstrate a phenotype for many tested lncRNAs (Gao et al., 2020). For instance, a recent study in zebrafish genetically deleted 25 separate lncRNAs, including several that had previously been ascribed functional roles, and failed to observe phenotypic effects on fitness (Goudarzi et al., 2019). At the same time, signatures of varying evolutionary constraint have been demonstrated for many vertebrate lncRNAs, including fishes, albeit these molecules on average evolve very rapidly both at the sequence level and in terms of gains and losses during evolution. Thus most detectable vertebrate lncRNAs are lineage specific and more than 70% were shown to lack detectable homologues at evolutionary distances > 50 million years (Hezroni et al., 2015). This is consistent with a recent study reporting that around one-third and two-thirds of the lncRNAs detected in coho salmon (*Oncorhynchus kisutch*) are shared with Atlantic salmon and rainbow trout (*O. mykiss*) (Leiva et al., 2020), reflecting ~30 and 10 million years of divergence, respectively. While the high evolutionary turnover of lncRNAs is hard to reconcile with broad-scale functionality, that stance fails to acknowledge many alternative views that exist, and this is a rapidly evolving field, where methods and concepts in evolution have a key role to play in shaping the direction of ongoing travel (Ulitsky, 2016). We will halt our coverage on lncRNAs here, but interested readers are encouraged to explore the literature further by building on the papers we have cited before.

7.3 Evolution of the epigenome

The epigenome plays a central role in determining how genes are expressed under different biological conditions (e.g., in different cell types/tissues or developmental stages, and in response to environmental stimuli) independently of variation in DNA sequence. The epigenome includes DNA biochemical modifications (including methylation), histone posttranslational modifications, and the occupancy and organization of DNA within nucleosomes (i.e., units of DNA wrapped around histone molecules). lncRNAs and other types of ncRNA are also considered important epigenetic regulators, via their roles in regulating gene expression and chromatin structure (reviewed in Peschansky and Wahlestedt, 2014; Han and Chang, 2015; Allis and Jenuwein, 2016). All the main components of epigenome machinery present in mammals are conserved and active in fish genomes, with some notable differences and expansions (Best et al., 2018). Together, these features govern chromatin accessibility, the interaction of DNA sequences across the genome in three dimensions, and the occupancy of DNA binding proteins (e.g., transcription factors) at regulatory elements. Variation in epigenome state occurs within life, involving dramatic changes during early fish development (e.g., Pálfy et al., 2020) and more focused changes influencing a subset of genes in response to environmental stimuli (see Chapter 3 and 5). For example, methylation changes were proposed to facilitate adaptation in advance of genetic selection in sticklebacks during the transition from marine to freshwater environments (Artemov et al., 2017). Methylation changes were reported to arise repeatedly to modify the expression of genes in hatchery-reared coho salmon, which may contribute to the reduced fitness of these populations when released into the wild for restocking (Le Luyer et al., 2017). Histone modifications and methylation were shown to be regulated by diet in liver of rainbow trout (Marandel et al., 2016; for further details on how diet might regulate epigenome, see Chapter 5). Epigenomic states may be transmitted across generations through a range of mechanisms reviewed elsewhere (Best et al., 2018; Perez and Lehner, 2019; for further details on epigenetic transmission, see also Chapter 3).

Macroevolutionary studies comparing epigenome state across species are rare compared to sequence-based comparative studies. The likely reasons include: (i) that genome-wide methods used to measure epigenomic features in vertebrates (see later) were developed for (and have been most applied in) model species (e.g., human and zebrafish), leading to a paucity of comparable datasets, and (ii) whereas frameworks for comparing sequence conservation across genomes are established, methods for comparative and evolutionary analysis of epigenomic data are underdeveloped. The latter point is linked to the challenge that epigenomic status and sequence level conservation are not always coupled for the same location in the genome (Xiao et al., 2012; Lowdon et al., 2016). For example, many active enhancers in vertebrate genomes may reside in nonconserved locations across species, potentially regulating orthologous genes in both comparable and distinct ways. Indeed, only one obvious example of comparative studies of epigenomic state across fish species exists in the literature: a comparison of zebrafish and medaka that identified conserved histone modification states during embryogenesis (Tena et al., 2014). More emphasis has been given to mammals, and the available studies offer insights that may transfer to fishes. A recent study compared liver histone modifications across the genome of 15 mammal species to distinguish active promoters (regions responsible for transcription initiation marked by transcription start sites in genes) from active enhancers (Berthelot et al., 2018). By generating matched RNA-Seq data, the authors were able to correlate evolutionary conservation in gene expression levels with the evolutionary dynamics of regulatory regions. They reported that slowly evolving, highly expressed genes are defined by more conserved promoter and enhancers, but also that newly evolved lineage-specific enhancers, in addition to measurably increasing gene expression, supported the maintenance of high expression levels during evolution, contributing "resilience" to gene expression (Berthelot et al., 2018). Another recent study identified conserved epigenomic states present across a range of farmed animals including chicken and mammals (Foissac et al., 2019).

Historical limitations to cross-species analyses of epigenomic data are being lifted as the gap between model and nonmodel organisms closes in genomics. Sequencing-based methods to probe chromatin accessibility (including

DNase-seq and ATAC-Seq), histone modifications (i.e., ChIP-Seq), long-range DNA-to-DNA interactions (e.g., Hi-C), and methylation status (e.g., bisulfite-based sequencing) are now being applied and promoted in many species through actions such as the Functional Annotation of Animal Genomes (FAANG) initiative (Giuffra et al., 2019), and related projects for zebrafish (https://danio-code.zfin.org/) and salmonids (Macqueen et al., 2017). A project funded recently by the European Commission called "AQUA-FAANG" (https://www.aqua-faang.eu/) is generating such datasets in six different commercial fish species across a range of standardized biological samples, providing a basis for comparing the evolution of epigenomic features across species and in relation to the Ts3R, Ss4R, and cyprinid WGD events. Novel bioinformatic frameworks are being developed with the aim of capturing evolutionary conservation among epigenomic states across species (e.g., Dukler et al., 2020) and will likely prove vital for future progress in this area for fishes and other animals.

7.4 Ohnologue regulatory evolution

It has long been proposed that gene duplication opens new directions in functional evolution, allowing novel functions to evolve that were not possible at a single locus (e.g., Ohno, 1970). This is due to the redundancy created when two gene copies replace a single copy in the genome, which happens on a massive scale after WGD. An important literature exists on this topic, which has focused on coding genes, and to some extent other features discussed before. Several classic reviews should be read to understand the various models of duplicated gene evolution proposed to date (Conant and Wolfe, 2008; Innan and Kondrashov, 2010; Kondrashov, 2012), including for fishes in the context of Ts3R (Glasauer and Neuhauss, 2014). We do not cover this field comprehensively, instead outlining recent genome-wide studies advancing our understanding on the evolution of ohnologue expression following WGD, using salmonids as a study system.

One of the key findings presented by Lien et al. (2016) during their analysis of the Atlantic salmon genome concerned the evolution of ohnologue expression. Prior to this work, it was already well established that a sizeable portion of Ss4R ohnologues from many gene families had evolved divergent expression, indicative of functional divergence (e.g., Macqueen et al., 2010, 2013). Adding to such studies, a global analysis of the rainbow trout genome showed that a similar number of Ss4R ohnologues retained either the same or more divergent expression patterns across a panel of tissues (Berthelot et al., 2014). What such work lacked was an indication of the evolutionary direction of changes in ohnologue expression, which requires an outgroup species to infer the ancestral condition. Lien et al. (2016) achieved this (and others have taken a similar approach after Ts3R; see Pasquier et al., 2017) by comparing expression for a panel of tissues shared by Atlantic salmon and northern pike *Esox lucius*, a member of Esociformes—the sister lineage to salmonids lacking Ss4R (Fig. 1).

Among a large set of gene trios (two salmonid ohnologues and their common pike orthologue), 42% showed conserved expression, while for 28%, one of the ohnologues had diverged in expression with the other being conserved with pike (Lien et al., 2016). The latter result points to salmonid-specific expression divergence in one of the ohnologues, but it was unclear if this was due to adaptation or neutral genetic drift (Sandve et al., 2018). In a follow-up analysis, deeper insights were gained by adding an additional salmonid, the European grayling (*Thymallus thymallus*) (Varadharajan et al., 2018), whose ancestor diverged from the Atlantic salmon lineage ~50 mya (Fig. 1). This study revealed that among thousands of analyzed ohnologue pairs, one-quarter showed conserved tissue expression in both salmonid species and pike, which interestingly were enriched for brain genes. However, a minor fraction (~5%) showed a pattern where one ohnologue evolved with a divergent expression in the ancestor to salmon and grayling that was maintained by selection in both species, while the other copy maintained the ancestral expression state; indicating adaptive expression evolution following Ss4R (Varadharajan et al., 2018).

A recent study extends on the earlier findings at mechanistic levels (Gillard et al., 2021). This study compared the evolution of ohnologue expression in four different salmonid species in comparison to single copy orthologues in pike, and more distantly related teleosts, using a novel modeling approach to identify adaptive shifts in gene expression along a phylogenetic tree. It was reported that adaptive shifts in tissue-specific expression level were rare in salmonid ohnologues, with the lion's share of expression divergence between ohnologues being present across multiple tissues. The results presented are consistent with adaptive changes in ohnologue expression to overcome several challenges posed by WGD events in newly formed polyploids. Such issues include problems with cell division, and in maintaining basic cellular functions such as translation and energy production (Gillard et al., 2021). Overall, salmonids are providing an excellent model for understanding the evolution of gene regulation following Ss4R, and ongoing work will extend these insights to additional levels of genome regulation, including the epigenome.

8. A primer on miRNA evolutionary characterization

In this final section, we provide a short introduction to miRNA characterization in fishes, with a step-by-step approach to infer the evolutionary relationships of miRNA families expanded by WGD events during teleost evolution (Fig. 3).

8.1 miRNAs and their role in fish biology and evolution

miRNAs are a highly conserved class of sRNA encoded within all animal genomes. They are transcribed much like coding genes within the nucleus into primary miRNA transcripts (pri-miRNAs), from which hairpin-shaped precursor miRNAs (pre-miRNAs) are processed by a conserved ribonuclease enzyme

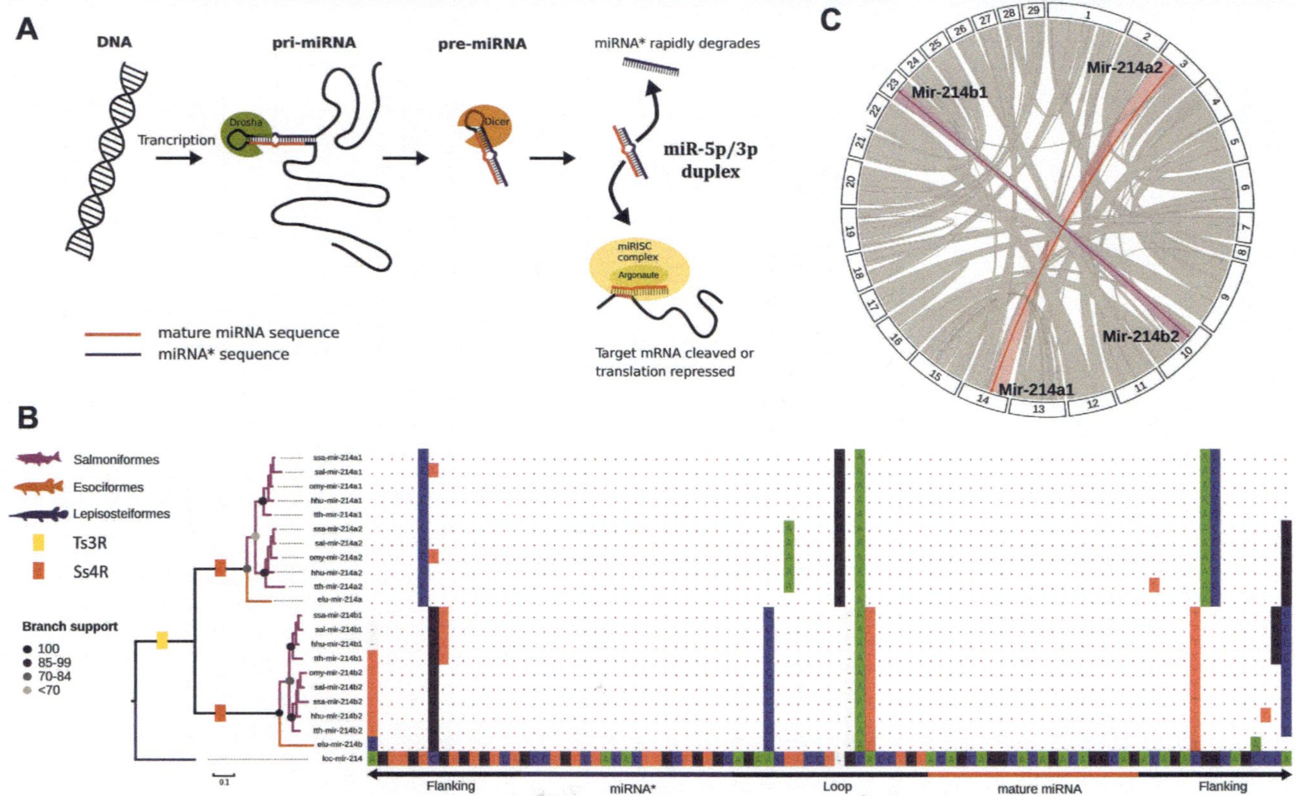

FIG. 3 microRNA evolutionary characterization in fish. (A) Overview of miRNA biogenesis and processing. (B) Left panel: Maximum likelihood phylogram of 22 miR-214 sequences analyzed in our example, annotated with bootstrap statistical support for the major nodes and WGD events. The scale bar shows the number of modeled substitutions per site. Species abbreviations are as follows: *elu*: *E. lucius*; *hhu*: *Hucho huchen*; *loc*: *L. oculatus*; *omy*: *O. mykiss*; *sal*: *Salvelinus alpinus*; *ssa*: *S. salar*; *tth*: *T. thymallus*. Also shown on the right panel is the sequence alignment used in the analysis, annotated with miRNA features. The dots show conserved bases with respect to the miRNA shown for L. *oculatus*. (C) Circos plot highlighting duplicated blocks in the Atlantic salmon genome (indicated by sweeping bands across 29 chromosomes) and the position of miR-214 ohnologues retained after Ss4R. The Circos plot was built in Circlize (Gu et al., 2014) with the coordinates for duplicated Ss4R regions defined in Lien et al. (2016).

called Drosha, before being exported to the cytoplasm and processed by Dicer (another ribonuclease) into a short (20–26 bp) sequence, and finally entering the miRNA-induced silencing complex (mi-RISC) to become a short mature miRNA (Fig. 3A). More detailed explanations on how miRNAs are generated and processed can be found elsewhere (e.g., Bizuayehu and Babiak, 2014; also see Chapter 4). miRNAs primarily act to target cytoplasmic mRNAs that contain a target site that is complementary in sequence to nucleotides 2–8 of the mature miRNA sequence, and usually located in the 3′ untranslated region of the mRNA. The mi-RISC causes gene silencing through the action of multiple proteins, notably the Argonaute family, resulting in mRNA degradation or translational repression, regulating how much protein is generated from the targeted mRNA (Fig. 3A).

miRNAs have been characterized in diverse fish taxa leading to a substantial body of literature spanning many species (reviewed in Herkenhoff et al., 2018) and different functional contexts, including accumulating evidence for their role in the regulation of fish immunity (Andreassen and Høyheim, 2017) and a range of other traits with commercial relevance (Herkenhoff et al., 2018). In an evolutionary context, a potential role for miRNA regulation in the rapid diversification of cichlid species has been proposed (Brawand et al., 2014; Franchini et al., 2019). A surprisingly limited amount of work has addressed the role of WGD events on miRNA evolution in fishes, though it is established that many miRNA ohnologues have been retained following Ts3R and Ss4R (Berthelot et al., 2014; Desvignes et al., 2021). For instance, following Ss4R, the rainbow trout genome has conserved 97% of miRNAs as ohnologues (Berthelot et al., 2014), which is much higher than for protein coding genes (~50%–60% ohnologue retention rate). Desvignes et al. (2021) reported a comprehensive analysis of miRNA evolution after Ts3R, revealing several potential mechanisms involved in duplicated miRNA retention and functional diversification.

As mentioned earlier (Section 6), phylogenetic analysis is an essential tool to infer the evolution of genetic characters in fish genomes and can be applied to explore miRNA gene family evolution. Here, we outline the key steps to achieve such an analysis using the miR-214 family as an example. The miR-214 family is highly conserved among vertebrates and has important roles in vertebrate skeletogenesis (Desvignes et al., 2014). For example, in zebrafish miR-214 overexpression hampers cranial cartilage formation during development (Roberto et al., 2018). miR-214 is also highly expressed in muscle and was reported to be duplicated in the spiny-rayed fish clade (Nachtigall et al., 2014).

8.2 Main steps in miRNA phylogenetic analysis

Four key steps underlie any phylogenetic analysis of gene family evolution. First, the collection of sequence data for the gene family and species of interest. Second, the alignment of the sequences, aiming to capture genetic differences that have been descended during evolution. Third, the selection of a suitable method to construct the phylogenetic tree through which to infer speciation and gene duplication events. Fourth, interpretation of the resulting data. These steps apply to other genetic characters (e.g., coding genes), though a couple of points specific to miRNAs should be noted. The median length of an animal pre-miRNA (Fig. 3A) is only 59 bp (Fromm et al., 2015), and thus significantly shorter than most coding genes, leaving less phylogenetic "signal" to inform the analysis. The most highly conserved regions of miRNAs are the miR-5p/3p duplex that is integrated into the mi-RISC complex (Fig. 3A), which are typically 22 bp in length but can fall in the range of 20–26 bp (Fromm et al., 2015). An alternative nomenclature denotes the two strands as the "mature" (miRNA) and "star" (miRNA*) strand (i.e., miRNA/miRNA* duplex). The mature strand is typically more highly expressed and biologically active, whereas the star strand is comparatively less expressed, less evolutionarily conserved and likely nonfunctional, normally rapidly degrading after processing (Fromm et al., 2015). While this nomenclature contains additional information, in some cases expression levels of both strands are similar, or the two strands show differential expression under distinct biological conditions, making it unclear which of the miR-5p/3p strands is functional (indeed, both may be functional) (Desvignes et al., 2015). This led miRBase (https://www.mirbase.org; the main database repository for miRNAs) to adopt the unbiased 5p/3p nomenclature (Kozomara and Griffiths-Jones, 2014). Due to strong selective pressure on these regions, perfect sequence conservation is often observed over large evolutionary distances, leading to zero phylogenetic signal (as seen for our example in Fig. 3B). For this reason, in addition to the short length of the precursor sequence, the flanking regions of miRNAs, which exhibit more genetic variation, are often used to expand the signal for analysis (Kenny et al., 2015) (e.g., see Fig. 3B). In addition, miRNAs are highly conserved over large evolutionary distances, showing far fewer losses during evolution than coding genes (Tarver et al., 2018), with little to no evidence for convergent evolution or horizontal gene transfer. Thus, despite being short in sequence, their high degree of conservation provides advantages that can directly address challenging evolutionary hypotheses (e.g., Heimberg et al., 2010).

A complicating factor when considering miRNA evolution is the coevolution of their target sites. Target site sequences evolve at a faster rate than miRNAs (Simkin et al., 2020), with mutations in these sequences leading to the potential reduction or elimination of regulation of that gene by the targeting miRNA. However, it is difficult to firmly establish the rate of evolutionary change of target sites, as many identified target sites are likely to be false positives (Pinzón et al., 2017) and therefore not under the same evolutionary pressures as true target sites. Indeed, while early studies estimated that each miRNA is likely to target hundreds of genes (Bartel and Chen, 2004; Friedman et al., 2009), more recent work suggests most miRNAs may actually have only a small number of true targets (e.g., Pinzón et al., 2017; Fridrich et al., 2019). Even for validated target sites, there is significant variation in the degree to which the target sites are conserved across species, with target sites for more recently evolved miRNA showing greater mutation rates than target sites of older miRNAs. There are also significant differences in how well target sites are conserved for ancient miRNA families showing similar ages, with miR-20 and miR-30 having highly conserved targets, while the targets of similarly ancient miR-146 showing much higher levels of divergence (Simkin et al., 2020).

8.2.1 Step 1: Data collection

Due to the challenges in dealing with complex transcriptomic data, it has been shown that large numbers of miRNAs deposited in online databases do not satisfy the gold-standard criteria for annotation (Kozomara et al., 2019). These criteria are based on a unique signature resulting from the biogenesis pathway (Fig. 3A) that results in high numbers of identical small RNA molecules being processed from a miRNA precursor. Remarkably, just 17% of animal miRNA families in miRBase v.21 are supported by sufficient evidence to be considered high-confidence miRNAs (Fromm et al., 2015). Therefore due care must be taken during data collection, and we opted to use sequences deposited on mirgenedb.org (Fromm et al., 2020), a manually curated database of high-confidence animal miRNAs.

All sequences from the miR-214 family were downloaded from https://mirgenedb.org, along with a genome assembly (from NCBI: https://www.ncbi.nlm.nih.gov/genome/) for five salmonid species, northern pike (lacking Ss4R), and spotted gar (lacking Ts3R) (see Fig. 1). The use of these species together allows us to make inferences about retention of duplicated miRNAs from Ts3R and Ss4R, while the spotted gar will be used to root the tree, as the sister to all teleosts (Fig. 1). A homology search for miR-214 was conducted using the alignment tool BLASTn (Altschul et al., 1990), employing all miR-214 sequences downloaded as the "query" sequences, allowing us to identify conserved sequences across each of the target fish genomes, with the assumption that a high level of sequence similarity is a consequence of common evolutionary origin. This was done locally after installing BLAST+ (https://www.ncbi.nlm.nih.gov/guide/howto/run-blast-local/), which is the best option if large numbers of sequences are analyzed, for example as part of a broader pipeline. However, as an alternative to downloading sequences and performing BLAST locally, NCBI provides online tools to analyze genomes on their servers (e.g., BLAST can be run at https://blast.ncbi.nlm.nih.gov/Blast.cgi for all the genomes used here). Results with an expectation value (e-value; a parameter indicating how probable it is for an observed BLAST result to be due to chance) of less than 1e-10 were retained. This is a more relaxed requirement than for most protein coding genes at the same evolutionary distances, reflecting the shorter sequence length of miRNAs.

Twenty-two unique miR-214 sequences were identified using this approach: one in spotted gar, two in northern pike, and up to four in each of the salmonids. The four copies identified in Atlantic salmon are located in regions of the genome that suggest ohnologue retention following Ss4R (Fig. 3C). As mentioned earlier, the Atlantic salmon genomes is composed of duplicated blocks where genes retain similar positional order along different chromosomes, often termed collinearity, reflecting the descent of the organization existing before WGD. In our example, it can be clearly seen that two pairs of miR-214 reside within duplicated blocks retained from Ss4R, which have been visualized in a Circos plot (Fig. 3C) created using the Circlize package in R (Gu et al., 2014; additional guidance available at https://jokergoo.github.io/circlize_book/book/). As mentioned before, due to the lack of phylogenetic signal available within the pre-miRNA sequence, we also collected an additional 500 bp either side of each sequence identified by BLAST.

8.2.2 Step 2: Sequence alignment

The accurate alignment of genetic sequences is crucial to the inference of evolutionary relationships in phylogenetic studies and can be a nontrivial problem. A range of alignment programs are available with different goals and approaches (see Kapli et al., 2020). MAFFT (Katoh and Standley, 2013) was used in our example to align the 22 miRNA sequences (Fig. 3A). As with the BLAST search, this can be done locally by downloading and installing the software, or through an online server (https://mafft.cbrc.jp/alignment/server/). We used a tool called TrimAl (Capella-Gutiérrez et al., 2009), which is available online (http://phylemon2.bioinfo.cipf.es/), to remove the alignment sites where more than 80% of the aligned miRNA sequences had gaps. It is an open debate over whether use of alignment trimming improves phylogenetic inference, but we chose to use it here as significant parts of the flanking regions displayed low levels of conservation.

8.2.3 Step 3: Phylogenetic tree reconstruction

With a set of miRNA family members aligned, the final step is to reconstruct their evolutionary history along a phylogenetic tree. Various methods are available for tree building based either on genetic distance, the most popular being neighbor joining, or various character-based methods (Kapli et al., 2020). The latter includes probabilistic approaches that search for the tree that best explains the observed sequence variation in the alignment considering the fit of the data to an evolutionary model of nucleotide substitution. The relative strengths and disadvantages of different tree building approaches are covered elsewhere (Kapli et al., 2020). We used IQ-Tree (Nguyen et al., 2015), a maximum likelihood method that automatically selects the most appropriate (i.e., best-fitting the data) evolutionary model of nucleotide substitution among a set of tested models. Fig. 3B includes a consensus tree for our data with statistical support for branching patterns generated by a bootstrapping method implemented by IQ-Tree (Hoang et al., 2018). Branch lengths represent genetic differences among the miRNA sequences measured as the number of substitutions per aligned site.

8.2.4 Step 4: Inferences on miRNA family evolution

Several results can be drawn from our analysis. The branching patterns indicate that miR-214 existed as a single copy in the common ancestor of all seven species, was duplicated in the lineage leading to pike and salmonids, and duplicated again in the salmonid ancestor (Fig. 3B). This is consistent with the retention of ohnologues following Ts3R and Ss4R, with a single miR-214 member retained in spotted gar, two in pike, and four members in all but one of the salmonids. In salmonids, we observed four clear "orthogroups" each matching expectations of the species relationships (Fig. 1), with two sister miRNAs in pike resolving these into two pairs of orthogroups. Establishing orthology in this manner allows the miRNAs to be named according to a convention consistent across species that accurately reflects evolutionary status as orthologues *and* ohnologues. This is important for several reasons. Comparative inferences on gene "functions" are founded on the idea that function is inherited from ancestral genes—if miR-214a and miR-214b had different functions in the salmonid ancestor, the descendants of each Ss4r ohnologue pair (miR-214a1 and a2; miR-214b1 and b2) are likely to retain these functions or diversify them independently of one other. Ignorance of the existence of these miRNA ohnologues or their evolutionary history could easily mislead conclusions in ongoing studies. Finally, the results are also consistent with Berthelot et al. (2014) in showing the high level of retention in salmonid miRNA ohnologues.

9. Concluding remarks

It is an exciting time to study the evolution of fish genomes, with resources, methodologies, and datasets available now that could not have been dreamed of one decade ago. The future in this field holds great immediate promise for delivering new findings on how fish lineages have diversified and adapted to different environments during their evolutionary history, and how this is linked to the diverse genomic features we have outlined in this chapter. Comparative studies of genome evolution will also increasingly support fish breeding and conservation efforts by helping to define conserved functional sequences and epigenomic features controlling traits of interest in aquaculture or that are selected or subject to drift in different wild and domesticated populations.

References

Alfaro, M.E., 2018. Resolving the ray-finned fish tree of life. Proc. Natl. Acad. Sci. U. S. A. 115 (24), 6107–6109. https://doi.org/10.1073/pnas.1807018115.

Alfaro, M.E., Faircloth, B.C., Harrington, R.C., et al., 2018. Explosive diversification of marine fishes at the Cretaceous-Palaeogene boundary. Nat. Ecol. Evol. 2 (4), 688–696. https://doi.org/10.1038/s41559-018-0494-6.

Allendorf, F.W., Bassham, S., Cresko, W.A., et al., 2015. Effects of crossovers between homeologs on inheritance and population genomics in polyploid-derived salmonid fishes. J. Hered. 106 (3), 217–227. https://doi.org/10.1093/jhered/esv015.

Allendorf, F.W., Thorgaard, B.W., 1984. Tetraploidy and the evolution of Salmonid fishes. In: Turner, B.J. (Ed.), Evolutionary Genetics of Fishes. Springer, Boston, MA, pp. 1–53.

Allis, C.D., Jenuwein, T., 2016. The molecular hallmarks of epigenetic control. Nat. Rev. Genet. 17 (8), 487–500. https://doi.org/10.1038/nrg.2016.59.

Altschul, S.F., Gish, W., Miller, W., et al., 1990. Basic local alignment search tool. J. Mol. Biol. 215 (3), 403–410. https://doi.org/10.1016/S0022-2836(05)80360-2.

Amarasinghe, S.L., Su, S., Dong, X., et al., 2020. Opportunities and challenges in long-read sequencing data analysis. Genome Biol. 21 (1), 30. https://doi.org/10.1186/s13059-020-1935-5.

Andreassen, R., Høyheim, B., 2017. miRNAs associated with immune response in teleost fish. Dev. Comp. Immunol. 75, 77–85. https://doi.org/10.1016/j.dci.2017.02.023.

Aparicio, S., Chapman, J., Stupka, E., et al., 2002. Whole-genome shotgun assembly and analysis of the genome of *Fugu rubripes*. Science 297 (5585), 1301–1310. https://doi.org/10.1126/science.1072104.

Artemov, A.V., Mugue, N.S., Rastorguev, S.M., et al., 2017. Genome-wide DNA methylation profiling reveals epigenetic adaptation of stickleback to marine and freshwater conditions. Mol. Biol. Evol. 34 (9), 2203–2213. https://doi.org/10.1093/molbev/msx156.

Bartel, D.P., Chen, C.Z., 2004. Micromanagers of gene expression: the potentially widespread influence of metazoan microRNAs. Nat. Rev. Genet. 5 (5), 396–400. https://doi.org/10.1038/nrg1328.

Betancur-R, R., Wiley, E.O., Arratia, G., et al., 2017. Phylogenetic classification of bony fishes. BMC Evol. Biol. 17 (1), 162. https://doi.org/10.1186/s12862-017-0958-3.

Berthelot, C., Brunet, F., Chalopin, D., et al., 2014. The rainbow trout genome provides novel insights into evolution after whole-genome duplication in vertebrates. Nat. Commun. 5, 3657. https://doi.org/10.1038/ncomms4657.

Berthelot, C., Villar, D., Horvath, J.E., et al., 2018. Complexity and conservation of regulatory landscapes underlie evolutionary resilience of mammalian gene expression. Nat. Ecol. Evol. 2 (1), 152–163. https://doi.org/10.1038/s41559-017-0377-2.

Bertolotti, A.C., Layer, R.M., Gundappa, M.K., et al., 2020. The structural variation landscape in 492 Atlantic salmon genomes. Nat. Commun. 11 (1), 1–6. https://doi.org/10.1038/s41467-020-18972-x.

Best, C., Ikert, H., Kostyniuk, D.J., et al., 2018. Epigenetics in teleost fish: from molecular mechanisms to physiological phenotypes. Comp. Biochem. Physiol. B Biochem. Mol. Biol. 224, 210–244. https://doi.org/10.1016/j.cbpb.2018.01.006.

Bizuayehu, T.T., Babiak, I., 2014. MicroRNA in teleost fish. Genome Biol. Evol. 6 (8), 1911–1937. https://doi.org/10.1093/gbe/evu151.

Bourque, G., Burns, K.H., Gehring, M., et al., 2018. Ten things you should know about transposable elements. Genome Biol. 19 (1), 199. https://doi.org/10.1186/s13059-018-1577-z.

Braasch, I., Peterson, S.M., Desvignes, T., et al., 2015. A new model army: Emerging fish models to study the genomics of vertebrate Evo-Devo. J. Exp. Zool. B Mol. Dev. Evol. 324 (4), 316–341. https://doi.org/10.1002/jez.b.22589.

Braasch, I., Gehrke, A.R., Smith, J.J., et al., 2016. The spotted gar genome illuminates vertebrate evolution and facilitates human-teleost comparisons. Nat. Genet. 48 (4), 427–437. https://doi.org/10.1038/ng.3526.

Brawand, D., Wagner, C.E., Li, Y.I., et al., 2014. The genomic substrate for adaptive radiation in African cichlid fish. Nature 513 (7518), 375–381. https://doi.org/10.1038/nature13726.

Capella-Gutiérrez, S., Silla-Martínez, J.M., Gabaldón, T., 2009. trimAl: a tool for automated alignment trimming in large-scale phylogenetic analyses. Bioinformatics 25 (15), 1972–1973. https://doi.org/10.1093/bioinformatics/btp348.

Carleton, K.L., Conte, M.A., Malinsky, M., et al., 2020. Movement of transposable elements contributes to cichlid diversity. Mol. Ecol. 29 (24), 4956–4969. https://doi.org/10.1111/mec.15685.

Chalopin, D., Volff, J.N., 2017. Analysis of the spotted gar genome suggests absence of causative link between ancestral genome duplication and transposable element diversification in teleost fish. J. Exp. Zool. B Mol. Dev. Evol. 328 (7), 629–637. https://doi.org/10.1002/jez.b.22761.

Chapman, E.J., Carrington, J.C., 2007. Specialization and evolution of endogenous small RNA pathways. Nat. Rev. Genet. 8 (11), 884–896. https://doi.org/10.1038/nrg2179.

Chen, Z., Omori, Y., Koren, S., et al., 2019. De novo assembly of the goldfish (*Carassius auratus*) genome and the evolution of genes after whole-genome duplication. Sci. Adv. 5 (6). https://doi.org/10.1126/sciadv.aav0547, eaav0547.

Cheng, P., Huang, Y., Lv, Y., et al., 2021. The American paddlefish genome provides novel insights into chromosomal evolution and bone mineralization in early vertebrates. Mol. Biol. Evol. 38 (4), 1595–1607. https://doi.org/10.1093/molbev/msaa326.

Chuong, E.B., Elde, N.C., Feschotte, C., 2017. Regulatory activities of transposable elements: from conflicts to benefits. Nat. Rev. Genet. 18 (2), 71–86. https://doi.org/10.1038/nrg.2016.139.

Clarke, J.T., Lloyd, G.T., Friedman, M., 2016. Little evidence for enhanced phenotypic evolution in early teleosts relative to their living fossil sister group. Proc. Natl. Acad. Sci. U. S. A. 113 (41), 11531–11536. https://doi.org/10.1073/pnas.1607237113.

Conant, G.C., Wolfe, K.H., 2008. Turning a hobby into a job: how duplicated genes find new functions. Nat. Rev. Genet. 9 (12), 938–950. https://doi.org/10.1038/nrg2482.

Cosby, R.L., Chang, N.C., Feschotte, C., 2019. Host-transposon interactions: conflict, cooperation, and cooption. Genes Dev. 33 (17–18), 1098–1116. https://doi.org/10.1101/gad.327312.119.

Davesne, D., Friedman, M., Schmitt, A.D., et al., 2021. Fossilized cell structures identify an ancient origin for the teleost whole-genome duplication. Proc. Natl. Acad. Sci. U. S. A. 118 (30). https://doi.org/10.1073/pnas.2101780118, e2101780118.

de Boer, J.G., Yazawa, R., Davidson, W.S., Koop, B.F., 2007. Bursts and horizontal evolution of DNA transposons in the speciation of pseudotetraploid salmonids. BMC Genomics 8, 422. https://doi.org/10.1186/1471-2164-8-422.

Dehal, P., Boore, J.L., 2005. Two rounds of whole genome duplication in the ancestral vertebrate. PLoS Biol. 3 (10). https://doi.org/10.1371/journal.pbio.0030314, e314.

Desvignes, T., Contreras, A., Postlethwait, J.H., 2014. Evolution of the miR199-214 cluster and vertebrate skeletal development. RNA Biol. 11 (4), 281–294. https://doi.org/10.4161/rna.28141.

Desvignes, T., Batzel, P., Berezikov, E., et al., 2015. miRNA nomenclature: a view incorporating genetic origins, biosynthetic pathways, and sequence variants. Trends Genet. 31 (11), 613–626. https://doi.org/10.1016/j.tig.2015.09.002.

Desvignes, T., Sydes, J., Montfort, J., et al., 2021. Evolution after whole-genome duplication: teleost microRNAs. Mol. Biol. Evol. 38 (8), 3308–3331. https://doi.org/10.1093/molbev/msab105.

Doolittle, W.F., Brunet, T.D.P., 2017. On causal roles and selected effects: our genome is mostly junk. BMC Biol. 15 (1), 116. https://doi.org/10.1186/s12915-017-0460-9.

Douglas, A.T., Hill, R.D., 2014. Variation in vertebrate cis-regulatory elements in evolution and disease. Transcription 5 (3). https://doi.org/10.4161/trns.28848, e28848.

Du, K., Stöck, M., Kneitz, S., et al., 2020. The sterlet sturgeon genome sequence and the mechanisms of segmental rediploidization. Nat. Ecol. Evol. 4 (6), 841–852. https://doi.org/10.1038/s41559-020-1166-x.

Dukler, N., Huang, Y.F., Siepel, A., 2020. Phylogenetic modeling of regulatory element turnover based on epigenomic Data. Mol. Biol. Evol. 37 (7), 2137–2152. https://doi.org/10.1093/molbev/msaa073.

Emms, D.M., Kelly, S., 2019. OrthoFinder: phylogenetic orthology inference for comparative genomics. Genome Biol. 20 (1), 238. https://doi.org/10.1186/s13059-019-1832-y.

Fan, G., Song, Y., Yang, L., et al., 2020. Initial data release and announcement of the 10,000 Fish Genomes Project (Fish10K). Gigascience 9 (8). https://doi.org/10.1093/gigascience/giaa080, giaa080.

Ferris, S.D., 1984. Tetraploidy and the evolution of the catostomid fishes. In: Turner, B.J. (Ed.), Evolutionary Genetics of Fishes, pp. 55–93.

Foissac, S., Djebali, S., Munyard, K., et al., 2019. Multi-species annotation of transcriptome and chromatin structure in domesticated animals. BMC Biol. 17 (1), 108. https://doi.org/10.1186/s12915-019-0726-5.

Franchini, P., Xiong, P., Fruciano, C., et al., 2019. MicroRNA gene regulation in extremely young and parallel adaptive radiations of crater lake cichlid fish. Mol. Biol. Evol. https://doi.org/10.1093/molbev/msz168, msz168.

Fridrich, A., Hazan, Y., Moran, Y., 2019. Too many false targets for MicroRNAs: challenges and pitfalls in prediction of mirna targets and their gene ontology in model and non-model organisms. BioEssays 41 (4). https://doi.org/10.1002/bies.201800169, e1800169.

Friedman, M., Feilich, K.L., Beckett, H.T., et al., 2019. A phylogenomic framework for pelagiarian fishes (Acanthomorpha: Percomorpha) highlights mosaic radiation in the open ocean. Proc. Biol. Sci. 286 (1910), 20191502. https://doi.org/10.1098/rspb.2019.1502.

Friedman, R.C., Farh, K.K., Burge, C.B., Bartel, D.P., 2009 Jan. Most mammalian mRNAs are conserved targets of microRNAs. Genome Res. 19 (1), 92–105. https://doi.org/10.1101/gr.082701.108.

Fromm, B., Billipp, T., Peck, L.E., et al., 2015. A uniform system for the annotation of vertebrate microRNA genes and the evolution of the human microRNAome. Annu. Rev. Genet. 49, 213–242. https://doi.org/10.1146/annurev-genet-120213-092023.

Fromm, B., Domanska, D., Høye, E., et al., 2020. MirGeneDB 2.0: the metazoan microRNA complement. Nucleic Acids Res. 48 (D1), D132–D141. https://doi.org/10.1093/nar/gkz885.

Garcia-Fernández, J., Holland, P.W., 1994. Archetypal organization of the amphioxus Hox gene cluster. Nature 370 (6490), 563–566. https://doi.org/10.1038/370563a0.

Gallagher, M.D., Macqueen, D.J., 2017. Evolution and expression of tissue globins in ray-finned fishes. Genome Biol. Evol. 9 (1), 32–47. https://doi.org/10.1093/gbe/evw266.

Gao, F., Cai, Y., Kapranov, P., Xu, D., 2020. Reverse-genetics studies of lncRNAs-what we have learnt and paths forward. Genome Biol. 21 (1), 93. https://doi.org/10.1186/s13059-020-01994-5. Published 2020 Apr 14.

Gillard, G.B., Grønvold, L., Røsæg, L.L., et al., 2021. Comparative regulomics supports pervasive selection on gene dosage following whole genome duplication. Genome Biol. 22 (1), 103. https://doi.org/10.1186/s13059-021-02323-0.

Gilbert, C., Cordaux, R., 2017. Viruses as vectors of horizontal transfer of genetic material in eukaryotes. Curr. Opin. Virol. 25, 16–22. https://doi.org/10.1016/j.coviro.2017.06.005.

Ghurye, J., Pop, M., 2019. Modern technologies and algorithms for scaffolding assembled genomes. PLoS Comput. Biol. 15 (6). https://doi.org/10.1371/journal.pcbi.1006994, e1006994.

Giuffra, E., Tuggle, C.K., FAANG consortium, 2019. Functional annotation of animal genomes (FAANG): current achievements and roadmap. Annu. Rev. Anim. Biosci. 7, 65–88. https://doi.org/10.1146/annurev-animal-020518-114913.

Glasauer, S.M., Neuhauss, S.C., 2014. Whole-genome duplication in teleost fishes and its evolutionary consequences. Mol. Gen. Genomics. 289 (6), 1045–1060. https://doi.org/10.1007/s00438-014-0889-2.

Goudarzi, M., Berg, K., Pieper, L.M., Schier, A.F., 2019. Individual long non-coding RNAs have no overt functions in zebrafish embryogenesis, viability and fertility. Elife 8. https://doi.org/10.7554/eLife.40815, e40815. Published 2019 Jan 8.

Graur, D., Zheng, Y., Price, N., Azevedo, R.B., Zufall, R.A., Elhaik, E., 2013. On the immortality of television sets: "function" in the human genome according to the evolution-free gospel of ENCODE. Genome Biol. Evol. 5 (3), 578–590. https://doi.org/10.1093/gbe/evt028.

Gu, Z., Gu, L., Eils, R., et al., 2014. Circlize implements and enhances circular visualization in R. Bioinformatics 30 (19), 2811–2812. https://doi.org/10.1093/bioinformatics/btu393.

Han, P., Chang, C.P., 2015. Long non-coding RNA and chromatin remodeling. RNA Biol. 12 (10), 1094–1098. https://doi.org/10.1080/15476286.2015.1063770.

Harmston, N., Ing-Simmons, E., Tan, G., et al., 2017. Topologically associating domains are ancient features that coincide with Metazoan clusters of extreme noncoding conservation. Nat. Commun. 8 (1), 441. https://doi.org/10.1038/s41467-017-00524-5. Published 2017 Sep 5.

Heimberg, A.M., Cowper-Sal-lari, R., Sémon, M., et al., 2010. MicroRNAs reveal the interrelationships of hagfish, lampreys, and gnathostomes and the nature of the ancestral vertebrate. Proc. Natl. Acad. Sci. U. S. A. 107 (45), 19379–19383. https://doi.org/10.1073/pnas.1010350107.

Helfman, G., Collette, B.B., Facey, D.E., Bowen, B.W., 2009. The Diversity of Fishes: Biology, Evolution and Ecology, second ed. Wiley-Blackwell.

Herkenhoff, M.E., Oliveira, A.C., Nachtigall, P.G., et al., 2018. Fishing into the MicroRNA transcriptome. Front. Genet. 9, 88. https://doi.org/10.3389/fgene.2018.00088.

Herpin, A., Braasch, I., Kraeussling, M., et al., 2010. Transcriptional rewiring of the sex determining dmrt1 gene duplicate by transposable elements. PLoS Genet. 6 (2). https://doi.org/10.1371/journal.pgen.1000844, e1000844.

Hezroni, H., Koppstein, D., Schwartz, M.G., et al., 2015. Principles of long noncoding RNA evolution derived from direct comparison of transcriptomes in 17 species. Cell Rep. 11 (7), 1110–1122. https://doi.org/10.1016/j.celrep.2015.04.023.

Hoang, D.T., Chernomor, O., von Haeseler, A., et al., 2018. UFBoot2: improving the ultrafast bootstrap approximation. Mol. Biol. Evol. 35 (2), 518–522. https://doi.org/10.1093/molbev/msx281.

Houston, R.D., Macqueen, D.J., 2019. Atlantic salmon (*Salmo salar* L.) genetics in the 21st century: taking leaps forward in aquaculture and biological understanding. Anim. Genet. 50 (1), 3–14. https://doi.org/10.1111/age.12748.

Houston, R.D., Bean, T.P., Macqueen, D.J., et al., 2020. Harnessing genomics to fast-track genetic improvement in aquaculture. Nat. Rev. Genet. 21 (7), 389–409. https://doi.org/10.1038/s41576-020-0227-y.

Houwing, S., Kamminga, L.M., Berezikov, E., et al., 2007. A role for Piwi and piRNAs in germ cell maintenance and transposon silencing in Zebrafish. Cell 129 (1), 69–82. https://doi.org/10.1016/j.cell.2007.03.026.

Howe, K., Clark, M.D., Torroja, C.F., et al., 2013. The zebrafish reference genome sequence and its relationship to the human genome. Nature 496 (7446), 498–503. https://doi.org/10.1038/nature12111.

Hughes, L.C., Ortí, G., Huang, Y., et al., 2018. Comprehensive phylogeny of ray-finned fishes (Actinopterygii) based on transcriptomic and genomic data. Proc. Natl. Acad. Sci. U. S. A. 115 (24), 6249–6254. https://doi.org/10.1073/pnas.1719358115.

Innan, H., Kondrashov, F., 2010. The evolution of gene duplications: classifying and distinguishing between models. Nat. Rev. Genet. 11 (2), 97–108. https://doi.org/10.1038/nrg2689.

Inoue, J., Sato, Y., Sinclair, R., et al., 2015. Rapid genome reshaping by multiple-gene loss after whole-genome duplication in teleost fish suggested by mathematical modeling. Proc. Natl. Acad. Sci. U. S. A. 112 (48), 14918–14923. https://doi.org/10.1073/pnas.1507669112.

Jaillon, O., Aury, J.M., Brunet, F., et al., 2004. Genome duplication in the teleost fish *Tetraodon nigroviridis* reveals the early vertebrate proto-karyotype. Nature 431 (7011), 946–957. https://doi.org/10.1038/nature03025.

Jones, F.C., Grabherr, M.G., Chan, Y.F., et al., 2012. The genomic basis of adaptive evolution in threespine sticklebacks. Nature 484 (7392), 55–61. https://doi.org/10.1038/nature10944. Published 2012 Apr 4.

Kapli, P., Yang, Z., Telford, M.J., 2020. Phylogenetic tree building in the genomic age. Nat. Rev. Genet. 21 (7), 428–444. https://doi.org/10.1038/s41576-020-0233-0.

Kasahara, M., Naruse, K., Sasaki, S., et al., 2007. The medaka draft genome and insights into vertebrate genome evolution. Nature 447 (7145), 714–719. https://doi.org/10.1038/nature05846.

Kassahn, K.S., Dang, V.T., Wilkins, S.J., et al., 2009. Evolution of gene function and regulatory control after whole-genome duplication: comparative analyses in vertebrates. Genome Res. 19 (8), 1404–1418. https://doi.org/10.1101/gr.086827.108.

Katoh, K., Standley, D.M., 2013. MAFFT multiple sequence alignment software version 7: improvements in performance and usability. Mol. Biol. Evol. 30 (4), 772–780. https://doi.org/10.1093/molbev/mst010.

Kenny, N.J., Sin, Y.W., Hayward, A., et al., 2015. The phylogenetic utility and functional constraint of microRNA flanking sequences. Proc. Biol. Sci. 282 (1803), 20142983. https://doi.org/10.1098/rspb.2014.2983.

Kon, T., Omori, Y., Fukuta, K., et al., 2020. The genetic basis of morphological diversity in domesticated goldfish. Curr. Biol. 30 (12), 2260–2274.e6. https://doi.org/10.1016/j.cub.2020.04.034.

Kondrashov, F.A., 2012. Gene duplication as a mechanism of genomic adaptation to a changing environment. Proc. Biol. Sci. 279 (1749), 5048–5057. https://doi.org/10.1098/rspb.2012.1108.

Koonin, E.V., 2005. Orthologs, paralogs, and evolutionary genomics. Annu. Rev. Genet. 39, 309–338. https://doi.org/10.1146/annurev.genet.39.073003.114725.

Kozomara, A., Griffiths-Jones, S., 2014. miRBase: annotating high confidence microRNAs using deep sequencing data. Nucleic Acids Res. 42 (Database issue), D68–D73. https://doi.org/10.1093/nar/gkt1181.

Kozomara, A., Birgaoanu, M., Griffiths-Jones, S., 2019. miRBase: from microRNA sequences to function. Nucleic Acids Res. 47 (D1), D155–D162. https://doi.org/10.1093/nar/gky1141.

Krabbenhoft, T.J., MacGuigan, D.J., Backenstose, N.J.C., et al., 2021. Chromosome-level genome assembly of Chinese sucker (*Myxocyprinus asiaticus*) reveals strongly-conserved synteny following a catostomid-specific whole genome duplication. Genome Biol. Evol. https://doi.org/10.1093/gbe/evab190, evab190.

Laurent, S., Salamin, N., Robinson-Rechavi, M., 2017. No evidence for the radiation time lag model after whole genome duplications in Teleostei. PLoS One 12 (4). https://doi.org/10.1371/journal.pone.0176384, e0176384.

Lee, A.P., Kerk, S.Y., Tan, Y.Y., et al., 2011. Ancient vertebrate conserved noncoding elements have been evolving rapidly in teleost fishes. Mol. Biol. Evol. 28 (3), 1205–1215. https://doi.org/10.1093/molbev/msq304.

Lee, H., Zhang, Z., Krause, H.M., 2019. Long noncoding RNAs and repetitive elements: junk or intimate evolutionary partners? Trends Genet. 35 (12), 892–902. https://doi.org/10.1016/j.tig.2019.09.006.

Leiva, F., Rojas-Herrera, M., Reyes, D., et al., 2020. Identification and characterization of miRNAs and lncRNAs of coho salmon (*Oncorhynchus kisutch*) in normal immune organs. Genomics 112 (1), 45–54. https://doi.org/10.1016/j.ygeno.2019.07.015.

Le Luyer, J., Laporte, M., Beacham, T.D., et al., 2017. Parallel epigenetic modifications induced by hatchery rearing in a Pacific salmon. Proc. Natl. Acad. Sci. U. S. A. 114 (49), 12964–12969. https://doi.org/10.1073/pnas.1711229114.

Lewin, H.A., Robinson, G.E., Kress, W.J., et al., 2018. Earth biogenome project: sequencing life for the future of life. Proc. Natl. Acad. Sci. U. S. A. 115 (17), 4325–4333. https://doi.org/10.1073/pnas.1720115115.

Li, X., Guo, B., 2020. Substantially adaptive potential in polyploid cyprinid fishes: evidence from biogeographic, phylogenetic and genomic studies. Proc. Biol. Sci. 287 (1920), 20193008. https://doi.org/10.1098/rspb.2019.3008.

Lien, S., Koop, B.F., Sandve, S.R., et al., 2016. The Atlantic salmon genome provides insights into rediploidization. Nature 533 (7602), 200–205. https://doi.org/10.1038/nature17164.

Lin, J.J., Wang, F.Y., Li, W.H., Wang, T.Y., 2017. The rises and falls of opsin genes in 59 ray-finned fish genomes and their implications for environmental adaptation. Sci. Rep. 7 (1), 15568. https://doi.org/10.1038/s41598-017-15868-7.

Linquist, S., Doolittle, W.F., Palazzo, A.F., 2020. Getting clear about the F-word in genomics. PLoS Genet. 16 (4). https://doi.org/10.1371/journal.pgen.1008702, e1008702.

Liu, D.W., Wang, F.Y., Lin, J.J., et al., 2019. The cone opsin repertoire of osteoglossomorph fishes: gene loss in mormyrid electric fish and a long wavelength-sensitive cone opsin that survived 3R. Mol. Biol. Evol. 36 (3), 447–457. https://doi.org/10.1093/molbev/msy241.

Ludwig, A., Belfiore, N.M., Pitra, C., et al., 2001. Genome duplication events and functional reduction of ploidy levels in sturgeon (*Acipenser*, *Huso* and *Scaphirhynchus*). Genetics 158 (3), 1203–1215. https://doi.org/10.1093/genetics/158.3.1203.

Luo, J., Gao, Y., Ma, W., et al., 2014. Tempo and mode of recurrent polyploidization in the *Carassius auratus* species complex (Cypriniformes, Cyprinidae). Heredity (Edinb.) 112 (4), 415–427. https://doi.org/10.1038/hdy.2013.121.

Luo, J., Chai, J., Wen, Y., et al., 2020. From asymmetrical to balanced genomic diversification during rediploidization: Subgenomic evolution in allotetraploid fish. Sci. Adv. 6 (22). https://doi.org/10.1126/sciadv.aaz7677, eaaz7677.

Lowdon, R.F., Jang, H.S., Wang, T., 2016. Evolution of epigenetic regulation in vertebrate genomes. Trends Genet. 32 (5), 269–283. https://doi.org/10.1016/j.tig.2016.03.001.

Macqueen, D.J., Kristjánsson, B.K., Johnston, I.A., 2010. Salmonid genomes have a remarkably expanded akirin family, coexpressed with genes from conserved pathways governing skeletal muscle growth and catabolism. Physiol. Genomics 42 (1), 134–148. https://doi.org/10.1152/physiolgenomics.00045.2010.

Macqueen, D.J., Garcia de la Serrana, D., Johnston, I.A., 2013. Evolution of ancient functions in the vertebrate insulin-like growth factor system uncovered by study of duplicated salmonid fish genomes. Mol. Biol. Evol. 30 (5), 1060–1076. https://doi.org/10.1093/molbev/mst017.

Macqueen, D.J., Johnston, I.A., 2014. A well-constrained estimate for the timing of the salmonid whole genome duplication reveals major decoupling from species diversification. Proc. Biol. Sci. 281 (1778), 20132881. https://doi.org/10.1098/rspb.2013.2881.

Macqueen, D.J., Primmer, C.R., Houston, R.D., et al., 2017. Functional annotation of all salmonid genomes (FAASG): an international initiative supporting future salmonid research, conservation and aquaculture. BMC Genomics 18 (1), 484. https://doi.org/10.1186/s12864-017-3862-8.

Marlétaz, F., Firbas, P.N., Maeso, I., et al., 2018. Amphioxus functional genomics and the origins of vertebrate gene regulation. Nature 564 (7734), 64–70. https://doi.org/10.1038/s41586-018-0734-6.

Marandel, L., Lepais, O., Arbenoits, E., et al., 2016. Remodelling of the hepatic epigenetic landscape of glucose-intolerant rainbow trout (*Oncorhynchus mykiss*) by nutritional status and dietary carbohydrates. Sci. Rep. 6, 32187. https://doi.org/10.1038/srep32187.

Martin, K.J., Holland, P.W., 2014. Enigmatic orthology relationships between Hox clusters of the African butterfly fish and other teleosts following ancient whole-genome duplication. Mol. Biol. Evol. 31 (10), 2592–2611. https://doi.org/10.1093/molbev/msu202.

McDonald, J.F., 1998. Transposable elements, gene silencing and macroevolution. Trends Ecol. Evol. 13 (3), 94–95. https://doi.org/10.1016/S0169-5347(97)01282-2.

Moriyama, Y., Ito, F., Takeda, H., et al., 2016. Evolution of the fish heart by sub/neofunctionalization of an elastin gene. Nat. Commun. 7, 10397. https://doi.org/10.1038/ncomms10397.

Nachtigall, P.G., Dias, M.C., Pinhal, D., 2014. Evolution and genomic organization of muscle microRNAs in fish genomes. BMC Evol. Biol. 14, 196. https://doi.org/10.1186/s12862-014-0196-x.

Nguyen, L.T., Schmidt, H.A., von Haeseler, A., Minh, B.Q., 2015. IQ-TREE: a fast and effective stochastic algorithm for estimating maximum-likelihood phylogenies. Mol. Biol. Evol. 32 (1), 268–274. https://doi.org/10.1093/molbev/msu300.

Nguyen, N.T.T., Vincens, P., Roest Crollius, H., Louis, A., 2018. Genomicus 2018: karyotype evolutionary trees and on-the-fly synteny computing. Nucleic Acids Res. 46 (D1), D816–D822. https://doi.org/10.1093/nar/gkx1003.

Ohno, S., Wolf, U., Atkin, N.B., 1968. Evolution from fish to mammals by gene duplication. Hereditas 59 (1), 169–187. https://doi.org/10.1111/j.1601-5223.1968.tb02169.x.

Ohno, S., 1970. Evolution by Gene Duplication. Springer-Verlag, New York.

Otto, S.P., 2007. The evolutionary consequences of polyploidy. Cell 131 (3), 452–462. https://doi.org/10.1016/j.cell.2007.10.022.

Palazzo, A.F., Lee, E.S., 2015. Non-coding RNA: what is functional and what is junk? Front. Genet. 6, 2. https://doi.org/10.3389/fgene.2015.00002. Published 2015 Jan 26.

Pálfy, M., Schulze, G., Valen, E., Vastenhouw, N.L., 2020. Chromatin accessibility established by Pou5f3, Sox19b and Nanog primes genes for activity during zebrafish genome activation. PLoS Genet. 16 (1). https://doi.org/10.1371/journal.pgen.1008546, e1008546.

Parey, E., Louis, A., Cabau, C., et al., 2020. Synteny-guided resolution of gene trees clarifies the functional impact of whole-genome duplications. Mol. Biol. Evol. 37 (11), 3324–3337. https://doi.org/10.1093/molbev/msaa149. 32556216.

Parhad, S.S., Theurkauf, W.E., 2019. Rapid evolution and conserved function of the piRNA pathway. Open Biol. 9 (1), 180181. https://doi.org/10.1098/rsob.180181.

Pasquier, J., Braasch, I., Batzel, P., et al., 2017. Evolution of gene expression after whole-genome duplication: New insights from the spotted gar genome. J. Exp. Zool. B Mol. Dev. Evol. 328 (7), 709–721. https://doi.org/10.1002/jez.b.22770.

Perez, M.F., Lehner, B., 2019. Intergenerational and transgenerational epigenetic inheritance in animals. Nat. Cell Biol. 21 (2), 143–151. https://doi.org/10.1038/s41556-018-0242-9.

Pertea, M., Shumate, A., Pertea, G., et al., 2018. CHESS: a new human gene catalog curated from thousands of large-scale RNA sequencing experiments reveals extensive transcriptional noise. Genome Biol. 19 (1), 208. https://doi.org/10.1186/s13059-018-1590-2.

Peschansky, V.J., Wahlestedt, C., 2014. Non-coding RNAs as direct and indirect modulators of epigenetic regulation. Epigenetics 9 (1), 3–12. https://doi.org/10.4161/epi.27473.

Pinzón, N., Li, B., Martinez, L., et al., 2017. microRNA target prediction programs predict many false positives. Genome Res. 27 (2), 234–245. https://doi.org/10.1101/gr.205146.116.

Putnam, N.H., Butts, T., Ferrier, D.E., et al., 2008. The amphioxus genome and the evolution of the chordate karyotype. Nature 453 (7198), 1064–1071. https://doi.org/10.1038/nature06967.

Ramat, A., Simonelig, M., 2021 Feb. Functions of PIWI proteins in gene regulation: new arrows added to the pirna quiver. Trends Genet. 37 (2), 188–200. https://doi.org/10.1016/j.tig.2020.08.011.

Ravi, V., Venkatesh, B., 2018. The divergent genomes of teleosts. Annu. Rev. Anim. Biosci. 6, 47–68. https://doi.org/10.1146/annurev-animal-030117-014821.

Roberto, V.P., Gavaia, P., Nunes, M.J., et al., 2018. Evidences for a new role of mir-214 in chondrogenesis. Sci. Rep. 8 (1), 3704. https://doi.org/10.1038/s41598-018-21735-w.

Robertson, F.M., Gundappa, M.K., Grammes, F., et al., 2017. Lineage-specific rediploidization is a mechanism to explain time-lags between genome duplication and evolutionary diversification. Genome Biol. 18 (1), 111. https://doi.org/10.1186/s13059-017-1241-z.

Roest Crollius, H., Weissenbach, J., 2005. Fish genomics and biology. Genome Res. 15 (12), 1675–1682. https://doi.org/10.1101/gr.3735805.

Rozenfeld, C., Blanca, J., Gallego, V., et al., 2019. De novo European eel transcriptome provides insights into the evolutionary history of duplicated genes in teleost lineages. PLoS One 14 (6). https://doi.org/10.1371/journal.pone.0218085, e0218085.

Sacerdot, C., Louis, A., Bon, C., et al., 2018. Chromosome evolution at the origin of the ancestral vertebrate genome. Genome Biol. 19 (1), 166. https://doi.org/10.1186/s13059-018-1559-1.

Sallan, L.C., 2014. Major issues in the origins of ray-finned fish (Actinopterygii) biodiversity. Biol. Rev. Camb. Philos. Soc. 89 (4), 950–971. https://doi.org/10.1111/brv.12086.

Sandve, S.R., Rohlfs, R.V., Hvidsten, T.R., 2018. Subfunctionalization versus neofunctionalization after whole-genome duplication. Nat. Genet. 50 (7), 908–909. https://doi.org/10.1038/s41588-018-0162-4.

Santini, F., Harmon, L.J., Carnevale, G., Alfaro, M.E., 2009. Did genome duplication drive the origin of teleosts? A comparative study of diversification in ray-finned fishes. BMC Evol. Biol. 9, 194. https://doi.org/10.1186/1471-2148-9-194.

Schartl, M., Schories, S., Wakamatsu, Y., et al., 2018. Sox5 is involved in germ-cell regulation and sex determination in medaka following co-option of nested transposable elements. BMC Biol. 16 (1), 16. https://doi.org/10.1186/s12915-018-0485-8.

Simakov, O., Marlétaz, F., Yue, J.X., et al., 2020. Deeply conserved synteny resolves early events in vertebrate evolution. Nat. Ecol. Evol. 4 (6), 820–830. https://doi.org/10.1038/s41559-020-1156-z.

Simkin, A., Geissler, R., McIntyre, A.B.R., Grimson, A., 2020. Evolutionary dynamics of microRNA target sites across vertebrate evolution. PLoS Genet. 16 (2). https://doi.org/10.1371/journal.pgen.1008285, e1008285.

Smith, J.J., Timoshevskaya, N., Ye, C., et al., 2018. The sea lamprey germline genome provides insights into programmed genome rearrangement and vertebrate evolution [published correction appears in Nat Genet. 2018 Apr 19;:] [published correction appears in Nat Genet. 2018 Nov;50(11):1617]. Nat. Genet. 50 (2), 270–277. https://doi.org/10.1038/s41588-017-0036-1.

Sotero-Caio, C.G., Platt, R.N., Suh, A., Ray, D.A., 2017. Evolution and diversity of transposable elements in vertebrate genomes. Genome Biol. Evol. 9 (1), 161–177. https://doi.org/10.1093/gbe/evw264#.

Star, B., Nederbragt, A.J., Jentoft, S., et al., 2011. The genome sequence of Atlantic cod reveals a unique immune system. Nature 477 (7363), 207–210. https://doi.org/10.1038/nature10342.

Stark, R., Grzelak, M., Hadfield, J., 2019. RNA sequencing: the teenage years. Nat. Rev. Genet. 20 (11), 631–656. https://doi.org/10.1038/s41576-019-0150-2.

Stein, C., Caccamo, M., Laird, G., Leptin, M., 2007. Conservation and divergence of gene families encoding components of innate immune response systems in zebrafish. Genome Biol. 8 (11), R251. https://doi.org/10.1186/gb-2007-8-11-r251.

Stephen, S., Pheasant, M., Makunin, I.V., Mattick, J.S., 2008. Large-scale appearance of ultraconserved elements in tetrapod genomes and slowdown of the molecular clock. Mol. Biol. Evol. 25 (2), 402–408. https://doi.org/10.1093/molbev/msm268.

Storz, J.F., Opazo, J.C., Hoffmann, F.G., 2013. Gene duplication, genome duplication, and the functional diversification of vertebrate globins. Mol. Phylogenet. Evol. 66 (2), 469–478. https://doi.org/10.1016/j.ympev.2012.07.013.

Tarver, J.E., Taylor, R.S., Puttick, M.N., et al., 2018. Well-annotated microRNAomes do not evidence pervasive miRNA loss. Genome Biol. Evol. 10 (6), 1457–1470. https://doi.org/10.1093/gbe/evy096.

Tena, J.J., González-Aguilera, C., Fernández-Miñán, A., et al., 2014. Comparative epigenomics in distantly related teleost species identifies conserved cis-regulatory nodes active during the vertebrate phylotypic period. Genome Res. 24 (7), 1075–1085. https://doi.org/10.1101/gr.163915.113.

Tong, C., Fei, T., Zhang, C., Zhao, K., 2017. Comprehensive transcriptomic analysis of Tibetan Schizothoracinae fish *Gymnocypris przewalskii* reveals how it adapts to a high altitude aquatic life. BMC Evol. Biol. 17 (1), 74. https://doi.org/10.1186/s12862-017-0925-z.

Ulitsky, I., 2016. Evolution to the rescue: using comparative genomics to understand long non-coding RNAs. Nat. Rev. Genet. 17 (10), 601–614. https://doi.org/10.1038/nrg.2016.85.

Vasil'ev, V.P., 2009. Mechanisms of polyploid evolution in fish: polyploidy in sturgeons. In: Carmona, R., Domezain, A., García-Gallego, M., Hernando, J.A., Rodríguez, F., Ruiz-Rejón, M. (Eds.), Biology, Conservation, and Sustainable Development of Sturgeons. Fish & Fisheries Series. vol. 29. Springer, Dordrecht, https://doi.org/10.1007/978-1-4020-8437-9_6.

Varadharajan, S., Sandve, S.R., Gillard, G.B., et al., 2018. The grayling genome reveals selection on gene expression regulation after whole-genome duplication. Genome Biol. Evol. 10 (10), 2785–2800. https://doi.org/10.1093/gbe/evy201.

Volff, J.N., 2005. Genome evolution and biodiversity in teleost fish. Heredity (Edinb.) 94 (3), 280–294. https://doi.org/10.1038/sj.hdy.6800635.

Wang, K., Shen, Y., Yang, Y., et al., 2019. Morphology and genome of a snailfish from the Mariana Trench provide insights into deep-sea adaptation. Nat. Ecol. Evol. 3 (5), 823–833. https://doi.org/10.1038/s41559-019-0864-8.

Wickham, H., 2016. ggplot2: Elegant Graphics for Data Analysis. Springer-Verlag New York, ISBN: 978-3-319-24277-4. https://ggplot2.tidyverse.org.

Wolfe, K., 2000. Robustness—it's not where you think it is. Nat. Genet. 25 (1), 3–4. https://doi.org/10.1038/75560.

Wolfe, K.H., 2001. Yesterday's polyploids and the mystery of diploidization. Nat. Rev. Genet. 2 (5), 333–341. https://doi.org/10.1038/35072009.

Wray, G.A., 2007. The evolutionary significance of cis-regulatory mutations. Nat. Rev. Genet. 8 (3), 206–216. https://doi.org/10.1038/nrg2063.

Xiao, S., Xie, D., Cao, X., et al., 2012. Comparative epigenomic annotation of regulatory DNA. Cell 149 (6), 1381–1392. https://doi.org/10.1016/j.cell.2012.04.029.

Xu, P., Xu, J., Liu, G., et al., 2019. The allotetraploid origin and asymmetrical genome evolution of the common carp Cyprinus carpio. Nat. Commun. 10 (1), 4625. https://doi.org/10.1038/s41467-019-12644-1.

Yates, A.D., Achuthan, P., Akanni, W., et al., 2020. Ensembl 2020. Nucleic Acids Res. 48 (D1), D682–D688. https://doi.org/10.1093/nar/gkz966.

Yi, M., Chen, F., Luo, M., et al., 2014. Rapid evolution of piRNA pathway in the teleost fish: implication for an adaptation to transposon diversity. Genome Biol. Evol. 6 (6), 1393–1407. https://doi.org/10.1093/gbe/evu105.

Yu, G., Lam, T.T., Zhu, H., Guan, Y., 2018. Two methods for mapping and visualizing associated data on phylogeny using ggtree. Mol. Biol. Evol. 35 (12), 3041–3043. https://doi.org/10.1093/molbev/msy194.

Zhang, J., 2003. Evolution by gene duplication: an update. Trends Ecol. Evol. 18 (6), 292–298. https://doi.org/10.1016/S0169-5347(03)00033-8.

Zhang, H.H., Peccoud, J., Xu, M.R., Zhang, X.G., Gilbert, C., 2020. Horizontal transfer and evolution of transposable elements in vertebrates. Nat. Commun. 11 (1), 1362. https://doi.org/10.1038/s41467-020-15149-4.

Chapter 11

Zebrafish xenotransplantation: A versatile platform for cancer and stem cell translational research

Nicole Melong[a], Kim Kobar[a,b], Nadine Azzam[a], and Jason N. Berman[a,b]
[a]*The Department of Pediatrics, Division of Hematology/Oncology, Children's Hospital of Eastern Ontario Research Institute (CHEO RI), Ottawa, ON, Canada,* [b]*Faculty of Medicine, University of Ottawa, Ottawa, ON, Canada*

1. Introduction

The field of medical oncology continues to use traditional nonspecific chemotherapeutic treatment options for cancer patients. However, over the last decade there has been a surge in novel anticancer agents, including a broad range of targeted therapies and immunotherapeutic agents that are more specific to treat certain diseases (National Cancer Institute, 2020). This ever-growing list of candidate drugs is exciting for patients and clinicians alike, but also creates a challenge for establishing the preclinical data that is needed to help prioritize these compounds. The decision of which treatment regimen or compound to incorporate into a patient's schedule remains difficult and complex for the oncology team (Fazio et al., 2020). The advent of these molecularly-based treatment strategies has created the concept of precision oncology, an emerging paradigm that focuses on features of the specific patient's disease rather than a more generalized chemotherapeutic treatment regimen.

One major approach to precision oncology is the xenotransplantation (XT) platform, whereby human cancer cells, obtained from tumor biopsies, are injected into a host animal and used as a patient avatar to study the unique disease progression and drug responses of that tumor. Recent advances in the zebrafish (*Danio rerio*) as a XT model have played a substantial role in the precision oncology and novel drug discovery fields. Both the larval and adult XT platforms have their advantages: larval zebrafish are well suited for preliminary experiments and timely drug screens, while adult zebrafish are better equipped for long-term studies, allowing for more accurate modeling of drug dosing schedules and timing as well as administration routes (Yan et al., 2019a,b). In zebrafish, the adaptive immune system is not fully functional

Cellular and Molecular Approaches in Fish Biology. https://doi.org/10.1016/B978-0-12-822273-7.00005-7

until 3–4 weeks postfertilization, allowing short-term transplantation studies in larvae without the need for immunosuppression (Willett et al., 1999; Lam et al., 2004; Haldi et al., 2006; Nicoli et al., 2007). Additionally, immunodeficient zebrafish lines have been engineered to support adult transplantation models and long-term engraftment studies (Hess et al., 2013; Tang et al., 2014; Yan et al., 2019a,b).

Furthermore, zebrafish possess the remarkable ability to repair complex tissues after injury, including tissues found in the heart, retina, hair cells, central nervous system, caudal fin, kidney, and liver (Marques et al., 2019). Therefore, not only are they an excellent host for human cancer cell XT studies, but their regenerative capacities have been increasingly studied to unveil the molecular mechanisms underlying stem cell biology and their therapeutic potential. Stem cell studies in zebrafish have been conducted to investigate evolutionarily conserved pathways for tissue regeneration across different species (Canaider et al., 2014). In this chapter, we will discuss the larval and adult zebrafish XT models that have been created to date and their roles in precision oncology, as well as for stem cell colony expansion and regenerative therapies.

2. Larval zebrafish XT models of cancer

At 28 °C, the larval stage in the zebrafish life cycle starts at 72h postfertilization (hpf), and the beginning of this stage is defined by the obvious protruding mouth, which extends anteriorly just beyond the eye (Kimmel et al., 1995). The larvae have hatched completely from their chorions and the blade of the pectoral fin continues to expand and extends posteriorly over the length of the diminishing yolk sac (Kimmel et al., 1995). Larval development is rapid and at this time point, the larvae already have functioning organs, including the circulatory system (heart, closed blood vessels, circulating blood cells), gastrointestinal tract, and metabolic system (Kimmel et al., 1995). Because of this rapid organogenesis, larvae have been used for human cancer cell line XT and as a model for patient-derived xenografts (PDX) to study cancer progression and migration. The animals are often injected with human cells at the embryonic stage (~48 hpf) but the majority of the experiment takes place during the larval stage. This platform has contributed to novel drug discovery, identifying targeted cancer therapies and aided in the screen for combination therapies (Cagan et al., 2019). However, it is known that tumors are heterogeneous and complex, making clinical treatment decisions difficult due to unknown mutations or treatment resistance. The zebrafish larval XT platform has shown promising preclinical potential to produce patient avatars of disease and provide essential drug response data in real time to help clinicians make the most informed treatment decisions possible. The ability of this model organism to host a PDX followed by medium to high-throughput drug screening is unparalleled in any other model organism.

2.1 Advantages and limitations of larval zebrafish XT models

One major advantage of the larval zebrafish model that is essential for XT assays, is the absence of a fully developed immune system during this stage of life, which means no immunosuppresion is required. It was proposed that zebrafish at 48 hpf are at the optimal developmental stage for transplantation (Haldi et al., 2006). At this stage, the nervous system is well developed, and the animal possesses a functional circulatory system and critical muscular and skeletal development for body mobility (Kimmel et al., 1995; Vittori et al., 2015). Along with rapid development, the high fecundity of the zebrafish allows for a large number of vetebrates to be studied per experiment. Prior to injection, human cancer cells can be transiently fluorescently labeled (using CellTracker cytoplasmic or membrane dyes) or stably transfected with fluorescent proteins and can be injected into different transplantation sites (Haldi et al., 2006; Veinotte et al., 2014). The transplantation sites vary depending on the cancer type and the assay of choice, but most often include the yolk sac, perivitelline space, common cardinal vein, Duct of Cuvier, or hindbrain ventricle (Fig. 1). Furthermore, the number of cells required to complete an experiment of this nature is relatively low (minimum of 2×10^6 cells) compared to the cell number requirements for mammalian XT studies. This is especially advantageous when working with precious patient-derived biopsy samples that are small in size and primarily acquired for clinical diagnostic purposes.

Over a decade ago, Richard White and Leonard Zon created a double pigment mutant known as the *casper* line (White et al., 2008) that is frequently employed across the zebrafish community given its optical clarity throughout the life of the animal. Once fluorescently labeled cancer cells are injected into the transparent larvae, the cells can be tracked in real time for proliferation and migration at single cell resolution using fluorescent microscopy and imaging. The larvae can be arrayed in a variety of isolated well plates, with the additional benefit of not requiring feeding as they rely on their yolk sac for nutrition for

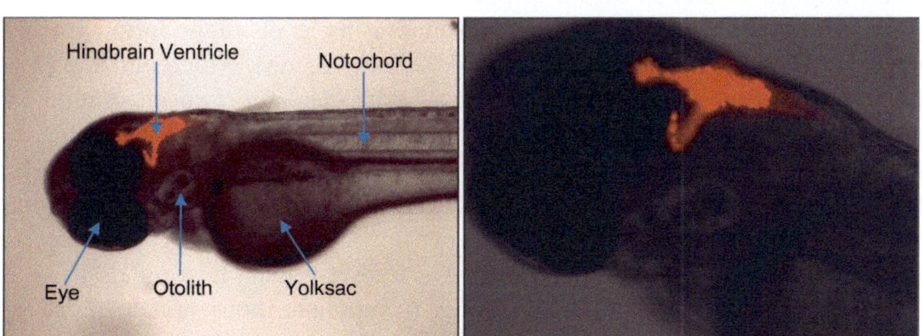

FIG. 1 Images of a 4-day postfertilization (dpf) zebrafish with CM-DiI labeled patient-derived Ewing sarcoma cells (labeled red) injected into the hindbrain ventricle at 5× and 10×.(Images courtesy of the Berman Zebrafish Lab, 2019.)

up to 5days of life (Vittori et al., 2015). Another advantage of the larval stage is their small size, allowing them to be permeable to small molecules through immersion therapy, facilitating moderate to high-throughput screening for drug efficacy and toxicity. This means drugs can be simply added to the larval water if the drug is water soluble, and absorbed across the body of the animal, removing the need for individual drug administration (Brown et al., 2017).

Like every animal model system, there are several drawbacks when using larval zebrafish for XT assays. Firstly, incubation temperature has been shown to be a critical factor impacting cancer cell proliferation (Lee et al., 2005). The ideal physiological temperature for human cells is 37°C and the optimal temperature for zebrafish growth and development is 28 °C; however, zebrafish can tolerate temperatures ranging from 6 °C to 38 °C (Kimmel et al., 1995; Spence et al., 2008; Detrich et al., 2010). Therefore XT experiments conducted at 28 °C are not ideal for human tumor cell proliferation. The commonly employed incubation temperature for larval zebrafish following XT is 35 °C, as a compromise for the normal growth of both larvae and human cancer cells (Haldi et al., 2006; Marques et al., 2009; Corkery et al., 2011; Bentley et al., 2015; Melong et al., 2017). Secondly, we know that accurate and specific dosing in zebrafish compared to humans has been a frequent topic of discussion. Adding the drug into larval water does not ensure equal or adequate absorption of the drug into each larva. Immersion therapy, albeit convenient and technically straightforward, cannot achieve accurate dosing per larva, which is important information when moving preclinical animal studies into human clinical trials. In contrast, although more time consuming and technically challenging, it is possible to gavage larval zebrafish or inject the drug directly into circulation to mimic oral or intravenous (IV) drug administration (Cocchiaro and Rawls, 2013; Cloney et al., 2018; Wertman et al., 2020). While we understand the drug doses used in zebrafish experiments cannot be directly applied to a human context, it is important to note that in our studies involving cisplatin, plasma levels measured in pediatric patients were within the range of our experimental larval doses. Studies of human serum cisplatin concentrations suggest that over the initial 24h period following administration, plasma levels range from approximately 1–10μg/mL, which converts to 0.0033–0.033mM, which encompasses the doses we used when treating our larvae with cisplatin (Lanvers-Kaminsky et al., 2006; Wertman et al., 2020). Another common issue concerning the zebrafish XT model is the conservation of molecular mechanisms and pathways, and more specifically drug absorption, distribution, metabolism, and excretion (ADME), in relation to humans. The pregnane X receptor (PXR) is a nuclear receptor that functions as a ligand-activated transcriptional factor and has vital functions in mammals, including glucose metabolism, bone homeostasis, reproduction, neuronal development, cognitive capacities, transport, and excretion of exogenous compounds (Marques et al., 2017; Ekins et al., 2008). PXRs show significant differences in ligand specificity across species, and although zebrafish do possess PXRs, they were shown to be more sterically constrained and were activated by fewer compounds, suggesting a smaller ligand-binding pocket compared to those in humans (Ekins et al., 2008). Zebrafish PXRs were activated by a smaller number of steroid hormones than human PXRs, but there was substantial overlap between the selectivity of the two PXRs (Ekins et al., 2008). Furthermore, it is widely known that drug-metabolizing enzymes play central roles in metabolism, elimination, and detoxification (de Souza Anselmo et al., 2018). Zebrafish have the ability to perform both phase I and phase II metabolism reactions, and the metabolic enzymes responsible for these reactions are highly conserved in relation to mammals (Chng et al., 2012; Diekmann and Hill, 2013; Brox et al., 2016). In humans, the five CYP gene families (CYP1, CYP2, CYP3, CYP4, and CYP7) are the main metabolizing enzymes responsible for xenobiotic metabolism and are associated with the oxidative biotransformation of approximately 70% of most clinically used drugs (de Souza Anselmo et al., 2018). Many human CYP enzymes have direct orthologs in zebrafish suggesting similar metabolic profiles between the two (Goldstone et al., 2010).

Lastly, as with any transplant model, the microenvironment provided by the zebrafish is not an exact replica of the human cell niche (Brown et al., 2017). Although zebrafish have a high level of genetic conservation with humans at the genomic and proteomic levels (Howe et al., 2013), the transplanted cells are exposed to a different host niche and environmental factors than what would naturally occur in a human host. As previously mentioned, the larval zebrafish does not possess an adaptive immune system, which enables the engraftment of human cancer cells. Thus, it does not recapitulate the mature immune cells present in a human host. Moreover, zebrafish also lack specific cytokines, which are known to be essential protein factors in the human microenvironment for the differentiation and maintenance of different cell types. To address this limitation and improve the engraftment efficiency of transplanted human cancer cells, our group generated a humanized transgenic zebrafish that expresses multiple human hematopoietic-specific cytokines (Rajan et al., 2020). The humanized zebrafish, termed the *GSS* zebrafish (an acronym for the cytokines expressed: granulocyte monocyte stimulating factor (GM-CSF), stem cell factor (SCF), and SDF1α), enhanced the proliferation and the hematopoietic niche-specific homing of patient-derived leukemia cells and umbilical cord-derived human hematopoietic stem and progenitor cells (HSPCs) (Rajan et al., 2020). The *GSS* fish provide a multiuse platform for the transplantation and expansion of HSPCs as well as human cancer cells to evaluate oncogenic potential in vivo.

2.2 Transplantation of human cancer cell lines in larval zebrafish

Larval zebrafish have been successfully utilized in many XT studies and have become popular in drug screening assays for the discovery of novel potential therapeutics. Before using larval zebrafish, cancer cells were first transplanted into zebrafish blastula-stage embryos where the survival and motility of cancer cells was observed (Lee et al., 2005). Researchers transplanted melanoma cells into the zebrafish blastula and observed that the cells survived and divided but did not form tumors (Lee et al., 2005). In a follow-up study, human melanoma cells were injected into zebrafish blastula-stage embryos and it was determined that aggressive melanoma cells secreted Nodal (Topczewska et al., 2006). Based on this work, Nodal has emerged as a diagnostic marker of disease progression and a possible target for development of an anticancer therapeutic (Topczewska et al., 2006). In one of the first studies that explored the development of larval zebrafish as a XT model, human metastatic melanoma, pancreatic carcinoma, and colon carcinoma cancer cells were transplanted into 48 hpf zebrafish. These cell lines were able to proliferate, migrate, and form masses in the zebrafish. Optimizations were made in this study, demonstrating the feasibility of using 48 hpf zebrafish for cancer research and as a XT host for human cancer cells, laying the foundation for future work in this area (Haldi et al., 2006).

2.2.1 Angiogenesis and the microenvironment

Other studies have employed the zebrafish larval XT model to examine the tumor microenvironment through the analysis of tumor-induced angiogenesis. Angiogenesis, the formation of new and recruitment of existing blood vessels, is an important factor in cancer progression and metastasis and is therefore a target for antitumor therapies. When tumorigenic fibroblast growth factor (FGF) 2-overexpressing mouse aortic endothelial cells were injected into 48 hpf zebrafish, they triggered a potent angiogenic response. Using different cancer cell lines, this study showed the capability of zebrafish to discriminate between highly angiogenic and poorly angiogenic tumor cell lines (Nicoli et al., 2007). The transplantation of melanoma cells into transgenic *Tg(flk1:EGFP)* 48 hpf zebrafish, which possess fluorescently labeled blood vessel endothelial cells, has shown the contribution of vascular endothelial cell growth factor receptor 2 positive ($VEGFR2^+$) to individual endothelial cells in the formation of the tumor vascular network. The larvae were treated with a specific VEGFR2 inhibitor, SU5416, which significantly inhibited vascularization and tumor growth, while having little effect on normal vessels (Zhao et al., 2011). Another study injected 48 hpf zebrafish with melanoma cells that overexpressed antiapoptotic bcl-xL, and then assessed the role of interleukin CXCL8, and bcl-xL on the dissemination and angiogenesis of cancer cells (Gabellini et al., 2018). Results suggested that the autocrine CXCL8/CXCR2 signaling pathway induces melanoma aggressiveness (Gabellini et al., 2018). These studies and others (Haldi et al., 2006; Vlecken and Bagowski, 2009; Stoletov et al., 2007) have shown the utility of the zebrafish XT model, when used in conjunction with transgenic zebrafish vascular reporter lines, as a tool for the discovery of antiangiogenic therapies.

2.2.2 Liquid tumor XTs

Leukemia cell xenografts in zebrafish larvae have been conducted to study antileukemic drug efficacy and toxicity. In a study examining the inhibition of human leukemia cell growth, leukemic cell lines and CD34$^+$ leukemic blast cells sorted from the blood of acute myeloid leukemia (AML) patients were injected into 48 hpf zebrafish (Pruvot et al., 2011). After transplantation and engraftment of the cells, antileukemic drugs were tested, and cancer cell growth was inhibited by imatinib mesylate and oxaphorines. Treatment decreased the leukemic burden and importantly, did not demonstrate any toxicity to the zebrafish (Pruvot et al., 2011). Our group injected human myeloid leukemia cell lines (K562 and NB4) into 48 hpf *casper* embryos, which were able to proliferate and survive for up to 7days posttransplant (dpt). Once the cells were injected and engrafted, specific molecularly targeted agents, imatinib mesylate (targeting BCR-ABL in chronic myeloid leukemia K562 cells) and all-trans retinoic acid (ATRA) (targeting PML-RARA in acute promyelocytic leukemia NB4 cells), were added to the water. An ex vivo cancer cell quantification assay was employed, showing a selective decrease in the number of leukemic cells following treatment with the appropriate targeted agent, in contrast to when that drug was used against the opposite cell line (Corkery et al., 2011).

As previously mentioned, the larval zebrafish does not possess a fully adaptive immune system until 3–4weeks after fertilization. However, the innate immune system is detectable and active at day one of zebrafish embryogenesis (Herbomel et al., 1999). An interesting approach to address the contribution of innate immunity to engraftment was recently undertaken by Gauert et al. (2020). Human B-cell precursor acute lymphoblastic leukemia (BCP-ALL) cell lines (SEM, Nalm-6, and RCH-ACV) were injected into 48 hpf zebrafish, which had been previously injected with morpholinos targeting innate immune cell genes, resulting in transient knockdown of neutrophils and monocytes. BCP-ALL survival was improved in morpholino-injected xenografts compared with controls. The xenografted zebrafish were then treated with venetoclax for 72h and human cancer cell viability was analyzed through flow cytometry (Fig. 2). Increasing doses of venetoclax showed decreased BCP-ALL cell viability in vivo (Gauert et al., 2020).

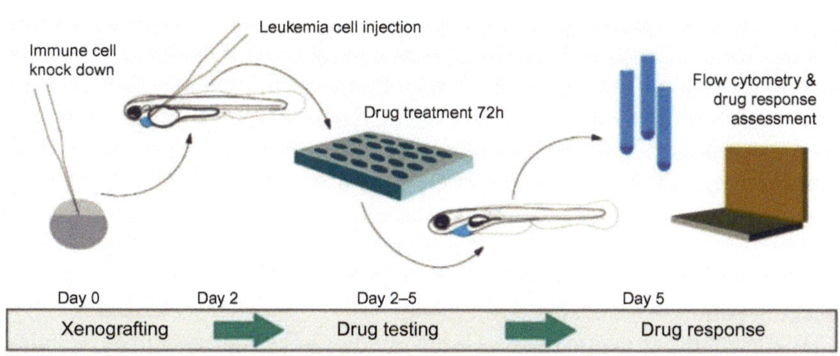

FIG. 2 Experimental schematic of Gauert et al. (2020) immunosuppressed larval zebrafish XT platform.

2.2.3 Solid tumor XTs

Although zebrafish lack mammary glands, prostate tissue, and lungs, studies have proven zebrafish to be a good model for studying human breast and prostate cancer by using XT. One study looked at metastasis and dissemination of breast, colorectal, and prostate cancer cell lines in zebrafish (Ghotra et al., 2012). The authors established that silencing E-cadherin accelerates breast carcinoma cell dissemination, supporting the inhibitory role of E-cadherin in cancer metastasis (Ghotra et al., 2012). In addition, findings from this study showed that cancer cells of mesenchymal origin displayed higher dissemination capacity than those of epithelial origin. Another study looked at the metastatic properties of different breast cancer lines following XT into larval zebrafish (Drabsch et al., 2013). The transforming growth factor beta (TGF-β) signaling pathway is known to contribute to the invasiveness of breast cancer (Wiercinska et al., 2011). Researchers showed that a TGF-β receptor kinase inhibitor prevented the invasion of cancer cells by inhibiting TGF-β, which was consistent with the results obtained from a mouse metastasis model (Drabsch et al., 2013). Triple-negative breast cancer (TNBC), a highly aggressive breast carcinoma, has limited treatment options. Using the zebrafish XT model, investigators found that human cancer cells expressing CXCR4 can respond to zebrafish cognate ligands at the metastatic site (Tulotta et al., 2016). They were able to block TNBC metastasis by using IT1t, a CXCR4 antagonist. Further, they proposed CXCR4 as a pharmacological target in TNBC (Tulotta et al., 2016).

In a zebrafish XT model of human prostate cancer, it was reported that the spleen kinase (SYK) affects the dissemination of prostate cancer cells (Ghotra et al., 2015). The authors showed the absence of SYK prevented cancer dissemination (Ghotra et al., 2015). In another study conducted by our group, the androgen-dependent LNCaP prostate cancer cell line was xenografted into larval zebrafish (Melong et al., 2017). Exogenous testosterone was administered to model the hormonal microenvironment in which prostate cancer develops in humans. The administration of testosterone increased the proliferation of LNCaP cells in comparison with controls and this effect was reversed by the antiandrogen receptor drug, enzalutamide. This study provided evidence that testosterone administration should be considered in future zebrafish XT studies of prostate cancer (Melong et al., 2017). Furthermore, like most cancers, the early diagnosis of prostate cancer is critical. A study utilizing the larval zebrafish XT platform established an improved in vivo model for the early diagnosis of prostate cancer. PC3-CTR cells (PC3 cells overexpressing the calcitonin receptor) were injected into zebrafish, and the cells successfully migrated and proliferated. The invasiveness of these cells provided a good model for both drug-targeted screening and for specific early markers to better detect prostate cancer in patients (Xu et al., 2018).

While zebrafish do not possess lungs, studies have been able to recapitulate cancer–tumor microenvironment interactions using XT models. A non-small-cell lung cancer (NSCLC) XT model was successfully designed by injecting

A549 lung cancer cells into zebrafish. This XT model was used to evaluate the anticancer effects of bevacizumab, endostar, and apatinib (Jin et al., 2018). The authors were able to observe inhibition of tumor growth in zebrafish (Jin et al., 2018). In collaboration with the Bally and Lockwood labs, our group examined the expression levels of 3′-phosphoadenosine 5′-phosphosulfate synthase 1 (PAPSS1) and cisplatin sensitivity in lung adenocarcinoma cell lines (Leung et al., 2017). PAPSS1 suppression in A549 cancer cells, in combination with low-dose cisplatin treatment, inhibited proliferation of lung tumor cells in zebrafish xenografts. Targeting PAPSS1 activity along with the administration of platinum-based chemotherapy was presented as an approach to improve treatment outcomes in patients whose lung cancers harbor this mutation (Leung et al., 2017). Lastly, NSCLC XTs were used to reveal and learn more about the role of autophagy in cancer (Chou et al., 2019). After the transplantation of human H1299 cells into 48 hpf zebrafish, authors investigated the effect of C2-ceramide and chloroquine (an autophagy inhibitor) on the growth of NSCLC XT cells. The treatment reduced the mass of the H1299-forming tumor in the zebrafish, suggesting that the use of autophagy inhibitors and a sublethal dose of C2-ceramide may be promising for the treatment of NSCLC (Chou et al., 2019).

2.3 Patient-derived xenografts (PDX) in larval zebrafish

The first study that utilized zebrafish to directly transplant patient-derived tumor cells was conducted to better understand the metastatic behavior of cancer cells (Marques et al., 2009). Primary human cancer cells from pancreas, colon, and gastric carcinomas were transplanted into 48 hpf zebrafish and both human tumor cell invasion and micrometastasis were observed within 24h of transplantation (Marques et al., 2009). Our group transplanted patient-derived T-cell acute lymphoblastic leukemia (T-ALL) into 48 hpf zebrafish to assess drug response in cancer cells from different individuals (Bentley et al., 2015). We were able to identify a mutation sensitive to γ-secretase inhibition through drug response assays and this mutation was confirmed as a gain-of-function *NOTCH1* mutation (Bentley et al., 2015). Patient-derived multiple myeloma (MM) cells are difficult to both establish in culture and grow in vivo using murine models, but were able to survive, engraft, and proliferate following transplantation into larval *casper* zebrafish (Lin et al., 2016). Furthermore, $CD138^+$ MM cells derived from patients that were transplanted into larval zebrafish, exhibited growth inhibition to the same drugs to which the patients had responded, providing evidence for the utility of this preclinical platform (Lin et al., 2016).

Patient-derived human colorectal cancer (CRC) cells were transplanted into 48 hpf zebrafish, and while the CRC cells maintained their general cellular characteristics, different patient samples exhibited different proliferation dynamics. The zebrafish xenografts were treated with two common chemotherapy regimens, 5- fluorouracil (5-FU)+oxaliplatin + folinic acid (FOLFOX) or 5-FU +irinotecan + folinic acid (FOLFIRI) and showed an overall higher sensitivity to FOLFIRI than to FOLFOX. These proof-of-concept experiments provided evidence of similar behavior responses to therapies between patients and zebrafish PDXs (Fior et al., 2017). These studies, and others (Gaudenzi, 2016; Mercatali, 2016; Wu et al., 2017), have shown that PDX studies and drug response, efficacy, and sensitivity data can be obtained in a rapid and actionable time frame in the larval zebrafish XT model. This model has proven to be valuable for drug discovery and has the potential to improve and accelerate the design of personalized cancer therapy options through the generation of patient avatars.

2.4 Drug screening, novel drug discovery, and radiotherapy enhancement using larval zebrafish

Zebrafish are an ideal choice for an unbiased novel compound testing platform, as mammalian models are expensive, laborious, and thus not well suited to screening large numbers of compounds for evidence of an efficacy signal. In the zebrafish larval model, drug treatments can be easily administered by simply adding the drug to the water in which the animals are maintained. Drug toxicity and antitumor effects can be tracked in real time with noninvasive imaging. Another advantage of the zebrafish XT platform is that it is ideal for providing drug responses in an actionable time frame. Transplantation of cells into zebrafish larvae and responses to drug treatments, including the evaluation of the inhibition of cell proliferation and migration, can be assessed within days.

Drug screening in zebrafish has become popular, as screening can be carried out at a medium to high-throughput rate. Drug screening studies have traditionally been performed in nonxenografted zebrafish embryos, but rather focused on targeting specific oncogenic molecular pathways. For example, a chemical screen conducted in transgenic zebrafish embryos (*Tg(flk1:EGFP)* line) tested for small molecules capable of the inhibition of angiogenesis (Wang et al., 2010). Rosuvastatin was identified among seven hits as an inhibitor of angiogenesis, and the drug was subsequently tested in XT nonobese diabetic severe combined immunodeficiency (NOD-SCID) mice and showed suppression of prostate tumor growth by inhibiting cancer cell proliferation (Wang et al., 2010). Another study screened 2000 chemicals to identify inhibitors of the neural crest lineage in zebrafish embryos (White et al., 2011). Leflunomide, an inhibitor of dihydro-orotate dehydrogenase, was found to inhibit the development of both neural crest cells and human melanoma cells (White et al., 2011). Furthermore, the larval zebrafish XT platform has shown the capacity to identify novel therapeutics used to treat acute leukemia. An example of novel drug discovery using this platform is a study that employed phloridzin (PZ), a flavonoid precursor found in apple peels, and docosahexaenoic acid (DHA), to yield a novel ester known as PZ-DHA (Arumuggam et al., 2017). PZ-DHA was shown to significantly inhibit the proliferation of the Jurkat human T-ALL cell line in zebrafish larvae (Arumuggam et al., 2017). In another study an antileukemic drug screen was performed, where more than 25,000 small compounds were tested. Among the compound hits, the authors identified Lenaldekar to have a generalized antileukemia effect on T-ALL cells (Ridges et al., 2012). There are other papers extensively reviewing zebrafish as a preclinical model in drug discovery (Parng et al., 2002; Peterson and MacRae, 2012; MacRae and Peterson, 2015; Kithcart and MacRae, 2017).

Not only can this model be used to discover new compounds, but it can also be employed to improve existing drugs that are currently used in the clinic. Chemotherapeutics are effective but highly toxic nonspecific therapies that can lead to lifelong complications and undesired side effects. The zebrafish XT model has served as a model to discover protective agents that mitigate unwanted side effects of chemotherapeutics while ensuring their chemotherapeutic efficacy and not interfering with their capacity to kill cancer cells. Standard induction regimens used for most patients with acute leukemia are based on a backbone of anthracyclines, an effective treatment option, but are highly cardiotoxic drugs shown to increase the risk of heart failure (Kang et al., 2019). A zebrafish model of doxorubicin-induced cardiomyopathy was used to screen for new cardioprotective compounds (Liu et al., 2014). Two compounds, visnagin (VIS) and diphenylurea (DPU), attenuated cardiac cell death and preserved cardiac function in larval zebrafish (Liu et al., 2014). This led to the discovery that VIS binds to mitochondrial malate dehydrogenase (MDH2), which plays a role in the tricarboxylic acid cycle. In so doing, this study identified VIS and DPU as potential cardioprotective compounds and MDH2 as a target for doxorubicin-induced cardiomyopathy. Importantly, the zebrafish XT model was employed to demonstrate that MDH2 inhibition is sufficient for cardioprotection and that the anticancer effects of doxorubicin were preserved (Liu et al., 2014). Our group recently completed a 1280 compound screen in larval zebrafish for otoprotective agents against cisplatin-induced damage to the hair cells of the lateral line neuromasts, an analogous structure to human hair cells of the inner ear (Wertman et al., 2020). Dopamine and L-mimosine were found to protect zebrafish neuromasts from cisplatin-induced damage and also preserve zebrafish larval glomerular filtration. This study provided preclinical evidence for the potential utility of dopamine and L-mimosine in the safer administration of cisplatin.

Zebrafish models have also been used in radiotherapy studies. U251 neuroblastoma cells transplanted into zebrafish embryos were treated with radiosensitizer 4′-bromo-3’nitropropiophenone (NS-123), to test the efficacy and toxicity of this treatment (Lally et al., 2007). Treatment reduced the number of cancer cells in zebrafish xenografts and no toxic effects were observed (Lally et al., 2007). Another study transplanted U251-RFP human glioma cells into zebrafish embryos and treated the XT fish with ionizing radiation and temozolomide (TMZ) (Geiger et al., 2008). Results showed TMZ enhanced the inhibitory effects of radiotherapy on the proliferation of these cells but had no effect on embryonic development (Geiger et al., 2008). Lastly, using a novel gold nanoparticle-mediated diamond target beam radiation platform, our group demonstrated the enhanced cell killing of human head and neck cancer cells in zebrafish xenografts compared to those receiving standard radiotherapy (Piccolo et al., 2019).

The power of this model lies in the numbers, as large sample sizes and clutches of larvae can be obtained and tested in a relatively short time frame, in comparison to mammalian chemical screenings, where this would not be feasible. The efficiency of embryo/larval screening can be combined with XT to enable unbiased in vivo cell viability screens of human cancers in a zebrafish host. This platform creates new opportunities for anticancer personalized therapies, as recent studies have shown the reliability of the zebrafish PDX model. From a translational perspective, cancer drug discovery using zebrafish PDXs could be a new scalable paradigm that provides fast and reliable assessments for personalized treatments for every patient.

3. Using the zebrafish for translational stem cell research

3.1 Nonmalignant stem cells

Stem cells have the ability to give rise to every tissue in the human body, and therefore have great potential for therapeutic uses involving tissue regeneration and repair. There are two essential characteristics of the stem cell: first, the cells must be able to self-renew and produce daughter cells identical to the original cell; second, the stem cell must be able to differentiate into a specialized mature cell type (Biehl and Russell, 2009). As such, stem cell transplants are a promising regenerative therapeutic option for patients with diseased or damaged cells, but further investigation in vivo is needed to improve clinical efficacy. Many current expansion protocols fail to produce clinically relevant numbers of functional stem cells or lack the ability to guide lineage-specific differentiation. Zebrafish, with their numerous advantages, have been exploited as a powerful tool in stem cell research for over three decades (De La Garza et al., 2017; Perlin et al., 2017; Cavalcante et al., 2019; Arjmand et al., 2020) and as a result, stem cell XT zebrafish models are emerging as a promising tool for preclinical stem cell research. As described for cancer cells, use of the transparent mutant *casper* zebrafish line similarly enables single-cell resolution imaging and observation of real-time stem cell interactions and behaviors in vivo. Mammalian stem cells (induced pluripotent stem cells (iPSCs), hematopoietic stem and progenitor cells (HSPCs), mesenchymal stem cells (MSCs), neural stem cells (NSCs), and their derivatives) transplanted into zebrafish can survive, differentiate, migrate to niches, and interact with zebrafish cells and signaling molecules (Patel et al., 2012; Orlova et al., 2014; Staal et al., 2016; Han and Hsu, 2017; Hamilton et al., 2018; Parada-Kusz et al., 2018). Zebrafish XT models are being developed to gain a better understanding of stem cell behavior and mechanisms, and once characterized, these models can serve as a drug discovery platform to find compounds that enhance stem cell numbers or target disease-causing stem cells.

Zebrafish stem cell XT models are being used to functionally assess alternative stem cell sources, including adipose-derived stem cells (ADSCs) and iPSCs as they are both abundant, can differentiate into many cell lineages, and can be used for autologous transplant therapies (Li et al., 2015; Xiao et al., 2016; Strnadel et al., 2018). Additionally, the zebrafish is a robust tool being used to better understand and improve hematopoietic stem cell transplants (HSCTs) and regenerative cellular therapies.

3.2 Hematopoietic stem cell transplants

Allogeneic HSCTs are widely used as a therapeutic option for patients with certain blood cancers, primary immune deficiencies, and some genetic disorders. For an HSCT to be successful, donor HSPCs must migrate to the bone marrow niche, engraft, and expand to repopulate the hematopoietic cell lineages (Sullivan et al., 2010). Ablation of the bone marrow prior to transplant is necessary to clear out diseased or damaged cells and to support the engraftment of transplanted cells (Hamilton et al., 2018). Additionally, there is a limited number of HSPCs that can be obtained from donor samples, so research efforts have been focused on expanding the HSPC population ex vivo to produce clinically relevant numbers (De La Garza et al., 2017). Further research is needed to better understand engraftment, expansion, and niche-homing mechanisms, so these processes can be optimized to decrease risks associated with HSCTs, including rejection, infection due to delayed engraftment, and death (Hamilton et al., 2018).

The zebrafish is a well-established model for hematopoiesis research and is emerging as a powerful tool for hematopoietic transplantation studies as the optical clarity of zebrafish allows for real-time tracking of transplanted cells to visualize cell interactions, engraftment success, and migration patterns. Importantly, there is a high degree of conservation between the mammalian and zebrafish hematopoietic systems that allows for the translation of findings from zebrafish to humans (Davidson and Zon, 2004; Clements and Traver, 2013). Although zebrafish and mammals have different anatomical hematopoietic niche sites (the caudal hematopoietic tissue (CHT) is akin to the fetal liver/placenta while the kidney marrow is analogous to the mammalian bone marrow), critical hematopoietic genes, pathways, and mechanisms are well conserved and common hematopoietic lineages give rise to the same major blood cell types (Traver et al., 2003; Murayama et al., 2006; Orkin and Zon, 2008; Tamplin et al., 2015). Recent studies have demonstrated that transplanted mammalian HSPCs home to the CHT and successfully engraft, respond to zebrafish signals, and are integrated into the developing hematopoietic and vasculature systems, further highlighting the high degree of hematopoietic conservation between mammals and zebrafish (Orlova et al., 2014; Staal et al., 2016; Parada-Kusz et al., 2018; Hamilton et al., 2018). Thus the zebrafish represents a robust drug discovery tool to identify compounds that increase expansion, promote engraftment, and direct migration of HSPCs to the hematopoietic niche to help reduce the risks associated with HSCTs.

In vivo chemical screens performed on zebrafish embryos have identified compounds that promote the expansion of endogenous zebrafish HSPCs, and several of these compounds also enhance functional expansion of mammalian HSPCs both in vivo and ex vivo (Arulmozhivarman et al., 2016; North et al., 2007). For example, valproic acid (VPA) was identified as a hit in one of these chemical screens and further validation revealed that treatment of human peripheral blood with VPA promoted the expansion and maintenance of isolated HSPCs ex vivo (Arulmozhivarman et al., 2016). Another compound identified in a different screen was 16,16-dimethyl-prostaglandin E2 (dmPGE2), a stable derivative of prostaglandin E2, which was the first compound found using the zebrafish platform that led to a human clinical trial (Cutler et al., 2013; North et al., 2007). Umbilical-cord blood-derived stem cells treated with dmPGE2 prior to transplant exhibited preferential engraftment in 10 out of 12 patients (Cutler et al., 2013). These compounds showed preclinical potential to increase numbers of human HSPCs in a Petri dish prior to transplant (Arulmozhivarman et al., 2016), in patients posttransplant, and for individuals with bone marrow failure syndromes (North et al., 2007). These studies provide further evidence that the knowledge gained from zebrafish is clinically relevant and that the zebrafish can be used as a drug discovery platform to find compounds that enhance human HSPC expansion and engraftment.

As human HSPCs are in short supply, mouse HSPCs are an alternative source for preliminary studies prior to validation in human. Mouse HSPCs transplanted into zebrafish embryos are able to interact with zebrafish cells, respond to zebrafish niche-homing signals to migrate to the hematopoietic niche, and mount a response to a bacterial infection (Parada-Kusz et al., 2018). The ability of mammalian HSPCs to respond to and interact with endogenous zebrafish signals and cells is key to study HSPC function in real time and to identify compounds that enhance engraftment and niche signaling for therapeutic applications (Parada-Kusz et al., 2018).

In one study, human HSPCs transplanted into the duct of Cuvier of 48 hpf zebrafish were able to survive for up to 6 dpt, respond to zebrafish stromal signals necessary for HSC maintenance, and differentiate into myeloid lineages (Staal et al., 2016). Additionally, human HSPCs transplanted into zebrafish embryos were able to engraft in the zebrafish hematopoietic niche and interact with the developing vascular tissue (Pozzoli et al., 2011; Hamilton et al., 2018). These models can be employed to further investigate cell fate and niche signaling and to identify compounds that enhance the expansion and engraftment of human HSPCs. Konantz et al. (2019) describe a method to transplant human HSPCs into transgenic zebrafish embryos to visualize the interactions of transplanted cells with the microenvironment. These studies encouraged the development of humanized zebrafish models to better represent the human context and further increase the clinical relevance of xenotransplantation studies.

In this regard, our group engineered humanized *GSS* fish (see above) that express human hematopoietic-specific cytokines (Rajan et al., 2020). Cytokines support the survival, proliferation, and differentiation of human cells, but there is poor evolutionary conservation of hematopoietic-specific cytokines between zebrafish and humans which may reduce the ability of human cells to thrive in zebrafish (Stachura et al., 2013; Rajan et al., 2020). Employing both fluorescent imaging and digital droplet PCR for HSPC and lineage markers, we conclusively determined that there was increased survival and differentiation of transplanted human HSPCs in the *GSS* fish compared with HSPCs transplanted into *casper* control larvae (Rajan et al., 2020). This observation suggests that these humanized zebrafish represent an improved platform with greater clinical relevance to aid in the identification of compounds that enhance human HSPC expansion, engraftment, and differentiation.

3.3 Regenerative therapies

Stem cell transplants are a regenerative therapeutic option for diseased or damaged tissue with some success, but further validation in vivo is needed to better understand mechanisms, enhance tissue engineering techniques, and to find compounds that promote the expansion and differentiation and target disease-causing stem cells (Kwon et al., 2018). The zebrafish xenograft platform is being used to investigate translational applications involving stem cells in the nervous system, vasculature system, and fibrotic diseases.

Firstly, adult NSCs show potential for neural regenerative therapies, but further optimization is needed, because NSCs are rare, need to be expanded prior to transplant, and have a low survival rate postengraftment (Takahashi et al., 2008; Han and Hsu, 2017). Alternatively, MSCs may be useful for neural regenerative applications as they can be induced to form functional neurons, are abundant in bone marrow and adipose tissue, and have established expansion protocols (Cho et al., 2005; Patel et al., 2012). There are some promising clinical results using MSCs for regenerative applications, but further investigation in vivo is needed to advance MSC cellular therapies to large-scale clinical trials for central nervous system (CNS) disorders and injuries (Momin et al., 2010; Patel et al., 2012).

To investigate the regenerative abilities of MSCs in the CNS, human MSCs that were induced to form functional neurons were transplanted into zebrafish larvae that had been subjected to a spinal cord injury (SCI) (Patel et al., 2012). Human MSC-transplanted fish displayed better wound healing, had human MSCs present at the SCI site, and both locomotion and sensory assays showed significant functional neural regeneration compared to untransplanted controls over time (Patel et al., 2012). This zebrafish model can be used to investigate

MSC repair mechanisms and be employed as a drug discovery platform to identify compounds that enhance MSC neural regeneration capabilities (Patel et al., 2012).

Coculturing NSCs with ADSCs, a type of MSC that serves as an alternate stem cell source as they are abundant and possess the capacity to differentiate into many cell lineages, encouraged the formation of cospheroids that increase NSC stemness and survival rates in vivo, which is another strategy being evaluated to improve NSC transplants (Haragopal et al., 2015; Han and Hsu, 2017). Han and Hsu (2017) transplanted either NSC mono-spheroids or NSC/ADSC cospheroids into the injury site of cerebellum-injured adult immunosuppressed zebrafish to investigate the impact of the spheroids on neural regeneration. Cell survival was increased in the cospheroid transplanted fish, and there was partial rescue of locomotion activity in both the mono- and cospheroid transplanted fish (Han and Hsu, 2017). Interestingly, the cospheroid transplanted fish had the same degree of functional rescue with half the number of NSCs and thus this represents a potential strategy to reduce the already limited numbers of available NSCs (Han and Hsu, 2017).

Secondly, iPSCs are a promising alternative source of the hard-to-produce endothelial cells (ECs) for both vascular tissue engineering and regenerative transplant therapies for patients with vascular diseases (Orlova et al., 2014; Chan et al., 2015; Olmer et al., 2018). Human iPSC-derived ECs integrate into the vasculature in developing zebrafish embryos, and the functionality of these cells can be assessed by quantifying the percentage and length of chimeric vessels (Orlova et al., 2014; Olmer et al., 2018). This method can be applied to assess the therapeutic potential of vascular cells in vascular diseases and for tissue engineering strategies. For example, human iPSCs derived from patients with type 1 diabetes (T1D) differentiate into ECs that form vascular networks in hydrogel and these cells successfully integrate into the zebrafish vasculature, indicating their functionality (Chan et al., 2015). As these cells are patient specific, they can be used for autologous vascular therapies to improve wound healing and treat acquired vascular diseases in patients with diabetes (Chan et al., 2015).

Finally, the zebrafish xenograft platform is being used to investigate therapeutic options for idiopathic pulmonary fibrosis (IPF), a progressive lung disease caused by fibrotic lesions that form a reticular network in the lungs, for which the only currently available effective treatment is a lung transplant (Xia et al., 2014). Mesenchymal progenitor cells (MPCs) have been shown to be the cells of origin of IPF-causing fibroblasts, as transplantation of MPCs from IPF patients form fibrotic lesions in zebrafish embryos (Benyumov et al., 2012; Xia et al., 2014). This zebrafish model can be leveraged for drug discovery to identify compounds that prevent the transformation of MPCs into fibroblasts and thus, abrogate the formation of fibrotic lesions (Xia et al., 2014). Overall, these studies highlight how the zebrafish XT platform will foster preclinical stem cell research through the identification of novel compounds to expand stem cell populations, target disease-causing cells, and as regenerative therapeutic options.

4. Adult zebrafish XT models of cancer

More recently, to address the limitations of both the mouse and larval zebrafish XT models, xenografts have been conducted in adult zebrafish through three different approaches: genetic alterations to induce immunopermissiveness, chemical immunosuppression, and/or ɣ-irradiation immunosuppression. The main limitation in developing a XT tumor model in any adult animal is the presence of a functional immune system, which requires adequate ablation prior to successful cell engraftment.

As previously described, many zebrafish xenograft cancer models have been developed when the animals are in their embryonic or larval stages (approximately 48–72 hpf), when the immune system is not fully developed, animal numbers are abundant, and the fish are transparent and easily manipulated. Using adult zebrafish for human cancer cell xenografts, although more laborious, is emerging as an innovative approach to study drug responses and other biologically relevant disease information in vivo. A zebrafish is officially classified in the adult stage when it reaches sexual maturity, at approximately 90 dpf (Kimmel et al., 1995). Immunocompromised mice are an invaluable model for xenograft transplantation studies and engraftment into nude, NOD-SCID or highly immune deficient NOD-SCID-gamma null (NSG) mice, represents the gold standard for human cancer avatars to study disease progression, metastasis, and as a platform for drug screening (Pearson et al., 2008; Bankert et al., 2011; Quintana et al., 2012; Radaelli et al., 2015). However, murine models are expensive and require both advanced technologies and techniques for imaging and surgeries for examining opaque and deep anatomic structures. Furthermore, propagation of a substantial number of immunocompromised mice is challenging, as mice have low fecundity rates and biologically develop at a relatively slow rate. The advantages and disadvantages of the XT mouse model, the larval zebrafish model, and the adult zebrafish model are annotated in Fig. 3.

In contrast, while the larval drug screening platform has tremendous utility as outlined before, the rapid generation of results is ultimately a trade-off for the accuracy of drug penetration and dosing. The major objective of the larval XT platform is to obtain drug response data in a sufficiently rapid time frame for the potential to have clinical implications. In this model, the larvae are frequently treated by immersion therapy as a group and therefore each fish is not individually dosed like a mouse or human. Larval zebrafish drug doses are used to identify an efficacy signal in the transplanted tumor cells and determine evidence of general and organ-specific toxicity. Short-term drug response data in XT larval models certainly has its place, especially in patients who present with hard-to-treat cancers that are refractory, relapsed, or metastatic. In these cases, any functional drug response data that can be generated may be valuable to help inform potential treatment strategies. This approach is currently being leveraged as part of the pan-Canadian PROFYLE (Precision Oncology for Young PeopLE) program designed to molecularly profile and model hard-to-treat pediatric cancers.

	 Immune compromized mice	 Larval zebrafish	 Adult immune deficient zebrafish
Clinically-relevant dosing using oral gavage	+++	-	+++
PK/PD studies possible	+++	-	+++
Long-term engraftment	+++	-	+++
Growth at 37C	+++	-	+++
Growth similar to human tumors	+++	+/-	+++
Facile imaging at single cell resolution	-	+++	+++
High throughput drug screening	-	+++	+
Cost	$$$	$	$$
Maximum cells implanted	1-5x10^6 cells	10-200 cells	1-5x10^6 cells

FIG. 3 Comparison of mouse and zebrafish XT models (Yan et al., 2019a,b). PK/PD, pharmacokinetics, pharmacodynamics.

4.1 Genetic approaches for adult immunosuppression and XT

The Langenau lab has generated a zebrafish line that is deficient in *protein kinase, DNA activated, catalytic polypeptide (prkdc)*, and *interleukin-2 receptor gamma a (il2rga)* (Tang et al., 2017). This specific line has been generated in the *casper* background, creating a lifelong optically clear homozygous compound mutant (*prkdc−/−, il2rga−/−*) into which foreign human tissues successfully engraft due to the lack of host T, B, and natural killer (NK) cells (Yan et al., 2019a,b). These models can be acclimatized to survive at the human physiological relevant temperature of 37°C and have been used to successfully transplant a variety of fluorescently labeled human cancer cells for more than 28days. High engraftment rates were achieved using 16 different cancer cell lines including melanoma, renal cell carcinoma, chronic myeloid leukemia, mantle cell lymphoma, triple-negative breast cancer, and several sarcomas (Yan et al., 2019a,b). It is important to note that the husbandry for these immunocompromised animals was complex, as they were maintained in the presence of a multitude of antibiotics, and were housed in individual sterile tanks or in a custom-built standalone recirculating system. Therefore feeding and maintenance schedules were more labor intensive compared to regular husbandry standard operating procedures to prevent infection and optimize animal survival. Nonetheless, an important highlight of this study is that XT animals were monitored for tumor progression for more than 28days and overall morbidity associated with transplanted cells was less than 15%. Once the cells had been transplanted into the adult zebrafish, histopathological analysis was conducted and cell morphology, diagnostic gene expression markers, and gene fusions were consistent with each respective cancer type (Yan et al., 2019a,b). These results demonstrate that these animals are a suitable host for human cancer cells, as there was no rejection of the transplanted cells and minimal host immune system interference. Furthermore, not only did the cells engraft and proliferate, but they also retained their original characteristics, which is critical when studying tumor cells, as those genetic characteristics determine cell behavior, microenvironment interactions, and drug responses.

4.1.1 PDXs using *prkdc*$^{-/-}$, *il2rga*$^{-/-}$ mutant zebrafish

The *prkdc−/−, il2rga−/−* mutant zebrafish have also been used as a preclinical model to study personalized medicine to evaluate specific patient drug responses by way of PDXs. The tumors that were transplanted included glioblastoma, embryonal rhabdomyosarcoma (RMS), metastatic BRAFV600E-induced melanoma, and therapy-resistant $ER^+/PR^+/Her2^-$ breast cancer (Yan et al., 2019a,b). Before the cells were transplanted into the immunocompromised adult zebrafish, they were cultured in sphere-inducing conditions (glioblastoma), passaged in mice (rhabdomyosarcomas), or obtained from circulating tumor cells and cultured for a short period of time (melanoma and breast cancer). Similar to the cell lines, all PDX samples successfully engrafted and progressed over 28days, until ultimately the host succumbed to the tumor burden (Fig. 4) (Yan et al., 2019a,b). In these transplant studies, tumor cells were labeled with transient fluorescent cytoplasmic dyes or by permanent fluorescent cell labeling using lentivirus. The two anatomical locations that were used for injection were the peritoneal cavity and the periocular space.

4.1.2 Single cell resolution imaging for XT studies

One of the major advantages of the zebrafish over the mouse XT model is the ability to image cancer cell progression in vivo without having to sacrifice the animal. This means that the same animal can be tracked in real time for multiple weeks with standard microscopy techniques. Furthermore, due to the high fecundity of the zebrafish, the sample size is much greater than that of a comparable murine XT experiment. A greater number of adult zebrafish can be followed over the course of the experiment, which allows for a more comprehensive sampling of the disease, as cancer is highly heterogeneous, ultimately providing greater power to determine statistical significance (Gupta and Massague, 2006). This model not only allows for the real-time visualization and movement of cancer cells within a live organism, but it also allows for a straightforward quantification method for fluorescent tumor cells. The transparency of the *casper* mutant

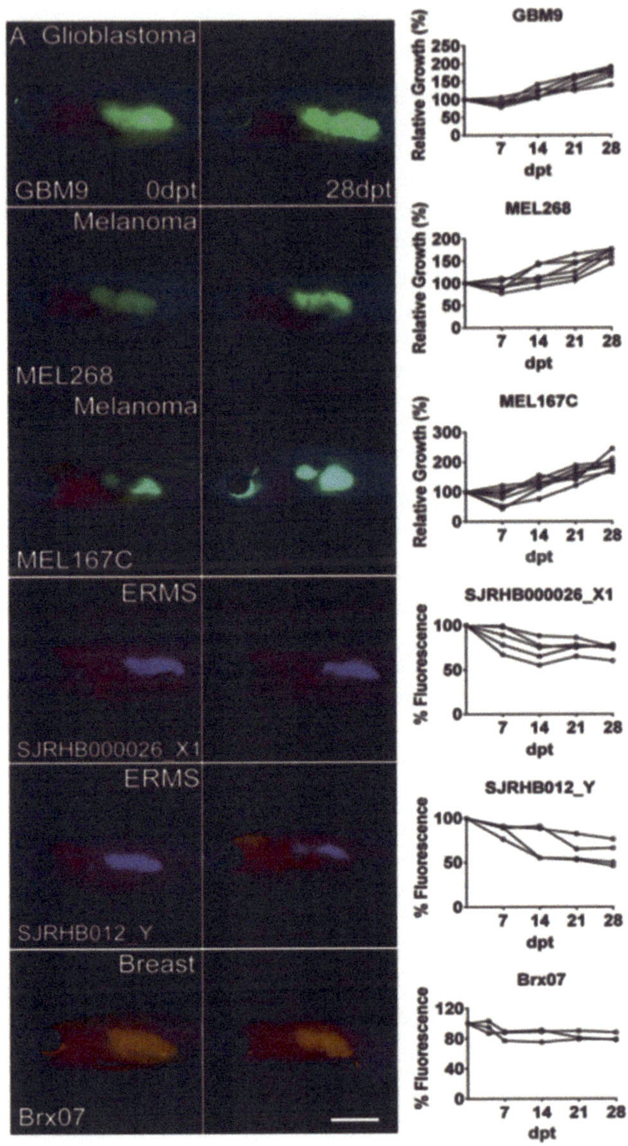

FIG. 4 Patient-derived cancer cells engraft and grow in *prkdc−/−, il2rga−/−* zebrafish. Adult fish were imaged at 0- and 28-days posttransplant (dpt) and graphs show relative tumor growth over the 28days (Yan et al., 2019a,b).

provides unparalleled access to quantify cancer cell growth at single cell resolution, which ultimately provides a practical way to assess drug efficacy and cancer cell killing capacity. Fluorescent cell readouts and numerical quantification of cancer cell numbers within an animal before and after treatment with the addition of one or more compound(s) result in valuable drug response data.

Moreover, with in vivo cancer cell tracking, researchers were able to identify several functionally distinct types of cancer cells (Yan et al., 2019a,b). Specifically for rhabdomyosarcoma cells, three cell populations were observed: (1) the highly migratory cells, (2) the actively proliferating cells, and (3) the bystander cells that did not divide or move over 7 days of daily evaluation. This was interesting as the analysis also revealed that the proliferation and migratory abilities of rhabdomyosarcoma cells are mutually exclusive, something posited in the cancer field, but for the first time directly observed in vivo through the "window" of a live adult zebrafish model (Gallaher et al., 2019; Yan et al., 2019a,b). This highlights the overall potential of the zebrafish XT model not only to provide response data but to also elucidate mechanistic insights at the single cell level.

4.1.3 Pharmacodynamic studies in zebrafish XT models

The second major advantage of the adult XT model that sets it apart from other XT models is the ability to conduct pharmacodynamic (PD) studies while undertaking drug response assays. After cancer cell transplantation and engraftment, researchers were able to individually treat adult zebrafish with drug by way of oral gavage (Yan et al., 2019a,b). This approach ensures each animal receives an exact dose of drug that can be better matched to the clinical context. Drug dosing schedules and administration routes can be more closely modeled in the adult zebrafish XT model resulting in a more clinically representative patient avatar. To investigate the PD consequences in this XT model, Yan et al. (2019a,b) created an RMS cell line that stably expressed the FUCCI4 cell cycle fluorescent reporter. This reporter system emits different fluorescent proteins (colors) during G1, S, G2, and M phases of the cell cycle (Bajar et al., 2016), which allows for the real-time tracking of proliferation and cellular state in live RMS cells that have been xenotransplanted into adult zebrafish. Researchers assessed for PD data by injecting FUCCI4$^+$ RMS cells into the periocular muscle of *prkdc−/−, il2rga−/−* fish, allowed for engraftment for 7days, and then orally gavaged fish with either vehicle control, single (TMZ or olaparib) or combination drug treatment (TMZ+olaparib) (Fig. 5). From these experiments, it was concluded that combination therapy (TMZ+olaparib) accelerated the time and numbers of G2-arrested cells, which precede apoptosis, compared to control or single agent therapy. This work identified an effective combination therapy for RMS while simultaneously providing important PD data at single cell resolution in vivo (Yan et al., 2019a,b).

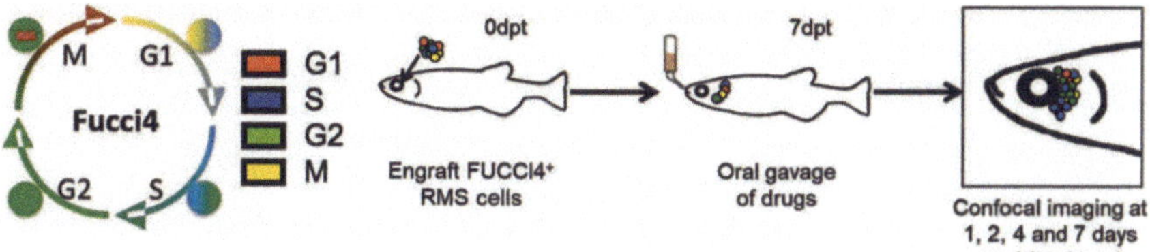

FIG. 5 Schematic and experimental design for the FUCCI4-color cell reporter assay in the adult XT platform (Yan et al., 2019a,b). dpt, days posttransplant; RMS, rhabdomyosarcoma.

4.2 Chemical approaches for adult immunosuppression and XT

An alternative approach to genetic immune system suppression is chemical suppression through the administration of drugs. Researchers have adapted drugs employed clinically for immune suppression prior to stem cell transplantation for use in adult zebrafish (Khan et al., 2019; Patel and Tadi, 2020). Busulfan is used in combination with fludarabine (BUFLU) as a fairly standard conditioning regimen for allogenic transplants in patients with acute myeloid leukemia (AML) (Sora et al., 2020). Busulfan has also been used in murine models for conditioning and subsequent successful long-term engraftment of multiple AML cell lines for the purpose of studying human disease (Saland et al., 2015).

Busulfan is an alkylating chemotherapeutic agent that works by substituting alkyl groups for hydrogen atoms on DNA, resulting in DNA crosslinking and transcription inhibition, which ultimately causes inhibition of protein synthesis and subsequent cytotoxicity (Patel and Tadi, 2020). In one study, busulfan was injected intraperitoneally in adult zebrafish two days prior to human cancer cell transplantation (Khan et al., 2019). Researchers injected 1×10^5 fluorescently labeled human cancer cells (U937, monocytic leukemia cell line) near the dorsal aorta of *casper* fish 2 days after busulfan administration (White et al., 2008; Khan et al., 2019). Once the cells were injected, fish were placed in water containing 1% penicillin and streptomycin for 24h to avoid infection. Injected fish were evaluated and monitored at 0, 3, 7, and 15days postinjection and researchers observed successful engraftment in approximately 87% of all the injected fish. Two weeks after XT, the overall survival of fish was ~45%. Further dissection of the fish showed substantial tumor growth around the gastrointestinal tract that emerged as a slight enlargement of the peritoneal area and appeared as a solid mass under a fluorescent filter (Fig. 6). To examine this cell homing further, the researchers assessed intracardiac injection of cells, as cells would be circulating and would need to actively leave the bloodstream to home to a certain anatomical location. Intracardiac injection of cells resulted in a high mortality rate of fish (~50%, n=5/10), but those that did survive showed cell

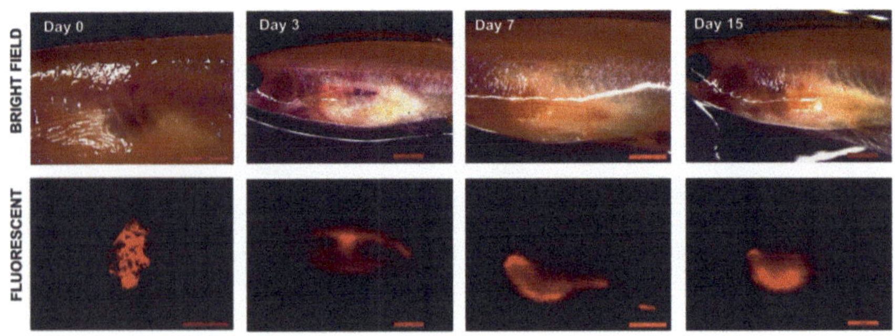

FIG. 6 AML tumors engraft and grow in immunosuppressed adult *casper* zebrafish. U937 cells (red fluorescence) were injected and monitored over 15days (Khan et al., 2019).

homing to the peritoneal cavity despite a remote initial injection site (Khan et al., 2019). This model presents a cost-effective platform to study the biology of cancer and the ability to evaluate potential anticancer agents.

As mentioned before, one of the major benefits of working with the zebrafish cancer model is real-time visualization and the ability to monitor transplanted cells in a living organism. This characteristic is especially valuable when studying the process of cancer cell metastasis, as understanding this process has been limited by the inability to visualize tumor cell behavior inside a mammalian model. Metastasis is the major cause of death in patients with solid tumors and there are limited treatment options and poor response rates for patients who present aggressive metastatic disease (Gupta and Massague, 2006). The zebrafish xenograft model platform can serve as a tool to help understand the knowledge gaps associated with the complex phenomenon of cancer cell migration and metastasis. Researchers have utilized the transgenic *Tg(fli:EGFP)* zebrafish line, in which endothelial cells express green fluorescent protein (GFP), to observe the process of active cancer cell movement from the initial injection site and ultimately the colonization of another location (Stoletov et al., 2007).

Zebrafish aged 25–35days were treated with dexamethasone for 2days for immunosuppression and were subsequently injected in the peritoneal cavity with up to 300 human cancer cells (breast adenocarcinoma MDA-MB-435, fibrosarcoma HT1080, or melanoma B16 cell lines) that were permanently labeled by stable fluorescent transfection (Stoletov et al., 2007). Dexamethasone is a corticosteroid that has been shown to suppress IL-2-mediated T-cell proliferation and cytokine production and it has also been shown to severely inhibit naïve T-cell proliferation and differentiation (Giles et al., 2018). The injected fish were then monitored and imaged daily for tumorigenesis and vascularization around the cancer cells. In these experiments, all tumor types displayed various levels of cell invasion, angiogenesis, and overall animal viability in contrast to animals injected with nontumorigenic cells or fluorescent beads, which showed no active movement from the initial injection site. With vasculature easily visualized in these transgenic animals, changes in angiogenesis before and after tumor inoculation were observed and recorded. Approximately 20% of developing MDA-MB-435 tumors showed blood vessel remodeling. In contrast, when VEGF-secreting engineered MDA-MB-435 were transplanted, 100% of tumors showed robust vessel sprouting and remodeling that completely infiltrated the tumor mass within 3–4days postinjection (Stoletov et al., 2007). This work showed integration of zebrafish transgenesis with human cancer cell XT in adult fish as a powerful and unparalleled platform for the development of potential therapeutic agents targeted at cancer progression and tumor-induced angiogenesis.

4.3 Irradiation approaches for adult immunosuppression and XT

Along with chemical approaches of immunosuppression, total body irradiation (TBI) has been used in the clinic as a component of some conditioning regimens for patients with acute leukemia who are undergoing an HSCT (Wong et al., 2018). Again, using the same ideology as the clinic, adult zebrafish have been irradiated prior to the transplantation of cancer cells to condition the organism for engraftment and the prevention of tumor cell rejection. The Zon lab was the first group to show the biological responses to ionizing irradiation in adult zebrafish using fish-to-fish cancer cell transplants. Similar to mammals, the minimum lethal dose ablates the zebrafish blood-forming system and sublethal irradiation doses transiently ablate lymphocytes, which allowed for the successful transplantation and engraftment of leukemia cells (Traver et al., 2004). Two days prior to injection, adult zebrafish were irradiated with 20Gy and injected intraperitoneally with 1–5 million zebrafish leukemia cells taken from the transgenic *rag2-EGFP-mMyc* fish. Zebrafish carrying a *rag2-EGFP-mMyc* fusion transgene developed a lethal T-cell leukemia that was harvested and transplanted into irradiated adult recipients and visualized by fluorescent microscopy. GFP^+ tumor cells were observed colonizing the thymus and subsequently spreading throughout the body of the animals. In comparison, unirradiated controls did not show any signs of disease at 25days after cell transplantation even at the highest concentration of tumor cell injection (5×10^6 cells). Fluorescent microscopy visualization was used to confirm these results and subsequently followed up with harvesting whole kidney marrow and FACS sorting for GFP^+ tumor cells (Traver et al., 2004).

As mentioned, efficient engraftment of donor cells generally requires ablation of the host hematopoietic and immune defense system. A similar study used six-month-old adult zebrafish that were irradiated with 25Gy of ɣ-irradiation 2days prior to cell transplant (Guo et al., 2015). This study, however, xenotransplanted a human pancreatic cancer cell line (PaCa-2 cells), labeled with a fluorescent lipophilic transient dye prior to injection into the adult fish (Guo et al., 2015). The cells were then injected into the heart or the intraperitoneal cavity of the adult fish, which was subsequently evaluated daily for up to 10 days by fluorescent microscopy (Guo et al., 2015). Researchers found that after intraperitoneal transplantation, the pancreatic cancer cells engrafted, proliferated, and migrated to the zebrafish pancreas (Guo et al., 2015).

In animal models of cancer, irradiation has proven an essential tool in the ablation of the immune response for the successful transplantation and engraftment of oncogenic cell populations. The techniques used in the clinic have been adapted and modified to be well suited for mammalian TBI/transplantation studies, which are essential preclinical tools for cancer and therapeutics research in mouse cancer models (Bailey et al., 1980; Huang et al., 1996; Kahn et al., 2012). It is evident that a similar approach can be used to prime adult zebrafish for human cancer cell transplantation, ultimately showing their suitability for such experiments.

5. Conclusion

The versatility of the zebrafish XT platform is unparalleled in other model organisms (Table 1). The evidence provided in this chapter has shown the potential of

TABLE 1 Summary of the models discussed in this chapter.

Model	Larvae—XT cancer cells 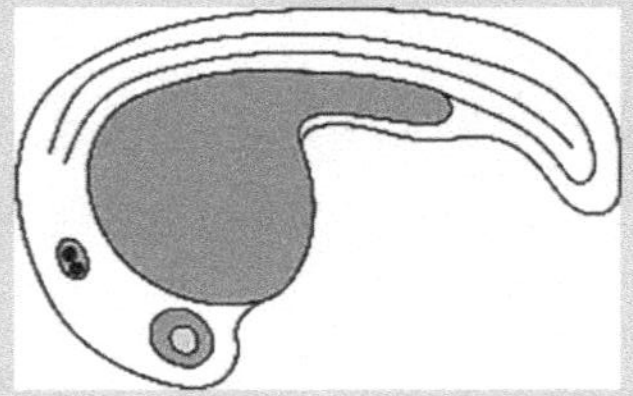	Larvae—XT SCs 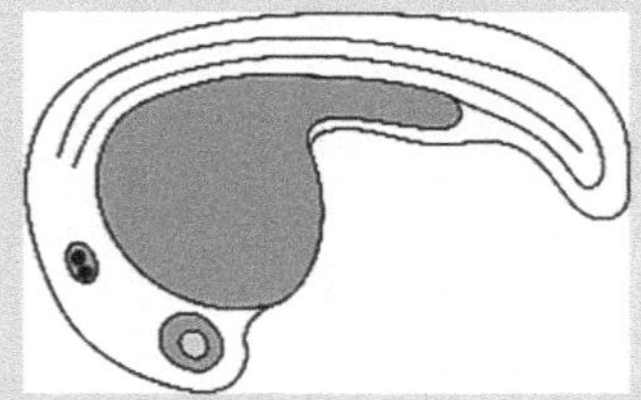	Adult—XT cancer cells 
Key features	- Established platform for liquid and solid mammalian tumors - Rapid organism development and organogenesis - Relatively low tumor cell number required for XT - High sample size (n>100) - No immunosuppression required - Straight forward drug treatment by immersion therapy - Rapid generation of results, drug response and efficacy data (within days)	- Can host mammalian iPSCs, HSPCs, MSCs, NSCs, with successful survival, differentiation, migration, and interactions - Particularly translational for HSPC studies due to high degree of genetic conservation and analogous structures of the hematopoietic systems - XT platform has been adapted to evaluate SC regenerative therapies in the nervous and vascular systems, and fibrotic diseases	- Established platform for liquid and solid mammalian tumors - More closely resembles the dosing schedule and administration routes of clinical therapeutics - Experiments are conducted at 37 °C - Can conduct PD studies - Drug response data can be obtained within weeks
Limitations	- Experiments are not conducted at 37 °C - Unequal dosing/uptake of drug per larva - Partial conservation of pathways and ADME mechanisms - Lacking important human proteins and cytokines		- Requires immunosuppression (genetically, chemically, or by radiation) - Using immunocompromised animals can be difficult to maintain, requires the use of antibiotics - Requires a standalone fish housing system, which can require a substantial footprint and can be costly
Alternative strategies and advantages	- All experiments can be conducted in *casper* fish, which allow for complete optical transparency - Quantitative cell assays are possible by fluorescent microscopy without sacrificing the animals - Zebrafish have direct orthologs of the CYP gene families for essential metabolizing enzymes - Ability to utilize transgenic fish (e.g., *GSS, flk1:EGFP, prkdc−/−, il2rga−/−*)		

the zebrafish to impact clinical decision-making and highlights the amenability of the model to provide preclinical drug response data in an actionable time frame necessary for a personalized medicine approach. The data obtained from both larval and adult zebrafish XT studies can bridge knowledge gaps and uncover crucial mechanisms and pathways in both the oncology and regenerative medicine fields. Animal models are an essential part of translational health research, and the zebrafish occupies a unique niche in the preclinical drug discovery pipeline. As the oncology and regenerative medicine fields continue to move toward precision therapeutic solutions, we anticipate additional opportunities, and advantages will be realized for incorporating the zebrafish model to improve the outcomes and quality of life for patients with cancer and degenerative disorders.

References

Arjmand, B., Tayanloo-Beik, A., Foroughi Heravani, N., Alaei, S., Payab, M., Alavi-Moghadam, S., Goodarzi, P., Gholami, M., Larijani, B., 2020. Zebrafish for personalized regenerative medicine; a more predictive humanized model of endocrine disease. Front. Endocrinol. 11, 396.

Arulmozhivarman, G., Kräter, M., Wobus, M., et al., 2016. Zebrafish *in vivo* screening for compounds amplifying hematopoietic stem and progenitor cells: preclinical validation in human CD34 + stem and progenitor cells. Sci. Rep. 7 (1), 12084.

Arumuggam, N., Melong, N., Too, C.K., Berman, J.N., Rupasinghe, H.V., 2017. Phloridzin docosahexaenoate, a novel flavonoid derivative, supresses growth and induces apoptosis in T-cell acute lymphoblastic leukemia cells. Am. J. Cancer Res. 7 (12), 2454–2462.

Bailey, M.J., Gazet, J.C., Peckham, M.J., 1980. Human breast-cancer xenografts in immune-suppressed mice. Br. J. Cancer 42 (4), 524–529.

Bajar, B.T., Lam, A.J., Badiee, R.K., Oh, Y.-H., Chu, J., Zhou, X.X., Kim, N., Kim, B.B., Chung, M., Yablonovitch, A.L., Cruz, B.F., Kulalert, K., Tao, J.J., Meyer, T., Su, X.-D., Lin, M.Z., 2016. Fluorescent indicators for simultaneous reporting of all four cell cycle phases. Nat. Methods 13 (12), 993–996.

Bankert, R.B., Balu-Iyer, S.V., Odunsi, K., Shultz, L.D., Kelleher Jr., R.J., Barnas, J.L., Simpson-Abelson, M., Parsons, R., Yokota, S.J., 2011. Humanized mouse model of ovarian cancer recapitulates patient solid tumor progression, ascites formation, and metastasis. PLoS One 6, e24420.

Bentley, V.L., Veinotte, C.J., Corkery, D.P., Pinder, J.B., LeBlanc, M.A., Bedard, K., Weng, A.P., Berman, J.N., Dellaire, G., 2015. Focused chemical genomics using zebrafish xenotransplantation as a pre-clinical therapeutic platform for T-cell acute lymphoblastic leukemia. Haematologica 100 (1), 70–76.

Benyumov, A.O., Hergert, P., Herrera, J., Peterson, M., Henke, C., Bitterman, P.B., 2012. A novel zebrafish embryo xenotransplantation model to study primary human fibroblast motility in health and disease. Zebrafish 9 (1), 38–43.

Biehl, J.K., Russell, B., 2009. Introduction to stem cell therapy. J. Cardiovasc. Nurs. 24 (2), 98–105.

Brown, H.K., Schiavone, K., Tazzyman, S., Heymann, D., Chico, T.J., 2017. Zebrafish xenograft models of cancer and metastasis for drug discovery. Expert Opin. Drug Discovery 12 (4), 379–389.

Brox, S., Seiwert, B., Haase, N., Küster, E., Reemtsma, T., 2016. Metabolism of clofibric acid in zebrafish embryos (*Danio rerio*) as determined by liquid chromatography-high resolution-mass spectrometry. Comp. Biochem. Physiol., Part C: Toxicol. Pharmacol. 185–186, 20–28.

Cagan, R.L., Zon, L.I., White, R.M., 2019. Modeling cancer with flies and fish. Dev. Cell 49. https://doi.org/10.1016/j.devcel.2019.04.013.

Canaider, S., Maioli, M., Facchin, F., Bianconi, E., Santaniello, S., Pigliaru, G., Ljungberg, L., Burigana, F., Bianchi, F., Olivi, E., Tremolada, C., Biava, P.M., Ventura, C., 2014. Human stem cell exposure to delopmental stage zebrafish extracts: a novel strategy for tuning stemness and senescence patterning. CellR4 2 (3), e1226.

Cavalcante, L., Toner, M., Uygun, K., Tessier, S.N., 2019. Leveraging the zebrafish to model organ transplantation. Curr. Opin. Organ Transplant. 24 (5), 613–619.

Chan, X.Y., Black, R., Dickerman, K., et al., 2015. Three-dimensional vascular network assembly from diabetic patient-derived induced pluripotent stem cells. Arterioscler. Thromb. Vasc. Biol. 35 (12), 2677–2685.

Chng, H.T., Ho, H.K., Yap, C.W., Lam, S.H., Chan, E.C.Y., 2012. An investigation of the bioactivation potential and metabolism profile of zebrafish versus human. J. Biomol. Screen. 17, 974–986.

Cho, K.J., Trzaska, K.A., Greco, S.J., McArdle, J., Wang, F.S., Ye, J.H., Rameshwar, P., 2005. Neurons derived from human mesenchymal stem cells show synaptic transmission and can be induced to produce the neurotransmitter substance P by interleukin-1 alpha. Stem Cells 23 (3), 383–391.

Chou, H.L., Lin, Y.H., Liu, W., Wu, C.Y., Li, R.N., Huang, H.W., Chou, C.H., Chiou, S.J., Chiu, C.C., 2019. Combination therapy of chloroquine and C(2)-ceramide enhances cytotoxicity in lung cancer H460 and H1299 cells. Cancer 11, 370.

Clements, W.K., Traver, D., 2013. Signalling pathways that control vertebrate haematopoietic stem cell specification. Nature reviews. Immunology 13 (5), 336–348.

Cloney, K., Steele, S.L., Stoyek, M.R., Croll, R.P., Smith, F.M., Prykhozhij, S.V., Brown, M.M., Midgen, C., Blake, K., Berman, J.N., 2018. Etiology and functional validation of gastrointestinal motility dysfunction in a zebrafish model of CHARGE syndrome. *FEBS* J. 285 (11), 2125–2140.

Cocchiaro, J.L., Rawls, J.F., 2013. Microgavage of zebrafish larvae. J. Vis. Exp. 72, e4434.

Corkery, D.P., Dellaire, G., Berman, J.N., 2011. Leukaemia xenotransplantation in zebrafish chemotherapy response assay *in vivo*. Br. J. Haematol. 153 (6), 786–789.

Cutler, C., Multani, P., Robbins, D., Kim, H.T., Le, T., Hoggatt, J., Pelus, L.M., Desponts, C., Chen, Y.B., Rezner, B., Armand, P., Koreth, J., Glotzbecker, B., Ho, V.T., Alyea, E., Isom, M., Kao, G., Armant, M., Silberstein, L., Hu, P., et al., 2013. Prostaglandin-modulated umbilical cord blood hematopoietic stem cell transplantation. Blood 122 (17), 3074–3081.

Davidson, A.J., Zon, L.I., 2004. The 'definitive' (and 'primitive') guide to zebrafish hematopoiesis. Oncogene 23 (43), 7233–7246.

De La Garza, A., Sinha, A., Bowman, T.V., 2017. Concise review: hematopoietic stem cell origins: lessons from embryogenesis for improving regenerative medicine. Stem Cells Transl. Med. 6 (1), 60–67.

de Souza Anselmo, C., Sardela, V.F., de Sousa, V.P., Pereira, H.M.G., 2018. Zebrafish (*Danio rerio*): a valuable tool for predicting the metabolism of xenobiotics in humans? Comp. Biochem. Physiol., Part C: Toxicol. Pharmacol. 212, 34–46.

Detrich, H.W., Westerfield, M., Zon, L.I., 2010. The zebrafish: cellular and developmental biology, part A. Preface. Methods Cell Biol. 100, xiii.

Diekmann, H., Hill, A., 2013. ADMETox in zebrafish. Drug Discov. Today Dis. Model. 10, e31–e35.

Drabsch, Y., He, S., Zhang, L., Snaar-Jagalska, B.E., tenDijke, P., 2013. Transforming growth factor-beta signalling controls human breast cancer metastasis in a zebrafish xenograft model. Breast Cancer Res. 15, R106.

Ekins, S., Reschly, E.J., Hagey, L.R., et al., 2008. Evolution of pharmacologic specificity in the pregnane X receptor. BMC Evol. Biol. 8, 103.

Fazio, M., Ablain, J., Chuan, Y., Langenau, D.M., Zon, L.I., 2020. Zebrafish patient avatars in cancer biology and precision cancer therapy. Nat. Rev. Cancer 20, 263–272.

Fior, R., Póvoa, V., Mendes, R.V., Carvalho, T., Gomes, A., Figueiredo, N., Ferreira, M.G., 2017. Single-cell functional and chemosensitive profiling of combinatorial colorectal therapy in zebrafish xenografts. Proc. Natl. Acad. Sci. 114 (39).

Gabellini, C., Gomez-Abenza, E., Ibanez-Molero, S., Tupone, M.G., Perez-Oliva, A.B., de Oliveira, S., Del Bufalo, D., Mulero, V., 2018. Interleukin 8 mediates bcl-xL-induced enhancement of human melanoma cell dissemination and angiogenesis in a zebrafish xenograft model. Int. J. Cancer 142, 584–596.

Gallaher, J.A., Brown, J.S., Anderson, A.R.A., 2019. The impact of proliferation-migration tradeoffs on phenotypic evolution in cancer. Sci. Rep. 9, 2425.

Gaudenzi, G, et al., 2016. Patient-derived xenograft in zebrafish embryos: A new platform for translational research in neuroendocrine tumors. Endocrine 57 (2), 214–219.

Gauert, A., Olk, N., Pimentel-Gutierrez, H., Astrahantseff, K., Jensen, L.D., Cao, Y., Eggert, A., Eckert, C., Hagemann, A.I.H., 2020. Fast, *in vivo* model for drug-response prediction in patients with B-cell precursor acute lymphoblastic leukemia. Cancer 12, 1883.

Geiger, G.A., Fu, W., Kao, G.D., 2008. Temozolomide- mediated radiosensitization of human glioma cells in a zebrafish embryonic system. Cancer Res. 68, 3396–3404.

Ghotra, V.P., He, S., de Bont, H., van der Ent, W., Spaink, H.P., van de Water, B., Snaar-Jagalska, B.E., Danen, E.H., 2012. Automated whole animal bio-imaging assay for human cancer dissemination. PLoS One 7, e31281.

Ghotra, V.P., He, S., van der Horst, G., Nijhoff, S., de Bont, H., Lekkerkerker, A., Janssen, R., Jenster, G., van Leenders, G.J., Hoogland, A.M., et al., 2015. SYK is a candidate kinase target for the treatment of advanced prostate cancer. Cancer Res. 75, 230–240.

Giles, A.J., Hutchinson, M.-K.N.D., Sonnemann, H.M., Jung, J., Fecci, P.E., Ratnam, N.M., Zhang, W., Song, H., Bailey, R., Davis, D., Reid, C.M., Park, D.M., Gilbert, M.R., 2018. Dexamethasone-induced immunosuppression: mechanisms and implications for immunotherapy. J. Immunother. Cancer 6, 51.

Goldstone, J.V., McArthur, A.G., Kubota, A., Zanette, J., Parente, T., Jönsson, M.E., Nelson, D.R., Stegeman, J.J., 2010. Identification and developmental expression of the full complement of cytochrome P450 genes in zebrafish. BMC Genomics 11, 643.

Guo, M., Wei, H., Hu, J., Sun, S., Long, J., Wang, X., 2015. U0126 inhibits pancreatic cancer progression via the KRAS signaling pathway in a zebrafish xenotransplantation model. Oncol. Rep. 34, 699–706.

Gupta, G.P., Massague, J., 2006. Cancer metastasis: building a framework. Cell 127 (4), 679–695.

Haldi, M., Ton, C., Seng, W.L., McGrath, P., 2006. Human melanoma cells transplanted into zebrafish proliferate, migrate, produce melanin, form masses and stimulate angiogenesis in zebrafish. Angiogenesis 9, 139–151.

Hamilton, N., Sabroe, I., Renshaw, S.A., 2018. A method for transplantation of human HSCs into zebrafish, to replace humanised murine transplantation models. F1000Res. 7, 594.

Han, H.W., Hsu, S.H., 2017. Chitosan derived co-spheroids of neural stem cells and mesenchymal stem cells for neural regeneration. Colloids Surf. B: Biointerfaces 158, 527–538.

Haragopal, H., Yu, D., Zeng, X., et al., 2015. Stemness enhancement of human neural stem cells following bone marrow MSC coculture. Cell Transplant. 24 (4), 645–659.

Herbomel, P., Thisse, B., Thisse, C., 1999. Ontogeny and behaviour of early macrophages in the zebrafish embryo. Development 126 (17), 3735–3745.

Hess, I., Iwanami, N., Schorpp, M., Boehm, T., 2013. Zebrafish model for allogeneic hematopoietic cell transplantation not requiring preconditioning. PNAS 110 (11), 4327–4332.

Howe, K., Clark, M.D., Torroja, C.F., et al., 2013. The zebrafish reference genome sequence and its relationship to the human genome. Nature 496 (7446), 498–503.

Huang, P., Taghian, A., Allam, A., Freeman, J., Duffy, M., Suit, H., 1996. The effect of whole-body irradiation of nude mice on the tumor transplantability and control probability of a human soft tissue sarcoma xenograft. Radiat. Res. 145 (3), 337–342.

Jin, Y., Wei, L., Jiang, Q., Song, X., Teng, C., Fan, C., Lv, Y., Liu, Y., Shen, W., Li, L., et al., 2018. Comparison of efficacy and toxicity of bevacizumab, endostar and apatinib in transgenic and human lung cancer xenograft zebrafish model. Sci. Rep. 8, 15837.

Kahn, J., Tofilon, P.J., Camphausen, K., 2012. Preclinical models in radiation oncology. Radiat. Oncol. 7, 223.

Kang, Y., Leal Assuncao, B., Denduluri, S., McCurdy, S., Luger, S., Lefebvre, B., Carver, J., Scherrer-Crosbie, M., 2019. Symptomatic heart failure in acute leukemia patients treated with anthracyclines. J. Am. Coll. Cardiol. 1 (2), 208–217.

Khan, N., Mahajan, N.K., Sinha, P., Jayandharan, G.R., 2019. An efficient method to generate xenograft tumor models of acute myeloid leukemia and hepatocellular carcinoma in adult zebrafish. Blood Cells Mol. Dis. 75, 48–55.

Kimmel, C.B., Ballard, W.W., Kimmel, S.R., Ullmann, B., Schilling, T.F., 1995. Stages of embryonic development of the zebrafish. Dev. Dyn. 203, 253–310.

Kithcart, A., MacRae, C.A., 2017. Using zebrafish for high-throughput screening of novel cardiovascular drugs. JACC: Basic to Transl. Sci. 2 (1), 1–12.

Konantz, M., Müller, J.S., Lengerke, C., 2019. Zebrafish xenografts for the *in vivo* analysis of healthy and malignant human hematopoietic cells. Methods Mol. Biol. 2017, 205–217.

Kwon, S.G., Kwon, Y.W., Lee, T.W., Park, G.T., Kim, J.H., 2018. Recent advances in stem cell therapeutics and tissue engineering strategies. Biomater. Res. 22, 36.

Lally, B.E., et al., 2007. Identification and biological evaluation of a novel and potent small molecule radiation sensitizer via an unbiased screen of a chemical library. Cancer Res. 67, 8791–8799.

Lam, S.H., Chua, H.L., Gong, Z., Lam, T.J., Sin, Y.M., 2004. Development and maturation of the immune system in zebrafish, *Danio rerio*: a gene expression profiling, in situ hybridization and immunological study. Dev. Comp. Immunol. 28 (1), 9–28.

Lanvers-Kaminsky, C., Krefeld, B., Dinnesen, A.G., Deuster, D., Seifert, E., Würthwein, G., Jaehde, U., Pieck, A.C., Boos, J., 2006. Continuous or repeated prolonged cisplatin infusions in children: a prospective study on ototoxicity, platinum concentrations, and standard serum parameters. Pediatr. Blood Cancer 2, 183–193.

Lee, L.M.J., Seftor, E.A., Bonde, G., Cornell, R.A., Hendrix, M.J.C., 2005. The fate of human malignant melanoma cells transplanted into zebrafish embryos: assessment of migration and cell division in the absence of tumor formation. Dev. Dyn. 233 (4), 1560–1570.

Leung, A.W.Y., Veinotte, C.J., Melong, N., Oh, M.H., Chen, K., Enfield, K.S.S., et al., 2017. In vivo validation of PAPSS1 (3'-phosphoadenosine 5' phosphosulfate synthase 1) as a cisplatin-sensitizing therapeutic target. Clin. Cancer Res. 23 (21), 6555–6566.

Li, J., Zeng, G., Qi, Y., Tang, X., Zhang, J., Wu, Z., Liang, J., Shi, L., Liu, H., Zhang, P., 2015. Xenotransplantation of human adipose-derived stem cells in zebrafish embryos. PLoS One 10 (4), e0123264.

Lin, J., Zhang, W., Zhao, J.J., Kwart, A.H., Yang, C., Ma, D., Ren, X., Tai, Y.T., Anderson, K.C., Handin, R.I., et al., 2016. A clinically relevant *in vivo* zebrafish model of human multiple myeloma to study preclinical therapeutic efficacy. Blood 128, 249–252.

Liu, Y., Asnani, A., Zou, L., Bentley, V.L., Yu, M., Wang, Y., et al., 2014. Visnagin protects against doxorubicin-induced cardiomyopathy through modulation of mitochondrial malate dehydrogenase. Sci. Transl. Med. 6 (266), 266ra170.

MacRae, C.A., Peterson, R.T., 2015. Zebrafish as tools for drug discovery. Nat. Rev. Drug Discov. 14, 721–731.

Marques, I.J., Weiss, F.U., Vlecken, D.H., Nitsche, C., Bakkers, J., Lagendijk, A.K., Partecke, L.I., Heidecke, C.D., Lerch, M.M., Bagowski, C.P., 2009. Metastatic behaviour of primary human tumours in a zebrafish xenotransplantation model. BMC Cancer 9, 14.

Marques, C., Roberto, V.P., Granadeiro, L., Trindade, M., Gavaia, P.J., Laizé, V., Cancela, M.L., Fernández, I., 2017. The xenobiotic sensor PXR in a marine flatfish species (*Solea senegalensis*): gene expression patterns and its regulation under different physiological conditions. Mar. Environ. Res. 130, 187–199.

Marques, I.J., Lupi, E., Mercader, N., 2019. Model systems for regeneration: zebrafish. Development 146 (18). https://doi.org/10.1242/dev.167692.

Melong, N., Steele, S., MacDonald, M., Holly, A., Collins, C.C., Zoubeidi, A., et al., 2017. Enzalutamide inhibits testosterone-induced growth of human prostate cancer xenografts in zebrafish and can induce bradycardia. Sci. Rep. 7 (1), 14698.

Mercatali, L., et al., 2016. Development of a patient-derived xenograft (PDX) of breast cancer bone metastasis in a zebrafish model. Int. J. Mol. Sci. 17 (8).

Momin, E.N., Mohyeldin, A., Zaidi, H.A., Vela, G., Quiñones-Hinojosa, A., 2010. Mesenchymal stem cells: new approaches for the treatment of neurological diseases. Curr. Stem Cell Res. Ther. 5 (4), 326–344.

Murayama, E., Kissa, K., Zapata, A., Mordelet, E., Briolat, V., Lin, H.F., Handin, R.I., Herbomel, P., 2006. Tracing hematopoietic precursor migration to successive hematopoietic organs during zebrafish development. Immunity 25 (6), 963–975. https://doi.org/10.1016/j.immuni.2006.10.015. Epub 2006 Dec 7.

National Cancer Institute. Targeted Cancer Therapies, 2020. https://www.cancer.gov/about-cancer/treatment/types/targeted-therapies/targeted-therapies-fact-sheet.

Nicoli, S., Ribatti, D., Cotelli, F., Presta, M., 2007. Mammalian tumor xenografts induce neovascularization in zebrafish embryos. Cancer Res. 67, 2927–2931.

North, T., Goessling, W., Walkley, C., et al., 2007. Prostaglandin E2 regulates vertebrate haematopoietic stem cell homeostasis. Nature 447, 1007–1011. https://doi.org/10.1038/nature05883.

Olmer, R., Engels, L., Usman, A., Menke, S., Malik, M., Pessler, F., Göhring, G., Bornhorst, D., Bolten, S., Abdelilah-Seyfried, S., Scheper, T., Kempf, H., Zweigerdt, R., Martin, U., 2018. Differentiation of human pluripotent stem cells into functional endothelial cells in scalable suspension culture. Stem Cell Rep. 10 (5), 1657–1672.

Orkin, S.H., Zon, L.I., 2008. Hematopoiesis: an evolving paradigm for stem cell biology. Cell 132 (4), 631–644.

Orlova, V.V., Drabsch, Y., Freund, C., Petrus-Reurer, S., van den Hil, F.E., Muenthaisong, S., Dijke, P.T., Mummery, C.L., 2014. Functionality of endothelial cells and pericytes from human pluripotent stem cells demonstrated in cultured vascular plexus and zebrafish xenografts. Arterioscler. Thromb. Vasc. Biol. 34 (1), 177–186.

Parada-Kusz, M., Penaranda, C., Hagedorn, E.J., Clatworthy, A., Nair, A.V., Henninger, J.E., Ernst, C., Li, B., Riquelme, R., Jijon, H., Villablanca, E.J., Zon, L.I., Hung, D., Allende, M.L., 2018. Generation of mouse-zebrafish hematopoietic tissue chimeric embryos for hematopoiesis and host-pathogen interaction studies. Dis. Model. Mech. 11 (11). https://doi.org/10.1242/dmm.034876.

Parng, C., Seng, W.L., McGrath, P., 2002. Zebrafish: a preclinical model for drug screening. Assay Drug Dev. Technol. 1 (1 Pt 1), 41–48.

Patel, R., Tadi, P., 2020. Busulfan. StatPearls Publishing, Treasure Island (FL).

Patel, N., Klassert, T.E., Greco, S.J., et al., 2012. Developmental regulation of TAC1 in peptidergic-induced human mesenchymal stem cells: implication for spinal cord injury in zebrafish. Stem Cells Dev. 21 (2), 308–320.

Pearson, T., Greiner, D.L., Shultz, L.D., 2008. Creation of "humanized" mice to study human immunity. Curr. Protoc. Immunol. 15 (21), 1–15.

Perlin, J.R., Robertson, A.L., Zon, L.I., 2017. Efforts to enhance blood stem cell engraftment: recent insights from zebrafish hematopoiesis. J. Exp. Med. 214 (10), 2817–2827.

Peterson, R.T., MacRae, C.A., 2012. Systematic approaches to toxicology in the zebrafish. Annu. Rev. Pharmacol. Toxicol. 52, 433–453.

Piccolo, O., Melong, N., Lincoln, J., Fernandez, N.R., Ha, M., Berman, J.N., Robar, J., 2019. Nanoparticle-mediated low Z radiotherapy as a potential treatment paradigm for radiosensitive cancers. In: Conference Publication, Zebrafish Disease Models Society Conference, Boston, MA, USA, July 2019.

Pozzoli, O., Vella, P., Iaffaldano, G., et al., 2011. Endothelial fate and angiogenic properties of human CD34+ progenitor cells in zebrafish. Arterioscler. Thromb. Vasc. Biol. 31 (7), 1589–1597.

Pruvot, B., Jacquel, A., Droin, N., Auberger, P., Bouscary, D., Tamburini, J., Muller, M., Fontenay, M., Chluba, J., Solary, E., 2011. Leukemic cell xenograft in zebrafish embryo for investigating drug efficacy. Haematologica 96, 612–616.

Quintana, E., Piskounova, E., Shackleton, M., Weinberg, D., Eskiocak, U., Fullen, D.R., Johnson, T.M., Morrison, S., 2012. Human melanoma metastasis in NSG mice correlates with clinical outcome in patients. Sci. Transl. Med. 4 (159), 149–159.

Radaelli, E., Hermans, E., Omodho, L., Francis, A., Vander Borght, S., Marine, J.-C., van den Oord, J., Amant, F., 2015. Spontaneous post-transplant disorders in NOD.Cg-*Prkdc*^{scid} *Il2rg*^{tm1Sug}/JicTac (NOG) mice engrafted with patient-derived metastatic melanomas. PLoS One 10 (5), e0124974.

Rajan, V., Melong, N., Hing Wong, W., King, B., Tong, S.R., Mahajan, N., Gaston, D., Lund, T., Rittenberg, D., Dellaire, G., Campbell, C.J.V., Druley, T., Berman, J.N., 2020. Humanized zebrafish enhance human hematopoietic stem cell survival and promote acute myeloid leukemia clonal diversity. Haematologica 105 (10), 2391–2399. https://doi.org/10.3324/haematol.2019.223040.

Ridges, S., Heaton, W.L., Joshi, D., Choi, H., Eiring, A., Batchelor, L., et al., 2012. Zebrafish screen identifies novel compound with selective toxicity against leukemia. Blood 119 (24), 5621–5631.

Saland, E., Boutzen, H., Castellano, R., Pouyet, L., Griessinger, E., Larrue, C., de Toni, F., Scotland, S., David, M., Danet-Desnoyers, Vergez, F., Barreira, Y., Collette, Y., Recher, C., Sarry, J.-E., 2015. A robust and rapid xenograft model to assess efficacy of chemotherapeutic agents for human acute myeloid leukemia. Blood Cancer J. 5, e297.

Sora, F., Di Grazia, C., Chiusolo, P., Raiola, A.M., Bregante, S., Mordini, N., Olivieri, A., Iori, A.P., Patriarca, F., Grisariu, S., Terruzzi, E., Rambaldi, A., Sica, S., Bruno, B., Angelucci, E., Bacigalupo, A., 2020. Allogeneic hemopoietic stem cell transplants in patients with acute myeloid leukemia (AML) prepared with busulfan and fludarabine (BUFLU) or thiotepa, busulfan, and fludarabine (TBF): a retrospective study. Biol. Blood Marrow Transplant. 26, 698–703.

Spence, R., Gerlach, G., Lawrence, C., Smith, C., 2008. The behaviour and ecology of the zebrafish, *Danio rerio*. Biol. Rev. 83, 13–34.

Staal, F.J., Spaink, H.P., Fibbe, W.E., 2016. Visualizing human hematopoietic stem cell trafficking *in vivo* using a zebrafish xenograft model. Stem Cells Dev. 25 (4), 360–365.

Stachura, D.L., Svoboda, O., Campbell, C.A., Espín-Palazón, R., Lau, R.P., Zon, L.I., Bartůněk, P., Traver, D., 2013. The zebrafish granulocyte colony-stimulating factors (Gcsfs): 2 paralogous cytokines and their roles in hematopoietic development and maintenance. Blood 122 (24), 3918–3928.

Stoletov, K., Montel, V., Lester, R.D., Gonias, S.L., Klemke, R., 2007. High-resolution imaging of the dynamic tumor cell–vascular interface in transparent zebrafish. PNAS 104 (44), 17406–17411.

Strnadel, J., Wang, H., Carromeu, C., et al., 2018. Transplantation of human-induced pluripotent stem cell-derived neural precursors into early-stage zebrafish embryos. J. Mol. Neurosci. 65 (3), 351–358.

Sullivan, K.M., Muraro, P., Tyndall, A., 2010. Hematopoietic cell transplantation for autoimmune disease: updates from Europe and the United States. Biol. Blood Marrow Transplant. 16 (1 Suppl), S48–S56.

Takahashi, K., Yasuhara, T., Shingo, T., et al., 2008. Embryonic neural stem cells transplanted in middle cerebral artery occlusion model of rats demonstrated potent therapeutic effects, compared to adult neural stem cells. Brain Res. 1234, 172–182.

Tamplin, O.J., Durand, E.M., Carr, L.A., Childs, S.J., Hagedorn, E.J., Li, P., Yzaguirre, A.D., Speck, N.A., Zon, L.I., 2015. Hematopoietic stem cell arrival triggers dynamic remodeling of the perivascular niche. Cell 160 (1–2), 241–252. https://doi.org/10.1016/j.cell.2014.12.032.

Tang, Q., Abdelfattah, N.S., Blackburn, J.S., et al., 2014. Optimized cell transplantation using adult rag2 mutant zebrafish. Nat. Methods 11 (8), 821–824.

Tang, Q., Iyer, S., Lobbardi, R., Moore, J.C., Chen, H., Lareau, C., Hebert, C., Shaw, M.L., Neftel, C., Suva, M.L., Ceol, C.J., Bernards, A., Aryee, M., Pinello, L., Drummond, I.A., Langenau, D.M., 2017. Dissecting hematopoietic and renal cell heterogeneity in adult zebrafish at single-cell resolution using RNA sequencing. J. Exp. Med. 214 (10), 2875–2887.

Topczewska, J.M., Postovit, L.-M., Margaryan, N.V., Sam, A., Hess, A.R., Wheaton, W.W., Nickoloff, B.J., Topczewski, J., Hendrix, M.J.C., 2006. Embryonic and tumorigenic pathways converge via Nodal signaling: role in melanoma aggressiveness. Nat. Med. 12 (8), 925–932.

Traver, D., Paw, B.H., Poss, K.D., Penberthy, W.T., Lin, S., Zon, L.I., 2003. Transplantation and *in vivo* imaging of multilineage engraftment in zebrafish bloodless mutants. Nat. Immunol. 4 (12), 1238–1246.

Traver, D., Winzeler, A., Stern, H.A., Mayhall, E.A., Langenau, D.M., Kutok, J.L., Look, T.A., Zon, L.I., 2004. Effects of lethal irradiation in zebrafish and rescue by hematopoietic cell transplantation. Blood 104 (5), 1298–1305.

Tulotta, C., Stefanescu, C., Beletkaia, E., Bussmann, J., Tarbashevich, K., Schmidt, T., Snaar-Jagalska, B.E., 2016. Inhibition of signaling between human CXCR4 and zebrafish ligands by the small molecule IT1t impairs the formation of triple-negative breast cancer early metastases in a zebrafish xenograft model. Dis. Model. Mech. 9, 141–153.

Veinotte, C.J., Dellaire, G., Berman, J.N., 2014. Hooking the big one: the potential of zebrafish xenotransplantation to reform cancer drug screening in the genomic era. Dis. Model. Mech. 7 (7), 745–754.

Vittori, M., Motain, H., Turnsek, T.L., 2015. The study of glioma by xenotransplantation in zebrafish early life stages. J. Histochem. Cytochem. 63 (10), 749–761.

Vlecken, D.H., Bagowski, C.P., 2009. LIMK1 and LIMK2 are important for metastatic behavior and tumor cell-induced angiogenesis of pancreatic cancer cells. Zebrafish 6, 433–439. https://doi.org/10.1089/zeb.2009.0602.

Wang, C., Tao, W., Wang, Y., Bikow, J., Lu, B., Keating, A., et al., 2010. Rosuvastatin, identified from a zebrafish chemical genetic screen for antiangiogenic compounds, suppresses the growth of prostate cancer. Eur. Urol. 58, 418–426.

Wertman, J.N., Melong, N., Stoyek, M.R., Piccolo, O., Langley, S., Orr, B., Steele, S.L., Razaghi, B., Berman, J.N., 2020. The identification of dual protective agents against cisplatin-induced oto- and nephrotoxicity using the zebrafish model. elife 9, e56235.

White, R.M., Sessa, A., Burke, C., Bowman, T., LeBlanc, T., Ceol, C., Bourque, C., Dovey, M., Goessling, W., Erter Burns, C., Zon, L.I., 2008. Transparent adult zebrafish as a tool for *in vivo* transplantation analysis. Cell Stem Cell 2 (2), 183–189.

White, R.M., Cech, J., Ratanasirintrawoot, S., Lin, C.Y., Rahl, P.B., Burke, C.J., et al., 2011. DHODH modulates transcriptional elongation in the neural crest and melanoma. Nature 471 (7339), 518–522.

Wiercinska, E., Naber, H.P.H., Pardali, E., van der Pluijm, G., van Dam, H., ten Dijke, P., 2011. The TGF-β/Smad pathway induces breast cancer cell invasion through the up-regulation of matrix metalloproteinase 2 and 9 in a spheroid invasion model system. Breast Cancer Research and Treatment 128 (3). https://doi.org/10.1007/s10549-010-1147-x.

Willett, C., Cortes, A., Zuasti, A., Zapata, A., 1999. Early hematopoiesis and developing lymphoid organs in the zebrafish. Dev. Dyn. 214, 323–336.

Wong, J.Y.C., Riccardo Filippi, A., Dabaja, B.S., Yahalom, J., Specht, L., 2018. Total body irradiation: guidelines from the international lymphoma radiation oncology group (ILROG). Int. J. Radiat. Oncol. Biol. Phys. 101 (3), 521–529.

Wu, J., Zhai, J., Li, C., Tan, A., Wei, P., Shen, L., He, M., 2017. Patient-derived xenograft in zebrafish embryos: A new platform for translational research in gastric cancer. J. Exp. Clin. Cancer Res. 36 (1).

Xia, H., Bodempudi, V., Benyumov, A., Hergert, P., Tank, D., Herrera, J., Braziunas, J., Larsson, O., Parker, M., Rossi, D., Smith, K., Peterson, M., Limper, A., Jessurun, J., Connett, J., Ingbar, D., Phan, S., Bitterman, P.B., Henke, C.A., 2014. Identification of a cell-of-origin for fibroblasts comprising the fibrotic reticulum in idiopathic pulmonary fibrosis. Am. J. Pathol. 184 (5), 1369–1383.

Xiao, C., Qian, M., Yin, C., Zhang, Y., Hu, H., Yao, S., 2016. A zebrafish mosaic assay to study mammalian stem cells in real time *in vivo*. J. Mol. Histol. 47 (5), 437–444.

Xu, W., Foster, B.A., Richards, M., Bondioli, K.R., Shah, G., Green, C.C., 2018. Characterization of prostate cancer cell progression in zebrafish xenograft model. Int. J. Oncol. 52, 252–260.

Yan, C., Brunson, D.C., Tang, Q., Do, D., Iftimia, N.A., Moore, J.C., Hayes, M.N., Welker, A.M., Garcia, E.G., Dubash, T.D., Hong, X., Drapkin, B.J., Myers, D.T., Phat, S., Volorio, A., Marvin, D.L., Ligorio, M., Dershowitz, L., McCarthy, K.M., Karabacak, M.N., Fletcher, J.A., Sgroi, D., Iafrate, J.A., Maheswaran, S., Dyson, N.J., Haber, D.A., Rawls, J.F., Langenau, D.M., 2019a. Visualizing engrafted human cancer and therapy responses in immunodeficient zebrafish. Cell 117 (7), 1903–1914.

Yan, C., Yang, Q., Do, D., Brunson, D.C., Langenau, D.M., 2019b. Adult immune compromised zebrafish for xenograft cell transplantation studies. EBio Med. 47, 24–26.

Zhao, C., Wang, X., Zhao, Y., Li, Z., Lin, S., Wei, Y., Yang, H., 2011. A novel xenograft model in zebrafish for high-resolution investigating dynamics of neovascularization in tumors. PLoS One 6, e21768.

Chapter 12

Status, challenges, and perspectives of fish cell culture—Focus on cell lines capable of in vitro mineralization

Vincent Laizé[a,b], Joana T. Rosa[a,b], Marco Tarasco[a], and M. Leonor Cancela[a,c,d]

[a]*CCMAR—Centre of Marine Sciences, Campus de Gambelas, Faro, Portugal,* [b]*S2 AQUA—Sustainable and Smart Aquaculture Collaborative Laboratory, avenida Parque Natural da Ria Formosa, Olhão, Portugal,* [c]*FMCB—Faculty of Medicine and Biomedical Sciences, University of Algarve, Campus de Gambelas, Faro, Portugal,* [d]*ABC—Algarve Biomedical Center, Campus de Gambelas, Faro, Portugal*

1. A growing interest for cell lines of fish origin

While the culture of mammalian cells was already a routine laboratory technique in the 1950s, the successful culture of fish cells was only reported in 1960 (Wolf et al., 1960) and the first teleost fish cell line—the rainbow trout (*Oncorhynchus mykiss*) gonad cell line RTG-2—was established in 1962 by Wolf and Quimby. Sixty years later, more than 700 fish cell lines have been developed from a wide range of species and tissues, a relatively high number in comparison to other vertebrate taxa such as amphibians and reptilians, but a small number in comparison to primates (92,668 cell lines only for human) and rodents (Fig. 1A). The steady increase in the provision of cell lines/cultures of fish origin observed in the last six decades (Fig. 1B) is the direct consequence of an increasing usage of fish in science, resulting from more research on fish species (e.g., those important for human socioeconomic activities or those that bring an evolutionary perspective to mechanisms of vertebrate development), but also on fish models (e.g., those important for biomedicine and human diseases) that represent a promising alternative or complement to classical mammalian models. As it is often the case, in vitro cell systems are necessary companions to in vivo systems to gain insights into cellular mechanisms and signaling pathways, but also to accelerate compound (e.g., drugs and toxicants) screening pipelines. An evolution of the ethical principles in science toward a reduction in animal experimentation and the implementation of regulations in many countries to preserve animal welfare, in particular fish (see directive

Cellular and Molecular Approaches in Fish Biology. https://doi.org/10.1016/B978-0-12-822273-7.00004-5

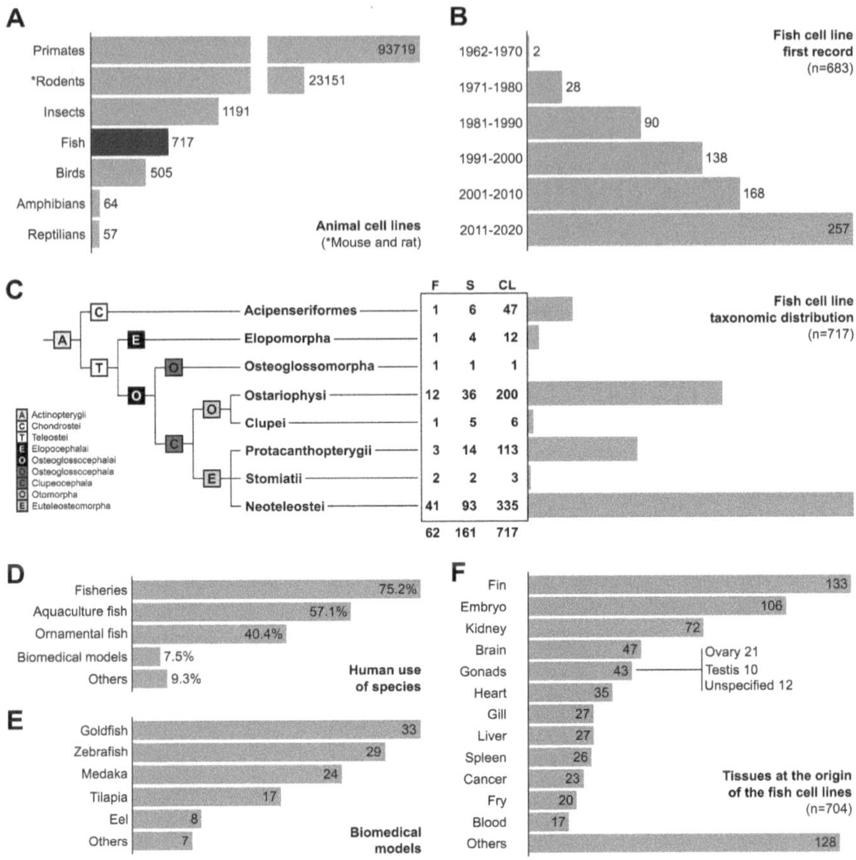

FIG. 1 Analysis of the data retrieved from the Cellosaurus cell line database (version 35). (A) Number of animal cell lines available from the major contributors. (B) Number of fish cell line publications since the first publication of RTG-2 in 1962 until today. (C) Taxonomic representation of the fish cell line (CL) diversity, including species (S) and family (F) numbers. (D) Human uses of the fish species represented with a cell line. (E) Number of cell lines for classical biomedical models. (F) Tissues that have originated the fish cell lines. Environment and human uses data were retrieved from the Fishbase (www.fishbase.org). Taxonomic information was retrieved from the Integrated Taxonomic Information System (www.itis.gov).

2010/63/EU in Europe), have also played a role in the rapid advance in the provision of fish cell lines. In fact, many research institutions and pharmaceutical companies have already validated in vitro cell systems as suitable alternatives to reduce or replace animal models (Goh et al., 2015).

The fast provision of fish cell lines and the successful implementation of fish cell culture worldwide largely benefited from the techniques, reagents, and medium developed for mammalian cell culture. Despite differences in several physiological parameters (e.g., osmoregulation, thermoregulation, and energy metabolism), fish cells adapted well to conditions used for mammalian cell cultivation, in particular the supplementation of culture medium with bovine (also horse, porcine, and lamb) serum as a source of growth factors and hormones. Although using mammalian serum may seem physiologically inappropriate to sustain fish cell growth, it represents an enormous financial benefit—the preparation of fish serum in similar quantity would require a massive investment in the farming of large size donor fish—and further confirm that cellular and molecular pathways of fish cells are evolutionarily not that far from those of mammalian cells. Trials to culture fish cells with fish serum have been reported (Rosa et al., 2010; Hashimoto et al., 1997; Couso et al., 2002) but remain marginal because they are economically not viable. The culture of fish cells in serum-free medium was also tried, mainly to overcome the masking effect of serum in cytotoxicity tests (Ackermann and Fent, 1998; Radošević et al., 2016; Shea and Berry, 1983). Despite the plasticity of the fish cell lines that allow them to adapt to suboptimal culture conditions, growth temperature had to be lowered and adjusted to the diversity of the (cold-blooded) fish species, e.g., 15–21 °C for cold-water species and 25–35 °C for warm-water species. Still it has been showed in various cases that fish cells can tolerate a wide range of temperatures (Bols et al., 2011), which is not the case for mammalian cells. It is worth to mention that, for reasons that remain to be fully determined but could be related to aneuploidy, high telomerase activity (Bryson et al., 2006), and loss of p53 protein (Liu et al., 2011), nearly all fish cell lines have suffered spontaneous transformation and have acquired the ability to grow continuously (Bols et al., 2005). Note that karyotypic information is lacking for many of the cell lines that originated from fish, thus it is difficult to conclude on whether aneuploidy is a critical parameter for the spontaneous immortalization of fish cell lines. This immortalization, which does not occur spontaneously in other animal cell lines (it must be induced following a variety of techniques), has greatly improved the success rate for the development of fish cell lines and could be advantageous in certain applications that require the timely production of cell systems such as cellular conservation (i.e., cell lines developed from endangered species). Not to mention that the immortalization of fish cell lines may also be used to gain knowledge on the mechanisms of cell tumorigenesis/transformation.

2. Current status of fish cell culture

Data available from the Cellosaurus database (Bairoch, 2018) indicates that the 717 cell lines developed from ray-finned fish (Actinopterygii) originated from 161 species distributed in 62 families (Fig. 1C). Four taxonomic groups—i.e., Acipenseriformes, Ostariophysi, Protacanthopterygii, and Neoteleostei—account for most of the species (93%) and most of the cell lines (97%), although to different extents. Neoteleosteans form a large clade of bony fish and are the major contributing species for cell lines (47%). It is important to mention that not all taxa are represented with a cell line. For example, no cell lines are available for Holosteans, in particular for the spotted gar (*Lepisosteus oculatus*), a fish that has recently emerged as an important species for evo-devo studies and a possible bridge species between human and teleostean biomedical models (Braasch et al., 2016). A deeper analysis of the fish species contributing to cell lines listed in the Cellosaurus database revealed that marine species are slightly more represented (52%) than freshwater species and that most of them are important to fisheries and in a lesser extent to the aquaculture industry and the ornamental fish trade (Fig. 1D). Interestingly, while teleost fish used mostly as biomedical models only represent 7.5% of the species (Fig. 1D), they contribute for 16.5% of the cell lines available, with goldfish (*Carassius auratus*), zebrafish (*Danio rerio*), and Japanese medaka (*Oryzias latipes*) being the most important contributors (Fig. 1E). Only few of these cell lines (less than 5%) are available from cell banks such as the American Type Culture Collection (ATCC; www.atcc.org), the RIKEN BioResource Research Center (RIKEN BRC; cell.brc.riken.jp), and the European Collection of Authenticated Cell Cultures (ECACC; www.phe-culturecollections.org.uk). Consequently, most of the fish cell lines are currently only available upon direct request to the developers, which is certainly a limitation to their wide distribution and usage. It is also more hazardous as it misses the quality controls that cell culture collections apply to the material they host and ship around the world (e.g., absence of contaminants such as mycoplasma, authentication of cell origin, and certified good laboratory practices).

In the absence of a homogenous and complete information on the type of cells that characterize the fish cell lines listed in the Cellosaurus, we looked at the tissues/organs from which these cell lines originated. Information retrieved from 704 cell lines (Fig. 1F) showed that they originated from 33 different tissues and that most of those are soft tissues (only 12 cell lines were developed from hard tissues; this will be further discussed in Section 4). Fins (mostly caudal fins but also dorsal, pectoral, and anal fins) were used to establish 133 fish cell lines, the highest number for a tissue/organ. While fins are probably the anatomical structures that best define fish, the biological interest that they sparked among cell culturists is much less evident. Their large usage is probably more related to technical advantages, as fins are of easy access and have the capacity to regenerate upon amputation in teleosts. It is probably safe to say that these cell lines were developed to be species specific and not organ specific, and a review of the literature associated with fish fin cell lines further confirmed that they are predominantly used to assess species-specific susceptibility to a variety of virus and, to a lesser extent, in ecotoxicological studies (Kalaiselvi Sivalingam et al., 2019; Yashwanth et al., 2020; Kumar et al., 2021; Swaminathan et al., 2015). Still, few of these cell lines were primarily developed to investigate fin-specific biological processes such as regeneration and wound healing.

The picture is slightly different for fish cell lines developed from embryos. While many of them were developed primarily for virus studies, many others were developed to study the pluripotency of embryonic stem (ES) cells, in particular those developed from blastula-stage embryos. The kidney, which hosts both renal and hematopoietic activity in fish, also contributes with many cell lines ($n=72$). Although the primary use for many kidney cell lines is also viral studies, several of these cell lines were developed to answer questions related to the immune system, inflammation, and osmotic stress. Gills are organs that are directly exposed to dissolved organic compounds and the liver is an organ that plays a central role in xenobiotic metabolism. Thus gill cell lines ($n=27$) and liver cell lines (n=27) have been commonly used for (eco)toxicological studies. A similar trend was observed for most of the other tissues, i.e., some cell lines have been developed to investigate mechanisms related to viral and ecotoxicological aspects, while others have been established to investigate mechanisms related to particular cell types and biological processes. In the first case, the origin of the species is more important than the tissue used to derive the cells, while in the second case, the tissue is probably more important. Cancer cell lines form a small ($n=23$) and atypic group. While they were among the first cell lines to be established (the goldfish GEM-81 and the walleye WC-1 were reported in 1980), no cancer cell lines have been developed since 2008 (according to the Cellosaurus database). The reasons for this lack of interest in fish cancer cell lines are unknown and the growing interest for fish, in particular zebrafish, as human cancer models (Hason and Bartůněk, 2019), is certainly a contradictory signal.

3. Applications for fish cell lines

As mentioned before, fish cell lines/cultures have been chiefly developed for the detection and isolation of fish viruses, but, although it remains a major application, fish cell lines are today more than a diagnostic tool in fish virology. They are indispensable research tools in a wide range of applications in many fields of biological interest. The most popular applications are listed as follows. The following works cited are examples and do not represent an exhaustive overview of the literature available in the field of these applications.

3.1 Applications in virology and immunology

Rearing conditions of farmed fish have drastically improved in the past decade, but they do not always favor fish health and resistance to bacteria and viruses. Disease outbreaks have occurred in many fish farms, causing massive economic losses and negatively impacting on animal welfare (Lafferty et al., 2015). As part of the common effort to find a solution to this issue, cell lines have been developed from many aquaculture species and tissues as diagnostic and research tools, being used as host to isolate, propagate, and detect fish viruses, to develop vaccines, but also to study host–pathogen interactions and fish immune response (Fierro-Castro et al., 2012; Villena, 2003; Lakra et al., 2011). Several fish cell lines have been genetically engineered and used to gain knowledge on fish diseases and fish immunology. Examples are cell reporter assays for type I interferon or Hsp70 chaperone, and stable cell lines silenced or overexpressing immune and pathogen genes (Collet et al., 2018).

3.2 Applications in ecotoxicology and toxicology

Because they are first line organisms to experience the toxicity of anthropogenic chemicals present in the aquatic environment and also because they share with humans many physiological and pathological processes, fish have been traditionally used as in vivo systems to assess the ecological risk of environmental pollutants or mixtures of chemicals from complex effluents, but also to assess the safety of a large variety of compounds and predict adverse effects on human health (Bols et al., 2005). For ethical reasons, but also for practical and financial considerations, fish cell lines are increasingly used in ecotoxicological and toxicological studies in place of whole animals, as they allow a precise assessment of the cytotoxicity or bioactivity of environmental pollutants and can give important clues on the cellular and molecular mechanisms underlying the toxicity of a large variety of chemicals, food additives, pharmaceuticals, veterinary drugs, biocides, and cosmetics (Bols et al., 2005; Castaño et al., 2003; Fent, 2007, 2001; Segner, 1998; Natsch et al., 2018; Rodrigues et al., 2020; Babich and Borenfreund, 1991). The following examples, recently published, are about contaminants of emerging concern—the nanoparticles—and illustrate both ecotoxicological and toxicological applications of fish cell lines. The presence of nanoparticles in the aquatic environment has raised many concerns in relation to their effects on wildlife. In this regard, polystyrene nanoplastics that originate from commercial products or result from the breakdown of the larger plastics polluting the oceans were recently shown to alter the cytotoxicity of human pharmaceuticals in marine fish cell lines (Almeida et al., 2019). Silica nanoparticles (used for drug delivery), silver nanoparticles (used in cosmetics, textiles, and disinfection products), and copper oxide nanoparticles (used in cosmetics, gas sensor, electronics, and in heat transfer fluids) were all shown to trigger cytotoxic effects in fish cell lines (Vo et al., 2014; Yue et al., 2015; Srikanth et al., 2016), confirming that their production should be tightly regulated and their release in the aquatic environment totally forbidden. To get deeper insights into the presence of specific chemicals in environmental samples, fish cell lines can be genetically engineered to express a reporter (a fluorescent protein such as the GFP or an enzyme such as the firefly luciferase) under the control of DNA regulatory elements that respond to ecotoxicants, e.g., the dioxin-response elements (DREs) (Richter et al., 1997) or the estrogen-response elements (EREs) (Cosnefroy et al., 2009), to genotoxicants, e.g., the *p21* promoter (Geng et al., 2012), or to stressors, e.g., the electrophile-response elements (EPREs) or the *hsp70* promoter (Molina et al., 2002). Fish cell lines transiently transfected with reporter vectors were also used (Ackermann et al., 2002; Carvan et al., 2005).

3.3 Biotechnological applications

Some were developed nearly three decades ago (e.g., transgenesis), others represent recent advances made to sustain the production of novel fish feeds and cell-based food to reduce fish farming and fisheries. Until the recent development of new techniques (see below), the **production of transgenic fish lines** was done using ES cells developed from a variety of fish species. ES cells have the capacity to differentiate into nearly all cell types and are amenable to genetic manipulation (e.g., the expression of a transgene that would favor a particular trait). Transgenic ES cells are then injected in the embryo to integrate the germ line and generate fish lines that stably express the transgene. In the field of aquaculture, transgenesis has the potential to greatly enhance fish production through the development of animals with increased disease resistance, improved reproduction rate, feed conversion efficiency, and growth performance (Alvarez et al., 2007). However, the use of transgenic animals for human consumption is a source of great concern and the production of transgenic farmed fish is still limited to few species and few countries. A recent milestone in biotechnological innovation in fish was the successful application of genome editing—the engineering of the genome through DNA modification, deletion, insertion, or replacement—using CRISPR/cas9 technology to cell lines from biomedical fish models such as the Japanese medaka (Liu et al., 2018), but also to fish cell lines from farmed fish such as the Chinook salmon (*Oncorhynchus tshawytscha*) (Gratacap et al., 2020; Dehler et al., 2019; Escobar-Aguirre et al., 2019), the grass carp (*Ctenopharyngodon idella*) (Ma et al., 2018), and the Mozambique tilapia (*Oreochromis mossambicus*) (Hamar and Kültz, 2021). These edited cell lines can not only provide critical data on the mechanisms underlying disease resistance, reproduction, and growth for posterior in vivo editing to establish new fish lines with improved immunity and enhanced performances, but also they can be engineered to produce viruses for vaccine development (Gratacap et al., 2019). **Cellular aquaculture** is a promising field of research to produce fish-based products for human consumption from fish cell lines/cultures rather than from whole animals. Besides an obvious interest in cell-based fish products to feed the growing world population, issues related to human health, fish welfare, and environment can also be addressed with the use of fish cell lines (Rubio et al., 2019). The availability of cell lines from a variety of fish species (in particular from key aquaculture species) and tissues (in particular from muscle), but also the availability of pluripotent cell lines that can be differentiated in multiple cell types (see below), should provide the basis for applications of fish cell lines in this field, providing that suitable scaffolds and serum-free culture media are developed (Potter et al., 2020). BlueNalu (www.bluenalu.com) and Finless Foods (www.finlessfoods.com) are two companies involved in cellular aquaculture and the development of products from mahi-mahi (*Coryphaena hippurus*) and the Pacific bluefin tuna (*Thunnus orientalis*), but more companies are being created in this very innovative field. Besides human nutrition, fish cell lines can also be used to produce **ingredients for animal feeds**, in particular farmed fish feeds. Fish meal and fish oil are traditionally used in these feeds, but demand will soon exceed the natural resources, thus alternative sources for fish meal and fish oil must be found. Fish cell lines can be a solution to this issue as they are a source of high-quality proteins and important omega-3 fatty acids (e.g., DHA). The Fraunhofer Research and Development Center for Marine and Cellular Biotechnology is among the institutions that believe that fish cell lines could be used to develop new feed formulations (see Technology Offer F52145 at www.emb.fraunhofer.de). Another recent biotechnological application of fish cell lines is the **screening of natural extracts** toward the discovery of bioactive compounds, in this case mineralogenic compounds that have the capacity to increase extracellular matrix (ECM) mineralization and could be applied to the development of the next-generation drugs for treating low bone mineral density disorders such as osteoporosis (Surget et al., 2017).

3.4 Applications in functional genomics

Genomics, transcriptomics, proteomics, and metabolomics are molecular biology approaches, collectively referred as functional genomics, used to assess gene, protein, and metabolite function, as well as their interactions and contribution to biological processes. Cell lines have commonly been used in functional genomics studies because they provide a simple well-defined context for a precise assessment of gene function, but also because they are amenable to genetic manipulation. There are numerous examples of functional genomics using fish cell lines: (i) gain-of-function studies, where a gene is forced/overexpressed using constitutive or inducible promoter constructs (Collet et al., 2018; Rafael et al., 2012); (ii) loss-of-function studies, where a gene is silenced (knock-out or knock-down expression) using mutagenesis, RNA interference, or gene editing (Collet et al., 2018; Dehler et al., 2019; Escobar-Aguirre et al., 2019; Ruiz et al., 2009); (iii) global analysis of gene expression using DNA-microarrays (Tiago et al., 2011a) and RNA-Seq (Kwok et al., 2020; Chaves-Pozo et al., 2017) to quantify the expression of thousands of genes at a time and identify those critical to fish development and physiology or to ecological, evolutionary, and environmental issues, but also to seek for genetic markers pivotal to conservation and fisheries management.

3.5 Application in cell biology

Stem cells—e.g., ES, induced pluripotent stem-like (iPS-like) and spermatogonial stem (SS) cells—are important tools to study cell self-renewal and differentiation, and pluripotency marker genes (Hong et al., 2011; Wang et al., 2011), but can also be used to produce transgenic fish lines through germline transmission or initiate muscle cell lines for cellular aquaculture (see earlier the paragraph on biotechnological applications). ES cell lines are available from a variety of fish including model laboratory species such as the zebrafish (Ciarlo and Zon, 2016) and the Japanese medaka (Hong et al., 2011), but also from important marine fish species such as the European sea bass (*Dicentrarchus labrax*) (Buonocore et al., 2006), the gilthead seabream (*Sparus aurata*) (Parameswaran et al., 2012; Béjar et al., 2002), the sea perch (*Lateolabrax japonicus*) (Chen et al., 2007), the Asian sea bass (*Lates calcarifer*) (Parameswaran et al., 2007), and the Atlantic cod (*Gadus morhua*) (Holen et al., 2010). iPS-like cell lines exhibiting features of ES cell lines have also been recently generated from adult zebrafish fibroblasts (Peng et al., 2019; Kato et al., 2015). Besides basic research on stem cells, fish cell lines have also been used to investigate specific biological processes/problems or environmental changes (e.g., temperature and salinity). The extensive literature on this matter will not be listed here, but an example—bone cell lines to study ECM mineralization—is presented in the next section.

3.6 Applications in species conservation

The preservation of genetic material from endangered species through cell culture is also a field of application for in vitro cell system that has gained momentum in the last decade. In the list released by the International Union for Conservation of Nature (IUCN; www.iucn.org), hundreds of fish species are critically endangered. The Cellosaurus database is providing basic information on cell lines developed from endangered species (see group name "Endangered species/breed cell line") and few of them are from ray-finned fish (e.g., Dussumier's catfish (*Clarias dussumieri*), Gunther's catfish (*Horabagrus brachysoma*), Japanese huchen (*Parahucho perryi*), Willow shiner (*Gnathopogon caerulescens*), Tor mahseer (*Tor tor*), and Yangtze sturgeon (*Acipenser dabryanus*)). Much remains to be done in this field of application and more cell lines/cultures should be developed from endangered fish species (e.g., primordial germ cell (PGC) cultures; see Chapters 1 and 2), cryopreserved and archived (e.g., in a cell repository such as the ATCC or the ECACC or in a biobank such as the frozen ark at www.frozenark.org) to preserve their genetic material and/or recover their populations through surrogate reproduction. Cell lines of endangered species can also be used to gain critical biological knowledge or to develop targeted therapeutical approaches (e.g., vaccines) to save them from extinction (Goswami et al., 2016).

4. Fish cell lines capable of in vitro mineralization

Because they bring an evolutionary perspective into mechanisms of vertebrate tissue mineralization, but also because they offer many technical advantages over the classical mammalian systems (see Laizé et al. (2014) and references therein), fish have been increasingly used to gain knowledge on the mechanisms underlying bone formation and mineralization, but also to assess compounds with osteogenic/toxic activity (Laizé et al., 2014; Fernández et al., 2018).

Among those fish, small teleosts such as the zebrafish and the Japanese medaka that are amenable to gene editing and laboratory domestication have been used to model human skeletal disorders and bone diseases (e.g., osteoporosis), and to discover new therapeutic targets or drugs toward an improvement of human health. Other fish such as the Atlantic salmon (*Salmo salar*), the Senegalese sole (*Solea senegalensis*), and the gilthead seabream have been used to study the cause of the high incidence of skeletal deformities in farmed fish, a major issue for the productivity of the aquaculture industry and for animal welfare. In this context, fish cell lines capable of ECM mineralization represent a valuable research tool to assess bone gene function and the corresponding signaling pathways, and to screen for pro/antimineralogenic compounds, but also to reduce animal experimentation for research purposes.

As previously mentioned, the survey of the Cellosaurus reveals that most of the fish cell lines listed in this database have been developed from soft-organ tissues (e.g., brain, heart, liver, gonads), while very few are available from hard tissues such as cartilage and bone or described for having mineralogenic activity. This situation is not specific to fish, as the same applies for other vertebrates and the reason, besides the fact that the diversity of soft organs is higher, is probably that the ECM in soft tissues is more amenable to disruption (i.e., less mechanical forces are required), leading to an easier and quicker release of the cells from the matrix. Hard tissues such as bone have a mineralized ECM and embedded cells must find their way out of the dense network of collagen fibers and interspersed crystals of hydroxyapatite. In this regard, most fish cell lines of bone origin were developed from explant culture following collagenase treatment to lose the bone matrix and facilitate the release of bone cells.

The first report in the literature of a fish cell line capable of in vitro mineralization traces back to 2004 (Pombinho et al., 2004), when, according to the Cellosaurus, a total of 288 fish cell lines were already available. In their publication, Pombinho and colleagues described two cell lines—VSa13 and VSa16—established from vertebrae of the gilthead seabream and capable of mineralizing their ECM when culture medium was supplemented with a mineralogenic cocktail containing a source of calcium (calcium chloride), a source of phosphate (β-glycerophosphate), and ascorbic acid. After several weeks of treatment, the formation of calcium phosphate crystals, whose composition was similar to hydroxyapatite, could be detected at the surface of the cell monolayer using alizarin red S or von Kossa staining (Pombinho et al., 2004). Based on their mineralogenic properties but also on the expression and activity of marker genes and enzymes, VSa13 and VSa16 were proposed to be osteochondroprogenitor cell lines.

The development of these two cell lines has given a momentum to fish bone cell culture, and today a total of 14 fish cell lines are available to study ECM mineralization (Table 1). Although they have been established from a limited number of fish species ($n=7$), these cell lines represent several key families of Actinopterygii (2 classes and 6 orders), different environments (saline, brackish,

TABLE 1 Fish cell lines to study in vitro mineralization.

Class	Order	Species	Bone type	Cell line	Tissue	Accession number[a]	References
Chondrostei	Acipenseriformes	Adriatic sturgeon *Acipenser naccarii*	C	BAAn1F	Branchial arch	CVCL_6E13	Viegas et al. (2010)
				VAn2H	Vertebra	CVCL_6E14	Viegas et al. (2010)
Teleostei	Spariformes	Gilthead seabream *Sparus aurata*	A	VSa13	Vertebra	CVCL_S952	Pombinho et al. (2004)
				VSa16	Vertebra	CVCL_S953	Pombinho et al. (2004)
				ABSa15	Branchial arch	CVCL_6E03; 13112201[b]	Marques et al. (2007)
				ESSA1	Embryo	CVCL_6E07	Parameswaran et al. (2012)
				BSa2	Blastema	CVCL_YU24	Vijayakumar et al. (2020)
				JSa1	Lower jaw	CVCL_6E04	Unpublished
	Salmoniformes	Atlantic salmon *Salmo salar*	C	SsV2	Vertebra	CVCL_6E06	Unpublished
				SsJ1	Lower Jaw	CVCL_6E05	Unpublished
	Cypriniformes	Zebrafish *Danio rerio*	C	ZFB1	Bone[c]	CVCL_6E12	Vijayakumar et al. (2013)
		Blunt snout bream *Megalobrama amblycephala*	C	MBCs	Bone[c]	CVCL_YD02	Guan et al. (2018)
	Cyprinodontiformes	Mosquitofish *Gambusia affinis*	A	BGA	Vertebra	CVCL_YP00	Jyotsna et al. (2020)
	Pleuronectiformes	Senegalese sole *Solea senegalensis*	A	SS1C	Vertebra	CVCL_6E61	Unpublished

[a] *Cellosaurus accession number.*
[b] *ECACC accession number.*
[c] *Information on bone tissues not available.*
A, acellular bone; *C*, cellular bone.

or fresh water) and, most importantly, different types of bone (osteocytic/cellular or anosteocytic/acellular bone) (Shahar and Dean, 2013).

With six cell lines, the gilthead seabream is the species that has concentrated most of the effort related to the provision of mineralogenic cell lines. The dominant position of the gilthead seabream versus model organisms, such as the zebrafish and the Japanese medaka, may be related to (i) the high incidence of skeletal deformities in a species that is critically important for the Mediterranean aquaculture industry (thus the need for in vitro cell systems to gain insights into the mechanisms underlying these deformities) and (ii) a higher rate of success in establishing new cell lines from this species, and in general from marine species (unpublished observation). Among the cell lines listed in Table 1, 12 have been established from skeletal elements, mostly vertebrae ($n=6$) but also branchial arches ($n=2$) and the lower jaw ($n=2$), and a mixture of tissues enriched in calcified structures ($n=2$) in the case of the zebrafish and mosquitofish (*Gambusia affinis*), because of their small size and the difficulty to individualize skeletal elements. The remaining two cell lines were developed from nonmineralized tissues—blastula stage embryo for ESSA1 and fin blastema for BSa2—and described as pluripotent and multipotent cell lines, respectively (Parameswaran et al., 2012; Vijayakumar et al., 2020). Under an appropriate stimulus, cells of both lines (ESSA1 and BSa2) can differentiate into osteoblast-like cells capable of mineralizing their ECM but also osteoclast-like cells capable of bone resorption. It is also worth to mention that the two cell lines developed from cartilaginous elements of the noncalcified skeleton of the Adriatic sturgeon (*Acipenser naccarii*) are not capable of in vitro mineralization (Viegas et al., 2010) and thus represent critical cellular tools to better understand the evolutionary mechanisms regulating matrix mineralization in vertebrates. The cell line ABSa15 (developed from mineralized branchial arches of the gilthead seabream) is the only one available in a cell bank (ECACC no. 13112201). All the cell lines listed in Table 1 form adherent monolayers, i.e., they attach to the surface of the culture dish/flask and their proliferation is contact inhibited. They are also subcultured and cryopreserved using techniques commonly used for other cell lines. For example, cells are detached from the substratum using 0.2% trypsin-ethylenediaminetetraacetic acid (EDTA) in phosphate-buffered saline (PBS) and cell cryopreservation is commonly achieved using 10% dimethyl sulfoxide (DMSO) in culture medium.

A review of the literature showed that these cell lines have already been used in many studies, with VSa13 being the most used and most versatile cell line. Table 2 gives an overview on their usage in different fields including (i) toxicology (to evidence the mineralogenic effect of aquatic pollutants or study their toxicity in a bone cell system), (ii) pharmacology (to evidence the mineralogenic effect of drugs, hormones, or nutrients, and to study the mechanisms underlying this effect), (iii) bioprospection (to screen marine extracts for mineralogenic activities and to identify bioactive compounds), (iv) functional genomics (to study genes/proteins function and interaction, and to understand their contribution to ECM mineralization), (v) transcriptional regulation (to study

TABLE 2 Fields of application for mineralogenic fish cell lines.

Field of application	Cell lines	Study objective	Compounds, genes	References
Toxicology	VSa13, VSa16, ABSa15	Mineralogenic effect	Azoxystrobin, 3-methylcholanthrene	Fernández et al. (2018), Laizé et al. (2018), Rodrigues et al. (2015)
Pharmacology	VSa13, VSa16	Mineralogenic effect	Vanadium, retinoic acid, insulin, IGF-1, PUFAs, minerals, vitamins	Tiago et al. (2008a,b, 2011a,b), Laizé et al. (2009), Viegas et al. (2012, 2020)
Bioprospection	VSa13	Mineralogenic effect	Marine algae extracts	Surget et al. (2017)
Functional genomics	VSa13, VSa16, ABSa15, ZFB1	Gene/protein function	*oc*, *mgp*, *sparc*, *spp1*, *bmp2*, *S100*, *fhl2*, *vkor*, *bhmt3*, *sdr*	Rafael et al. (2006, 2012), Pombinho et al. (2004), Laizé et al. (2005), Fonseca et al. (2011), Fernández et al. (2015)
Transcriptional regulation	VSa13, SS1C, ABSa15, ZFB1	Gene transcription	Fibroblast growth factor, retinoic acid, bone morphogenetic proteins	Marques et al. (2014, 2016), Conceição et al. (2008)
Epigenetics	ABSa15	Antimineralogenic	microRNA miR-20a	Tiago et al. (2014)

gene transcription in a mineralogenic context or a bone cell system), and (vi) epigenetics (to study microRNA-associated posttranscriptional mechanisms in a mineralogenic context or in a bone cell system). These cell lines have been also used in studies not directly related to biomineralization, e.g., the capacity of pluripotent/multipotent cells to differentiate into a mineralogenic cell phenotype (Parameswaran et al., 2012; Vijayakumar et al., 2020) and the plasticity of unipotent cells to adapt to new culture conditions (Rosa et al., 2010).

Efficient DNA transfer into a cell line (transfection) is an important factor for in vitro cell systems as it allows studies on gene function and transcription but also to develop cell clones expressing specific transgenes. Bone-derived cells are notoriously hard-to-transfect cells—i.e., transfection efficiency (the percentage of cells that are transfected in an entire population) is usually low—and those of fish origin are no exception. While this may not be an issue for developing stable cell clones, it could hamper the use of fish bone-derived cell lines in studies aiming at assessing transcriptional regulation through the transient transfection of reporter constructs. In 2006, Braga and colleagues evaluated the efficiency of several nonviral methods and reagents to transfect DNA into VSa13 and VSa16 cells (Braga et al., 2006). They found that a cationic polymer, the polyethylenimine (PEI), could achieve a transfection efficiency of up to 40% in VSa13 cells, twice as efficient as the best reagent for lipofection, while much cheaper. DNA transfection into VSa16 cells was also higher using PEI but efficiency was only 8%. Other methods/reagents for DNA transfer have been evaluated since then, and nucleofection, a nonchemical transfection method, has achieved slightly better results than PEI, i.e., a transfection efficiency of 42% in VSa13 and 21% in VSa16 (unpublished results). While transfection efficiency was consistently higher in VSa13 cells than in VSa16 cells, independently of the method used, highest efficiency of DNA transfer into a fish bone-derived cell line was achieved in ABSa15 cell line, where 62% of the cells could be transfected by nucleofection (unpublished results). A nucleofection protocol optimized for fish mineralogenic cell lines (buffer composition and electrical settings may need to be adapted as each cell line has different tolerance to nucleofection) is available in the toolbox of the ASSEMBLE project (Association of European Marine Biological Laboratories) at www.assemblemarine.org.

5. Development of bone-derived cells

Most of the fish cell lines established from skeletal tissues (mineralized or not) were developed from explant cultures following a simple procedure originally described by Marques and colleagues in 2007 (Marques et al., 2007), and updated as follows (see also Fig. 2).

a. Anesthetize the fish (prefer juveniles as they are actively developing) with isoeugenol diluted in fish water, then place it in iced water until cessation of the opercular movements and sacrifice by decapitation. Check literature to use the amount of anesthetic adapted to fish species and ontogenetic stages (tricaine methanesulfonate is also used as anesthetic);

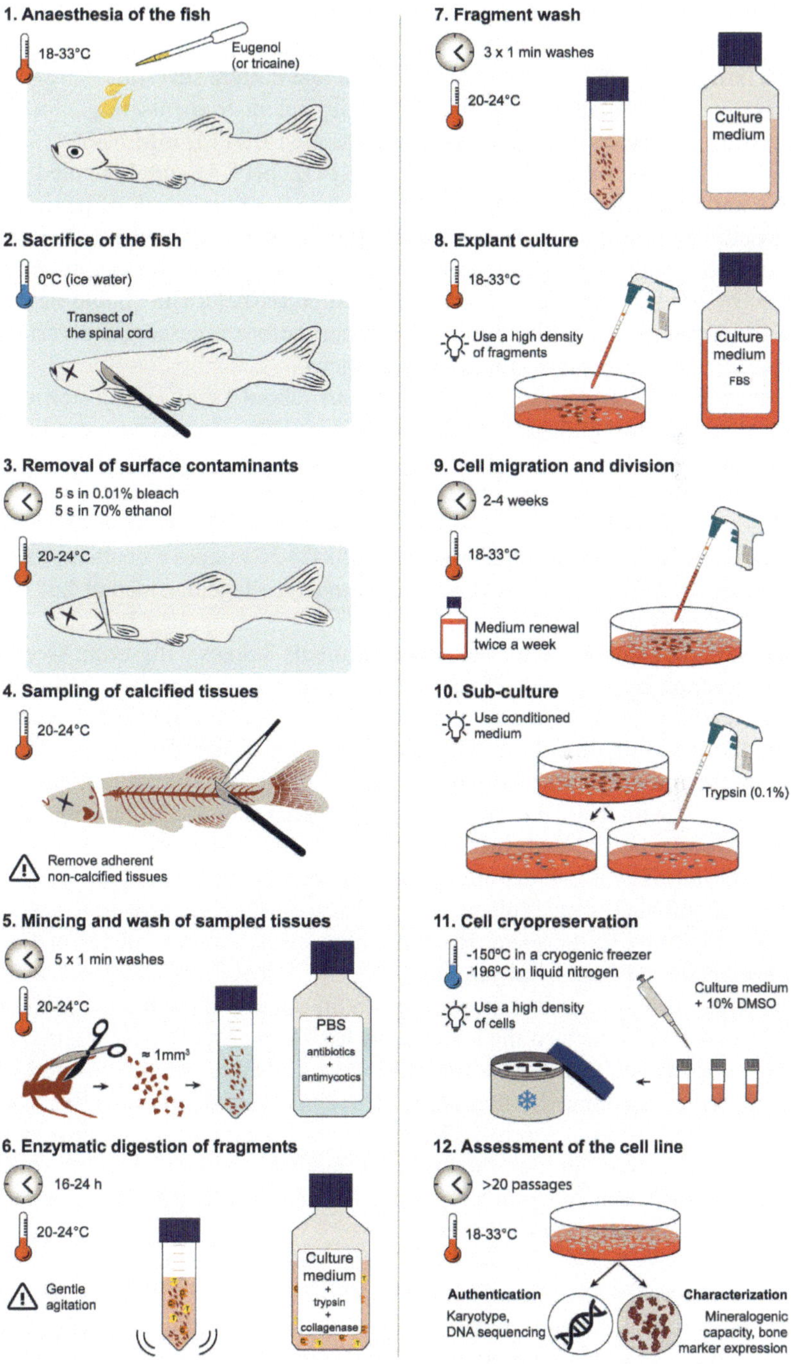

FIG. 2 Step-by-step protocol to develop primary cell cultures from fish calcified tissues.

b. Quickly wash fish body with bleach diluted 1:100 in distilled water, then with 70% ethanol to remove surface microorganisms, as skin hosts many bacteria and fungi that may jeopardize explant culture survival;

c. Collect calcified tissues using sterile surgical instruments (e.g., forceps, scissors, and scalpels) and soak them in sterile PBS (137 mM NaCl, 2.7 mM KCl, 15.8 mM Na_2HPO_4, 1.23 mM KH_2PO_4, pH 7.4) supplemented with 5% antibiotics and 1% antimycotic (usually 10,000 units/mL of penicillin–streptomycin and 250 µg/mL of amphotericin B, respectively). Clean them from adherent tissues using the surgical instruments (a sterile toothbrush can also be used to clean large pieces of calcified tissues). Gut should be carefully removed without exposing the lumen before accessing the vertebral column as it hosts many microorganisms;

d. Mince calcified tissues into small fragments (about 1 mm^3) using sterile surgical instruments and wash them 5 times with PBS. Collagenase used in the next step will digest more efficiently the collagen matrix in small fragments and cells may also leave more easily from explants of smaller size;

e. Digest fragments with 0.125% collagenase type II (in serum-free medium) for 16–24 h at room temperature (20–24 °C) under a gentle agitation. Trypsin (0.2%) may be used to complement collagenase action but for a much shorter time (15–30 min);

f. Remove digestion solution and wash fragments 3 times with serum-free medium supplemented with 5% antibiotics and 1% antimycotic. Leibovitz's L15 medium and Dulbecco's modified Eagle's medium (DMEM) have been used successfully to culture fish mineralogenic cell lines;

g. Place fragments into a cell culture vessel containing culture medium supplemented with 15% fetal bovine serum (FBS), 1% antibiotics, and 0.2% antimycotic. Use enough medium for the fragments to stay in contact with the substratum without drifting. Keep fragment quantity and density high as factors diffusing from explants are needed to sustain the growth of the first cells migrating to the plate. Routine uncoated cell culture plates or flasks have been used successfully;

h. Incubate the explants in appropriate conditions and allow the cells to leave fragments and attach to the substratum for 2–4 weeks. Renew the medium of the explant culture twice a week using fresh medium complemented with conditioned medium (spent medium harvested from cultured cells). Using conditioned medium (up to 50%) will avoid the culture medium to be depleted of essential growth factors. Cell incubator may need CO_2 to buffer medium pH and incubation temperature should be adapted to each fish species and may range from 18 to 33 °C;

i. Transfer the cells that have attached to the substratum into new plates/flasks using trypsin–EDTA (1.1 mM EDTA and 0.2% trypsin in PBS). Renew the medium of the primary cell cultures using fresh medium complemented with conditioned medium (up to 25%). The trypsinization of cells in explant cultures is done following routine protocols and may use culture cylinders to isolate cell colonies. Explant may be maintained in culture as remaining cells may form new primary cell cultures;

j. Divide (typically 1:2) subconfluent cell cultures into larger plates/flasks and reduce gradually the proportion of conditioned medium, FBS, and antibiotics/antimycotic. Most fish mineralogenic cell lines grow well in the presence of 10% FBS. It is advisable to culture cells without antibiotics and antimycotic, although it may not always be possible;
k. Cryopreserve cells that have successfully survived several passages and show signs of spontaneous immortalization. Fish mineralogenic cells have been cryopreserved and successfully revived using 10% DMSO. Regularly test cell cultures/lines for mycoplasma;
l. Test the bone origin of the cells and their mineralogenic capacity. Assess the expression of bone marker genes by qPCR or RNA-Seq, and ECM mineralization through alizarin red S or von Kossa staining (Tiago et al., 2011a; Marques et al., 2007).

Critical points to a successful culture of skeletal explants are (1) a rapid handling and processing of the explants to limit the extent of cellular damages; (2) a high density of fragments to provide (through diffusion in the culture medium) enough growth factors (e.g., BMPs, TGFs, PDGFs, and IGFs) to sustain the growth of the first cells that will leave the explants and colonize the culture plate; (3) a medium that can support fish cell growth, e.g., commercial solutions such as L15 medium and DMEM; and (4) the supplementation of the saline solution for washes and culture medium with antibiotics (e.g., penicillin and streptomycin) and antimycotics (e.g., amphotericin B) to decrease the risks of contamination and thus, an early termination of the explant culture. The timing for cell migration from explants is variable (from days to weeks) and largely depends on the type of skeletal tissue and the efficiency of the enzymatic treatment to lose the cells in the matrix. Once a primary cell culture is established from the explant culture, it will be subcultured until cell eventually die or get immortalized (see earlier the paragraph on the spontaneous transformation of fish cell lines). As mentioned previously, a critical aspect for experiments evaluating ECM mineralization is the source of the serum used to culture the cells, as its mineralogenic capacity may vary according to suppliers (country of origin and processing may be different) but also according to batches of the same references (unpublished observation). Therefore it is important to carefully evaluate the serum mineralogenic capacity prior to purchase and reserve a large quantity of the selected batch, at least to cover the duration of ECM mineralization experiments.

6. Perspectives for fish cell culture

Following the growing interest of the scientific community for practical and cost-effective study models but also for experimental systems that could reduce animal experimentation and fulfill the 3Rs principle (reduction, refinement, and replacement), the number of cell lines of fish origin has steadily increased in the last 60 years. Still, an effort should be made to increase the provision of in vitro cell systems from important biomedical models (e.g., small teleosts such as the zebrafish, medaka, and mosquitofish) or emerging and economically important aquaculture species (e.g., marine fish such as the meager (*Argyrosomus regius*), the sardine (*Sardina pilchardus*), and the longfin yellowtail (*Seriola rivoliana*)), in order to gain novel insights into mechanisms of human diseases and also into physiological processes that have been overlooked in farmed fish in the absence of suitable cellular tools. Cell lines should be also developed from endangered fish species to participate in the worldwide action aiming at preserving the genetic material of these threatened species. An effort should also be made to increase the storage of fish cell lines in cell banks, where they can be properly preserved, authenticated, assessed for contaminants, and distributed worldwide. The scarce availability of fish cell lines in cell bank catalogs is clearly a limitation to a wider usage by the scientific community. In what concerns the use of fish cell lines to study mechanisms of bone formation, screen osteoactive compounds, or assess molecule osteotoxicity, efforts should be made toward the better characterization of available cell lines but also toward the development of cell lines representing all bone cell types. In this regard, lines of cells from the osteoclast/macrophage lineage (typically present in fish head–kidney) are urgently needed for in vitro studies on mechanisms of matrix resorption. While the provision of bone cell lines must increase for biomedical models (only one is available for zebrafish), in vitro cell systems should also be developed from nonmodel species that have demonstrated some usefulness to better understand mechanisms regulating bone mineral density, e.g., icefishes (Notothenioidei), which have evolved an osteopenic skeleton to adapt to their environment (Albertson et al., 2009) and could be used to screen compounds with mineralogenic activity, and the gar (Holostei), that has evolved a weakly ossified skeleton that places it as a bridge species between Chondrostei (e.g., sturgeon; unossified skeleton) and Teleostei (e.g., zebrafish; ossified skeleton) but also between zebrafish and human, as it did not suffer the whole genome duplication event that occurred in teleosts (Braasch et al., 2016). Gene editing methodologies (e.g., CRISPR/Cas9) have been successfully applied to fish and fish cell lines, although not yet used to assess bone gene function in fish bone-derived cell lines. In this regard, fish mineralogenic cell lines hold a great potential for genetic and epigenetic studies aiming at the identification of genes and signaling pathways involved in bone formation and mineralization.

While the provision of fish cell lines is certainly a major challenge for the next decade, there is room for the optimization of the reagents and techniques used for fish cell culture, which have been developed for mammalian cells. For example, medium composition and osmolality could be adjusted to better fit fish physiological parameters (e.g., saline versus freshwater fish), serum-free medium could be developed to boost the production of cell-based fish product by cellular aquaculture, and cryopreservation and transfection protocols could be improved to increase cell survival and enhance DNA delivery. Niels Bols and Lucy Lee have recently pointed out the importance of consolidating the past discoveries and of providing direction for future research in fish cell culture (Bols et al., 2017). In that sense, they have reviewed the literature available on rainbow trout cell lines and collected information on their development (*invitromatics*), established a catalog (*invitrome*), and listed their use (*invitroomics*). Applying this idea to the 717 fish cell lines listed in the Cellosaurus would unify the scattered information on fish cell culture, but it represents a colossal task.

Acknowledgments

The authors acknowledge the financial support of the Portuguese Foundation for Science and Technology (FCT) through project UIDB/04326/2020 (CCMAR) and the contribution of former and present members of the BIOSKEL laboratory (CCMAR) to the development of fish mineralogenic cell lines, in particular Natércia Conceição, Cindy Fazenda, Ignacio Fernández Monzón, Paulo Gavaia, Cátia Marques, Vijayakumar Parameswaran, Jorge Pinto, António Pombinho, Marta Rafael, Daniel Tiago, Carla Viegas, and Michael Viegas.

References

Ackermann, G.E., Fent, K., 1998. The adaptation of the permanent fish cell lines PLHC-1 and RTG-2 to FCS-free media results in similar growth rates compared to FCS-containing conditions. Mar. Environ. Res. 46, 363–367.

Ackermann, G.E., Brombacher, E., Fent, K., 2002. Development of a fish reporter gene system for the assessment of estrogenic compounds and sewage treatment plant effluents. Environ. Toxicol. Chem. 21, 1864–1875.

Albertson, R.C., Cresko, W., Detrich, H.W., Postlethwait, J.H., 2009. Evolutionary mutant models for human disease. Trends Genet. 25, 74–81.

Almeida, M., Martins, M.A., Soares, A.M.V., Cuesta, A., Oliveira, M., 2019. Polystyrene nanoplastics alter the cytotoxicity of human pharmaceuticals on marine fish cell lines. Environ. Toxicol. Pharmacol. 69, 57–65.

Alvarez, M.C., Béjar, J., Chen, S., Hong, Y., 2007. Fish ES cells and applications to biotechnology. Mar. Biotechnol. 9, 117–127.

Babich, H., Borenfreund, E., 1991. Cytotoxicity and genotoxicity assays with cultured fish cells: a review. Toxicol. In Vitro 5, 91–100.

Bairoch, A., 2018. The Cellosaurus, a cell-line knowledge resource. J. Biomol. Tech. 29, 25–38.

Béjar, J., Hong, Y., Alvarez, M.C., 2002. An ES-like cell line from the marine fish *Sparus aurata*: characterization and chimaera production. Transgenic Res. 11, 279–289.

Bols, N.C., Dayeh, V.R., Lee, L.E.J., Schirmer, K., 2005. Use of fish cell lines in the toxicology and ecotoxicology of fish. Piscine cell lines in environmental toxicology. In: Mommsen, T.P., Moon, T.W. (Eds.), Biochemistry and Molecular Biology of Fishes. vol. 6. Elsevier B.V, pp. 43–84.

Bols, N.C., Kawano, A., Lee, L.E.J., 2011. Cellular, molecular, genomics, and biomedical approaches: culture of fish fell lines. In: Encyclopedia of Fish Physiology. vol. 3. Elsevier Inc, pp. 1965–1970.

Bols, N.C., Pham, P.H., Dayeh, V.R., Lee, L.E.J., 2017. Invitromatics, invitrome, and invitroomics: introduction of three new terms for *in vitro* biology and illustration of their use with the cell lines from rainbow trout. In Vitro Cell. Dev. Biol. Anim. 53, 383–405.

Braasch, I., Gehrke, A.R., Smith, J.J., Kawasaki, K., Manousaki, T., Pasquier, J., Amores, A., Desvignes, T., Batzel, P., Catchen, J., Berlin, A.M., Campbell, M.S., Barrell, D., Martin, K.J., Mulley, J.F., Ravi, V., Lee, A.P., Nakamura, T., Chalopin, D., Fan, S., Wcisel, D., Caestro, C., Sydes, J., Beaudry, F.E.G., Sun, Y., Hertel, J., Beam, M.J., Fasold, M., Ishiyama, M., Johnson, J., Kehr, S., Lara, M., Letaw, J.H., Litman, G.W., Litman, R.T., Mikami, M., Ota, T., Saha, N.R., Williams, L., Stadler, P.F., Wang, H., Taylor, J.S., Fontenot, Q., Ferrara, A., Searle, S.M.J., Aken, B., Yandell, M., Schneider, I., Yoder, J.A., Volff, J.N., Meyer, A., Amemiya, C.T., Venkatesh, B., Holland, P.W.H., Guiguen, Y., Bobe, J., Shubin, N.H., Di Palma, F., Alföldi, J., Lindblad-Toh, K., Postlethwait, J.H., 2016. The spotted gar genome illuminates vertebrate evolution and facilitates human-teleost comparisons. Nat. Genet. 48, 427–437.

Braga, D., Laizé, V., Tiago, D.M., Cancela, M.L., 2006. Enhanced DNA transfer into fish bone cells using polyethylenimine. Mol. Biotechnol. 34, 51–54.

Bryson, S.P., Joyce, E.M., Martell, D.J., Lee, L.E.J., Holt, S.E., Kales, S.C., Fujiki, K., Dixon, B., Bols, N.C., 2006. A cell line (HEW) from embryos of haddock (*Melanogrammus aeglefinius*) and its capacity to tolerate environmental extremes. Mar. Biotechnol. 8, 641–653.

Buonocore, F., Libertini, A., Prugnoli, D., Mazzini, M., Scapigliati, G., 2006. Production and characterization of a continuous embryonic cell line from sea bass (*Dicentrarchus labrax* L.). Mar. Biotechnol. 8, 80–85.

Carvan, M.J.I., Heiden, T.K., Tomasiewicz, H., 2005. The utility of zebrafish as a model for toxicological research. In: Mommsen, T.P., Moon, T.W. (Eds.), Biochemistry and Molecular Biology of Fishes. vol. **6**. Elsevier B.V, pp. 3–41.

Castaño, A., Bols, N., Braunbeck, T., Dierickx, P., Halder, M., Isomaa, B., Kawahara, K., Lee, L.E.J., Mothersill, C., Pärt, P., Repetto, G., Riego Sintes, J., Rufli, H., Smith, R., Wood, C., Segner, H., 2003. The use of fish cells in ecotoxicology: the report and recommandation of ECVAM workshop 47. ATLA Altern. Lab. Anim. 31, 317–351.

Chaves-Pozo, E., Valero, Y., Esteve-Codina, A., Gómez-Garrido, J., Dabad, M., Alioto, T., Meseguer, J., Esteban, M.Á., Cuesta, A., 2017. Innate cell-mediated cytotoxic activity of European sea bass leucocytes against nodavirus-infected cells: a functional and RNA-seq study. Sci. Rep. 7, 15396.

Chen, S.-L., Sha, Z.-X., Ye, H.-Q., Liu, Y., Tian, Y.-S., Hong, Y., Tang, Q.-S., 2007. Pluripotency and chimera competence of an embryonic stem cell line from the sea perch (*Lateolabrax japonicus*). Mar. Biotechnol. 9, 82–91.

Ciarlo, C.A., Zon, L.I., 2016. Embryonic cell culture in zebrafish. In: Methods in Cell Biology. vol. 133. Elsevier Ltd, pp. 1–10.

Collet, B., Collins, C., Lester, K., 2018. Engineered cell lines for fish health research. Dev. Comp. Immunol. 80, 34–40.

Conceição, N., Laizé, V., Simões, B., Pombinho, A.R., Cancela, M.L., 2008. Retinoic acid is a negative regulator of matrix Gla protein gene expression in teleost fish *Sparus aurata*. Biochim. Biophys. Acta 1779, 28–39.

Cosnefroy, A., Brion, F., Guillet, B., Laville, N., Porcher, J.M., Balaguer, P., Aït-Aïssa, S., 2009. A stable fish reporter cell line to study estrogen receptor transactivation by environmental (xeno) estrogens. Toxicol. in Vitro 23, 1450–1454.

Couso, N., Castro, R., Noya, M., Obach, A., Lamas, J., 2002. Formation of short-lived multinucleated giant cells (MGCS) from cultured gilthead seabream macrophages. Anat. Rec. 267, 204–212.

Dehler, C.E., Lester, K., Della Pelle, G., Jouneau, L., Houel, A., Collins, C., Dovgan, T., Machat, R., Zou, J., Boudinot, P., Martin, S.A.M., Collet, B., 2019. Viral resistance and IFN signaling in STAT2 knockout fish cells. J. Immunol. 203, 465–475.

Escobar-Aguirre, S., Arancibia, D., Escorza, A., Bravo, C., Andrés, M., Zamorano, P., Martínez, V., 2019. Development of a bicistronic vector for the expression of a CRISPR/Cas9-mCherry system in fish cell lines. Cell 8, 75.

Fent, K., 2001. Fish cell lines as versatile tools in ecotoxicology: assessment of cytotoxicity, cytochrome P4501A induction potential and estrogenic activity of chemicals and environmental samples. Toxicol. in Vitro 15, 477–488.

Fent, K., 2007. Permanent fish cell cultures as important tools in ecotoxicology. ALTEX 24, 26–28.

Fernández, I., Vijayakumar, P., Marques, C., Cancela, M.L.L., Gavaia, P.J.P.J., Laizé, V., 2015. Zebrafish vitamin K epoxide reductases: expression *in vivo*, along extracellular matrix mineralization and under phylloquinone and warfarin *in vitro* exposure. Fish Physiol. Biochem. 41, 745–759.

Fernández, I., Gavaia, P.J., Laizé, V., Cancela, M.L., 2018. Fish as a model to assess chemical toxicity in bone. Aquat. Toxicol. 194, 208–226.

Fierro-Castro, C., Barrioluengo, L., López-Fierro, P., Razquin, B.E., Carracedo, B., Villena, A.J., 2012. Fish cell cultures as *in vitro* models of pro-inflammatory responses elicited by immunostimulants. Fish Shellfish Immunol. 33, 389–400.

Fonseca, V.G., Rosa, J., Laizé, V., Gavaia, P.J., Cancela, M.L., 2011. Identification of a new cartilage-specific S100-like protein up-regulated during endo/perichondral mineralization in gilthead seabream. Gene Expr. Patterns 11, 448–455.

Geng, D., Zhang, Z., Guo, H., 2012. Development of a fish cell biosensor system for genotoxicity detection based on DNA damage-induced trans-activation of *p21* gene expression. Biosensors 2, 318–340.

Goh, J.Y., Weaver, R.J., Dixon, L., Platt, N.J., Roberts, R.A., 2015. Development and use of *in vitro* alternatives to animal testing by the pharmaceutical industry 1980-2013. Toxicol. Res. 4, 1297–1307.

Goswami, M., Mishra, A., Ninawe, A., Trudeau, V., Lakra, W., 2016. Bio-banking: an emerging approach for conservation of fish germplasm. Poult. Fish. Wildl. Sci. 4, 1–8.

Gratacap, R.L., Wargelius, A., Edvardsen, R.B., Houston, R.D., 2019. Potential of genome editing to improve aquaculture breeding and production. Trends Genet. 35, 672–684.

Gratacap, R.L., Regan, T., Dehler, C.E., Martin, S.A.M., Boudinot, P., Collet, B., Houston, R.D., 2020. Efficient CRISPR/Cas9 genome editing in a salmonid fish cell line using a lentivirus delivery system. BMC Biotechnol. 20, 35.

Guan, N., Lan, T., Chen, Y., Liu, L., Zhou, F., Gao, Z., 2018. Establishment and characterization of bone cell line for blunt snout bream (*Megalobrama amblycephala*). J. Fish. China 42, 513–521.

Hamar, J., Kültz, D., 2021. An efficient vector-based CRISPR/Cas9 system in an *Oreochromis mossambicus* cell line using endogenous promoters. Sci. Rep. 11, 7854.

Hashimoto, H., Toyohara, H., Yokoyama, Y., Sakagushi, M., Ozato, K., Wakamatsu, Y., 1997. Effects of carp serum on the growth of goldfish fin cells in early passage. J. Fish Biol. 50, 201–207.

Hason, M., Bartůněk, P., 2019. Zebrafish models of cancer—new insights on modeling human cancer in a non-mammalian vertebrate. Genes 10, 935.

Holen, E., Kausland, A., Skjærven, K., 2010. Embryonic stem cells isolated from Atlantic cod (*Gadus morhua*) and the developmental expression of a stage-specific transcription factor ac-Pou2. Fish Physiol. Biochem. 36, 1029–1039.

Hong, N., Li, Z., Hong, Y., 2011. Fish stem cell cultures. Int. J. Biol. Sci. 7, 392–402.

Jyotsna, Vijayakumar, P., Mani, R., Raja, S., Dhas, T.S., 2020. Susceptibility of betanodavirus in a newly established vertebra-derived cell line from mosquitofish (*Gambusia affinis*). J. Fish Dis. 43, 263–273.

Kalaiselvi Sivalingam, N.N., Seepoo, A.M., Gani, T., Selvam, S., Azeez Sait, S.H., 2019. Zebrafish fin-derived fibroblast cell line: a model for *in vitro* wound healing. J. Fish Dis. 42, 573–584.

Kato, H., Abe, K., Yokota, S., Matsuno, R., Mikekado, T., Yokoi, H., Suzuki, T., 2015. Establishment of *oct4:gfp* transgenic zebrafish line for monitoring cellular multipotency by GFP fluorescence. In Vitro Cell. Dev. Biol. Anim. 51, 42–49.

Kumar, M.S., Soni, P., Kumar, R., Singh, N., Srivastava, S., Mishra, A.K., Singh, V.K., Kushwaha, B., 2021. Development of caudal fin cell line from Hill trout *Barilius bendelisis* (Hamilton, 1807) for cytotoxicity and transfection studies. Croat. J. Fish. 79, 15–24.

Kwok, M.L., Hu, X.L., Meng, Q., Chan, K.M., 2020. Whole-transcriptome sequencing (RNA-seq) analyses of the zebrafish liver cell line, ZFL, after acute exposure to Cu^{2+} ions. Metallomics 12, 732–751.

Lafferty, K.D., Harvell, C.D., Conrad, J.M., Friedman, C.S., Kent, M.L., Kuris, A.M., Powell, E.N., Rondeau, D., Saksida, S.M., 2015. Infectious diseases affect marine fisheries and aquaculture economics. Annu. Rev. Mar. Sci. 7, 471–496.

Laizé, V., Pombinho, A.R., Cancela, M.L., 2005. Characterization of *Sparus aurata* osteonectin cDNA and *in silico* analysis of protein conserved features: evidence for more than one osteonectin in Salmonidae. Biochimie 87, 411–420.

Laizé, V., Tiago, D.M., Aureliano, M., Cancela, M.L., 2009. New insights into mineralogenic effects of vanadate. Cell. Mol. Life Sci. 66, 3831–3836.

Laizé, V., Gavaia, P.J., Cancela, M.L., 2014. Fish: a suitable system to model human bone disorders and discover drugs with osteogenic or osteotoxic activities. Drug Discov. Today Dis. Model. 13, 29–37.

Laizé, V., Gavaia, P.J., Tarasco, M., Viegas, M.N., Caria, J., Luis, N., Cancela, M.L., 2018. Osteotoxicity of 3-methylcholanthrene in fish. Ecotoxicol. Environ. Saf. 161, 721–728.

Lakra, W.S., Swaminathan, T.R., Joy, K.P., 2011. Development, characterization, conservation and storage of fish cell lines: a review. Fish Physiol. Biochem. 37, 1–20.

Liu, M., Tee, C., Zeng, F., Sherry, J.P., Dixon, B., Bols, N.C., Duncker, B.P., 2011. Characterization of p53 expression in rainbow trout. Comp. Biochem. Physiol., Part C: Toxicol. Pharmacol. 154, 326–332.

Liu, Q., Yuan, Y., Zhu, F., Hong, Y., Ge, R., 2018. Efficient genome editing using CRISPR/Cas9 ribonucleoprotein approach in cultured Medaka fish cells. Biol. Open **7**, bio035170.

Ma, J., Fan, Y., Zhou, Y., Liu, W., Jiang, N., Zhang, J., Zeng, L., 2018. Efficient resistance to grass carp reovirus infection in JAM-A knockout cells using CRISPR/Cas9. Fish Shellfish Immunol. 76, 206–215.

Marques, C.L., Rafael, M.S., Cancela, M.L., Laizé, V., 2007. Establishment of primary cell cultures from fish calcified tissues. Cytotechnology 55, 9–13.

Marques, C.L., Fernández, I., Rosa, J., Viegas, M.N., Cancela, M.L., Laizé, V., 2014. Spatiotemporal expression and retinoic acid regulation of bone morphogenetic proteins 2, 4 and 16 in Senegalese sole. J. Appl. Ichthyol. 30, 713–720.

Marques, C.L., Fernández, I., Viegas, M.N., Cox, C.J., Martel, P., Rosa, J., Cancela, M.L., Laizé, V., 2016. Comparative analysis of zebrafish bone morphogenetic proteins 2, 4 and 16: molecular and evolutionary perspectives. Cell. Mol. Life Sci. 73, 841–857.

Molina, A., Carpeaux, R., Martial, J.A., Muller, M., 2002. A transformed fish cell line expressing a green fluorescent protein-luciferase fusion gene responding to cellular stress. Toxicol. In Vitro 16, 201–207.

Natsch, A., Laue, H., Haupt, T., von Niederhäusern, V., Sanders, G., 2018. Accurate prediction of acute fish toxicity of fragrance chemicals with the RTgill-W1 cell assay. Environ. Toxicol. Chem. 37, 931–941.

Parameswaran, V., Shukla, R., Bhonde, R., Hameed, A.S.S., 2007. Development of a pluripotent ES-like cell line from Asian sea bass (*Lates calcarifer*)—an oviparous stem cell line mimicking viviparous ES cells. Mar. Biotechnol. 9, 766–775.

Parameswaran, V., Laizé, V., Gavaia, P.J., Leonor Cancela, M., 2012. ESSA1 embryonic stem like cells from gilthead seabream: a new tool to study mesenchymal cell lineage differentiation in fish. Differentiation 84, 240–251.

Peng, L., Zhou, Y., Xu, W., Jiang, M., Li, H., Long, M., Liu, W., Liu, J., Zhao, X., Xiao, Y., 2019. Generation of stable induced pluripotent stem-like cells from adult zebra fish fibroblasts. Int. J. Biol. Sci. 15, 2340–2349.

Pombinho, A.R., Laizé, V., Molha, D.M., Marques, S.M.P., Cancela, M.L., 2004. Development of two bone-derived cell lines from the marine teleost *Sparus aurata*; evidence for extracellular matrix mineralization and cell-type-specific expression of matrix Gla protein and osteocalcin. Cell Tissue Res. 315, 393–406.

Potter, G., Smith, A.S.T., Vo, N.T.K., Muster, J., Weston, W., Bertero, A., Maves, L., Mack, D.L., Rostain, A., 2020. A more open approach is needed to develop cell-based fish technology: it starts with zebrafish. One Earth 3, 54–64.

Radošević, K., Dukić, B., Andlar, M., Slivac, I., Gaurina Srček, V., 2016. Adaptation and cultivation of permanent fish cell line CCO in serum-free medium and influence of protein hydrolysates on growth performance. Cytotechnology 68, 115–121.

Rafael, M.S., Laizé, V., Cancela, M.L., 2006. Identification of *Sparus aurata* bone morphogenetic protein 2: molecular cloning, gene expression and *in silico* analysis of protein conserved features in vertebrates. Bone 39, 1373–1381.

Rafael, M.S., Laizé, V., Florindo, C., Ferraresso, S., Bargelloni, L., Cancela, M.L., 2012. Overexpression of four and a half LIM domains protein 2 promotes epithelial-mesenchymal transition-like phenotype in fish pre-osteoblasts. Biochimie 94, 1128–1134.

Richter, C.A., Tieber, V.L., Denison, M.S., Giesy, J.P., 1997. An *in vitro* rainbow trout cell bioassay for aryl hydrocarbon receptor-mediated toxins. Environ. Toxicol. Chem. 16, 543–550.

Rodrigues, E.T., Pardal, M.Â., Laizé, V., Cancela, M.L., Oliveira, P.J., Serafim, T.L., 2015. Cardiomyocyte H9c2 cells present a valuable alternative to fish lethal testing for azoxystrobin. Environ. Pollut. 206, 619–626.

Rodrigues, E.T., Varela, A.T., Pardal, M.A., Sardão, V.A., 2020. Cell-based assays as an alternative for the study of aquatic toxicity of pharmaceuticals. Environ. Sci. Pollut. Res. 27, 7145–7155.

Rosa, J., Tiago, D.M., Dias, J., Cancela, M.L., Laizé, V., 2010. Serum-specific stimulation of proliferation and mineralization of fish bone-derived cells. J. Appl. Ichthyol. 26, 251–256.

Rubio, N., Datar, I., Stachura, D., Kaplan, D., Krueger, K., 2019. Cell-based fish: a novel approach to seafood production and an opportunity for cellular agriculture. Front. Sustain. Food Syst. 3, 43.

Ruiz, S., Schyth, B.D., Encinas, P., Tafalla, C., Estepa, A., Lorenzen, N., Coll, J.M., 2009. New tools to study RNA interference to fish viruses: fish cell lines permanently expressing siRNAs targeting the viral polymerase of viral hemorrhagic septicemia virus. Antivir. Res. 82, 148–156.

Segner, H., 1998. Fish cell lines as a tool in aquatic toxicology. In: Braunbeck, T., Hinton, D.E., Streit, B. (Eds.), Fish Ecotoxicology. Birkhauser Verlag, Basel, pp. 1–38.

Shahar, R., Dean, M.N., 2013. The enigmas of bone without osteocytes. Bonekey Rep. 2, 343.

Shea, T.B., Berry, E.S., 1983. A serum-free medium that supports the growth of piscine cell cultures. In Vitro 19, 818–824.

Srikanth, K., Pereira, E., Duarte, A.C., Rao, J.V., 2016. Evaluation of cytotoxicity, morphological alterations and oxidative stress in Chinook salmon cells exposed to copper oxide nanoparticles. Protoplasma 253, 873–884.

Surget, G., Roberto, V.P., Le Lann, K., Mira, S., Guérard, F., Laizé, V., Poupart, N., Cancela, M.L., Stiger-Pouvreau, V., 2017. Marine green macroalgae: a source of natural compounds with mineralogenic and antioxidant activities. J. Appl. Phycol. 29, 575–584.

Swaminathan, T.R., Basheer, V.S., Kumar, R., Kathirvelpandian, A., Sood, N., Jena, J.K., 2015. Establishment and characterization of fin-derived cell line from ornamental carp, *Cyprinus carpio* koi, for virus isolation in India. In Vitro Cell. Dev. Biol. Anim. 51, 705–713.

Tiago, D.M., Cancela, M.L., Aureliano, M., Laizé, V., 2008a. Vanadate proliferative and anti-mineralogenic effects are mediated by MAPK and PI-3K/Ras/Erk pathways in a fish chondrocyte cell line. FEBS Lett. 582, 1381–1385.

Tiago, D.M., Laizé, V., Cancela, M.L., Aureliano, M., 2008b. Impairment of mineralization by metavanadate and decavanadate solutions in a fish bone-derived cell line. Cell Biol. Toxicol. 24, 253–263.

Tiago, D.M., Laizé, V., Bargelloni, L., Ferraresso, S., Romualdi, C., Cancela, M.L., 2011a. Global analysis of gene expression in mineralizing fish vertebra-derived cell lines: new insights into anti-mineralogenic effect of vanadate. BMC Genomics 12, 310.

Tiago, D.M., Cancela, M.L., Laizé, V., 2011b. Proliferative and mineralogenic effects of insulin, IGF-1, and vanadate in fish osteoblast-like cells. J. Bone Miner. Metab. 29, 377–382.

Tiago, D.M., Marques, C.L., Roberto, V.P., Cancela, M.L., Laizé, V., 2014. Mir-20a regulates *in vitro* mineralization and BMP signaling pathway by targeting BMP-2 transcript in fish. Arch. Biochem. Biophys. 543, 23–30.

Viegas, C.S.B., Conceição, N., Fazenda, C., Simes, D.C., Cancela, M.L., 2010. Expression of Gla-rich protein (GRP) in newly developed cartilage-derived cell cultures from sturgeon (*Acipenser naccarii*). J. Appl. Ichthyol. 26, 214–218.

Viegas, M.N., Dias, J., Cancela, M.L., Laizé, V., 2012. Polyunsaturated fatty acids regulate cell proliferation, extracellular matrix mineralization and gene expression in a gilthead seabream skeletal cell line. J. Appl. Ichthyol. 28, 427–432.

Viegas, M.N., Laizé, V., Salgado, M.A., Aguiar, C., Dias, J., 2020. Effect of trace minerals and B vitamins on the proliferation/cytotoxicity and mineralization of a gilthead seabream bone-derived cell line. Biol. Trace Elem. Res. 196, 629–638.

Vijayakumar, P., Laizé, V., Cardeira, J., Trindade, M., Cancela, M.L., 2013. Development of an *in vitro* cell system from zebrafish suitable to study bone cell differentiation and extracellular matrix mineralization. Zebrafish 10, 500–509.

Vijayakumar, P., Cardeira, J., Laizé, V., Gavaia, P.J., Cancela, M.L., 2020. Cells isolated from regenerating caudal fin of *Sparus aurata* can differentiate into distinct bone cell lineages. Mar. Biotechnol. 22, 333–347.

Villena, A.J., 2003. Applications and needs of fish and shellfish cell culture for disease control in aquaculture. Rev. Fish Biol. Fish. 13, 111–140.

Vo, N.T.K., Bufalino, M.R., Hartlen, K.D., Kitaev, V., Lee, L.E.J., 2014. Cytotoxicity evaluation of silica nanoparticles using fish cell lines. In Vitro Cell. Dev. Biol. Anim. 50, 427–438.

Wang, D., Manali, D., Wang, T., Bhat, N., Hong, N., Li, Z., Wang, L., Yan, Y., Liu, R., Hong, Y., 2011. Identification of pluripotency genes in the fish medaka. Int. J. Biol. Sci. 7, 440–451.

Wolf, K., Quimby, M., 1962. Established eurythermic line of fish cells *in vitro*. Science 135, 1065–1066.

Wolf, K., Quimby, M.C., Pyle, E.A., Dexter, R.P., 1960. Preparation of monolayer cell cultures from tissues of some lower vertebrates. Science 132, 1890–1891.

Yashwanth, B.S., Goswami, M., Kooloth Valappil, R., Thakuria, D., Chaudhari, A., 2020. Characterization of a new cell line from ornamental fish *Amphiprion ocellaris* (Cuvier, 1830) and its susceptibility to nervous necrosis virus. Sci. Rep. 10, 20051.

Yue, Y., Behra, R., Sigg, L., Fernández Freire, P., Pillai, S., Schirmer, K., 2015. Toxicity of silver nanoparticles to a fish gill cell line: role of medium composition. Nanotoxicology 9, 54–63.

Index

Note: Page numbers followed by *f* indicate figures, *t* indicate tables, and *b* indicate boxes.

E

F

N

V

W

Z

Kolibri 360 GmbH
Ferdinand-Jühlke-Straße 7
99095 Erfurt, Deutschland
produktsicherheit@kolibri360.de